Lecture Notes in Computer Science 15411

Founding Editors

Gerhard Goos
Juris Hartmanis

AF172974

The series Lecture Notes in Computer Science (LNCS), including its subseries Lecture Notes in Artificial Intelligence (LNAI) and Lecture Notes in Bioinformatics (LNBI), has established itself as a medium for the publication of new developments in computer science and information technology research, teaching, and education.

LNCS enjoys close cooperation with the computer science R & D community, the series counts many renowned academics among its volume editors and paper authors, and collaborates with prestigious societies. Its mission is to serve this international community by providing an invaluable service, mainly focused on the publication of conference and workshop proceedings and postproceedings. LNCS commenced publication in 1973.

Shin-ichi Nakano · Mingyu Xiao
Editors

WALCOM: Algorithms and Computation

19th International Conference and Workshops
on Algorithms and Computation, WALCOM 2025
Chengdu, China, February 28 – March 2, 2025
Proceedings

Editors
Shin-ichi Nakano 🆔
Gunma University
Maebashi, Japan

Mingyu Xiao 🆔
University of Electronic Science
and Technology of China
Chengdu, China

ISSN 0302-9743   ISSN 1611-3349 (electronic)
Lecture Notes in Computer Science
ISBN 978-981-96-2844-5   ISBN 978-981-96-2845-2 (eBook)
https://doi.org/10.1007/978-981-96-2845-2

Preface

Welcome to the proceedings of the 19th International Conference and Workshops on Algorithms and Computation (WALCOM 2025), held in Chengdu, China, from February 28 to March 2, 2025. This conference served as a prominent international forum for researchers and practitioners in the fields of algorithms and computation to share their latest findings, exchange ideas, and foster collaboration.

WALCOM 2025 covered a wide range of topics, including approximation algorithms, algorithmic graph theory and combinatorics, algorithmic and combinatorial aspects of logic, combinatorial algorithms, combinatorial optimization, combinatorial reconfiguration, computational biology, computational complexity, computational geometry, discrete geometry, data structures, experimental algorithm methodologies, graph algorithms, graph drawing, parallel and distributed algorithms, parameterized algorithms and complexity, network optimization, online algorithms, randomized algorithms, and string algorithms. These topics reflect the diverse and rapidly evolving nature of the field.

The conference received 71 submissions from 23 countries and regions, demonstrating the global reach and interest in WALCOM. The program committee, with the assistance of external referees, rigorously reviewed each submission with a minimum of three reviews per submission, ultimately selecting 26 papers for presentation at the conference. This acceptance rate was 37%.

We were honored to have three eminent researchers deliver invited talks at WALCOM 2025: Venkatesan Guruswami from the University of California, Berkeley, USA, Ken-ichi Kawarabayashi from the National Institute of Informatics, Japan, and Daniel Lokshtanov from the University of California, Santa Barbara, USA. Their insightful and inspiring presentations added greatly to the conference's intellectual richness.

Our heartfelt thanks go to all who contributed to the success of WALCOM 2025. We are grateful to the authors for their valuable submissions, the program committee members and external referees for their meticulous reviews, the invited speakers for their engaging talks, and the local organizers for their tireless efforts in ensuring the smooth running of the conference. We also acknowledge the Steering Committee members for their continuous support and guidance.

The conference was generously sponsored by the Algorithms and Logic Group at the University of Electronic Science and Technology of China, as well as the Theoretical Computer Science Committee of the China Computer Federation (CCF). Their support was instrumental in making WALCOM 2025 a reality.

Finally, we would like to express our sincere gratitude to Springer for publishing the proceedings of WALCOM 2025 in their esteemed LNCS series. This publication will ensure that the research presented at the conference reaches a wide audience and continues to contribute to the advancement of the field.

We hope that the proceedings will serve as a lasting testament to the vibrant and innovative research conducted in the areas of algorithms and computation, and that it will inspire further exploration and discovery in these crucial fields.

March 2025

Shin-ichi Nakano
Mingyu Xiao

Organization

General Chair

Bakh Khoussainov   University of Electronic Science and Technology of China, China

PC Co-chairs

Shin-ichi Nakano   Gunma University, Japan
Mingyu Xiao   University of Electronic Science and Technology of China, China

Steering Committee

Tamal Dey   Ohio State University, USA
Seok-Hee Hong   University of Sydney, Australia
Costas S. Iliopoulos   King's College London, UK
Giuseppe Liotta   University of Perugia, Italy
Petra Mutzel   University of Bonn, Germany
Shin-ichi Nakano   Gunma University, Japan
Subhas C. Nandy   Indian Statistical Institute, India
Md. Saidur Rahman   Bangladesh University of Engineering and Technology, Bangladesh
Ryuhei Uehara   Japan Advanced Institute of Science and Technology, Japan

Program Committee

Reyan Ahmed   University of Arizona, USA
Kristóf Bérczi   ELTE, Hungary
Sang Won Bae   Kyonggi University, South Korea
Zhipeng Cai   Georgia State University, USA
Marek Chrobak   University of California, Riverside, USA
Gautam K. Das   Indian Institute of Technology Guwahati, India
Petr Golovach   University of Bergen, Norway

Christian Komusiewicz	Friedrich Schiller University Jena, Germany
Bo Li	Hong Kong Polytechnic University, China
Yi Li	Nanyang Technological University, Singapore
Chun-Cheng Lin	National Yang Ming Chiao Tung University, Taiwan
Giuseppe Liotta	University of Perugia, Italy
Debajyoti Mondal	University of Saskatchewan, Canada
Krishnendu Mukhopadhyaya	Indian Statistical Institute, India
Shin-ichi Nakano	Gunma University, Japan
Subhas C. Nandy	Indian Statistical Institute, India
Rahnuma Islam Nishat	Brock University, Canada
Yoshio Okamoto	University of Electro-Communications, Japan
Maurizio Patrignani	Università Roma Tre, Italy
Pan Peng	University of Science and Technology of China, China
Md. Saidur Rahman	Bangladesh University of Engineering and Technology, Bangladesh
Adele A. Rescigno	University of Salerno, Italy
Toshiki Saitoh	Kyushu Institute of Technology, Japan
Frank Stephan	National University of Singapore, Singapore
Darren Strash	Hamilton College, USA
Ryuhei Uehara	Japan Advanced Institute of Science and Technology, Japan
Mingyu Xiao	University of Electronic Science and Technology of China, China
Chao Xu	University of Electronic Science and Technology of China, China
Katsuhisa Yamanaka	Iwate University, Japan
Meirav Zehavi	Ben-Gurion University of the Negev, Israel
Chihao Zhang	Shanghai Jiao Tong University, China
Jialin Zhang	Chinese Academy of Sciences, China

Local Organizing Committee

Mingyu Xiao	University of Electronic Science and Technology of China, China
Chao Xu	University of Electronic Science and Technology of China, China
Yi Zhou	University of Electronic Science and Technology of China, China
Dong Hao	University of Electronic Science and Technology of China, China

Toru Takisaka	University of Electronic Science and Technology of China, China
Mengxiao Zhang	University of Electronic Science and Technology of China, China
Yuting Fang	University of Electronic Science and Technology of China, China
Ting Gou	University of Electronic Science and Technology of China, China
Zimo Sheng	University of Electronic Science and Technology of China, China
Junqiang Peng	University of Electronic Science and Technology of China, China
Kangyi Tian	University of Electronic Science and Technology of China, China

Additional Reviewers

Taehoon Ahn
Narmina Baghirova
Arijit Bishnu
Nicolas Bousquet
Susanna Caroppo
Sourav Chakraborty
Shengminjie Chen
Gennaro Cordasco
Giordano Da Lozzo
Minati De
Hiroshi Eto
Jaroslav Garvardt
Barun Gorain
Sheikh Azizul Hakim
Yuchen He
Tomohiro I.
Naonori Kakimura
Csaba Király
Naoki Kitamura
Fabian Klute
JanMatyáš Křišvtan
Shaohua Li
Honghao Lin
Simon Mackenzie
Ratnadip Mandal
Gopinath Mishra

Takaaki Mizuki
Atsuki Nagao
Supantha Pandit
Daniel Paul-Pena
Tommaso Piselli
Li Qian
Jakub Radoszewski
Bastien Rivier
Zimo Sheng
Pranavi Sudireddy
Biaoshuai Tao
Xuan Wu
Muhammad Yanhaona
Ryo Yoshinaka
Yuhao Zhang
Carlos Alegría
Tian Bai
Greg Bodwin
Franz Brandenburg
Ruoxu Cen
Michael C. Chavrimootoo
Yu Cong
Gergely Csáji
Swagatam Das
Mark Dukes
Fabrizio Frati

Arijit Ghosh
Fabrizio Grosso
Dong Hao
Duc A. Hoang
Chuzo Iwamoto
Liana Khazaliya
Tamás Király
Sandi Klavzar
Yusuke Kobayashi
Jaegun Lee
Nutan Limaye
Mengfan Ma
Subhamoy Maitra
Toshimitsu Masuzawa
Pawan K. Mishra

Nils Morawietz
Masaaki Nishino
Irene Parada
Daniel Perz
Solon Pissis
Guoliang Qiu
Linus Richter
Tamas Schwarcz
Luca Pascal Staus
Asahi Takaoka
Francesco Verciani
Kuan Yang
Sang Duk Yoon
Rocco Zaccagnino
Jingyang Zhao

Sponsoring Institutions

Algorithms and Logic Group, University of Electronic Science and Technology of China
Theoretical Computer Science Committee of China Computer Federation (CCF)

Abstract of Invited Talks

When and Why do Efficient Algorithms Exist (for Constraint Satisfaction and Beyond)?

Venkatesan Guruswami

University of California, Berkeley, USA
venkatg@berkeley.edu

Abstract. Computational problems exhibit a diverse range of behaviors in terms of how quickly and effectively they can be solved. What underlying mathematical structure (or lack thereof) in a computational problem leads to an efficient algorithm for solving it (or dictates its intractability)? Given the vast landscape of problems and algorithmic approaches, it would be too much to hope for a universal theory explaining the underpinnings of easiness/hardness. Yet, in the realm of constraint satisfaction problems (CSP), the algebraic dichotomy theorem gives a definitive answer: a polynomial time algorithm exists when there are non-trivial local operations called polymorphisms under which the solution space is closed; otherwise the problem is NP-complete.

Inspired by this, one might speculate a polymorphic principle in more general contexts, with appropriate notions of symmetries governing the existence of efficient algorithms. Beginning with some background on CSP and the polymorphic approach to understand their complexity, the talk will discuss some extensions beyond CSP where the polymorphic principle seems promising (yet far from understood). In particular, we will discuss "promise CSP" where one is allowed to satisfy a relaxed version of the constraints, a framework that includes problems such as approximate graph coloring. We will provide glimpses of the emerging theory characterizing the (in)tractability of interesting classes of promise CSP, touch upon connections to optimization (linear and affine relaxations), and highlight some of the many challenges that remain.

Three-Edge-Coloring Cubic Graphs on Surfaces of Low Genus

Ken-ichi Kawarabayashi

National Institute of Informatics, Japan
k_keniti@nii.ac.jp

Abstract. It is well-known that every bridgeless planar cubic graph admits a three-edge coloring. Indeed, it has been known since 1890 that this is equivalent to the four color theorem.

Our recent work extended this result to graphs on the projective plane and the torus. On the projective plane there is, essentially, a single exception of the Petersen graph, while on the torus one obtains an infinite collection of "Petersenlike" exceptions. As a corollary, we obtain a strengthening of the well-known Grunbaum's conjecture from 1968 that every cubic graph with a polyhedral embedding in the torus is 3-edge-colorable.

We report more progress on this line of work.

Structure and Quasi-Polynomial Time Algorithms

Daniel Lokshtanov

University of California, Santa, USA
Barbaradaniello@ucsb.edu

Abstract. A substantial number of prominent computational problems in algorithmic graph theory still resist classification into polynomial time solvable or NP-hard. For example, is the Independent Set problem polynomial time solvable on every hereditary graph class that excludes some path as an induced subgraph? What about the 3-coloring problem on the same classes of graphs? What about Independent Set and Coloring on even-hole-free graphs? While we are still far away from being able to answer these questions, in the past 5 years many of these problems have been shown to admit quasi-polynomial time algorithms. This implies that the problems are very unlikely to be NP-hard. The new quasi-polynomial time algorithms are built on structural insights which are specifically tailored to the design of quasi-polynomial, instead of polynomial time algorithms. In this talk I will survey some of the recent quasi-polynomial time algorithms and the related developments in structural graph theory.

Contents

Parameterized Voter Relevance in Facility Location Games with Tree-Shaped Invitation Graphs

Ryoto Ando, Kei Kimrua, Taiki Todo$^{(\boxtimes)}$, and Makoto Yokoo

Graduate School of ISEE, Kyushu University, Motooka, Fukuoka 819-0395, Japan
todo@inf.kyushu-u.ac.jp

Abstract. Diffusion mechanism design, which investigates how to incentivise agents to invite as many colleagues to a multi-agent decision making process as possible, is a new research paradigm at the intersection between microeconomics and computer science. In this paper, we extend traditional facility location games into the model of diffusion mechanism design. Our objective is to completely understand to what extent of anonymity/voter-relevance we can achieve, along with strategy-proofness and Pareto efficiency when voters strategically invite colleagues. We define a series of anonymity properties applicable to the diffusion mechanism design model, as well as parameterized voter-relevance properties for guaranteeing reasonably-fair decision making. We obtained two impossibility theorems and two existence theorems, which partially answer the question we have raised at the beginning of the paper.

Keywords: Mechanism Design · Facility Location Games · Information Diffusion · Single-Peaked Preferences · Anonymity · Voter-Relevance

1 Introduction

Social choice theory is one of the mathematical foundations of multi-agent decision making. We assume there exists a number of individuals, usually called *agents* or *voters*. Each voter has a preference over a set of alternatives/outcomes, and a social choice function chooses, by taking into account the preferences of voters, an alternative as a final outcome. As many impossibility theorems have been obtained in the literature [18], designing social choice functions for making an appropriate social decision has been an important research topic.

Facility location games are a well-studied problem in the literature of social choice and are known as a special case of voting [17]. In the problem, each agent (voter) is located at a point on an interval that represents the set of social alternatives. Under the realization of a social alternative as the outcome of a social choice function, a voter's cost is defined as the distance between the outcome and her location. The domain restriction on such *single-peaked* preferences guarantees the existence of a Condorcet winner. Actually the alternative most preferred by the *median voter*, i.e., the voter whose location is the $\lfloor (n+1)/2 \rfloor$-th smallest among n voters, is a Condorcet winner.

In the literature of mechanism design, truthfulness, also known as *strategy-proofness*, is one of the most important properties that a social choice function should preserve. It requires that telling a preference honestly to the social choice function is a dominant strategy for every voter. Clarifying the necessary and sufficient condition for a social choice function to be strategy-proof has greatly attracted considerable attention from researchers. The *median voter schemes*, which contains the above median rule as a special case, are the only social choice functions that satisfy strategy-proofness, ontoness, and anonymity [17].

In practice, a mechanism designer, also called a *moderator*, struggles to observe the set of voters who potentially participate in the decision making process and to directly advertise it to them. Instead, voters can get the information about a decision making process from their peers/colleagues, through their shared *social networks*, sometimes also called as *invitation graphs*. It is therefore important to incentivize voters to invite as many colleagues as possible. Such a new paradigm of mechanism design is called *diffusion mechanism design* [14]. However, as far as the authors know, any existing work on social choice theory and facility location games, has never considered diffusion mechanism design.

According to these situations, in this paper, we consider social choice under single-peaked preferences over networks. The main goal of this research direction is to completely understand which social choice functions satisfy/do not satisfy strategy-proofness and ontoness, as well as some anonymity properties. As a first step, we define a class of anonymity properties applicable to the model of mechanism design via social networks. Specifically, we define three anonymity properties, namely anonymity on structures (AN-S), anonymity on distances (AN-D), and anonymity on structure-distance pairs (AN-SD), each of which considers the structure of social networks in their own definitions.

To understand whether socially-fair decisions can be made, we also define a parameterized class of *voter-relevance* properties and discuss which combinations between anonymity and voter-relevance are compatible with SP and onto. A social choice function is d-distance voter-relevant (VR-d in short) if any voter who is at distance $d \in \mathbb{N}_{\geq 0}$ or less from the moderator has *some chance* to change the final outcome by changing her own action. Obviously, achieving $d = n$ is the best when n voters participate; any voter is at distance d or less then, i.e., every voter has some chance to change the outcome. We obtained two impossibility and two existence theorems when the social networks among voters are tree-shaped. These results give us a complete understanding on to what extent of voter-relevance is achievable when AN-S/AN-D are mandatory, as shown in the two center columns of Table 1.

The rest of this paper is organized as follows. Section 2 reviews the related literature of social choice theory and diffusion mechanism design. Section 3 defines the standard model of facility location games and its extension with strategic

Table 1. Summary of existence of SCFs satisfying strategy-proofness over social networks and Pareto efficiency, along with corresponding anonymity and voter-relevance properties. Requirements basically get weaker by moving from the top-left corner to the bottom-right corner, while there is one exception in that there is no implication relation between the two center columns, AN-S and AN-D.

	AN	AN-S	AN-D	AN-SD
VR-n	✗	✗	✗	Open
VR-3 ... VR-n-1	✗	✗	✗	Open
VR-2	✗	✗	✗ (Theorem 3)	✓ (Theorem 5)
VR-1	✗	✗	✓ (Theorem 4)	✓
VR-0	✗	✗ (Theorem 2)	✓	✓

information diffusion. Section 4 shows an impossibility theorem on anonymity on structures (AN-S), corresponding to the AN-S column of the table. Section 5 shows an impossibility theorem and an existence theorem on anonymity on distances (AN-D), corresponding to the AN-D column. Section 6 shows an existence theorem on anonymity on structure-distance pairs (AN-SD), corresponding to the AN-SD column. Section 7 concludes the paper.

2   Literature

Under the single-peaked preferences, Moulin [17] investigated strategy-proof and Pareto efficient social choice functions on a continuous line and proposed a class of such social choice functions, so-called generalized median voter schemes. Indeed, it is the only class of deterministic, strategy-proof, Pareto efficient and anonymous social choice functions. Procaccia and Tennenholtz [21] initiated the research on approximation mechanism design, for which a facility location problem was chosen as a case study. Recently, some new facility location models have been investigated, including dynamic locations [11,28] and multiple facilities [3,7,23]. Some other research considered strategy-proof facility locations on discrete structures, such as grids [6,19,25], cycles [1,2,5], and wheel graphs [20].

The research of diffusion mechanism design, also known as *mechanism design over social networks*, was initiated by Li et al. [14], which considered single-item auctions and proposed a strategy-proof mechanism. After that, several works investigated strategy-proof resource allocation mechanisms with monetary compensations, e.g., multi-unit auctions and redistributions [8,9,13,31,32]. On the other hand, there is limited research on decision making without money from the perspective of diffusion mechanism design. Some recent works have house allocation problems without monetary compensation [10,30] and two-sided matching (also known as school choice) problem [4]. However, neither papers have addressed voting/social choice functions from the perspective of diffusion mechanism design.

3   Model

We begin with defining the traditional model of social choice with single-peaked preferences, which is followed by the extended model with information diffusion.

There is a line segment $\mathcal{O} := [0, 1]$, from which a social choice function chooses a point/outcome. Since in our model the set of participating voters depends on the information diffusion by voters, we need to define both *potential* and *participating* voters. Let $\mathcal{N}$ be the set of potential voters, and $N \subseteq \mathcal{N}$ be a set of participating voters. Each voter $i \in N$ has a *preference* $\succ_i \in \mathcal{P}$, which is a complete and transitive binary relation over $\mathcal{O}$. Let P denote the set of possible preferences over $\mathcal{O}$. In this paper, we assume preferences are *single-peaked*, which is one of the well-studied preference restrictions in the literature.

Definition 1 (Single-Peaked Preferences). *A domain $\mathcal{R} \subseteq \mathcal{P}$ of preferences over the set of (possibly infinite) alternatives $\mathcal{O}$ is* single-peaked *if there is a strict ordering $\rhd$ over $\mathcal{O}$ s.t. for any preference $\succ \in \mathcal{R}$, there is an associated ideal alternative $p \in \mathcal{O}$, and for any two distinct alternatives $x, y \in \mathcal{O}$, $x \succ y$ if and only if either $p \rhd x \rhd y$ or $y \rhd x \rhd p$ holds.*

That is, a voter who has preference $\succ_i$ associated with peak p_i prefers outcome x, which is strictly closer to (resp. farther from) p_i than another outcome y. When voter i has preference $\succ_i$ associated with the peak $p_i \in \mathcal{O}$, we sometimes say that voter i is *located at p_i*. Let $\succ := (\succ_i)_{i \in N}$ denote a profile of the voters' preferences, and let $\succ_{-i} := (\succ_{i'})_{i' \neq i}$ denote the profile without i's. Also, given preference $\succ_i$, let $\succsim_i$ denote the weak preference corresponding to $\succ_i$, where $x \succsim_i y$ means that x is weakly better than y.

A (deterministic) social choice function is a mapping from the set of possible profiles to the set of outcomes. Since the number of participating voters varies with regard to the voters' actions, a social choice function must be defined for different-sized profiles. To describe this feature, we define a social choice function $f = (N)_{N \subseteq \mathcal{N}}$ as a family of functions, where each f^N is a mapping from $\mathcal{R}^{|N|}$ to $\mathcal{O}$. When a set N of voters participates, the social choice function f uses function f^N to determine the outcome. The function f^N takes profile $\succ$ of preferences jointly reported by N as an input, and returns $f^N(\succ)$ as an outcome. We denote f^N as f if it is clear from the context.

Definition 2 (Social Choice Function (SCF)). *A social choice function (SCF) $f = (f^N)_{N \subseteq \mathcal{N}}$ is a family of mappings f^N from $\mathcal{R}^n$ to $\mathcal{O}$, where $|N| = n$, s.t. each f^N takes n preferences as an input and returns an alternative $o \in \mathcal{O}$.*

Strategy-proofness, which is one of the most important properties that SCFs should satisfy, requires that for each voter, reporting her true preference to the social choice function is a dominant strategy, i.e., one of the best actions regardless of the action profile of other voters.

Definition 3 (Strategy-Proofness). *An SCF f is* strategy-proof *if, for any $N \subseteq \mathcal{N}$ s.t. $|N| = n$, any voter $i \in N$, any preference profile $\succ'_{-i}$ of other voters, any preference $\succ_i$ of voter i, and any preference misreport $\succ'_i$, $f(\succ_i, \succ'_{-i}) \succsim_i f(\succ'_i, \succ'_{-i})$.*

Ontoness is a minimum requirement to guarantee that the decision making process is fair for outcomes, which requires that for any outcome, there is at least one profile of preferences under which it is chosen. If ontoness is not satisfied, i.e., there is an outcome that cannot be chosen under any profile, then it is reasonable to remove it from the set of outcomes.

Definition 4 (Ontoness). *An SCF f is onto if, for any $o \in \mathcal{O}$, there is at least one profile $\succ$ s.t. $f(\succ) = o$.*

Anonymity on preferences, traditionally just called anonymity, is, on the other hand, a requirement on fairness among voters. Under an SCF that is anonymous on preferences, a permutation of voters' names/identities does not affect the chosen outcome.

Definition 5 (Anonymity). *An SCF f is* anonymous *if, for any profile $\succ$ and any permutation $\succ'$ of $\succ$, $f(\succ) = f(\succ')$ holds.*

Moulin [17] proposed a class of SCFs satisfying strategy-proofness and ontoness. Furthermore, he showed that any SCF that satisfies both of these properties can be represented as an instance in this class.

Definition 6 (Generalized Median Voter Schemes (GMVS)). *A generalized median voter scheme (GMVS) f is an SCF is defined as follows: $\forall N \subseteq \mathcal{N}$ s.t., $|N| = n$, there is a profile $\alpha^N = (\alpha_S^N)_{S \subseteq N} \in \mathcal{O}^{2^n}$ of 2^n parameters s.t.*

1. *$\alpha_\emptyset^N = 0$, $\alpha_N^N = 1$*
2. *$\alpha_S^N \leq \alpha_T^N$ for any $S \subseteq T \subseteq N$, and*
3. *$f^N(\succ) = \max_{S \subseteq N} \min\{(p_i)_{i \in S}, \alpha_S^N\}$ for any input $\succ \in \mathcal{R}^n$.*

Theorem 1 (Moulin [17]). *Under the single-peaked preference domain, an SCF satisfies strategy-proofness and ontoness if and only if it is a generalized median voter scheme.*

3.1 Mechanism Design via Social Network

In our model of facility location games, voters are distributed over a social network (or an *invitation graph*), which is assumed to be a directed tree with a single source vertex. There is a special agent called *moderator*, represented as symbol m and corresponds to the source vertex. Let $r_m \subseteq \mathcal{N}$ be the set of the children of moderator m, which are also called the *direct children* of m. For each voter $i \in \mathcal{N}$, let $r_i \subseteq \mathcal{N} \setminus \{i\}$ denote i's children. Given $r_\mathcal{N} := (r_i)_{i \in \mathcal{N}}$ and r_m, all the parent-child relations are defined, specifying the *social network* $G(r_\mathcal{N}, r_m)$ among voters and the moderator.

Next we give some additional notations to formalize our model as a mechanism design problem. Let $\theta_i = (\succ_i, r_i)$ denote the *true type* of voter i, and let $\theta = (\theta_i)_{i \in \mathcal{N}}$ denote a type profile of all the voters. Let θ_{-i} denote a profile of the types owned by the voters except i. Let $R(\theta_i) = \{\theta_i' = (\succ_i', r_i') \mid r_i' \subseteq r_i\}$ denote the set of *reportable types* by voter i with true type θ_i, assuming that each i cannot pretend to be a parent of any voter of which i is not really a parent. When i reports r_i' as her children, we say i *invites* r_i'. Let $\theta' = (\theta_i')_{i \in \mathcal{N}} \in \times_{i \in \mathcal{N}} R(\theta_i) = R(\theta)$ denote a reportable type profile.

Given type profile θ', let $\hat{N}(\theta') \subseteq \mathcal{N}$ denote the set of *participating voters*, to whom a path exists from m in $G(r_{\mathcal{N}}', r_m)$. Given true type profile θ (which is not observable), a social choice function f maps each reported profile $\theta' \in R(\theta)$ into an outcome $o \in \mathcal{O}$, while f can use $(\succ_i)_{i \in \hat{N}}$ and $(r_i)_{i \in \hat{N}}$ as parameters. When the meaning is clear from the context, we will use slightly different notations such as $f(\theta)$ and $f(\succ)$, for the ease of understanding.

Strategy-proofness over networks [14] is a refinement of strategy-proofness for diffusion mechanism design. It requires that, for each voter, inviting as many children as possible and reporting her true preference is a dominant strategy.

Definition 7 (Strategy-Proofness over Networks (SP) [14]). *An SCF f is* strategy-proof over networks *(or satisfy SP in short) if for any $N \subseteq \mathcal{N}$ s.t. $|N| = n$, any $i \in N$, any $\theta_{-i} = ((\succ_j, r_j))_{j \in N \setminus \{i\}}$, any $\theta_{-i}' \in R(\theta_{-i})$, any $\theta_i = (\succ_i, r_i)$, and any $\theta_i' \in R(\theta_i)$, $f(\theta_i, \theta_{-i}') \succsim_i f(\theta_i', \theta_{-i}')$.*

Since one of the main objectives in our paper is to analyze the effect of strategic information diffusion in facility location games, we also formally define *strategy-proofness over networks on information diffusion*, SP-D in short. This weaker property requires that for any voter, who is assumed to report her preference truthfully, inviting as many colleagues as possible is a dominant strategy. In practice, such an assumption might be reasonable for some cases related to social choice and facility location. For example, if citizens are expected to prefer locations closer to their living addresses, it might be enough for the decision maker to ask citizens to invite as many colleagues as possible. SP-D is therefore a reasonable incentive property for such a situation.

Definition 8 (Strategy-Proofness over Networks on Information Diffusion (SP-D)). *An SCF is* strategy-proof over networks on information diffusion *(or SP-D in short) if for any $N \subseteq \mathcal{N}$ s.t. $|N| = n$, any $i \in N$, any $\theta_{-i} = ((\succ_j, r_j))_{j \in N \setminus \{i\}}$, any $\theta_{-i}' \in R(\theta_{-i})$, any $\theta_i = (\succ_i, r_i)$, and any $\theta_i' = (\succ_i, r_i') \in R(\theta_i)$, $f(\theta_i, \theta_{-i}') \succsim_i f(\theta_i', \theta_{-i}')$.*

In our model of mechanism design via social network, the original anonymity property requires that any permutation of preferences among all the voters never changes the outcome, which we sometimes call *full-anonymity*. Now we define further three variants of anonymity properties below for SCFs, namely anonymity on structures (AN-S), anonymity on distances (AN-D), and anonymity on structure-distance pairs (AN-SD). Briefly speaking, AN-S requires that any permutation of preferences among those who have the same number of children never changes the outcome, AN-D requires that any permutation of preferences among those who are in the same distance from the moderator never changes the outcome, AN-SD requires that any permutation of preferences among those who have the same number of children and are in the same distance from the moderator never changes the outcome. By definition, AN implies both AN-S and AN-D, and both AN-S and AN-D imply AN-SD.

Definition 9. *Given social network $G(r'_\mathcal{N}, r_m)$, let $N_S(k)$ be the set of participating voters who have k children, for each $k \in \mathbb{N}_{\geq 0}$. Also, given social network $G(r'_\mathcal{N}, r_m)$, let $N_D(d)$ be the set of participating voters who is at distance d from the moderator, for each $d \in \mathbb{N}_{\geq 0}$. An SCF f satisfies AN-S if, for any θ, any $k \in \mathbb{N}_{\geq 0}$, and any θ' in which any subset of voters in a certain set $N_S(k)$ permutes their preferences from θ, $f(\theta) = f(\theta')$ holds. An SCF f satisfies AN-D if, for any θ, any $d \in \mathbb{N}_{\geq 0}$, and any θ' in which any subset of voters in a certain set $N_D(k)$ permutes their preferences from θ, $f(\theta) = f(\theta')$ holds. An SCF f satisfies AN-SD if, for any θ, any pair $(k, d) \in \mathbb{N}_{\geq 0}^2$, and any θ' in which any subset of voters in a certain set $N_S(k) \cap N_D(d)$ permutes their preferences from θ, $f(\theta) = f(\theta')$ holds.*

Note that there is a naïve idea for achieving those anonymity properties; applying any fully-anonymous GMVS for the direct-children of the moderator. Indeed, such an SCF satisfies SP and onto as well. However, it has an obvious drawback that all the other voters have no effect *at all* on the final outcome.

To avoid this, we also define a parameterized class of *voter-relevance* properties based on the distance from the moderator, which enables us to quantify how far an SCF is from a desirable decision process in which every voter has some chance to change the final outcome[1]. We say an SCF is *d-distance voter-relevant* (or satisfies *VR-d* in short) if, for any voter i who is at distance d or less from the moderator, there exists at least one situation (i.e., a fixed profile of the other voters' actions) θ_{-i} in which i can change the outcome by her own action, e.g., two types θ'_i and θ''_i (see the formal definition below). In such a case, we say voter i is relevant. From the viewpoint of public decision making, having a larger $d (\leq n)$ is better. While satisfying VR-0 is meaningless since there is no voter with distance zero from the moderator, we define VR-0 as well for the completeness of the discussion.

Definition 10. *An SCF f is d-distance voter-relevant (or satisfies VR-d) for some $d \in \mathbb{N}_{\geq 0}$ if, for any voter $i \in \bigcup_{1 \leq d' \leq d} N_D(d')$ and any θ_i, there exists θ'_{-i} s.t. $f(\theta'_i, \theta'_{-i}) \neq f(\theta''_i, \theta'_{-i})$ holds for some $\theta'_i, \theta''_i \in R(\theta_i)$.*

4 Analyzing Anonymity on Structures

From now on, we will present our contributions. Each section corresponds to a column in Table 1. We skip AN and begin with AN-S, since just requiring AN-S without any voter-relevance is impossible, along with SP and onto.

---

[1] This relevance property discusses to what extent voters' actions affect the outcome. Similar concepts also exist in the literature, such as *null player* [24] and *player decisiveness* [12], while these concepts do not quantify the level of violation from some desirable property.

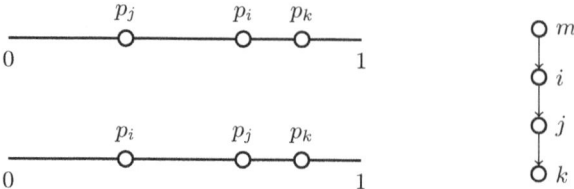

Fig. 1. An example showing that AN-S is incompatible with combination of SP and onto, explained in the proof of Theorem 2. The right diagram indicates the social network among three voters. The left top diagram shows the case where i's peak is the median among the three voters, and the left bottom shows the case where i's peak is the leftmost among the three voters. Both outcomes result in i's peak p_i.

Theorem 2. *There is no SCF that simultaneously satisfies SP, onto, and AN-S, even without any requirement on the voter-relevance property.*

Proof. Assume, for the sake of contradiction that there are three voters i, j, k whose parent-child relations are given in the right figure of Fig. 1. We also assume that $p_i \neq p_j$ holds, e.g., see the left top figure of Fig. 1.

SP implies SP-D by definition, and SP-D implies, for this example, that the facility must be located at the peak of voter i; otherwise i has an incentive to exclude voter j, which also removes k. This argument is true even when i and j swap their preferences. In Fig. 1, the middle peak is chosen in the left top figure, and the leftmost peak is chosen in the left bottom figure.

Here, two voters i and j has the same structure; both have one child. Therefore, AN-S requires that swapping their preference does not change the outcome when these two voters participate, which violates the above argument.   □

This theorem also has a bit surprising implication; the *target rule*, which is a well-known GMVS instance and robust to fake-votes [26], also violates the property of SP in our model. Indeed, the theorem shows that, when we require AN-S, any voter-relevance property is not achievable along with SP and onto. The intuition behind it is that treating two voters having the same number of children equally, even if their distances from the moderator are different, is quite unrealistic from the perspective of diffusion mechanism design. Indeed, to guarantee SP-D in general, we should give some priority to those voters closer to the moderator[2]. Since we further require AN-S, we end up choosing a fixed alternative as a final outcome, which violates onto. In the next section, we will consider AN-D, a different anonymity concept, and propose a class of SCFs satisfying it along with SP, onto, and VR-2, while AN-S and AN-D have no inclusion relation.

---

[2] Li et al. [14] handle this issue by proposing a concept of *diffusion critical tree* for the case of auctions.

5   Analyzing Anonymity on Distances

We now turn to show our analysis on AN-D. As we have mentioned, AN-D is independently defined from AN-S, and thus the result obtained in the previous section does not carry over in this section. Indeed, this section shows a positive result; AN-D is compatible with a certain level of voter-relevance.

We first show a proposition, stating that, when we require SP, onto, and AN-D, the outcome must be Pareto efficient even only for the direct-children of the moderator. This fact somewhat corresponds to giving a priority to those voters closer to the moderator, as we have mentioned at the end of the previous section.

Proposition 1. *Assume that an SCF f satisfies SP, onto, and AN-D. Then, for any input θ, the following holds:*

$$\min_{i \in N_D(1)} p_i \leq f(\theta) \leq \max_{j \in N_D(1)} p_j$$

Proof. We show the statement by mathematical induction on the set of participating voters $\hat{N}$. As a baseline, let us consider the case where $\hat{N} = N_D(1)$, i.e., only the set of direct-children of the moderator is participating. Let

$$p^1_{\min} := \min_{i \in N_D(1)} p_i, \qquad p^1_{\max} := \max_{j \in N_D(1)} p_j.$$

From the assumption that f satisfies onto, it clearly holds that $p^1_{\min} \leq f(\theta) \leq p^1_{\max}$ for arbitrarily chosen input θ under which only $N_D(1)$ is participating.

We then consider another profile θ' under which another voter $i' \notin N_D(1)$ is participating. Let $i^* \in N_D(1)$ be the unique ancestor of i'. Assume for the sake of contradiction that $f(\theta') < p^1_{\min}$.

Since f satisfies AN-D, the outcome does not change when we change the preferences of voter i^* and voter j such that $p_j := p^1_{\max}$. If $i^* = j$, i.e., the voter i^* is originally having the largest peak among $N_D(1)$, we do nothing.

After this operation, voter i^* whose current peak is $p^1_{\max}$ has an incentive to exclude i' (by not inviting i^*'s child who is an ancestor of i'); it is guaranteed from the assumption of induction that the outcome is in the range $[p^1_{\min}, p^1_{\max}]$ when i^* exclude i', while the outcome $f(\theta')$ under her sincere invitation is strictly less than $p^1_{\min}$. This violates the assumption that f satisfies SP-D. A similar argument works when $p^1_{\max} < f(\theta')$.   □

This proposition might be of independent interest to readers. Once we fall into a situation where some voters/agents are specially-treated with some sort of higher priorities, it says that we have to achieve a socially-optimal outcome only for those agents; otherwise, they have an incentive to exclude some of their colleagues from the decision making.

Now we are ready to show our impossibility theorem, which is on a combination of an anonymity property AN-D and a voter-relevance property VR-2.

Theorem 3. *There is no SCF that simultaneously satisfies SP, onto, AN-D, and VR-2.*

Proof. For the sake of contradiction, assume there exists an SCF f satisfying SP, onto, AN-D, and VR-2. From VR-2, there exists a voter $i \in N_D(2)$ such that

$$\exists \theta'_{-i}, \exists \theta'_i, \exists \theta''_i, f(\theta'_i, \theta'_{-i}) \neq f(\theta''_i, \theta'_{-i})$$

holds. From the above Proposition 1, both $f(\theta'_i, \theta'_{-i})$ and $f(\theta''_i, \theta'_{-i})$ are in the range $[p^1_{\min}, p^1_{\min}]$.

Since $f(\theta'_i, \theta'_{-i}) \neq f(\theta''_i, \theta'_{-i})$ holds, at least one of these two outcomes differs from $f(\theta'_{-i})$. Let us assume without loss of generality that $f(\theta'_i, \theta'_{-i}) \neq f(\theta'_{-i})$. Note that, again from Proposition 1, $f(\theta'_{-i})$ is also in the range $[p^1_{\min}, p^1_{\min}]$.

Now let $i^* \in N_D(1)$ be the parent of i, who has the right to exclude i. If $f(\theta'_i, \theta'_{-i}) < f(\theta'_{-i})$ holds, consider swapping preferences of voter i^* and voter j who has the largest peak among $N_D(1)$. We then apply the same argument with the proof of Proposition 1 and derive a contradiction. A similar argument applies for the case where $f(\theta'_i, \theta'_{-i}) > f(\theta'_{-i})$ holds. $\square$

What if we could require a weaker notion of voter-relevance? The answer is affirmative; if we just require VR-1 as a voter relevance property, we can achieve all the requirements, as the following theorem shows.

Theorem 4. *There is an SCF that satisfies SP, onto, AN-D, and VR-1.*

Proof. Consider applying an arbitrarily chosen anonymous GMVS only for direct-children of the moderator. For example, just choosing the median peak among the reported peaks by the direct-children of the moderator is fine.

SP trivially holds; for every voter, any invitation strategy gives her the same happiness level, which implies the definition of SP-D. Furthermore, for the direct-children of the moderator, it is well-known that telling their preference truthfully is a best strategy in any generalized median voter scheme. All the other voters' preferences have no effect on the outcome. Therefore SP holds.

The outcome is Pareto efficient for $N_D(1)$, and thus Pareto efficient for the whole society $\hat{N}$. Therefore the SCF satisfies onto.

Since we apply an anonymous GMVS, any preference permutation among $N_D(1)$ never changes the outcome. Furthermore, it entirely ignores the preferences of all the other voters. Thus, any preference permutation among $N_D(d)$ for each $d \geq 2$ also never changes the outcome. Therefore the SCF satisfies AN-D.

Finally, by definition of anonymous GMVSs, any direct-children of the moderator have an instance in which she has a right to choose at least two outcomes. Thus the SCF satisfies VR-1. $\square$

Given Theorems 3 and 4, we have found a *tight* parameter $d = 1$ so that, along with SP, onto, and AN-D, we can achieve VR-d but cannot achieve VR-$d + 1$. This is represented in the AN-D column of Table 1 in the introduction. While this is still quite a negative, there exists some sort of flexibility compared to the impossibility on AN-S presented in the previous section.

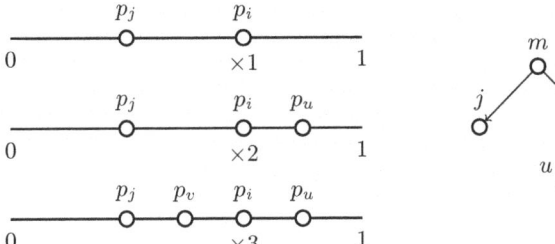

Fig. 2. The right diagram denotes the social network among the four voters. The left top diagram shows the case where voter i never invites any child, the left middle shows the case where i only invites a child u, and the left bottom shows the case where i invites both children u and v. The outcome monotonically gets closer to i's peak p_i.

6   Analysing Anonymity on Structure-Distance Pairs

We now consider a further weaker variant of anonymity, namely anonymity on structure-distance pairs (AN-SD), which is implied by both AN-S and AN-D. The SCF mentioned in the proof of Theorem 4 satisfies AN-SD, but violates VR-2, as any voter who is not directly-connected to the moderator has no effect on the outcome. Here, we propose an SCF based on an *weighted median method*, which satisfies SP, onto, AN-SD, and VR-2.

Theorem 5. *There is an SCF that satisfies SP, onto, AN-SD, and VR-2.*

Proof (Sketch). Consider the following SCF f; for each voter $i \in \hat{N}$, we first give a weight $w_i \in \mathbb{N}_{\geq 1}$ as

$$w_i = \begin{cases} |r'_i| + 1 & \text{if } i \in N_D(1) \\ 1 & \text{if } i \in N_D(2) \\ 0 & \text{otherwise.} \end{cases}$$

Then the SCF f takes the median among all the voters peaks, where voter i's peak has w_i copies. Formally,

$$f(\theta) = \text{med}(\underbrace{p_1, \ldots, p_1}_{w_1}, \underbrace{p_2, \ldots, p_2}_{w_2}, \ldots, \underbrace{p_n, \ldots, p_n}_{w_n}),$$

where p_i is the peak of voter i who reports preference $\succ_i$ and the median operator $\text{med}(\cdots)$ chooses the $\lceil m/2 \rceil$-th smallest value among m input values.

  Here we show that the above SCF satisfies all the properties in the theorem. onto can be easily shown by the same argument with the proof of Theorem 4.

  SP holds for voters in $N_D(2)$, with the same argument with Theorem 4. For those in $N_D(1)$, SP-D is guaranteed from the carefully chosen assignment of weights; each voter $i \in N_D(1)$ has the incentive to invite as many children as possible, since having a new child gives her an additional one unit of weight, while the invited child just have one unit of weight. For example, see Fig. 2, which explains an intuition why SP-D holds. Even in the worst case for voter $i \in N_D(1)$, where she has a peak at one extreme and all her children are at the other, inviting a child never affects the outcome by the definition of the median operator. Moreover, her weight is still one unit larger than the sum of the weights of her children. SP is then guaranteed based on the analysis by Moulin [17], since we can easily show that the above weighted median is an instance of GMVS.

About AN-SD, observe that any two voters, who are at the same distance from the moderator and invite the same number of children, are assigned the same weight. Therefore, by the definition of f, these two voters contribute the outcome in exactly the same way, which guarantees that AN-SD is satisfied.

We finally show that the SCF satisfies VR-2. It is obvious by definition that any direct-child of the moderator is relevant. For each voter i in $N_D(2)$, we can easily find an input in which all direct-children do not invite any of their children except for i, and half of their peaks are 0 while the other half are 1. In such an input, the facility is located at the voter i's peak, which satisfies the condition of relevance; the outcome changes when i reports a different peak location.   □

Note that the SCF mentioned in the above proof does not satisfy any stronger requirements on both anonymity and voter-relevance. On anonymity, it violates AN-S from the Theorem 2. Furthermore, it also violates AN-D, since the weights can be different among $N_D(0)$, according to the number of children. On voter-relevance, it violates VR-3, since any report by those who are at distance 3 or more from the moderator is completely ignored.

What if we can totally ignore anonymity properties? More specifically, to what extent of voter-relevance levels can we achieve, besides SP and onto? As we have already observed, the SCF mentioned in the proof of Theorem 5 satisfies AN-SD and VR-2, which implies that at least VR-2 is achievable if we totally ignore anonymity. One of our future directions examines whether there exist SCFs achieving the full voter-relevance (i.e., VR-n) along with SP and onto.

7   Discussions and Concluding Remarks

In this paper, we focused on deterministic social choice functions and discussed to what extent of anonymity and voter-relevance we can achieve, along with SP and onto. We provide two impossibility theorems and two existence theorems, summarized in Table 1. We still have open questions in the table; completing them is one of our future directions, as mentioned in the previous section.

There are various ways to extend our discussions. For example, what if we can also consider randomized social choice functions? One naïve way is to randomly choose a dictator. However, some voter having many children with completely opposite preferences then would remove them to increase the probability that she is chosen as the dictator, which seems to violate SP-D. We believe that, under some well-designed probability assignment, we can guarantee the incentive to invite as many children as possible, satisfying SP-D (and also SP) in expectation.

Extending our discussion to other outcome spaces, such as a tree-metric, a circle metric, an Euclidean space, and/or discrete graphs, is also an interesting direction. As many existing works discussed [1, 16, 22, 27], different spaces might have different properties, and the existence of desirable social choice functions strictly depends on the complexity of the outcome spaces, e.g., the number of vertices/dimensions. Furthermore, considering weakened incentive properties in facility location games with information diffusion would also be a promising direction, such as non-obvious manipulability [15, 29].

Last but not least, we strongly believe that both the anonymity properties and the voter-relevance properties defined in this paper are applicable to a more general diffusion mechanism design framework. The literature on diffusion mechanism design still lacks such a normative analysis. In other words, they have not discussed which decisions are reasonable/applicable and why, and just focused on how to incentivise agents to invite colleagues. As far as we observed, even the reason to increase the market by information diffusion has not been well-justified. Further normative analysis would make the diffusion mechanism design more appealing to the real-life decision making situations in the new era.

Acknowledgments. This work is partially supported by JST ERATO Grant Number JPMJER2301, and JSPS KAKENHI Grant Numbers JP21H04979 and JP20H00587.

References

1. Alon, N., Feldman, M., Procaccia, A.D., Tennenholtz, M.: Strategyproof approximation of the minimax on networks. Math. Oper. Res. **35**(3), 513–526 (2010)
2. Alon, N., Feldman, M., Procaccia, A.D., Tennenholtz, M.: Walking in circles. Discret. Math. **310**(23), 3432–3435 (2010)
3. Anastasiadis, E., Deligkas, A.: Heterogeneous facility location games. In: Proceedings of the AAMAS 2018, pp. 623–631 (2018)
4. Cho, S.H., Todo, T., Yokoo, M.: Two-sided matching over social networks. In: Proceedings of the IJCAI-ECAI 2022, pp. 186–193 (2022)
5. Dokow, E., Feldman, M., Meir, R., Nehama, I.: Mechanism design on discrete lines and cycles. In: Proceedings of the ACM-EC 2012, pp. 423–440 (2012)
6. Escoffier, B., Gourvès, L., Kim Thang, N., Pascual, F., Spanjaard, O.: Strategyproof mechanisms for facility location games with many facilities. In: Proceedings of the the Second International Conference on Algorithmic Decision Theory (ADT 2011), pp. 67–81 (2011)
7. Fong, C.K.K., Li, M., Lu, P., Todo, T., Yokoo, M.: Facility location game with fractional preferences. In: Proceedings of the AAAI 2018, pp. 1039–1046 (2018)
8. Jeong, S.E., Lee, J.: The groupwise-pivotal referral auction: core-selecting referral strategy-proof mechanism. Games Econom. Behav. **143**, 191–203 (2024)

9. Kawasaki, T., Barrot, N., Takanashi, S., Todo, T., Yokoo, M.: Strategy-proof and non-wasteful multi-unit auction via social network. In: Proceedings of the AAAI 2020, pp. 2062–2069 (2020)
10. Kawasaki, T., Wada, R., Todo, T., Yokoo, M.: Mechanism design for housing markets over social networks. In: Proceedings of the AAMAS 2021, pp. 692–700 (2021)
11. de Keijzer, B., Wojtczak, D.: Facility reallocation on the line. In: Proceedings of the IJCAI 2018, pp. 188–194 (2018)
12. Lavi, R., Mu'alem, A., Nisan, N.: Two simplified proofs for Roberts' theorem. Soc. Choice Welf. **32**(3), 407–423 (2009)
13. Li, B., Hao, D., Zhao, D.: Incentive-compatible diffusion auctions. In: Proceedings of the IJCAI 2020, pp. 231–237 (2020)
14. Li, B., Hao, D., Zhao, D., Zhou, T.: Mechanism design in social networks. In: Proceedings of the AAAI 2017, pp. 586–592 (2017)
15. Li, S.: Obviously strategy-proof mechanisms. Amer. Econ. Rev. **107**(11), 3257–87 (2017)
16. Lu, P., Sun, X., Wang, Y., Zhu, Z.A.: Asymptotically optimal strategy-proof mechanisms for two-facility games. In: Proceedings of the ACM-EC 2010, pp. 315–324 (2010)
17. Moulin, H.: On strategy-proofness and single peakedness. Public Choice **35**(4), 437–455 (1980)
18. Muller, E., Satterthwaite, M.A.: Strategy-proofness: the existence of dominant-strategy mechanisms. In: Hurwicz, L., Schmeidler, D., Sonnenschein, U. (eds.) Social Goals and Social Organization, pp. 131–171 (1985)
19. Okada, N., Todo, T., Yokoo, M.: SAT-based automated mechanism design for false-name-proof facility location. In: Proceedings of the the 22nd International Conference on Principles and Practice of Multi-Agent Systems (PRIMA 2019), pp. 321–337 (2019)
20. Osoegawa, K., Todo, T., Yokoo, M.: False-name-proof facility location on wheel graphs. In: Proceedings of the the 24th International Conference on Principles and Practice of Multi-Agent Systems (PRIMA 2022), pp. 139–155 (2022)
21. Procaccia, A.D., Tennenholtz, M.: Approximate mechanism design without money. ACM Trans. Econ. Comput. **1**(4), 18 (2013)
22. Schummer, J., Vohra, R.V.: Strategy-proof location on a network. J. Econ. Theory **104**(2), 405–428 (2002)
23. Serafino, P., Ventre, C.: Heterogeneous facility location without money on the line. In: Proceedings of the the 21st European Conference on Artificial Intelligence (ECAI 2014), pp. 807–812 (2014)
24. Shapley, L.S.: A value for n-person games, pp. 31–40. Cambridge University Press (1988)
25. Sui, X., Boutilier, C., Sandholm, T.: Analysis and optimization of multidimensional percentile mechanisms. In: Proceedings of the IJCAI 2013, pp. 367–374 (2013)
26. Todo, T., Iwasaki, A., Yokoo, M.: False-name-proof mechanism design without money. In: Proceedings of the AAMAS 2011, pp. 651–658 (2011)
27. Todo, T., Okada, N., Yokoo, M.: False-name-proof facility location on discrete structures. In: Proceedings of the the 24th European Conference on Artificial Intelligence (ECAI 2020), pp. 227–234 (2020)
28. Wada, Y., Ono, T., Todo, T., Yokoo, M.: Facility location with variable and dynamic populations. In: Proceedings of the AAMAS 2018, pp. 336–344 (2018)

29. Yoshida, K., Kimura, K., Todo, T., Yokoo, M.: Analyzing incentives and fairness in ordered weighted average for facility location games. In: Proceedings of the 27th European Conference on Artificial Intelligence (ECAI 2024), pp. 3380–3387 (2024)
30. You, B., Dierks, L., Todo, T., Li, M., Yokoo, M.: Strategy-proof house allocation with existing tenants over social networks. In: Proceedings of the AAMAS 2022, pp. 1446–1454 (2022)
31. Zhang, W., Zhao, D., Chen, H.: Redistribution mechanism on networks. In: Proceedings of the AAMAS 2020, pp. 1620–1628 (2020)
32. Zhao, D., Li, B., Xu, J., Hao, D., Jennings, N.R.: Selling multiple items via social networks. In: Proceedings of the AAMAS 2018, pp. 68–76 (2018)

Proportionally Dense Subgraphs: Parameterized Hardness and Efficiently Solvable Cases

Narmina Baghirova[1]([⊠])[ID] and Antoine Castillon[2]

[1] Department of Informatics, University of Fribourg, Fribourg, Switzerland
`narmina.baghirova@unifr.ch`
[2] Univ. Lille, CNRS, Centrale Lille, UMR 9189 CRIStAL, 59000 Lille, France
`antoine.castillon@ens-lyon.fr`

Abstract. A *proportionally dense subgraph (PDS)* of a graph is an induced subgraph of size at least two such that every vertex in the subgraph has proportionally as many neighbors inside as outside of the subgraph. Then, MAXPDS is the problem of determining a PDS of maximum size in a given graph. In this paper, we show that MAXPDS is FPT parameterized by $\Delta + \text{tw}$, where Δ is the maximum degree and tw is the treewidth of the input graph. This algorithm implies that the problem is polynomial-time solvable on graphs with degen ≤ 1, where degen represents the degeneracy. Moreover, the result implies that MAXPDS is polynomial-time solvable on graphs with $h \leq 2$ and graphs G such that $h(\overline{G}) \leq 2$, where h represents the h-index and $\overline{G}$ is the complement of G. Given the aforementioned results, we then show that MAXPDS is NP-hard parameterized by Δ. More specifically, we show that MAXPDS is NP-hard on graphs with $\Delta = 4$, $h = 4$ and degen = 2. Then, we show that MAXPDS is NP-hard on graphs G such that $\overline{G}$ is planar and $\Delta(\overline{G}) \leq 6$. We also show MAXPDS is NP-hard on graphs G such that degen($\overline{G}$) ≤ 2 and $\overline{G}$ is bipartite. Finally, we show that MAXPDS remains NP-hard on planar graphs.

Keywords: communities · dense subgraph · parameterized complexity · treewidth · complexity · algorithms

1 Introduction

1.1 Motivation and Known Results

In this paper, we study the MAXIMUM PROPORTIONALLY DENSE SUBGRAPH (MAXPDS) problem, introduced in [4] and motivated by community detection in networks. Community detection has diverse applications, ranging from developing social media algorithms for identifying individuals with shared interests to discovering a set of functionally associated proteins in protein-protein interaction networks in the field of bioinformatics (for instance, see [20]). By representing networks as graphs, with vertices representing individuals/elements and

S. Nakano and M. Xiao (Eds.): WALCOM 2025, LNCS 15411, pp. 16–31, 2025.
https://doi.org/10.1007/978-981-96-2845-2_2

edges representing connections between them, we can analyze the network from a structural perspective. In 2013, Olsen (see [18]) argued that the notion of *proportionality* in the definition of community is both intuitive and supported by observations in real-world networks. The definition studied in this paper captures the proportion of neighbors for each vertex within the subgraph and, most importantly, compares it with the proportion of neighbors outside of the subgraph. In this way, important properties of a community are captured.

Let us formally define the notion of a PDS studied in this paper. Given a graph $G = (V, E)$, $X \subseteq V$, and $u \in V$, we denote by $d_X(u)$ the number of neighbors of u in X.

Definition 1 (Bazgan et al. [4]). *Let $G = (V, E)$ be a graph and $S \subset V$, such that $2 \leq |S| < |V|$. The subgraph induced by S, which we denote by S for simplicity, is a* proportionally dense subgraph (PDS) *if for each $u \in S$:*

$$\frac{d_S(u)}{|S| - 1} \geq \frac{d_{\overline{S}}(u)}{|\overline{S}|}, \tag{1}$$

which is equivalent to $\dfrac{d_S(u)}{|S| - 1} \geq \dfrac{d(u)}{|V| - 1}$ and to $\dfrac{d(u)}{|V| - 1} \geq \dfrac{d_{\overline{S}}(u)}{|\overline{S}|}$, where $\overline{S} = V \setminus S$.

Notice that the proof of the first equivalence can be found in [6]. Then, since $d_S(u) = d(u) - d_{\overline{S}}(u)$ and $|S| = |V| - |\overline{S}|$, the second equivalence follows.

In other words, a PDS is an induced subgraph of size at least two such that every vertex of the subgraph is adjacent to proportionally at least as many vertices in the subgraph as to vertices outside. Then, MAXPDS consists in finding a PDS of maximum size, with respect to the number of vertices, in a given graph.

Below, we give definitions of problems of a similar flavour in the literature that fail to capture *natural* properties of community.

Two examples of such problems are, for instance, DENSEST K-SUBGRAPH and γ-QUASI-CLIQUES. DENSEST K-SUBGRAPH (see [11]) is the problem of finding a subgraph of maximum density containing exactly k vertices, where the density of a subgraph $S \subseteq V$ is commonly defined as $\frac{|E(S)|}{|S|}$, where $E(S)$ is the set of edges of G with both endpoints in S. Consider the example in which a vertex is part of the densest k-subgraph according to the definition, however, the vertex is adjacent to 10 out of 20 vertices in the subgraph but to 20 out of 25 vertices outside of the subgraph. Intuitively, considering the proportionality, it seems natural that, since the vertex is adjacent to a smaller proportion of vertices in the subgraph than outside, it should not be a part of the subgraph. Then, γ-QUASI-CLIQUES for some given constant $\gamma \in \,]0, 1[$, is the problem of finding an induced subgraph of maximum size with edge density at least γ. Even when γ is close to 1, the example used earlier applies similarly to the γ-Quasi-Cliques. There are other related problems in the literature, see for instance [17], for which we can similarly deduce that they may not necessarily capture the intuitive characteristics of the notion of community.

Let us give more details on the background of the problem. In 2013, Olsen (see [18]) introduced the so-called community structure problem, which is defined as the problem of partitioning vertices of a given graph into at least two sets (called *communities*), each containing at least two vertices, such that every vertex is adjacent to proportionally at least as many vertices in its own community as to vertices in any other community. Notice that in the definition introduced by Olsen in [18], the exact number of communities is not given, i.e. the only restriction is that there are at least two communities. In [6,10], the notion of *k-community structure* was first used to fix the number of communities to some integer $k \geq 2$. This problem has been intensively studied as well (see for instance [1,3,6,10]).

Based on the concept of k-community structures, the authors in [4] introduced the notion of a *proportionally dense subgraph (PDS)* and studied the problem of finding a PDS of maximum size. While closely linked to the definition of a k-community structure, this definition shifts from a partitioning problem to a maximization problem.

In [4], the authors show that MAXPDS is APX-hard on split graphs and NP-hard on bipartite graphs. In the same paper, the authors also show that deciding if a PDS is inclusion-wise maximal is co-NP-complete on bipartite graphs. Also, the authors present a polynomial-time $(2-\frac{2}{\Delta+1})$-approximation algorithm for the problem, where Δ is the maximum degree of the graph. Finally, the authors show that a PDS of maximum size can be found in linear time in Hamiltonian cubic graphs if a Hamiltonian cycle is given in the input. Recently, the authors of [2] introduced a framework, which allows solving a special family of partitioning problems in polynomial time in classes of graphs of bounded clique-width. As an application, they showed that the problem of finding a PDS of maximum size can be solved in polynomial time in classes of graphs of bounded cliquewidth. Notice that the algorithm presented in [2] is XP parameterized by clique-width, which implies an XP algorithm parameterized by treewidth. On the other hand, the algorithm that we introduce in this paper is FPT parameterized by $\Delta + \mathrm{tw}$.

Finally, we would like to mention that from a theoretical perspective, it is interesting to study problems where every vertex needs to satisfy a property which encompasses both global and local properties as in MAXPDS. This unique paradigm is not very common in the field of graph theory.

1.2   Our Contribution

In this paper, we study the parameterized complexity of MAXPDS with respect to parameters related to the density and the degree of the vertices of the given graph, and its complement. We present an FPT algorithm, parameterized by $\Delta + \mathrm{tw}$, where tw is the treewidth and Δ represents maximum degree, for solving MAXPDS. Then, we explain how the algorithm implies a polynomial-time algorithm on graphs with degen ≤ 1, where degen is the degeneracy of the graph. Moreover, the FPT algorithm implies that MAXPDS is polynomial-time solvable on graphs with $h \leq 2$ and graphs G such that $h(\overline{G}) \leq 2$, where h represents the h-index and $\overline{G}$ is the complement of G.

Given the existence of an FPT algorithm parameterized by $\Delta +$ tw, we investigate the complexity of MaxPDS when parameterized by Δ alone. Moreover, we explore the complexity for larger values of h and degen for graphs and their complements. We show that MaxPDS is para-NP-hard parameterized by Δ. More specifically, we show that MaxPDS is NP-hard on graphs with $\Delta = 4$, degen $= 2$ and $h = 4$. This result implies that the FPT algorithm parameterized by $\Delta +$ tw can not be improved to an FPT algorithm parameterized by Δ alone. Moreover, this shows that MaxPDS remains NP-hard when restricted to sparse graphs. Further, we show that MaxPDS is NP-hard when restricted to dense graphs as well. More specifically, we prove the following results. MaxPDS is NP-hard on graphs G such that $\overline{G}$ is planar and $\Delta(\overline{G}) \leq 6$. Then, we show that MaxPDS remains NP-hard on graphs G such that $\text{degen}(\overline{G}) = 2$ and $\overline{G}$ is bipartite. Finally, as mentioned previously, we prove NP-hardness on graphs whose complement is planar. Thus, we show that MaxPDS remains NP-hard on planar graphs.

In Table 1, we summarize the parameterized complexity results that we present in this paper. Notice that since degen $\leq h \leq \Delta$, if there exists a polynomial-time algorithm for a parameter for solving MaxPDS on a corresponding entry in the table, it implies polynomial-time complexity for the parameters corresponding to the entries above and on the right. For example, a polynomial time algorithm for $h \leq 2$ implies polynomial-time complexity for $h \leq 1$ and $\Delta \leq 2$. Conversely, NP-hardness extends to entries below and to the left.

Table 1. Summary of the complexity of MaxPDS when parameterized by degen, h and Δ of the graph G and its complement $\overline{G}$. Blue indicates polynomial-time solvability, red indicates NP-hardness, and a question mark means that the complexity is unknown.

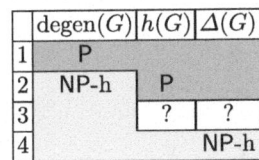
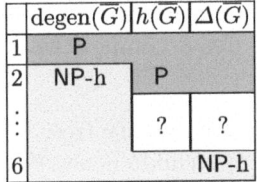

2   Preliminaries

2.1   Relevant Notions

In this paper, all graphs are simple and undirected. We use standard graph notations which can be found in [9]. We denote the *closed neighborhood* of a vertex $v \in V$ by $N[v] := N(v) \cup \{v\}$. The *neighborhood* of $v \in V$ *in* $V' \subseteq V$, denoted by $N_{V'}(v)$, is the set of vertices that are both in V' and adjacent to v, i.e. $N_{V'}(v) := N(v) \cap V'$.

For $S \subseteq V$, we denote by $\overline{S}$ the complement of S in V, i.e. $\overline{S} := V \setminus S$. We denote by $[\![a, b]\!]$, with $a, b \in \mathbb{N}$, the set $\{a, a + 1, \ldots, b\}$.

We denote a tree $T = (V, E)$ rooted at vertex r by (T, r). We denote by T_v the subtree of (T, r) rooted at v for every $v \in V(T)$. In a rooted tree, the *parent* of a vertex v is the vertex adjacent to v on the path to the root. Every vertex has a unique parent (notice, that root is the only vertex that has no parent). A *child* of a vertex v is a vertex of which v is the parent.

A graph is k-degenerate if its vertices can be successively deleted, according to an order, so that when deleted, each vertex has degree at most k. Such order is called a k-*elimination order*. The *degeneracy* of a graph is the smallest k such that it is k-degenerate.

Given a graph G, we define the *h-index* of G as the largest integer $h(G)$ such that there are at least $h(G)$ vertices of degree at least $h(G)$. Similarly, if the graph is clear from the context, we simply write h.

Further, let us define the notion of the treewidth of a graph. Given a graph $G = (V, E)$, a *tree-decomposition* of G is a tuple $(T, (X_t)_{t \in V(T)})$ where T is a tree and X_t is a subset of V called bag for every $t \in V(T)$, satisfying the following properties:

1. $\displaystyle\bigcup_{t \in V(T)} X_t = V$,

2. For all $v \in G$, if $v \in X_t \cap X_{t'}$, then all $X_{t''}$ with t'' on the unique path from t to t' contain v as well. Equivalently, the vertices t of T such that $v \in X_t$ form a connected subtree of T,

3. For every $uv \in E$, there exists X_t, for some $t \in V(T)$, that contains both u and v. Equivalently, two vertices are adjacent in G only if the corresponding subtrees have a vertex in common in T.

Let t be a node of T, such that $|X_t| = \max\{|X_{t'}| : t' \in V(T)\}$. The *width* of a tree-decomposition equals $|X_t| - 1$. The *treewidth* of a graph G, denoted by $\mathrm{tw}(G)$, is the minimum width among all possible tree-decompositions of G.

A *nice* tree-decomposition of G is a tree-decomposition $(T, (X_t)_{t \in V(T)})$ such that:

1. T is a rooted binary tree,
2. if t has two children, say t' and t'', then $X_t = X_{t'} = X_{t''}$,
3. if t has one child, say t', then either
   (a) $|X_t| = |X_{t'}| - 1$ and $X_t \subset X_{t'}$ or
   (b) $|X_t| = |X_{t'}| + 1$ and $X_{t'} \subset X_t$.

We call the node t in item 2 a *join node* and nodes in item 3a and item 3b *forget* and *introduce node*, respectively. The authors in [16, Lemma 13.1.3] show that from a given tree-decomposition of a graph G, we can construct a nice tree-decomposition of G with the same width in linear time.

In this paper, we use the notion of an *easy* tree-decomposition, which is a nice tree-decomposition with the additional property that the bags associated with the root and leaves of the corresponding tree are empty sets (see [8, Definition 2.5.8]).

Given a nice tree-decomposition of width k and $\mathcal{O}(n)$ nodes, one can construct in $\mathcal{O}(kn)$ time an easy tree-decomposition of width k and $\mathcal{O}(kn)$ nodes (see [8, Page 57]).

Further, as mentioned in Sect. 1, MaxPDS consists in finding a PDS of maximum size in a given graph. The corresponding decision problem is, given a graph $G = (V, E)$ and an integer k, to determine whether G contains a PDS of size at least k.

We say that a vertex $v \in S$ is *satisfied with respect to S* if it satisfies (1). If the PDS S is clear from the context, we may simply say that a vertex $v \in S$ is *satisfied*. Let $G = (V, E)$ be a graph and let C be a connected component of G. Then, when we say that $C \subseteq S$ when every vertex of $G[C]$ is in S.

The authors in [4] showed the following two results, which we frequently refer to in this paper.

Theorem 1. *[4, Theorem 4]* *For any given graph $G = (V, E)$, a proportionally dense subgraph of size $\lceil \frac{|V|}{2} \rceil$ or $\lceil \frac{|V|}{2} \rceil + 1$ can be found in linear time.*

The following lemma gives an upper bound on the size of a PDS depending on the maximum degree of a vertex in a given graph.

Lemma 1. *[4, Lemma 8]* *Let $G = (V, E)$ be a graph and $S \subseteq V$ such that $G[S]$ is a PDS. Then, $|S| \leq \lfloor \frac{(\Delta - 1)|V| + 1}{\Delta} \rfloor$.*

Due to lack of space, proofs of results marked by ♠ are omitted.

2.2 Properties

In this subsection, we derive properties of PDS's that we use later in the paper. Let $G = (V, E)$ be a graph. Let us note the following. Let $v \in V$ be such that $d(v) = 0$. Then, it is easy to see that $S = V \setminus \{v\}$ forms a PDS of a maximum size in G. Hence, as of now, we assume that $d(v) \geq 1$, for all $v \in V$.

Below, we give an upper bound on the size of a PDS to ensure that every vertex of G is satisfied with respect to the PDS.

Lemma 2 (♠). *Let $G = (V, E)$ be a graph and let S be a PDS in G. Then, for every $u \in S$ we have*

$$|S| \leq \frac{d_S(u)}{d(u)}(|V| - 1) + 1.$$

In the lemma below, we show that if a PDS S in G is sufficiently large and the degree of some vertex u is relatively small, then the inclusion of vertex u in S implies the inclusion of its neighborhood $N(u)$ in S.

Lemma 3 (♠). *Let $G = (V, E)$ be a graph and S a PDS in G and let $u \in S$. If $d(u) < \frac{|V| - 1}{|V| - |S|}$, or equivalently if $\frac{d(u) - 1}{|S| - 1} < \frac{1}{|V| - |S|}$, then we have $N(u) \subseteq S$.*

The lemma introduced above implies the following.

Corollary 1 (♠). *Let $G = (V, E)$ be a graph, S be a PDS in G, $X \subseteq V$ be such that $G[X]$ is connected and for all $u \in X$, $d(u) < \frac{|V| - 1}{|V| - |S|}$ for all $u \in X$. Then, either $N[X] \subseteq S$ or $N[X] \subseteq \overline{S}$.*

3   Graphs of Bounded Treewidth and Bounded Maximum Degree

In this subsection, we show how to find a PDS of maximum size in graphs of bounded treewidth and bounded maximum degree. More specifically, we prove the following theorem.

Theorem 2. MAXPDS *is* FPT *parameterized by* tw $+\Delta$.

To prove the theorem above, we introduce a dynamic programming algorithm which traverses an easy tree decomposition of a given graph in a bottom-up fashion, i.e. starting with the leaves and progressing upwards towards the root. Let $G = (V, E)$ be the input graph and let $(T, (X_t)_{t \in V(T)})$ be an easy tree decomposition of G. For $t, t' \in V(T)$, we say that $t' \leq t$ if t' is contained in the subtree of T rooted at t. For $t \in V(T)$, we denote $Y_t = (\bigcup_{t' \leq t} X_{t'}) \setminus X_t$. Informally, Y_t represents the set of forgotten vertices in the subtree of T rooted at a node t.

For every node $t \in V(T)$, we define tuples associated with t. A tuple (Σ, κ, δ) associated with t is such that:

– $\Sigma \subseteq X_t$;
– $\kappa \in \{0, \ldots, |Y_t|\}$;
– $\delta : X_t \to [\![0, \Delta]\!]$ is a function.

Recall that we may assume without loss of generality that $d(v) \geq 1$, for all $v \in V$ (see Sect. 2.2).

Let $t \in V(T)$ and $Z_t = X_t \cup Y_t$. For $S \subseteq Z_t$, we say that S is $(\Sigma, \kappa, \delta)_t$-*compatible* if $\Sigma = S \cap X_t$, $\kappa = |S \cap Y_t|$, and $\delta(v) = d_{S \cap Y_t}(v)$, for all $v \in X_t$. We say that S is $(\Sigma, \kappa, \delta)_t$-*optimal* if S is $(\Sigma, \kappa, \delta)_t$-compatible and $\min\limits_{v \in S \cap Y_t} \dfrac{d_S(v)}{d(v)} = \max\limits_{S'}\{\min\limits_{v \in S' \cap Y_t} \dfrac{d_{S'}(v)}{d(v)}\}$ among all $(\Sigma, \kappa, \delta)_t$-compatible sets S'; when $S' = \emptyset$, we set the value of the ratio to $+\infty$.

With each node $t \in V(T)$, we associate a function denoted by up_t. The input of the function is a tuple (Σ, κ, δ) associated with node t, as defined above, and the output is a value U. This value U represents an upper bound, derived from Lemma 2, on the size of a PDS Γ in G such that $\Gamma \cap Z_t = S$, where S is any $(\Sigma, \kappa, \delta)_t$-optimal set.

We want to ensure that the following property holds for each $t \in V(T)$.

Property 1. Let $t \in V(T)$ and (Σ, κ, δ) a tuple associated with t.

1. Let S be a $(\Sigma, \kappa, \delta)_t$-optimal set. Then $up_t(\Sigma, \kappa, \delta) = U$, where $U = (n - 1) \min\limits_{v \in S \cap Y_t} \dfrac{d_S(v)}{d(v)}$.
2. $up_t(\Sigma, \kappa, \delta) = -\infty$ if and only if there does not exist a $(\Sigma, \kappa, \delta)_t$-compatible set.

The algorithm introduced below updates the values $up_t(\Sigma, \kappa, \delta)$ so that Property 1 remains satisfied when going up the tree decomposition. Eventually, at the root of the tree decomposition, Property 1 allows us to derive the size of a PDS of maximum size in G.

We initialize the values of up_t to $-\infty$, for all $t \in V(T)$. In the following, we consider each node type in an easy tree decomposition independently. For each type, we describe the behavior of the algorithm and prove that Property 1 holds at each step. In particular, this shows that Property 1 remains true when going up the tree decomposition, i.e. if it holds for every child of a node t, then it also holds for node t.

First, let us consider the leaf nodes. Remember that, by the definition of an easy tree decomposition, the leaf nodes correspond to empty sets. Hence, by definition, there is no restriction on the value U from Property 1, and we set the upper bound to $+\infty$.

[**Leaf node**] When a node t is a leaf node, we proceed as follows:

$$up_t(\emptyset, 0, \varnothing) = +\infty.$$

Lemma 4 (♠). *Property 1 is satisfied by leaf nodes.*

Let us now explain how to proceed when the algorithm encounters an introduce node t. Informally, when a new vertex v_0 is introduced, we need to distinguish two cases: either v_0 belongs to the set Σ that we consider or it does not. We consider both cases and modify the values accordingly. Let $f : E \to F$ be a function and $A \subseteq E$. Then, *the restriction of f to A*, denoted by $f_{|A}$, is a function $f_{|A} : A \to F$ such that $f_{|A}(x) = f(x)$, for all $x \in A$.

[**Introduce node**] Let t be an introduce node, t' its child and $X_t = X_{t'} \cup \{v_0\}$. For all tuples (Σ, κ, δ) associated with t with $\delta(v_0) = 0$:

$$up_t(\Sigma, \kappa, \delta) = up_{t'}(\Sigma \setminus \{v_0\}, \kappa, \delta_{|X_{t'}}).$$

Lemma 5 (♠). *Property 1 is satisfied by introduce nodes.*

Let us now consider the case when the algorithm encounters a forget node t. In this case, for each tuple $(\Sigma', \kappa', \delta')$ associated with the child t' of t, the forgotten vertex v_0 is either contained in the set Σ' or not. If v_0 is contained in Σ', then the new corresponding values κ, δ, and U are computed accordingly; otherwise, the same values can be used.

[**Forget node**] Let t be a forget node, t' be its child, and $X_{t'} = X_t \cup \{v_0\}$. For all tuples (Σ, κ, δ) associated with t, we set

$$up_t(\Sigma, \kappa, \delta) = \max \begin{cases} up_{t'}(\Sigma, \kappa, \delta'), \\ \text{for each function } \delta' \text{ such that } \delta'(v) = \delta(v) \text{ for all } v \in X_t, \\ \\ \min\{up_{t'}(\Sigma \cup \{v_0\}, \kappa - 1, \delta'), (n-1)\frac{d_{\Sigma'}(v_0) + \delta'(v_0)}{d(v_0)}\}, \\ \text{for each function } \delta' \text{ such that } \delta'(v) = \delta(v) - 1 \text{ for all } v \in N(v_0) \cap X_t \\ \text{and } \delta'(v) = \delta(v) \text{ for all } v \notin N(v_0). \end{cases}$$

Lemma 6 (♠). *Property 1 is satisfied by forget nodes.*

[Join node] Let t be a join node, t' and t'' be its children. For all tuples (Σ, κ, δ) associated with t:

$$up_t(\Sigma, \kappa, \delta) = \max_{\kappa' + \kappa'' = \kappa} \max_{\delta' + \delta'' = \delta} \min\{up_{t'}(\Sigma, \kappa', \delta'), up_{t''}(\Sigma, \kappa'', \delta'')\}.$$

Intuitively, since $Y_t = Y_{t'} \cup Y_{t''}$ and $Y_{t'} \cap Y_{t''} = \emptyset$ by definition of a tree decomposition, the κ (resp. δ) values represent the size of disjoint sets of vertices, thus the new κ (resp. δ) value will be the sum of the two previous ones.

Lemma 7 (♠). *Property 1 is satisfied by join nodes.*

We are now ready to prove the main result of this section, namely Theorem 2.

Proof of Theorem 2. The algorithm consists of traversing a given easy tree decomposition in a bottom-up fashion and applying one of Lemma 4 to 7 depending on the node type. Finally, at the root node r, the algorithm returns the largest value of κ such that the tuple $(\emptyset, \kappa, \varnothing)$ associated with r satisfies the inequality $up_r(\emptyset, \kappa, \varnothing) \geq \kappa$.

Correctness: By Lemmas 4 to 7 the root node satisfies Property 1. Moreover, at the root node, we have $X_{root} = \emptyset$ and $Y_{root} = V$. Let S be a PDS of maximum size in G. Notice, that S is $(\emptyset, |S|, \varnothing)_r$-compatible and by the definition of a PDS we know that $up_r(\emptyset, |S|, \varnothing) = U$, otherwise there exists $u \in S$ which is not satisfied with respect to S.

Conversely, let S be $(\emptyset, \kappa, \varnothing)_r$-optimal set. Then, $up_r(\emptyset, \kappa, \varnothing) = U$, where $U = (n-1) \min_{v \in S \cap Y_t} \dfrac{d_S(v)}{d(v)}$, which implies that all vertices in S are satisfied with respect to S and it forms a PDS.

Complexity: First, notice that the lemmas above are applied at most $\mathcal{O}(n \cdot \text{tw})$ times and the complexity of the four lemmas above are the following:

- The complexity of Lemma 4 is clearly $\mathcal{O}(1)$.
- The complexity of Lemma 5 is $\mathcal{O}(2^{tw} \cdot n \cdot \Delta^{tw})$.
- The complexity of Lemma 6 is $\mathcal{O}(2^{tw} \cdot n^2 \cdot \Delta^{tw})$.
- The complexity of Lemma 7 is $\mathcal{O}((2^{tw} \cdot n \cdot \Delta^{tw})^2)$.

Then, the overall complexity of the algorithm is $\mathcal{O}(4^{tw} \cdot \Delta^{2tw} \cdot n^3 \cdot \text{tw})$.

Note that graphs with degen ≤ 1 are forests. Hence the complexity of the algorithm introduced above is $\mathcal{O}(n^5)$ on such graphs. In other words, the algorithm above implies a polynomial-time complexity on graphs with degen ≤ 1. Notice, that a PDS of a maximum size can easily be determined by backtracking the dynamic programming table.

Moreover, the algorithm described above implies that the MaxPDS problem is solvable in polynomial time for graphs with $h \leq 2$. More specifically, such graphs contain at most two vertices with a degree higher than 2, while the remaining vertices have a degree of at most 2. Graphs with degree at most 2 are a union of cycles and paths and have treewidth at most 2. Thus, graphs with $h \leq 2$ have treewidth at most 4 and the following result follows.

Corollary 2. MAXPDS *is polynomial-time solvable in graphs with* $h \leq 2$.

Below we show that the FPT algorithm implies polynomial-time solvability on graphs G such that $h(\overline{G}) \leq 2$.

Corollary 3. MAXPDS *can be solved in polynomial time in graphs* G *such that* $h(\overline{G}) \leq 2$.

Proof. Let $G = (V, E)$ be a graph such that $h(\overline{G}) \leq 2$. Let u^*, v^* be the two vertices of maximum degree in $\overline{G}$. Notice that any set $S \subseteq V$ with $|S| \geq \frac{|V|}{2} + 1$ is a PDS in G if and only if the vertices in $\{u^*, v^*\} \cap S$ are satisfied and $V \setminus S$ dominates all non-isolated vertices of $\overline{G}$.

Indeed, for all $u \in S$, if all neighbors of u in $\overline{G}$ are also contained in S, then the following holds

$$\frac{|S| - 1 - d_{\overline{G}}(u)}{|S| - 1} < 1 = \frac{|V \setminus S|}{|V \setminus S|}.$$

Thus, u is not satisfied with respect to S.

Conversely, let $u \in S$ be such that $d_{\overline{G}}(u) \leq 2$. If u has at least one neighbor in $\overline{G}$ contained in $V \setminus S$, then since $|S| \geq \frac{|V|}{2} + 1 \geq |V \setminus S|$, the following holds

$$\frac{|S| - 2}{|S| - 1} = 1 - \frac{1}{|S| - 1} \geq 1 - \frac{1}{|V \setminus S|} = \frac{|V \setminus S| - 1}{|V \setminus S|}.$$

Thus, u is satisfied.

Since, $\overline{G}$ has h-index at most 2, as discussed earlier, it has treewidth at most 4. Hence, a dominating set of minimum size can be found in polynomial time in such graphs [19]. Then, a PDS of maximum size in G can be found in polynomial time by combining the algorithm presented above for u^* and v^* and the algorithm for finding a dominating set of minimum size in $\overline{G}$.

4   Parameterized Hardness

4.1   Maximum Degree and Degeneracy

In this subsection, we establish the NP-hardness results of MAXPDS, parameterized by the degeneracy and maximum degree. More specifically, we prove that MAXPDS is NP-hard on 2-degenerate graphs in which each vertex has a degree at most 4. Note that due to space limitations, for all results in this section, we only provide proof sketches that omit details.

Theorem 3. MAXPDS *is NP-hard on graphs with* degen $= 2$ *and* $\Delta = 4$.

Proof Sketch: We provide a polynomial-time reduction from INDEPENDENT SET on cubic graphs. This problem is known to be NP-hard [14].

Let A, B be two integers such that $A, B >> n$, A and B are polynomials in n and $A \approx 3B$. The exact values of A and B are provided in the appendix.

Construction: Given an instance of INDEPENDENT SET on cubic graphs, that is, a cubic graph $G = (V, E)$ and a positive integer k, we construct $G' = (V', E')$, a positive integer k' an instance of MAXPDS as follows.

– The vertex set V' is constructed as follows.
  • We create a copy of vertices in V. We denote by V_G the set formed by such vertices.
  • We create a vertex e_{uv} for each uv in E. We denote the corresponding set of vertices by V_E.
  • We create two sets of A and B new vertices, which we denote by V_A and V_B, respectively.
– The edge set E' is constructed as follows.
  • By properly choosing the values of A and B, we have $B > n$. Then, for every $u \in V_G$, there exists a unique $u' \in V_B$ such that $uu' \in E'$.
  • By properly choosing the values, we also have that $A > 2m$. Then, each vertex in V_E is adjacent to two unique vertices in V_A, i.e. for every $u \in V_E$ there exists two vertices $w, w' \in V_A$ such that $wu, w'u \in E'$.
  • Each vertex $e_{uv} \in V_E$ is adjacent to u and v in the set V_G.
  • The vertices of set V_A form a path.
  • The vertices of set V_B form a path.

Let $k' = A + m + (n - k)$. See Fig. 1 for an illustration of the construction. Also, since N is a polynomial in n, G' can obviously be constructed in polynomial time in n.

Further, to complete the proof, it remains to show that $\Delta(G') = 4$, $\mathrm{degen}(G') = 2$ and prove the following. Let I be an independent set of size k in G. Let I' be the set of vertices in G' corresponding to the set I. It can be shown that $S = V_A \cup V_E \cup (V_G \setminus I')$ is a PDS of size k' in G'. Conversely, let S be a PDS of size at least k' in G'. Let $I' = V' \cap (V_G \setminus S)$ and I be the set of vertices in G corresponding to the set I'. It remains to show that I is an independent set of size k in G.   □

Theorem 3 implies the following.

Corollary 4. MAXPDS *is para-NP-hard parameterized by* degen *and* Δ.

Note that, under the classical complexity assumptions, this result implies that the FPT algorithm parameterized by $\Delta + \mathrm{tw}$ (presented in Sect. 3) cannot be improved to either an FPT or an XP algorithm parameterized by Δ solely. Also, since $h \leq \Delta$, the NP-hardness on graphs with $\Delta \leq 4$ implies the NP-hardness on graphs with $h \leq 4$.

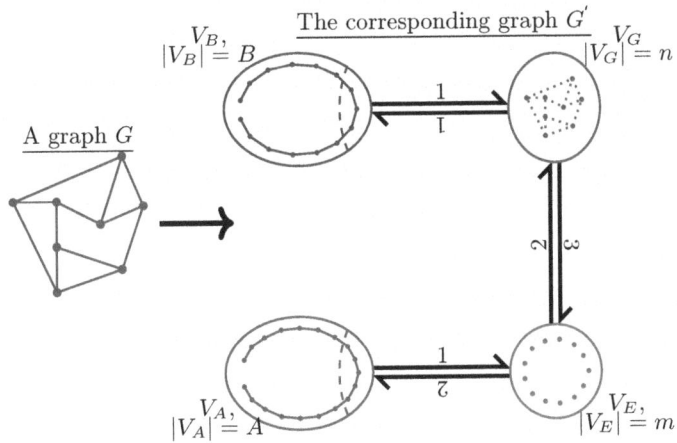

Fig. 1. On the left we present an example of a cubic graph G on 8 vertices. On the right, we present the construction explained above. A double arrow $X \overset{x}{\underset{y}{\rightleftharpoons}} Y$ represents that the vertices in X have exactly x neighbors in Y and the vertices in Y have exactly y neighbors in X.

Parameterization of the Complement Graph Hereafter, we show that MAXPDS is NP-hard on graphs G such that $\overline{G}$ is planar and $\Delta(\overline{G}) = 6$. Moreover, we show that MAXPDS remains NP-hard on graphs G, such that $\text{degen}(\overline{G}) = 2$ and $\overline{G}$ is bipartite.

Theorem 4. MAXPDS *is NP-hard on graphs G such that $\overline{G}$ is planar and* $\Delta(\overline{G}) = 6$.

Proof Sketch: We provide a polynomial-time reduction from INDEPENDENT SET on cubic graphs. This problem is known to be NP-hard [14].

Construction: Given an instance of INDEPENDENT SET on cubic graphs, that is, a cubic graph $G = (V, E)$, with $|V| = n$ and $|E| = m$, and a positive integer k, we construct $G' = (V', E')$, a positive integer k' an instance of MAXPDS as follows.

– The vertex set V' is constructed as follows.
  - We create a copy of vertices in V. We denote by V_G the set formed by such vertices.
  - We create a vertex e_{uv} for each $uv \in E$. We denote the corresponding set of vertices by V_E.
– The edge set E' is constructed as follows.
  - Let $v \in V$ and w, u, x be its neighbors in G. Let v', w', u', x' be the corresponding vertices in G'. Then, v' is adjacent to all vertices other than $w', u', x', e_{vw}, e_{vu}$, and e_{vx} in G'. Notice, that since G is a cubic graph, v is not adjacent to four vertices in V_G and 3 vertices in V_E.

- The vertices of V_E form a clique.

Note that the graph G' can be obtained from G by subdividing each edge of the graph and then taking its complement. Thus if the input graph G is planar, then the complement of G' constructed as explained above is planar as well. INDEPENDENT SET is NP-hard on planar graphs (see [15]).

Let $k' = m + k$. Further, to complete the proof, it remains to show that $\Delta(\overline{G'}) = 6$ and prove the following. Let I be an independent set of size k in G and I' be the corresponding set of vertices in G'. It can be shown that $S = I' \cup V_E$ is a PDS of size k' in G'. Conversely, let S be a PDS of size at least k' in G'. It remains to show that the set I in G corresponding to the set $I' = S \cap V_G$ in G' is an independent set in G of size at least k.   □

By using a similar reduction, we obtain an NP-hardness results on graphs that are complements of 2-degenerate bipartite graphs.

Theorem 5 (♠). MAXPDS *is NP-hard even when restricted to graphs G such that* $\operatorname{degen}(\overline{G}) = 2$ *and* $\overline{G}$ *is bipartite.*

4.2   Planar Graphs

As discussed in Sect. 4.1, MAXPDS remains NP-hard on graphs that are complements of planar graphs. In this section, we demonstrate that MAXPDS remains NP-hard on planar graphs. To achieve this, we reduce from INDEPENDENT SET, restricting it to instances $(G = (V, E), k)$ where G is planar, has no isolated vertices, $\Delta(G) \leq 6$, $|V| = n > 2k$, $|E| = m > 6k$. Indeed, INDEPENDENT SET remains NP-hard on such instances [12].

Before we move on to the main result, we introduce a preliminary result.

Lemma 8 (♠).   *Given a graph $G = (V, E)$, with $|V| = n$ and $|E| = m$ and such that $n \leq m \leq n^2$, without isolated vertices and a positive integer k such that $2k < n$ and $6k < m$, there exist integers N, k', n_x, n_y such that:*

1. $N = mn_x + m + nn_y$,
2. $k' = mn_x + m + (n - k)n_y = N - kn_y$,
3. $n_x > 5$,
4. $\frac{n_x + 1}{k' - 1} \geq \frac{1}{kn_y}$,
5. $\frac{2}{kn_y} > \frac{n_x}{k' - 1}$,
6. $\frac{1}{(k-1)n_y} > \frac{n_x + 1}{N - (k-1)n_y - 1}$,
7. $m(n_x + 1) > kn_y$,
8. N *is polynomial in* n.

Now, let us proceed to the main result of this section. Due to space limitations, we provide only the proof sketch.

Theorem 6. MAXPDS *is NP-hard on planar graphs.*

Proof Sketch: We provide a polynomial-time reduction from INDEPENDENT SET, restricted to the instances specified at the beginning of this section.

Sketch of Construction: Let $G = (V, E)$ be a planar graph such that $\Delta(G) \leq 6$, and a positive integer k, an instance of INDEPENDENT SET, with $|V| = n > 2k$ and $|E| = m > 6k$. Moreover, we assume that $m \geq n$. We fix a planar representation of G. Let k', n_x, n_y be values that satisfy conditions from Lemma 8. We construct an equivalent instance $G' = (V', E')$ of MAXPDS as follows.

– The vertex set V' is constructed as follows.
   - Copy of the vertices V (represented in blue).
   - For each $u \in V$, we create $n_y - 1$ new vertices $y_u^{(2)}, ..., y_u^{(n_y)}$ (represented in purple). We denote $u = y_u^{(1)}$.
   - For each $uv \in E$, we create a vertex e_{uv} in V' (represented in red).
   - For each $uv \in E$, we create n_x new vertices $x_{uv}^{(1)}, ..., x_{uv}^{(n_x)}$ (represented in green).
– The edge set E' is constructed as follows.
   - For each $uv \in E$, we create edges ue_{uv} and $e_{uv}v$.
   - For every $u \in V$, $y_u^{(1)}, ..., y_u^{(n_y)}$ induce a path.
   - For each $uv \in E$, we create the edges between e_{uv} and $x_{uv}^{(i)}$, for every $i \in \{1, ..., n_x\}$.
   - For each face F of G, the vertices of X contained in F are connected by a path.

See Fig. 2 for a graphical illustration of such construction.

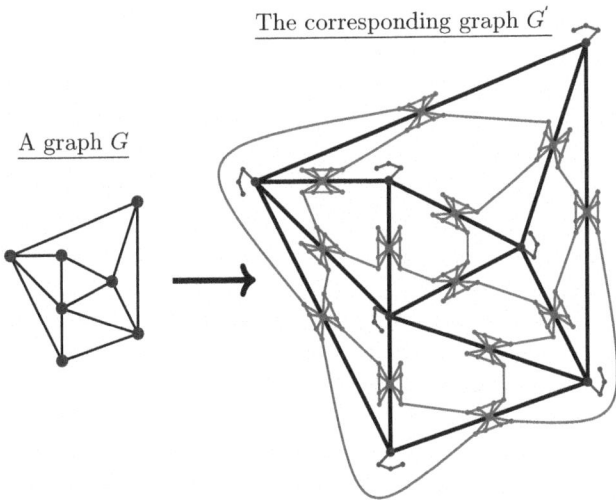

Fig. 2. On the left we present an example of a graph G on 7 vertices. On the right, we present the construction G' explained above. (Color figure online)

To complete the proof, we need to show that the reduction introduced above can be performed in polynomial time with respect to n and prove the following. Note that we use the color code used in Fig. 2.

Let I be an independent set of size k in G and let I' be the corresponding set of vertices in G'. Then, it can be shown that $S = V' \setminus (\bigcup_{u \in I'} Y_u)$, i.e., the set formed by all red, green, blue, and purple vertices corresponding to vertices outside of I' in G', is a PDS of G' of size k'. Conversely, let S be a PDS in G' of size at least k'. It remains to show that the blue vertices outside S, let us denote those vertices by I', form an independent set I of G, where I is the set of vertices in G corresponding to the set I' in G'.   □

5   Conclusion

There remain several interesting open questions, some of which are presented hereafter.

- As mentioned in Table 1, the complexity of finding a (connected) PDS of maximum size in graphs with $\Delta(G) = 3$, $h(G) = 3$, $3 \leq \Delta(\overline{G}) \leq 5$ or $3 \leq h(\overline{G}) \leq 5$ remain open problems.
- We show that the MaxPDS is para-NP-hard parameterized by Δ. However, we give an FPT algorithm parameterized by $\Delta + \mathrm{tw}$ for finding PDS of maximum size. Note that this algorithm implies an XP algorithm parameterized by treewidth, a result which was already introduced by the authors in [2]. It would be interesting to determine if XP algorithms, introduced in this paper or by the authors in [2], parameterized by treewidth could be improved to FPT algorithms.

  For instance, the authors in [13], showed that a related problem called a BALANCED SATISFACTORY PARTITION (a problem of finding a partition of the vertex set into two equal-sized subsets where each vertex in a subset has at least as many neighbors within its own subset as outside), introduced in [7], is $W[1]$-hard parameterized by treewidth.
- Finally, in this paper, we show that MaxPDS remains NP-hard on graphs whose complement is planar, as well as on planar graphs. As mentioned in Sect. 1, the authors in [5] showed that MaxPDS is NP-hard on bipartite graphs, while in this paper we show that MaxPDS is NP-hard on graphs whose complement is bipartite. This leads to a natural question: knowing the complexity of MaxPDS for a given hereditary graph class $\mathcal{G}$, what can we infer about the complexity of MaxPDS for the graph class consisting of the complements of graphs in $\mathcal{G}$?

References

1. Baghirova, N., Dallard, C., Ries, B., Schindl, D.: Finding k-community structures in special graph classes. Discret. Appl. Math. **359**, 159–175 (2024). https://doi.org/10.1016/j.dam.2024.07.033

2. Baghirova, N., Gonzalez, C.L., Ries, B., Schindl, D.: Locally checkable problems parameterized by clique-width. In: 33rd International Symposium on Algorithms and Computation (ISAAC 2022). Leibniz International Proceedings in Informatics (LIPIcs), vol. 248, pp. 31:1–31:20 (2022)
3. Bazgan, C., Chlebíková, J., Dallard, C.: Graphs without a partition into two proportionally dense subgraphs. Inf. Process. Lett. **155**, 105877 (2020). https://doi.org/10.1016/j.ipl.2019.105877
4. Bazgan, C., Chlebíková, J., Dallard, C., Pontoizeau, T.: Proportionally dense subgraph of maximum size: complexity and approximation. J. Comb. Algorithms Inform. Comput. Sci. **270**, 25–36 (2019). https://doi.org/10.1016/j.dam.2019.07.010
5. Bazgan, C., Chlebikova, J., Dallard, C., Pontoizeau, T.: Proportionally dense subgraph of maximum size: complexity and approximation. Discret. Appl. Math. **270** (2019). https://doi.org/10.1016/j.dam.2019.07.010
6. Bazgan, C., Chlebikova, J., Pontoizeau, T.: Structural and algorithmic properties of 2-community structures. Algorithmica **80**(6), 1890–1908 (2018). https://doi.org/10.1007/s00453-017-0283-7
7. Bazgan, C., Tuza, Z., Vanderpooten, D.: The satisfactory partition problem. Discret. Appl. Math. **154**(8), 1236–1245 (2006). https://doi.org/10.1016/j.dam.2005.10.014
8. Bonomo-Braberman, F., Gonzalez, C.L.: A new approach on locally checkable problems. Discret. Appl. Math. **314**, 53–80 (2022). https://doi.org/10.1016/j.dam.2022.01.019
9. Diestel, R.: Graph Theory. Springer, Cham (2005)
10. Estivill-Castro, V., Parsa, M.: On connected two communities. In: Proceedings of the Thirty-Sixth Australasian Computer Science Conference, ACSC 2013, vol. 135, pp. 23–30. Australian Computer Society, Inc. (2013)
11. Feige, U., Peleg, D., Kortsarz, G.: The dense k-subgraph problem. Algorithmica **29**, 410–421 (2001). https://doi.org/10.1007/s004530010050
12. Fleischner, H., Sabidussi, G., Sarvanov, V.I.: Maximum independent sets in 3- and 4-regular Hamiltonian graphs. Discret. Math. **310**(20), 2742–2749 (2010). https://doi.org/10.1016/j.disc.2010.05.028. graph Theory - Dedicated to Carsten Thomassen on his 60th Birthday
13. Gaikwad, A., Maity, S., Tripathi, S.K.: Parameterized complexity of satisfactory partition problem. Theor. Comput. Sci. **907**, 113–127 (2022). https://doi.org/10.1016/j.tcs.2022.01.022
14. Garey, M.R., Johnson, D.S.: Computers and Intractability; A Guide to the Theory of NP-Completeness. W. H. Freeman & Co. (1990)
15. Garey, M.R., Johnson, D.S., Stockmeyer, L.: Some simplified NP-complete graph problems. Theor. Comput. Sci. **1**, 237–267 (1976). https://doi.org/10.1016/0304-3975(76)90059-1
16. Kloks, T.: Treewidth. Lecture Notes in Computer Science, vol. 842 (1994). https://doi.org/10.1007/BFb0045375
17. Komusiewicz, C.: Multivariate algorithmics for finding cohesive subnetworks. Algorithms **9**(1) (2016). https://doi.org/10.3390/a9010021
18. Olsen, M.: A general view on computing communities. Math. Soc. Sci. **66**(3), 331–336 (2013). https://doi.org/10.1016/j.mathsocsci.2013.07.002
19. Vatshelle, M.: New width parameters of graphs. Ph.D. thesis (2012)
20. Zhu, Y., Li, Y., Liu, J., Qin, L., Yu, J.X.: Discovering large conserved functional components in global network alignment by graph matching. BMC Genomics **19**(Suppl 7), 670 (2018)

Computing Conforming Partitions with Low Stabbing Number for Rectilinear Polygons

Therese Biedl[1], Stephane Durocher[2], Debajyoti Mondal[3], Rahnuma Islam Nishat[4], and Bastien Rivier[4(✉)]

[1] University of Waterloo, Waterloo, Canada
biedl@uwaterloo.ca
[2] University of Manitoba, Winnipeg, Canada
stephane.durocher@umanitoba.ca
[3] University of Saskatchewan, Saskatoon, Canada
dmondal@cs.usask.ca
[4] Brock University, St. Catharines, Canada
{rnishat,brivier}@brocku.ca

Abstract. A *conforming partition* of a rectilinear n-gon P is a partition of P into rectangles without using Steiner points (i.e., all corners of all rectangles must lie on the boundary of P). The stabbing number of such a partition is the maximum number of rectangles intersected by an axis-aligned segment lying in the interior of P. In this paper, we examine the problem of computing conforming partitions with low stabbing number. We show that computing a conforming partition with stabbing number at most 4 is $\mathcal{NP}$-hard, which strengthens a previously known hardness result [Durocher & Mehrabi, Theor. Comput. Sci. 689: 157–168 (2017)] and eliminates the possibility for fixed-parameter-tractable algorithms parameterized by the stabbing number unless $\mathcal{P} = \mathcal{NP}$. In contrast, we give (i) an $O(n \log n)$-time algorithm to decide whether a conforming partition with stabbing number 2 exists, (ii) a fixed-parameter-tractable algorithm parameterized by both the stabbing number and treewidth of the pixelation of the polygon, and (iii) a fixed-parameter-tractable algorithm parameterized by the stabbing number for polygons without holes in general position.

Keywords: Stabbing Number · Partition · Rectilinear Polygon · NP-Hard · Fixed-Parameter Tractability · Treewidth

1 Introduction

Partitioning an n-gon P with nice properties is a fundamental paradigm in computational geometry. We are interested in the *stabbing number* of a partition,

This paper is dedicated to the memory of our friend Saeed, whose work inspired the project. This work is funded in part by the Natural Sciences and Engineering Research Council of Canada (NSERC). The *full version* of this article is at https://doi.org/10.48550/arXiv.2411.11274.

i.e., the maximum number of elements of the partition that are intersected by a line segment that lies in the interior of P. Consider a partition of P into triangles. Such a partition yields a data structure to efficiently process a ray shooting query starting inside P: compute the first intersection with the boundary of P by traversing the sequence of triangles that are stabbed by the ray. Since the running time is proportional to the number of stabbed triangles, it is desirable to find a triangular partition such that no ray intersects too many triangles, or in other words, to minimize the stabbing number. Hershberger and Suri [12] showed that every simple (without self intersections) polygon without holes has a triangular partition with stabbing number $O(\log n)$ and there exist polygons where any triangular partition has stabbing number $\Omega(\log n)$. There is also an $O(1)$-approximation algorithm for minimizing the stabbing number of triangular partitions [1].

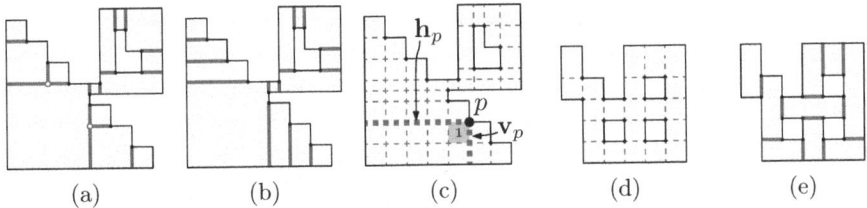

                    (a)   (b)   (c)   (d)   (e)

Fig. 1. (a) An optimal rectangular partition of a polygon P_1 in general position with one hole using two Steiner points (tiny hollow circles) with stabbing number 3. The portion of the edges of partition rectangles which are not on the boundary of P_1 are plain bold (red). (b) An optimal conforming partition of P_1 with stabbing number 4. (c) The pixelation of P_1. The reflex vertices are tiny (black) discs and the reflex segments are dotted (red). The horizontal and vertical reflex segments $\mathbf{h}_p$ and $\mathbf{v}_p$ from the reflex vertex p are bold. The wedge-pixel of p is shaded (in orange) and labeled 1. (d) The pixelation of a thin polygon P_2 with three holes (not in general position). (e) An optimal conforming partition of P_2 with stabbing number 3. (Color figure online)

In this paper, we restrict the attention to a problem motivated by orthogonal ray-shooting. We partition rectilinear polygons (i.e., polygons whose edges are axis-aligned) into rectangles, and for the stabbing number only consider line segments that are in the interior of the polygon and axis-aligned (we call these *stabbing segments*). More precisely, we study the following problem for a rectilinear n-gon P, possibly with holes: partition P into rectangles while minimizing the *stabbing number of the partition*, that is, the maximum over all stabbing segments **s** of the number of partition rectangles intersected by **s**. We say that such a rectangular partition is *optimal*, and call its stabbing number the *(minimum) stabbing number of P*. Figure 1(a) shows an example of an optimal partition. We often describe such a partition via the inserted segments. In this paper, a *Steiner point* of a partition is a corner of a partition rectangle that lies in the interior of P. Similar to triangular partitions, every rectilinear polygon has stabbing number $O(\log n)$, and there exist polygons of arbitrary size with stabbing number $\Omega(\log n)$ [8]. However, there also exist arbitrary-size polygons with stabbing number $O(1)$. To this end, Abam et al. [1] gave a 3-approximation algorithm for computing the stabbing number of simple rectilinear polygons without holes. An interesting open problem in this context is to determine the computational complexity of computing the stabbing number for simple polygons without holes. Although this question remains open in general, there has been some progress on a variant of rectangular partition called conforming partition.

A *conforming partition* of a rectilinear polygon P is a rectangular partition without Steiner points. Put differently, the partition is obtained by using internally disjoint axis-aligned segments that are *maximal* (i.e., both endpoints lie on the boundary of P). To minimize the stabbing number, it suffices to restrict the attention to partitions that use only *reflex segments*, i.e., maximal axis-aligned open segments where one endpoint is a reflex vertex of P (Fig. 1(b)). Again, we say that a conforming partition is *optimal* if its stabbing number is minimum among all the conforming partitions, and we call this stabbing number the *conforming stabbing number of P*. Durocher and Mehrabi [9,10] showed that computing an optimal conforming partition is $\mathcal{NP}$-hard for rectilinear polygons with holes, and gave a 2-approximation algorithm for computing the conforming stabbing number (see also [13] for experimental results). However, the complexity of the problem remains open for simple polygons without holes.

Contributions. In this paper, we investigate the problem of computing an optimal conforming partition of rectilinear polygons (possibly with holes) from the perspective of designing *fixed-parameter tractable (FPT) algorithms*, i.e., algorithms with a running time of the form $f(k)n^{O(1)}$ for some chosen parameter k and some computable function $f(\cdot)$ that is independent of n. A natural question in the context of asking for (conforming) partitions with stabbing number at most k is to search for an FPT algorithm parameterized by k. We show that such an algorithm does not exist unless $\mathcal{P} = \mathcal{NP}$. Specifically, deciding whether the conforming stabbing number (and in fact the stabbing number) of a polygon is at most 4 remains $\mathcal{NP}$-hard (Sect. 3). This strengthens the $\mathcal{NP}$-hardness result of Durocher and Mehrabi [10], who show it is $\mathcal{NP}$-hard to determine whether the conforming stabbing number is $\Theta(\sqrt{n})$.

Our hardness result puts forward two interesting questions. First, is it decidable whether a rectilinear polygon admits a conforming partition with stabbing number at most 2 or 3 in polynomial time? Second, are there other natural parameters for FPT algorithms to compute optimal conforming partitions? For the former, we give an $O(n \log n)$-time algorithm to decide whether a conforming partition with stabbing number 2 exists (Sect. 4); this leaves the case of stabbing number 3 open. For the latter, we give two FPT algorithms to test whether a polygon P has conforming stabbing number at most k (Sect. 5). One is parameterized by the sum of k plus the *treewidth* of P, the other is specific to polygons without holes in *general position* and is parameterized by k alone.

2   Preliminaries

Throughout the article, the polygons we consider are all simple (i.e., without self intersections), rectilinear (i.e., the edges are axis-aligned) and may have holes. A polygon is in *general position* if no three vertices lie on one axis-aligned line (Fig. 1(a–c)). A polygon is *thin* if no pair of its reflex segments intersect (Fig. 1(d)–(e)).

The *pixelation* of a polygon P (possibly with holes) is the partition of P obtained by adding for each reflex vertex p its horizontal and vertical reflex segments; these segments are denoted $\mathbf{h}_p$ and $\mathbf{v}_p$ (Fig. 1(c)–(d)). A *pixel* is a maximal region of P that does not intersect a reflex segment. For a reflex vertex p of P, the *wedge-pixel of p* is the pixel incident to the wedge defined by the reflex segments of p, i.e., the pixel that is incident to p and to $\mathbf{h}_p$ and $\mathbf{v}_p$ (Fig. 1(c)).

Recall that a *stabbing segment* of a rectilinear polygon P is an axis-aligned line segment that lies in the interior of P; for purposes of the stabbing number we only need to consider segments of maximal length, and we consider them to be open segments. We say that two stabbing segments are *equivalent* if they intersect the same set of pixels; there are $O(n)$ equivalence classes of stabbing segments. For instance, in Fig. 1(c), there are 26 equivalence classes. Given a rectilinear polygon P, by k-STAB (respectively k-CSTAB) we denote the problem of deciding whether P admits a partition (respectively conforming partition) into rectangles such that all stabbing segments intersect at most k rectangles.

We will reduce from an $\mathcal{NP}$-hard problem called *rectilinear planar monotone 3-SAT* (RPM-3-SAT) [3] to prove the hardness results. The RPM-3-SAT problem is a variant of 3-SAT where every clause is either negative or positive, i.e., consists of either three negated or three non-negated variables. Furthermore, the bipartite graph constructed from the variable-clause incidences admits a planar drawing such that all vertices are drawn as rectilinear rectangles, the *variable rectangles* (i.e., rectangles of vertices corresponding to variables) lie along the x-axis, the *positive (respectively negative) clause rectangles* (i.e., rectangles of vertices corresponding to such clauses) lie above (respectively below) the x-axis, and edges are represented by vertical lines of visibility between the rectangles of their endpoints. Fig. 2(a) illustrates such an instance where the rectangles are shaded in gray.

3   Intractability of Stabbing Number 4 or More

In this section, we sketch a proof of $\mathcal{NP}$-completeness.

Theorem 1. *For all integer $k \geq 4$, the decision problems k-STAB and k-CSTAB are $\mathcal{NP}$-complete. Moreover, k-CSTAB remains $\mathcal{NP}$-complete even for thin polygons and for polygons in general position.*

Proof Structure. It is straightforward to verify that k-STAB and k-CSTAB are in $\mathcal{NP}$ for any integer k. We thus concentrate on proving $\mathcal{NP}$-hardness. First, we prove that 4-CSTAB is $\mathcal{NP}$-hard, even if only thin polygons are considered. In a thin polygon, any optimal partition is conforming, so in consequence 4-STAB is also $\mathcal{NP}$-hard. However, the gadgets take advantage of not being in general position. Second, we provide an alternative version of this proof, this time for polygons in general position (but the gadgets take advantage of not being thin). As an aside, we note here that it is not possible to make the gadgets both thin and in general position; the problem is actually polynomial in this case, see Theorem 2.

As a third step, using a similar approach and with a similar alternative version for polygons in general position, we prove that 5-CSTAB, and thus 5-STAB, is $\mathcal{NP}$-hard. Finally, we show how to modify our constructions for ℓ-CSTAB ($\ell \in \{4, 5\}$) to work for ($\ell + 2\,m$)-CSTAB for any $m \geq 1$. Therefore k-CSTAB is $\mathcal{NP}$-hard for thin polygons for all $k \geq 4$, which implies hardness for k-STAB.

Proof Sketch of the $\mathcal{NP}$-Hardness of 4-CSTAB *for Thin Polygons.* We reduce RPM-3-SAT (defined in the preliminaries) to 4-STAB in polynomial time. We transform an instance ϕ of RPM-3-SAT (shaded in the background of Fig. 2(a)) into an instance $P\,(\phi)$ of 4-STAB (the polygon in Fig. 2(a)).

The polygon $P(\phi)$ consists of *variable gadgets* (drawn inside the variable rectangles), *split gadgets* (drawn above and/or below the variable gadgets and still inside the variable rectangles), and *clause gadgets* (drawn inside the clause rectangles). Crucial to our construction are *forcer gadgets*, indicated by a square labeled F in Fig. 2(a) and shown in detail in Fig. 2(b). A forcer gadget is designed to force the presence (in any conforming partition with stabbing number at most 4) of a certain pair of reflex segments in the pixel to which it is attached. Details of these gadgets and proofs of their properties can be found in the full version.

We now describe the properties of the gadgets and show at the same time that any conforming partition $\mathcal{R}$ of $P\,(\phi)$ implies a satisfying assignment for ϕ. Without loss of generality, $\mathcal{R}$ is *minimal*, i.e., no reflex segment can be removed while retaining a conforming partition.

We propagate information between gadgets along certain vertical stabbing segments that each intersect two gadgets. We say that such a stabbing segment $\mathbf{s}$ *propagates* 0 (standing for "false") if it intersects three segments of $\mathcal{R}$ within the intersected gadget which is the closest (in terms of vertical distance) to the x-axis, and that $\mathbf{s}$ propagates 1 (standing for "true") if it intersects only two segments of $\mathcal{R}$.

In the gadget of a variable x, there are two *out-stabs*, i.e., two vertical stabbing segments, which are assigned to the literals x and $\bar{x}$. We design variable gadgets such that not both out-stabs propagates 1, but all other combinations of propagated values are possible. We then read from the partition a value for x, namely, x is assigned the value propagated by the out-stab of the literal x. Note that we set $x = 0$ (by convention) if both out-stabs propagate 0 (the convention $x = 1$ would have worked as well). In the partition of $P(\phi)$ in Fig. 2(a), $x_1, x_2, x_3, x_4 = 0, 1, 1, 0$ (even though the out-stab of the literal $\overline{x_4}$ propagates 0).

A split gadget has an *in-stab* and two *out-stabs*. The in-stab is an out-stab of a variable gadget or of another split gadget. A split gadget "splits the propagation"

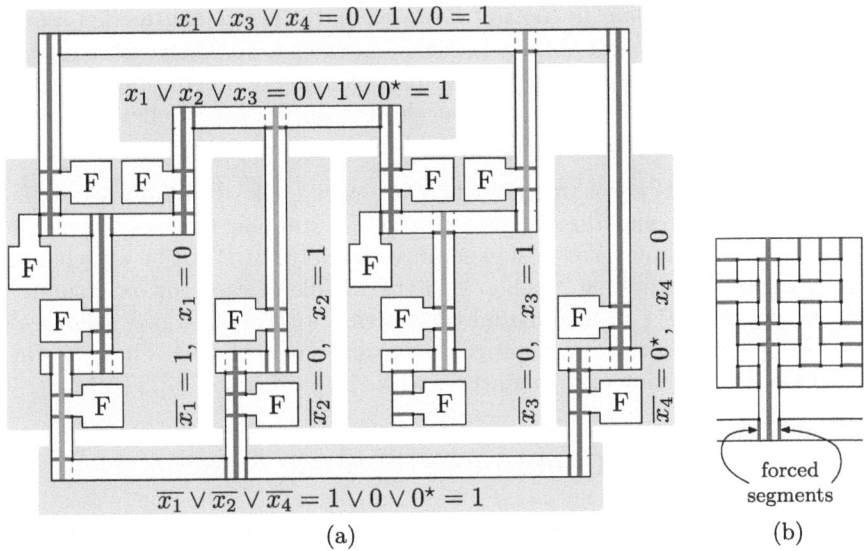

Fig. 2. Gadgets used in the proof of Theorem 1 for $k = 4$ (using thin polygons). (a) The polygon $P(\phi)$ (not to scale) of the RPM-3-SAT drawing of $\phi = (x_1 \vee x_3 \vee x_4) \wedge (x_1 \vee x_2 \vee x_3) \wedge (\overline{x_1} \vee \overline{x_2} \vee \overline{x_4})$. Forcer gadgets are represented by squares labeled F. The reflex segments of a partition with stabbing number 4 are solid bold (in red), the other reflex segments are dotted (in red). Vertical stabbing segments propagating 0 (respectively 1) are thick purple (respectively thick green). We use $0^\star$ for a value that is 1 in the variable assignment but that has been decreased by a variable gadget or by a split gadget and is propagated as 0. (b) The forcer gadget. (Color figure online)

in the sense that the value propagated by the two out-stabs is at most the value propagated by the in-stab. In Fig. 2(a), the split gadget in x_1 splits the in-stab's value 0 into two out-stabs propagating 0 as well, whereas the split gadget in x_3 splits the in-stab's value 1 into a left out-stab propagating 0 and a right one propagating 1.

A clause gadget has three *in-stabs* each of which is an out-stab (of a variable gadget or a split gadget) propagating the value (possibly decreased) from a variable gadget. We design the clause gadget such that there exists a conforming partition where the horizontal stabbing segment within a clause gadget intersects at most four rectangles if and only if at least one of the three in-stabs of the clause gadget propagates 1. This in turn is possible only if one of the literals of the clause corresponds to an out-stab propagating 1, which in turn implies that we have assigned 1 to this literal, since propagated values do not increase. Therefore, a solution to 4-CSTAB implies a satisfying assignment for ϕ.

The other direction (i.e., proving that a satisfying assignment to ϕ gives a solution to 4-CSTAB) is similar and even easier and the reduction is hence complete. We prove that 5-CSTAB is $\mathcal{NP}$-hard using the exact same reduction idea with slightly modified gadgets.

Proof Sketch for Polygons in General Position. In the previous reduction, we use aligned reflex vertices most notably in the forcer gadget. To achieve the reduction for polygons in general position, we design a completely different forcer gadget based on a staircase, and we show that shifting all the other reflex vertices in the other gadgets does not affect the proof.

Proof Sketch for $k > 4$. For the case when k is even, i.e. $k = 4 + 2m$ where m is a positive integer, we generalize the forcer gadget for stabbing number 4, by adding m rows and m columns. We then generalize the polygon P (ϕ) by attaching m forcer gadgets for stabbing number k to the middle of each row or column of adjacent pixels of P (ϕ). The hardness reduction for k-CSTAB now follows the same technique that we used to prove the hardness of 4-CSTAB. The case when k is odd is handled similarly by starting with the hardness of 5-CSTAB.

4   Tractability of Conforming Stabbing Number 2

The tractability of 2-CSTAB is very easy to show by phrasing the problem as a 2-SAT problem. The running time depends on the maximum number of reflex segments intersected by a stabbing segment.

Lemma 1. *There exists an algorithm that, for a rectilinear n-gon P where every stabbing segment intersects at most ℓ reflex segments, decides 2-CSTAB and provides a solution (if any) in $O(\ell n)$ time.*

Proof. Declare a boolean variable $x(\mathbf{s})$ for every reflex segment $\mathbf{s}$, with the intent that $\mathbf{s}$ is used in the solution if and only if $x(\mathbf{s})$ is true. To ensure that we have a conforming partition, we hence require

- $x(\mathbf{h}_p) \vee x(\mathbf{v}_p)$ for every reflex vertex p, as well as
- $\neg x(\mathbf{h}_p) \vee \neg x(\mathbf{v}_q)$ for any two intersecting reflex segments $\mathbf{h}_p, \mathbf{v}_q$.

To ensure that the conforming stabbing number is at most 2, we force that every stabbing segment intersects at most one chosen reflex segment. In other words, we require $\neg x(\mathbf{s}_1) \vee \neg x(\mathbf{s}_2)$ for any two reflex segments $\mathbf{s}_1, \mathbf{s}_2$ intersected by a common stabbing segment.

    All these restrictions only involve two variables, so this gives a 2-SAT instance that has $O(n)$ variables. For every reflex segment $\mathbf{s}$, variable $x(\mathbf{s})$ belongs to at most two clauses of the first kind, and at most ℓ clauses each of the second and the third kind. So the number of clauses is $O(\ell n)$. Since 2-SAT can be solved in linear time [2], the result follows.   □

    In an arbitrary polygon there could be stabbing segments that intersect $\Theta(n)$ reflex segments, so the running time of the 2-SAT approach is $O(n^2)$ in the worst case. Our main contribution in this section is to give a faster algorithm, with $O(n \log n)$ running time.

    We call a reflex segment *impossible* if no conforming partition with stabbing number 2 contains it, and *fixed* if any such conforming partition must contain it.

The idea of our algorithm is to determine via some rules that some segments are impossible or fixed, from which we deduce other segments to be impossible or fixed. Repeated applications either provide an answer to 2-CSTAB, or end with a situation where the *undecided* segments (i.e., the ones where we did not derive that they are fixed or impossible) are in very restricted positions; we then find a conforming partition easily. We start with three obvious rules:

(R1) If, at some reflex vertex p, both reflex segments are impossible, then there is no conforming partition.

(R2) If, at some reflex vertex p, one reflex segment is impossible, then the other one is fixed.

(R3) If a stabbing segment $\mathbf{s}$ intersects a fixed segment, then all other reflex segments intersected by $\mathbf{s}$ are impossible.

Two non-trivial rules, (R4) and (R5), which trigger the entire process, are in the following lemmas:

Lemma 2. (R4) *If two (distinct) reflex segments (open by definition) intersect, then both are impossible.*

Proof. Let $\mathbf{h}_p$ and $\mathbf{v}_q$ be a pair of horizontal and vertical reflex segments that intersect. Assume for contradiction that some conforming partition used $\mathbf{h}_p$ (the argument is similar for $\mathbf{v}_q$). Then we cannot use $\mathbf{v}_q$ (since partition segments must not intersect), so we must use $\mathbf{h}_q$. Let χ be the common point of $\mathbf{h}_p$ and $\mathbf{v}_q$. Up to symmetry, we may assume that the wedge-pixel of q lies to the right of $\mathbf{v}_q$. Then for small enough ε, the vertical stabbing segment through $\chi + (\varepsilon, \varepsilon)$ intersects both $\mathbf{h}_p$ and $\mathbf{h}_q$ (see also Fig. 3(a)) and the conforming partition has stabbing number 3 or more.   □

To explain (R5) we need a definition. A *gate* of a polygon is an axis-aligned segment $\overline{pq}$ that connects two reflex vertices p, q such that the wedge-pixels of p and q lie on the same side of $\overline{pq}$. Figure 3(b) shows a gate, while segment $\overline{pq}$ in Fig. 3(d) is not a gate since the wedge-pixels are not on the same side of $\overline{pq}$.

Lemma 3. (R5) *Any gate is fixed.*

Proof. Up to symmetry we may assume that gate $\overline{pq}$ is horizontal, so $\overline{pq} = \mathbf{h}_p = \mathbf{h}_q$. Since the wedge-pixels lie on the same side of $\overline{pq}$, we may assume up to symmetry that $\mathbf{v}_p$ and $\mathbf{v}_q$ both go upward from p and q. The horizontal stabbing segment through $p + (0, \varepsilon)$ (for a small enough ε) then intersects both $\mathbf{v}_p$ and $\mathbf{v}_q$ since it runs parallel to $\mathbf{h}_p = \mathbf{h}_q$ (see also Fig. 3(b)). Thus, by (R3), any conforming partition with stabbing number 2 does not include both $\mathbf{v}_p$ and $\mathbf{v}_q$, which means by (R2) that the segment $\overline{pq}$ is included instead.   □

Recall that our approach is to apply the above rules, and to keep track (by storing them in two lists L_{fixed} and $L_{\text{impossible}}$) of all reflex segments that we determine to be fixed or impossible. (There may be other fixed or impossible segments that we do not find.) This clearly can be done in polynomial time; we

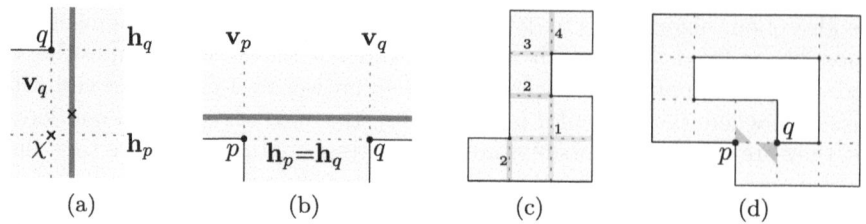

Fig. 3. (a) (R4) illustrated. (b) (R5) illustrated. (c) An example of the propagation. Both segments at 1 are impossible by (R4). This fixes the two segments labeled 2 by (R2). This makes the segment labeled 3 impossible by (R3), which in turn fixes the segment labeled 4 by (R2). (d) An example where no reflex segments get fixed. There is a horizontal segment connecting two reflex vertices p, q, but it is not a gate.

show in the full version how to implement it in $O(n \log n)$ time by applying line-sweep and ray-shooting techniques. If some segment $\mathbf{s}$ belongs to both L_{fixed} and $L_{\text{impossible}}$, then we conclude that there is no conforming partition with stabbing number 2.

We are left with three possible outcomes: We find that there is no conforming partition with stabbing number at most 2, or L_{fixed} defines a conforming partition, or neither. We are done in the first outcome. We are also done in the second outcome: If L_{fixed} defines a conforming partition, then by (R3) (and since L_{fixed} is disjoint from $L_{\text{impossible}}$) every stabbing segment intersects at most one segment of L_{fixed}, and so the stabbing number is 2. In the third outcome, we provide below an algorithm to test in linear time whether there exists a solution. (In fact, one can prove that there *always* is a solution. This proof is omitted here.)

Add the segments of L_{fixed} into P to obtain a partition of P into rectilinear polygons $P_1, \ldots, P_\ell$ that we call the *pieces* of P. The idea is now to solve the problem for each piece of P and to put the solutions together. Next, we make two useful observations.

Observation 4. *For every piece P_i, every reflex segment $\mathbf{s}$ of P_i is a reflex segment of P that was undecided (i.e., neither in L_{fixed} nor in $L_{\text{impossible}}$).*

Proof. Since $\mathbf{s}$ is a reflex segment of P_i, one endpoint of $\mathbf{s}$, say p, is a reflex vertex of P_i, hence also a reflex vertex of P. The other endpoint of $\mathbf{s}$ lies on the boundary of P_i. If this other endpoint were not on the boundary of P, then it would be on an open segment $\mathbf{s}' \in L_{\text{fixed}}$. But then rule (R4) would have been applied to $\mathbf{s}'$ and the reflex segment of P containing $\mathbf{s}$. This would have added $\mathbf{s}'$ to $L_{\text{impossible}}$, contradicting that L_{fixed} and $L_{\text{impossible}}$ are disjoint. Thus the other endpoint of $\mathbf{s}$ also lies on the boundary of P, and $\mathbf{s}$ is a reflex segment of P.

To see that $\mathbf{s}$ is undecided, observe first that p does not have an incident reflex segment in L_{fixed} since it is reflex in the piece P_i. Thus $\mathbf{s} \notin L_{\text{fixed}}$, and also $\mathbf{s} \notin L_{\text{impossible}}$, since otherwise rule (R2) would have added the other reflex segment at p to L_{fixed}. Therefore $\mathbf{s}$ is undecided. □

Observation 5. *P has a solution to* 2-CSTAB *if and only if each of the pieces* $P_1, \ldots, P_\ell$ *of P has a solution to* 2-CSTAB.

Proof. Any solution for P includes all segments of L_{fixed}, thereby yielding a solution for each piece. Vice versa, assume that each piece P_i of P admits a solution $\mathcal{R}_i$ of reflex segments to 2-CSTAB. We show that $\mathcal{R} := L_{\text{fixed}} \cup \bigcup_i \mathcal{R}_i$ is a solution for P. To see that $\mathcal{R}$ is a conforming partition, observe that it only contains reflex segments of P by Observation 4, and assigns at least one reflex segment to each reflex vertex of P. Since the pieces $P_1, \ldots, P_\ell$ are interior-disjoint, the reflex segments in $\bigcup_i \mathcal{R}_i$ do not intersect each other. They do not intersect a segment of L_{fixed} either, by Observation 4, so $\mathcal{R}$ yields a conforming partition.

To show that $\mathcal{R}$ has stabbing number at most 2, consider any stabbing segment $\mathbf{s}$ of P. If $\mathbf{s}$ intersects no segment of L_{fixed}, then it is also a stabbing segment for one piece P_i, and so will intersect at most one segment of $\mathcal{R}$. Now assume that $\mathbf{s}$ intersects a segment of L_{fixed}. Since rule (R3) was applied, all other reflex segments of P intersected by $\mathbf{s}$ were added to $L_{\text{impossible}}$, so were not reflex segments of any pieces, and hence are not used by $\mathcal{R}$. Therefore, stabbing segment $\mathbf{s}$ intersects at most one segment of $\mathcal{R}$. $\qquad\square$

It remains to show how to solve the problem for each piece efficiently. Here, our previous 2-SAT approach comes to the rescue since the pieces are not arbitrary polygons. Specifically, since rule (R4) does not apply to piece P_i (for $i \in \{1, \ldots, \ell\}$), it has no intersecting reflex segments, so it is thin. Since rule (R5) does not apply to P_i, it has no gate. We now prove a statement that holds for any thin gate-free polygon.

Lemma 6. *Let P be a thin rectilinear polygon that has no gates. Then every stabbing segment $\mathbf{s}$ intersects at most two reflex segments.*

Proof. Assume for contradiction that $\mathbf{s}$ intersects three reflex segments, say $\mathbf{s}$ intersects $\mathbf{s}_1, \mathbf{s}_2, \mathbf{s}_3$, in this order and with no other reflex segments in between. See also Fig. 4. Up to symmetry $\mathbf{s}$ is horizontal, so $\mathbf{s}_1, \mathbf{s}_2, \mathbf{s}_3$ are vertical, and up to renaming $\mathbf{s}_1$ is on the left of $\mathbf{s}_2$. Let p be the reflex vertex of P with $\mathbf{s}_2 = \mathbf{v}_p$.

Up to symmetry, the wedge-pixel ξ of p is to the left of $\mathbf{v}_p$ and below $\mathbf{h}_p$. Since P is thin, pixel ξ spans the entire length of $\mathbf{v}_p$, and in particular includes the point common to $\mathbf{s}$ and $\mathbf{v}_p$. Since there are no vertical reflex segments between $\mathbf{s}_1$ and $\mathbf{s}_2$ along $\mathbf{s}$, pixel ξ extends to the point common to $\mathbf{s}_1$ and $\mathbf{s}$, and therefore spans the entire length of $\mathbf{s}_1$. It also includes the entire length of $\mathbf{h}_p$. Thus the top left corner of ξ is a point q common to $\mathbf{s}_1$ and $\mathbf{h}_p$, hence q is a reflex vertex that lies on a horizontal line with p. Since both $\mathbf{v}_q$ and $\mathbf{h}_q$ bound sides of ξ, this makes $\overline{pq}$ a gate. $\qquad\square$

Lemma 6 has two consequences:

Theorem 2. *There exists an algorithm that, for a rectilinear n-gon P that is thin and has no gate (which is the case if P is in general position), computes the stabbing number of P in $O(n)$ time.*

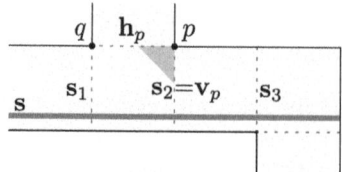

Fig. 4. If stabbing segment s intersects three reflex segments, and no two reflex segments intersect, then the polygon is not in general position. The upper right corner of the wedge-pixel ξ of p (and q) is shaded (in orange). (Color figure online)

Proof. The stabbing number of P is 1 if P is a rectangle, and at least 2 otherwise. By Lemma 6, any stabbing segment of P intersects at most two reflex segments. Thus, the stabbing number of P is at most 3. Lemma 1 gives an algorithm to test whether it is 2 or 3 in $O(n)$ time.   □

Theorem 3. *There exists an algorithm that, for any rectilinear n-gon P, decides* 2-CSTAB *and provides a solution (if any) in $O(n \log n)$ time.*

Proof. Apply all rules; this takes $O(n \log n)$ time. Then compute the pieces $P_1, \ldots, P_\ell$ in $O(n)$ time. Using Lemma 6 and Theorem 2, test in $O(|P_i|)$ time whether a piece P_i has a solution for 2-CSTAB. By Observation 5, this information is enough to decide 2-CSTAB for P and compute the solution in case of an affirmative answer. Since $\sum_i |P_i| \in O(n)$, the result follows.   □

5   Polygons with Small Treewidth

We now turn towards FPT algorithms, and in particular, study polygons with bounded treewidth. We recall first a few definitions. A *tree decomposition* of a graph $G = (V, E)$ is a tree $\mathcal{T}$ and an assignment β from the nodes of $\mathcal{T}$ to subsets of V (called *bags*) with the following properties: (a) For every vertex v of G, the bags containing v form a non-empty connected subtree of $\mathcal{T}$. (b) For every edge e of G, there exists a bag that contains both endpoints of e. The *width* of a tree decomposition is the maximum bag-size minus one, and the *treewidth* $tw(G)$ of G is the minimum width of a tree decomposition of G.

The treewidth has frequently been used for FPT algorithms for graph problems, but can also be used for solving problems on polygons, see e.g. [4]. Recall that the *pixelation* of a polygon P is obtained by inserting all reflex segments. This gives rise to a planar graph (the *pixelation graph* G_P) that has vertices at every vertex of P, every crossing, and every endpoint of a reflex segment, and edges as implies by the edges of P and the reflex segments (see Fig. 5(a)). The *treewidth* of P is the treewidth $tw(G_P)$ of the pixelation graph.

Our algorithm for polygons with small treewidth uses not only the pixelation graph, but also its *radial graph* R_P, which is defined as follows. The vertices of R_P are the vertices of G_P (we denote them by V_P), as well as one vertex for every pixel (we denote these by Ξ_P). We add an edge (ξ, v) between $\xi \in \Xi_P$

and $v \in V_P$ if and only if vertex v is incident to pixel ξ. See Fig. 5(b). Using the techniques of Borradaile et al. [6], one can easily show that R_P has treewidth $O(tw(G_P))$, since pixels are incident to at most four vertices of G_P.

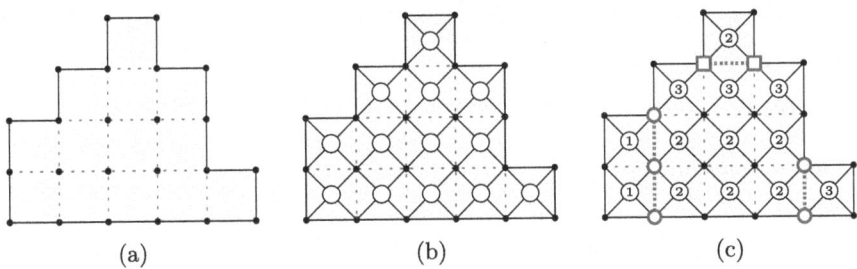

(a)   (b)   (c)

Fig. 5. (a) A polygon P with its pixelation graph G_P. (b) The radial graph R_P contains the thin solid edges. (c) One possible solution to the MSOL formula ϕ for $k = 3$. We only show parts of this solution: Partition segments are dotted, bold, and red. Vertices in $\mathcal{V}$ ($\mathcal{H}$) are bold hollow red circles (squares). Vertices in Ξ_i^{hor} are indicated by $\textcircled{i}$. (Color figure online)

So we now show how to exploit small treewidth of R_P to find the conforming stabbing number of P. To this end, we use Courcelle's theorem [7], which states that if a graph property can be expressed in monadic second-order logic (MSOL) as a formula ϕ, then testing whether a graph G with a tree decomposition of width w satisfies the property can be done in time that is linear in the number of vertices and fixed-parameter tractable in $|\phi| + tw(G)$. To express the conforming stabbing number of P via MSOL, we use the following ideas illustrated in Fig. 5(c):

- We have vertex sets $\mathcal{V}$ and $\mathcal{H}$, with the intended meaning that these are the vertices of G_P that lie on vertical/horizontal reflex segments used by a conforming partition. With easy formulas that only rely on adjacencies of graph R_P, we can express that $\mathcal{V}$ and $\mathcal{H}$ indeed correspond to reflex segments of the appropriate orientation, and that we have a conforming partition: every reflex vertex of P belongs to at least one of these vertex sets, and no reflex segments intersect. (See the full version for details.)
- We partition Ξ_P into $\Xi_1^{\text{hor}} \cup \cdots \cup \Xi_k^{\text{hor}}$, with the intended meaning that if $\xi \in \Xi_P$ belongs to Ξ_i^{hor}, then the horizontal stabbing segment through ξ, when traversed left-to-right, has encountered at most i rectangles when it reaches ξ. Since we require $i \leq k$, this enforces that all horizontal stabbing segments hit at most k rectangles. With easy formulas that only rely on adjacencies of graph R_P, we can express that indices of the sets Ξ_i^{hor} indeed express rectangle-counts. Namely, if ξ, ξ' are two pixels that share a vertical edge (say with ξ left), and $\xi \in \Xi_i^{\text{hor}}$, then we require $\xi' \in \Xi_{i+1}^{\text{hor}}$ or $\xi' \in \Xi_i^{\text{hor}}$ depending to whether the segment corresponding to the shared edge is in the partition or not.

– Symmetrically we can force that all vertical stabbing segments hit at most k rectangles.

The length of the resulting formula is linear in k and independent of the size of graph R_P. If polygon P has n vertices, then G_P has $O(n^2)$ vertices, and so does R_P. Therefore, with Courcelle's theorem, we obtain the first FPT algorithm.

Theorem 4. *There exists an algorithm that, for a rectilinear n-gon P with treewidth ℓ, decides k-CSTAB in $O(f(k, \ell)n^2)$ time, for some function $f(\cdot)$ that does not depend on n.*

The function $f(\cdot)$ that falls out of Courcelle's theorem is rather large (it could be a tower of exponents). It is possible to decide k-CSTAB directly by doing bottom-up dynamic programming in a tree decomposition of R_P of minimum width ℓ' (which we know to be in $O(\ell)$). Each pixel needs to keep track of which of the sets $\Xi_1^{\text{hor}}, \ldots, \Xi_k^{\text{hor}}, \Xi_1^{\text{ver}}, \ldots, \Xi_k^{\text{ver}}$ it belongs to, and each vertex of V_P needs to keep track whether it is in $\mathcal{V}$ or $\mathcal{H}$. Since bags contain up to $\ell' + 1$ vertices, this gives at most $k^{2\ell'+2}$ possible configurations per bag, and with (not difficult but tedious to write) update-formulas one can therefore show how to solve k-CSTAB in $O(k^{2\ell'+2}n^2)$ time. We leave the details as an exercise. It is also not hard to modify the MSOL formulations so that it permits arbitrary partitions, rather than restricting to conforming ones. In other words, k-STAB is also fixed-parameter tractable in $k+tw(G_P)$. Details are also left as an exercise.

Now we give a second FPT algorithm, which makes a different assumption on the polygon P. We require P to have no holes and no gates (the latter holds in particular if P is in general position), but in exchange we no longer need to bound the treewidth. The idea for this theorem is to distinguish by the maximum number of reflex segments intersected by a stabbing segment; if it is small then the treewidth is small and Theorem 4 applies, and if it is large enough then (as one shows) the conforming stabbing number is bigger than k. Details are in the full version.

Theorem 5. *There exists an algorithm that, for a gate-free rectilinear n-gon P without holes, decides k-CSTAB in $O(f'(k)n^2)$ time, for some function $f'(\cdot)$ that does not depend on n.*

6   Conclusion

In this paper, we show that computing a conforming partition of a rectilinear polygon with stabbing number k is $\mathcal{NP}$-hard for all $k \geq 4$. Since the reduction uses only thin polygons, the hardness result follows even if we omit the conforming constraint. The polygons used in our reduction have holes. Therefore, determining the time complexity of computing an optimal (conforming) partition for polygons without holes remains open.

On the positive side, we provide an $O(n \log n)$-time algorithm to decide whether a polygon admits a conforming partition with stabbing number 2. Since the problem is $\mathcal{NP}$-hard already for conforming stabbing number 4, only the case of conforming stabbing number 3 remains open. For polygons (possibly with holes) with bounded treewidth and bounded conforming stabbing number, we give a quadratic-time algorithm to compute the minimum stabbing number. An exciting direction would be to design fixed-parameter tractable algorithms for polygons without holes parameterized by the conforming stabbing number, which would complement the hardness result for polygons with holes. Interestingly, for polygons without holes that are in general position, we already gave such a fixed-parameter-tractable algorithm. But we also proved that general position does not help the case of polygons with holes: computing a conforming partition with stabbing number at most k (for $k \geq 4$) remains $\mathcal{NP}$-hard for polygons in general position.

Extending all these results to higher dimensions would be interesting, even for the restricted class of orthogonal 3D-histograms where previous results focus on minimizing the number of partitions into rectangular boxes [5,11].

References

1. Abam, M.A., Aronov, B., de Berg, M., Khosravi, A.: Approximation algorithms for computing partitions with minimum stabbing number of rectilinear and simple polygons. In: Proceedings of the 27th ACM Symposium on Computational Geometry (SCG), pp. 407–416 (2011). https://doi.org/10.1145/1998196.1998263
2. Aspvall, B., Plass, M.F., Tarjan, R.E.: A linear-time algorithm for testing the truth of certain quantified Boolean formulas. Inf. Process. Lett. **8**(3), 121–123 (1979)
3. de Berg, M., Khosravi, A.: Optimal binary space partitions for segments in the plane. Int. J. Comput. Geom. Appl. **22**(03), 187–205 (2012). https://doi.org/10.1142/S0218195912500045
4. Biedl, T., Mehrabi, S.: On orthogonally guarding orthogonal polygons with bounded treewidth. Algorithmica **83**, 641–666 (2021). https://doi.org/10.1007/s00453-020-00769-5
5. Biedl, T., Derka, M., Irvine, V., Lubiw, A., Mondal, D., Turcotte, A.: Partitioning orthogonal histograms into rectangular boxes. In: Proceedings of the 13th Latin American Symposium on Theoretical Informatics (LATIN), pp. 146–160. Springer (2018)
6. Borradaile, G., Erickson, J., Le, H., Weber, R.: Embedded-width: a variant of treewidth for plane graphs. CoRR abs/1703.07532 (2017). http://arxiv.org/abs/1703.07532
7. Courcelle, B.: The monadic second-order logic of graphs. I. Recognizable sets of finite graphs. Inf. Comput. **85**(1), 12–75 (1990)
8. de Berg, M., van Kreveld, M.: Rectilinear decompositions with low stabbing number. Inf. Process. Lett. **52**(4), 215–221 (1994). https://doi.org/10.1016/0020-0190(94)90129-5
9. Durocher, S., Mehrabi, S.: Erratum to: computing partitions of rectilinear polygons with minimum stabbing number. https://citeseerx.ist.psu.edu/document?repid=rep1&type=pdf&doi=0093663d76ea6411084216000c1c61b7c0b0f49d
10. Durocher, S., Mehrabi, S.: Computing conforming partitions of orthogonal polygons with minimum stabbing number. Theor. Comput. Sci. **689**, 157–168 (2017). https://doi.org/10.1016/j.tcs.2017.05.035

11. Floderus, P., Jansson, J., Levcopoulos, C., Lingas, A., Sledneu, D.: 3D rectangulations and geometric matrix multiplication. Algorithmica **80**, 136–154 (2018)
12. Hershberger, J., Suri, S.: A pedestrian approach to ray shooting: shoot a ray, take a walk. J. Algorithms **18**(3), 403–431 (1995). https://doi.org/10.1006/jagm.1995.1017
13. Piva, B., de Souza, C.C.: Minimum stabbing rectangular partitions of rectilinear polygons. Comput. Oper. Res. **80**, 184–197 (2017)

On the Approximability of Graph Visibility Problems

Davide Bilò[D], Alessia Di Fonso[(✉)][D], Gabriele Di Stefano[D],
and Stefano Leucci[D]

Department of Information Engineering, Computer Science, and Mathematics,
University of L'Aquila, L'Aquila, Italy
{davide.bilo,alessia.difonso,gabriele.distefano,
stefano.leucci}@univaq.it

Abstract. Visibility problems have been investigated for a long time
under different assumptions as they pose challenging combinatorial problems and are connected to robot navigation problems. The *mutual-
visibility* problem in a graph G of n vertices asks to find the largest set
of vertices $X \subseteq V(G)$, also called μ-set, such that for any two vertices
$u, v \in X$, there is a shortest u, v-path P where all internal vertices of P
are not in X. This means that u and v are visible w.r.t. X. Variations
of this problem are known as *total*, *outer*, and *dual* mutual-visibility
problems, depending on the visibility property of vertices inside and/or
outside X. The mutual-visibility problem and all its variations are known
to be NP-complete on graphs of diameter 4.
  In this paper, we design a polynomial-time algorithm that finds a μ-
set with size $\Omega\left(\sqrt{n/D}\right)$, where D is the average distance between any
two vertices of G. Moreover, we show inapproximability results for all
visibility problems on graphs of diameter 2 and strengthen the inapproximability ratios for graphs of diameter 3 or larger. More precisely, for
graphs of diameter at least 3 and for every constant $\varepsilon > 0$, we show that
mutual-visibility and dual mutual-visibility problems are not approximable within a factor of $n^{1/3-\varepsilon}$, while outer and total mutual-visibility
problems are not approximable within a factor of $n^{1/2-\varepsilon}$, unless $\mathsf{P} = \mathsf{NP}$.
Furthermore, in the extended version of this paper we study the rela-
tionship between the mutual-visibility number and the general position
number in which no three distinct vertices u, v, w of X belong to any
shortest path of G.

1 Introduction

Mutual-visibility problems on the Euclidean plane involve determining if a set
of points or entities can see each other without any obstacles blocking their line

---

The work been supported in part by the Italian National Group for Scientific
Computation (GNCS-INdAM) and by the Italian Ministry of Economic development (MISE) under the project "SICURA - Casa intelligente delle tecnologie per la
sicurezza", CUP C19C200005200004.

S. Nakano and M. Xiao (Eds.): WALCOM 2025, LNCS 15411, pp. 47–61, 2025.
https://doi.org/10.1007/978-981-96-2845-2_4

of sight. These problems have been investigated for a long time under different assumptions. The root of visibility problems dates back to the end of the 1800s when Dudeney first introduced the famous *no-three-in-line* problem in [21]: given an $n \times n$ grid, find the maximum number of points such that there are no three points on a line and it is still an open problem. The notion of visibility can be also defined in discrete spaces like graphs requiring that a set of entities see each other along the shortest paths connecting them without any obstacles. Visibility problems on networks pose interesting theoretical problems both in graph theory and combinatorics but are also of practical importance in research areas like distributed computing by mobile entities in connection to robot navigation problems [3,4,10,11,19,26,28].

For robots moving in the Euclidean plane, achieving a configuration of mutual visibility is crucial. A robot whose visibility is obstructed by others, might not be able to complete its task. Conversely, when robots are mutually visible, they can all see each other and collaborate to solve problems. In communication or social networks, a subset of agents located on some nodes of the network may need to communicate in an efficient (using the shortest paths) and confidential way, that is, in such a way that the exchanged messages do not pass through other agents in the subset.

The concept of mutual-visibility in graphs has been recently introduced and studied in [20]. Given a set of vertices of a graph, two vertices u, v are in *mutual-visibility* if there exists a shortest u, v-path without further vertices of the set. A set of vertices is a *mutual-visibility* set if each pair of vertices in the set is in mutual-visibility. This graph-based mutual-visibility concept generated significant interest within the research community since its introduction (see e.g., [6,7,11–18,23,29]). In the *mutual-visibility problem* the goal is to find the maximum number of vertices that can be in mutual visibility on a graph G. This problem is NP-complete [20] on general graphs, whereas there exist exact formulas for special graph classes like paths, cycles, trees, block graphs, cographs, grids [16,20] and for both the Cartesian and the strong product of graphs [16].

Formally, given a connected graph G and a set of vertices $X \subseteq V(G)$, two vertices $x, y \in V(G)$ are said to be *X-visible* if there is a shortest x, y-path whose internal vertices do not belong to X. If every two vertices from X are X-visible, then X is a *mutual-visibility* set (or μ-set). Let $\overline{X} = V(G) \setminus X$. A set X is said to be an *outer mutual-visibility set* (or μ_o-set) if every two vertices $x, y \in X$ are X-visible, and every two vertices $x \in X$ and $y \in \overline{X}$ are X-visible. A set X is a *dual mutual-visibility set* (or μ_d-set) if every two vertices $x, y \in X$ are X-visible, and every two vertices $x, y \in \overline{X}$ are X-visible. Finally a set X is said a *total mutual-visibility set* (or μ_t-set) if every two vertices $x, y \in V(G)$ are X-visible. If $\tau \in \{\mu, \mu_d, \mu_o, \mu_t\}$, then the cardinality of the largest τ-set is called the τ-*number* of G and is denoted by $\tau(G)$. A τ-set X such that $|X| = \tau(G)$ is called a *maximum τ-set* of G. For each of the above variants, the respective optimization problem asks to find the maximum τ-set in a given graph G. One can observe that any μ_t-set is both a μ_d- and a μ_o-set, while any μ_d- or μ_o-set is also a μ-set. The converse relationships are not true in general. Moreover, there

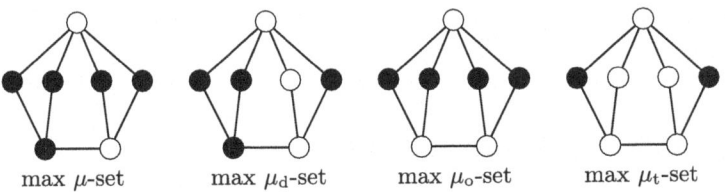

max μ-set   max μ_d-set   max μ_o-set   max μ_t-set

Fig. 1. Examples of maximum τ-sets on a graph G. Note that the μ_d-set is neither a μ_t-set nor a μ_o-set. The μ_o-set is neither a μ_t-set nor a μ_d-set. The μ_t-set is a feasible μ-set, μ_d-set and μ_o-set that is not maximum.

is no general relationship between μ_d- and μ_o-sets. As a consequence, for any graph G, we have $\mu_t(G) \leq \mu_d(G) \leq \mu(G)$ and $\mu_t(G) \leq \mu_o(G) \leq \mu(G)$. See Fig. 1 for some examples.

The mutual-visibility problem is connected to classical topics in combinatorics. For example, solving this problem in the Cartesian product of complete graphs is equivalent to solving an instance of Zarankiewicz's problem (see [16]). Finding the smallest maximal mutual-visibility set of a graph has been proven to be closely related to a classical Bollobás-Wessel theorem (see [6]). Also, the optimization visibility problems introduced above can be reformulated as Turán-type problems on hypergraphs and line graphs (see [7,18]). The mutual-visibility problem is also related to the *general position problem* in graphs that asks to determine a largest set X of vertices of a graph G such that no three vertices of X lie on a common shortest path [9,24,25]. Such a set is called *general position set* (or *gp-set*) and the cardinality of a maximum *gp*-set is called the *gp-number* of G and denoted by $gp(G)$. Clearly, any *gp*-set is a μ-set and then $gp(G) \leq \mu(G)$.

Our Results. For a given graph G of n vertices, the problem of finding the *gp*-number $gp(G)$, as well as the problems of finding the τ-number $\tau(G)$, for $\tau \in \{\mu, \mu_d, \mu_o, \mu_t\}$, have been proved to be NP-complete, respectively in [25] and in [15]. All the NP-completeness results hold for graphs of diameter 4 or larger. Then many other works restricted their attention to the study of these quantities for special classes of graphs (e.g., see [2,7,13,14,16,18,22,23,25,29,30]). However, these problems have not yet been considered from the approximability point of view. In this paper, we provide an algorithm that finds a μ-set with size $\Omega\left(\sqrt{n/D}\right)$, where D is the average distance between any two vertices of the graph (which is trivially a $O(\sqrt{n \cdot \text{diam}(G)})$-approximation algorithm), and we present strong inapproximability results about the computation of $gp(G)$, $\mu(G)$, $\mu_d(G)$, $\mu_o(G)$, and $\mu_t(G)$ on graphs of diameter 2 or larger, as summarized in Table 1. In [5], we also study the relationship between the general position number and the mutual-visibility number in graphs of diameter 2.

Structure of the Paper. The next section provides some preliminary notions. In Sect. 3 we present the algorithm for approximating a maximum mutual-visibility set. Then, we study the computational complexity of finding maximum τ-sets,

Table 1. Summary of our results. Here $\varepsilon > 0$ denotes a positive constant of choice, and D is the average distance in G.

Measure(s)	Result	Notes	Reference
μ	μ-set of size $\Omega(\sqrt{n/D})$ in poly time		Theorem 1
μ, μ_d, μ_o, μ_t	APX-Hard	$\mathrm{diam}(G) = 2$	Theorem 2
μ_t	Not approximable within $n^{1/3-\varepsilon}$	$\mathrm{diam}(G) = 2$	Theorem 3
μ, μ_d	Not approximable within $n^{1/3-\varepsilon}$	$\mathrm{diam}(G) = 3$	Theorem 4
μ_o, μ_t	Not approximable within $n^{1/2-\varepsilon}$	$\mathrm{diam}(G) = 3$	Theorem 4
gp	Not approximable within $n^{1-\varepsilon}$	$\mathrm{diam}(G) = 2$	[5]

for $\tau \in \{\mu, \mu_d, \mu_o, \mu_t\}$ in Sects. 4 and 5. Finally, Sect. 6 discusses some open problems. Due to space limitation, the inapproximability result of the general position problem and the relationship between the general position number and the mutual-visibility number can be found in [5].

2   Preliminaries

We consider undirected graphs and unless otherwise stated, all graphs in the paper are connected. Given a graph G, $V(G)$ and $E(G)$ are used to denote its vertex set and its edge set, respectively. The order of G, that is $|V(G)|$, is denoted by $n(G)$, and its size $|E(G)|$ is denoted by $m(G)$. We remove the argument G when it is clear from the context. If $X \subseteq V(G)$, then $G[X]$ denotes the subgraph of G induced by X. For a natural number k, we set $[k] = \{1, \ldots, k\}$.

An *independent set* is a set of vertices of G, no two of which are adjacent. The size of a largest independent set is the *independence number* $\alpha(G)$ of G.

The *complete graph* (or *clique*) K_n, is the graph with n vertices where each pair of distinct vertices are adjacent. A *subcubic graph* is a graph where each vertex has degree at most 3. The distance between two vertices u, v in a graph G is denoted $d(u, v)$ and is the number of edges in a shortest u, v-path. The *diameter* of G is the maximum distance between pairs of vertices of the graph.

The *Cartesian product* $G \square H$ of graphs G and H both have the vertex set $V(G) \times V(H)$. In $G \square H$, vertices (g, h) and (g', h') are adjacent if either $g = g'$ and $(h, h') \in E(H)$, or $h = h'$ and $(g, g') \in E(G)$. A *layer* in $G \square H$ is a subgraph induced by the vertices in which one of the coordinates is fixed. Note that each layer is isomorphic either to G or H.

3   A Polynomial-Time Algorithm for Computing Mutual-Visibility Sets

We consider graphs G with $n \geq 7$ vertices as for graphs with at most 6 vertices we can compute a μ-set of maximum size in a brute-force manner. We denote by $\binom{V(G)}{2}$ the set of all unordered pairs of distinct vertices in $V(G)$. For any

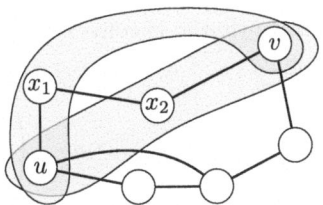

Fig. 2. A sample graph G and the hyperedges (shown as sets of vertices) added to the hypergraph H for the pair of vertices u, v when the chosen shortest path is $\langle u, x_1, x_2, v \rangle$.

$\{u, v\} \in \binom{V(G)}{2}$, fix a shortest path $\langle u, x_1, x_2, \ldots, x_k, v \rangle$ from u to v in G (break ties arbitrarily) and let $B(\{u, v\}) = \{x_1, \ldots, x_k\}$. Notice that it might be $k = 0$, in which case $B(\{u, v\}) = \emptyset$. This can be done in time $O(nm)$ by computing all-pairs shortest paths.

We build a 3-uniform hypergraph H (i.e., a hypergraph in which each hyperedge contains exactly 3 vertices) as follows: the set of vertices of H is $V(G)$ and there exists a hyperedge $\{u, v, x\}$ for each $\{u, v\} \in \binom{V(G)}{2}$ and $x \in B(\{u, v\})$. See Fig. 2 for an example.

An independent set of a hypergraph H is a subset S of vertices $V(G)$ such that, there exists no hyperedge e such that $e \subseteq S$.

Observe that an independent set S of H is a μ-set of G. Indeed, for every two distinct vertices u and v such that $u, v \in S$, it must be the case that no vertex $x \in B(\{u, v\})$ is in S as H contains the hyperedge $\{u, x, v\}$.

Let $D = \frac{2}{n(n-1)} \sum_{\{u,v\} \in \binom{V(G)}{2}} d(u, v)$ be the average distance in G, and let $m(H)$ be the number of hyperedges in H. We compute an independent set S of H in time $O(n + m(H)) = O(n^2 D)$ using the linear-time algorithm of [8]. For an integer $\ell \geq 0$ and a real $r \geq 0$, we define $\binom{r}{\ell} = \frac{1}{\ell!} \prod_{i=0}^{\ell-1}(r - i)$. It is known [27] that $|S|$ can be lower bounded as:

$$|S| \geq \sum_{v \in V(G)} \frac{1}{\binom{\delta(v)+\frac{1}{2}}{\delta(v)}} = \Theta\left(\sum_{v \in V(G)} \frac{1}{\sqrt{1 + \delta(v)}}\right),$$

where $\delta(v)$ denotes the number of hyperedges incident to v and we used $\binom{\delta(v)+\frac{1}{2}}{\delta(v)} = \Theta(\sqrt{1 + \delta(v)})$.

Using the fact that H is 3-uniform we have:

$$\sum_v \delta(v) = \frac{1}{3} m(H) \leq \frac{1}{3} \sum_{\{u,v\} \in \binom{V(G)}{2}} |B(\{u, v\})| = \frac{1}{3} \sum_{\{u,v\} \in \binom{V(G)}{2}} (d(u, v) - 1)$$

$$= \frac{n(n-1)}{6}(D - 1) \leq \frac{n^2}{6} D - n.$$

Let $\varphi(x) = \frac{1}{\sqrt{1+x}}$ and notice that φ is convex and monotonically decreasing for $x \geq 0$. By Jensen's inequality:

$$\sum_{v \in V(G)} \frac{1}{\sqrt{1 + \delta(v)}} = \sum_{v \in V(G)} \varphi(\delta(v)) \geq n \cdot \varphi \left(\frac{1}{n} \cdot \sum_{v \in V(G)} \delta(v) \right)$$

$$\geq n \cdot \varphi \left(\frac{n}{6} D - 1 \right) = \sqrt{\frac{6n}{D}},$$

hence $|S| = \Omega\left(\sqrt{n/D}\right)$. We have therefore shown:

Theorem 1. *Given an input graph G on n vertices, it is possible to find, in time $O(n(m + nD)) = O(n^3)$, a μ-set of G having size $\Omega\left(\sqrt{n/D}\right)$, where $D = \frac{2}{n(n-1)} \sum_{\{u,v\} \in \binom{V(G)}{2}} d(u, v)$ is the average distance in G.*

We observe that the approximation ratio guaranteed by the algorithm of this section is trivially $O(\sqrt{n \cdot D}) = O(\sqrt{n \cdot \operatorname{diam}(G)})$.

4   Inapproximability of Visibility Problems on Graphs of Diameter 2

In this section we show inapproximability results for the problem of computing τ-set of maximum size, where $\tau \in \{\mu, \mu_d, \mu_o, \mu_t\}$, for graphs with diameter 2 via reductions from the MAX-INDEPENDENT-SET problem.

Given an undirected graph H, the MAX-INDEPENDENT-SET problem asks to compute an independent set of H of maximum cardinality.

The MAX-INDEPENDENT-SET problem on subcubic graphs is not approximable in polynomial time within a factor of c' for a suitable constant $c' > 1$, unless P = NP, see [1]. Moreover, for general graphs, it cannot be approximated within a factor of $O(n(H)^{1-\delta})$ in polynomial time, for any constant $\delta > 0$, see [31].

The reduction we are going to show is from the MAX-INDEPENDENT-SET problem on connected graphs. Given a connected graph H with $n(H) \geq 3$ vertices which is the input instance of the MAX-INDEPENDENT-SET problem, we construct the graph G as a function of H and an additional integer parameter $L \geq 1$. The set of vertices of G consists of the union of (i) all edges in $E(H)$, (ii) L copies $v_1, \ldots, v_L$ of each vertex $v \in V(H)$, and (iii) two new vertices y and z. The set of edges of G contains (i) all edges $(u_1, v_1), \ldots, (u_L, v_L)$ for each pair of distinct vertices $u, v \in V(H)$ (ii) all edges $(e, v_1), \ldots, (e, v_L)$ for each $e \in E(H)$ and each $v \in V(H)$ such that v is an endvertex of e, (iii) all edges $(y, v_1), \ldots, (y, v_L)$ for all $v \in V(H)$, and finally (iv) all edges (z, e) for $e \in E(H)$. Observe that G has diameter 2. Figure 3 shows the construction of a graph G corresponding to a particular graph H.

We first present some technical lemmas that will be instrumental to proving our inapproximability results.

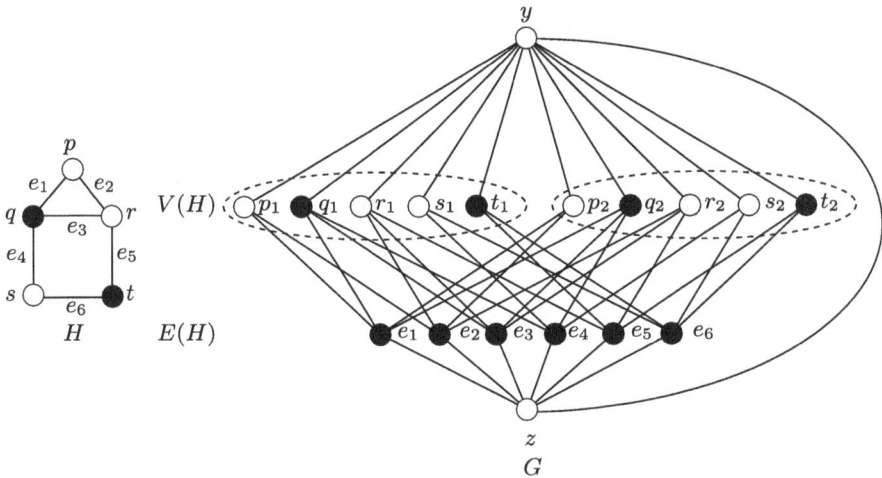

Fig. 3. Construction of graph G from graph H; $L = 2$. Vertices inside the dashed ellipses induce complete graphs.

Lemma 1. *Consider the graph G obtained from H when $L = 1$. Given a μ-set M of G, it is possible to compute, in polynomial time, an independent set S of H of size at least $|M| - m(H) - 4$.*

Proof. Let S' be the set of all vertices $v \in V(H)$ such that $v_1 \in M$. If $|S'| \leq 2$ we choose $S = \emptyset$, and the claim follows from $|M| \leq m(H) + |S'| + |\{y, z\}| \leq m(H) + 4$, i.e., $|S| = 0 \geq |M| - m(H) - 4$. Hence, in the rest of the proof we only consider the case $|S'| \geq 3$.

Choose an arbitrary end vertex r_e of each edge $e = (u, v) \in E(H)$ such that $e \notin M$ and define $R = \{r_e \mid e \in V(H) \setminus M\}$. We choose $S = S' \setminus R$.

To see that S is an independent set of H, consider an arbitrary edge $e = (u, v) \in E(H)$, let $w \in S' \setminus \{u, v\}$ and notice that it cannot be the case that $u, v \in S$ since this would imply $e \in M$ and e would not be in mutual visibility with w_1 in G. Indeed, the only two shortest paths connecting w_1 to e are $\langle w_1, u, e \rangle$ and $\langle w_1, v, e \rangle$.

Using $|S'| + |M \cap E(H)| \geq |M| - 2$, we can write:

$$|S| \geq |S'| - |R| \geq |S'| - |E(H) \setminus M| = |S'| - |E(H) \setminus (M \cap E(H))|$$
$$= |S'| - (m(H) - |M \cap E(H)|) \geq |M| - 2 - m(H).$$

$\square$

Lemma 2. *Given a μ_t-set X of G, it is possible to compute, in polynomial time, an independent set S of H of size at least $\frac{|X| - m(H) - 2}{L}$.*

Proof. Consider any $e = (u, v) \in E(H)$, and let $w \in V(H) \setminus \{u, v\}$. For any $i \in [L]$, the only two shortest paths between w_i and e in G are $\langle w_i, u_i, e \rangle$ and $\langle w_i, v_i, e \rangle$, hence at least one of u_i and v_i is not in X.

Let S_i be the set containing all vertices $v \in V(H)$ such that $v_i \in X$. By the above discussion we have that each S_i, with $i \in [L]$, is an independent set of H and we choose S as any of the sets S_i of maximum cardinality. Since $\sum_{i=1}^{L} |S_i| \geq |X| - m(H) - 2$, we have:

$$|S| = \max_{i \in [L]} |S_i| \geq \frac{1}{L} \sum_{i=1}^{L} |S_i| \geq \frac{|X| - m(H) - 2}{L}.$$

$\square$

Lemma 3. $\mu_t(G) \geq L \cdot \alpha(H) + m(H)$.

Proof. Let S be a maximum independent set of H. We define X as the subset of vertices of G that contains all $e \in E(H)$ plus all copies $v_1, \ldots, v_L$ for each $v \in S$. Clearly, $|X| = L \cdot |S| + m(H) = L \cdot \alpha(H) + m(H)$. We now argue that X is a μ_t-set of G.

Each vertex $v_i \in V(G)$ for $v \in V(H)$ and $i \in [L]$ is in X-visibility with y (via the edge (u, y)), with z (via the shortest path $\langle u, y, z \rangle$), with any other vertex u_i for $u \in V(H)$ (via the shortest path consisting of the sole edge (v_i, u_i)), and with any other vertex u_j with $u \in V(H)$ and $j \in [L] \setminus \{i\}$ (via the shortest path $\langle v_i, y, u_j \rangle$).

Similarly, each vertex $e \in E(H)$ is in X-visibility with z (via the edge (e, z)), with y (via the shortest path $\langle e, z, y \rangle$) and with any other vertex $f \in E(H)$ (via the shortest path $\langle e, z, f \rangle$).

It remains to argue that each vertex x_i with $x \in V(H)$ and $i \in [L]$ is in X-visibility with any other vertex $e = (u, v) \in E(H)$. If $x \in \{u, v\}$ this is clearly the case, since G contains the edge (x_i, e). If $x \notin \{u, v\}$ then let w be an arbitrary vertex in $\{u, v\} \setminus S$ (such a w exists since S is an independent set of H) and observe that $\langle x_i, w_i, e \rangle$ is a shortest path between x_i and e in G with $w_i \notin X$. $\square$

We can now prove the two main inapproximability results of this section.

Theorem 2. *For every* $\tau \in \{\mu, \mu_d, \mu_o, \mu_t\}$, *the problem of computing a maximum-size τ-set of an input graph G with diameter 2 is* APX-Hard.

Proof. Let $c \in \left(1, 1 + \frac{1}{12}\right]$ be a constant whose exact value will be given later. We show how to transform any c-approximation algorithm A for the problem of computing a τ-set into a $3c$-approximation algorithm for the minimum independent set problem on subcubic graphs.

Given an instance H of the MAX-INDEPENDENT-SET problem, where H is a subcubic graph, we can assume w.l.o.g. that H is connected (otherwise we can apply the following arguments on each connected component of H), that $m(H) \geq n(H)$ (otherwise H is a tree and we can find a maximum independent set in polynomial time), and that $n(H) \geq 104$ (otherwise we can find a maximum independent set in constant time by brute force).

Since H is subcubic, we must have $\alpha(H) \geq \frac{n(H)}{4}$.[1] We apply our reduction to H with $L = 1$ to obtain a graph G. Since $\mu_t(G) \leq \tau(G)$, Lemma 3 ensures that an optimal τ-set for G has size at least $\alpha(H) + m(H)$, and hence the c-approximate solution M computed by running algorithm A on G has size at least $\frac{\alpha(H)+m(H)}{c}$. Since M is always a mutual-visibility set, Lemma 1 allows us to compute, in polynomial time, an independent set S of H that satisfies:

$$|S| \geq |M| - m(H) - 4 \geq \frac{\alpha(H) + m(H)}{c} - m(H) - 4 = \frac{\alpha(H) - (c-1)m(H) - 4c}{c}.$$

Such a set S is exactly the one returned by our approximation algorithm for the independent set problem on subcubic graphs. The achieved approximation ratio is:

$$\frac{\alpha(H)}{|S|} \leq \frac{c\alpha(H)}{\alpha(H) - (c-1)m(H) - 4c} = c\left(1 + \frac{(c-1)m(H) + 4c}{\alpha(H) - (c-1)m(H) - 4c}\right)$$

$$\leq c\left(1 + \frac{(c-1)\frac{3n(H)}{2} + 4c}{\frac{n(H)}{4} - (c-1)\frac{3n(H)}{2} - 4c}\right) \leq c\left(1 + \frac{n(H) + \frac{104}{3}}{n(H) - \frac{104}{3}}\right) \leq 3c,$$

where we used $m(H) \leq \frac{3n(H)}{2}$ (since H is subcubic), $c \leq \frac{13}{12}$, and the fact that $\frac{x + \frac{104}{3}}{x - \frac{104}{3}} \leq 2$ for all $x \geq 104$.

As the MAX-INDEPENDENT-SET problem is not approximable in polynomial time within a factor of c' for a suitable constant $c' > 1$, unless $\mathsf{P} = \mathsf{NP}$, the claim follows by choosing $c = \min\left\{1 + \frac{1}{12}, \frac{c'}{3}\right\}$. $\square$

Theorem 3. *The problem of computing a maximum-size μ_t-set of an input graph G having diameter 2 is not approximable within $n^{\frac{1}{3}-\varepsilon}$, for any constant $\varepsilon > 0$, unless $\mathsf{P} = \mathsf{NP}$.*

Proof. Let $c > 0$ be a constant of choice. We show how to transform any $n(G)^c$-approximation algorithm A for the problem of computing a μ_t-set of a graph G with $n(G)$ vertices into a $O(n(H)^{3c})$-approximation algorithm for the problem of computing minimum independent of a graph H with $n(H)$ vertices.

Consider an instance H of the MAX-INDEPENDENT-SET problem, let $n(H)$ (resp. $m(H)$) be the number of vertices (resp. edges) of H, assume w.l.o.g. that $n(H) \geq 3$ (otherwise a maximum independent set of H can be found in constant time by brute force), and apply our reduction with $L = n(H)^2$ to construct G. Note that $n(G) = L \cdot n(H) + m(H) + 2 = \Theta(n(H)^3)$.

From Lemma 3, we have $\mu_t(G) \geq n(H)^2\alpha(H) + m(H)$, hence the μ_t-set X computed by running algorithm A on G has size at least $\frac{n(H)^2\alpha(H)+m(H)}{n(G)^c}$.

---

[1] A independent set S of H having size at least $\frac{n(H)}{4}$ can be computed by the greedy algorithm that starts from $S = \emptyset$ and iteratively (i) adds an arbitrary vertex v to S, and (ii) deletes v and all its (at most 3) neighbors form H until no vertices are left.

As shown by Lemma 2, we can convert X into an independent set S of H having size $\frac{|X|-m(H)-2}{n(H)^2}$. Then:

$$|S| \geq \frac{|X|-m(H)-2}{n(H)^2} \geq \frac{\frac{n(H)^2\alpha(H)}{n(G)^c}-m(H)-2}{n(H)^2} = \frac{\alpha(H)}{n(G)^c} - \frac{m(H)+2}{n(H)^2}$$

$$\geq \frac{\alpha(H)}{\Theta(n(H)^{3c})} - 1 = \Omega\left(\frac{\alpha(H)}{n(H)^{3c}}\right).$$

Hence, for any constant $\varepsilon > 0$, no polynomial-time $n(G)^{\frac{1}{3}-\varepsilon}$-approximation algorithm can exist for the problem of computing a maximum μ_t-set, unless P = NP, since it would imply the existence of a polynomial-time $O(n(H)^{1-3\varepsilon})$-approximation algorithm for the MAX-INDEPENDENT-SET problem.   □

5   Inapproximability of Visibility Problems on Graphs of Diameter 3

In this section we show stronger inapproximability results of visibility problems for graphs of diameter 3. More precisely, given a graph G of n vertices and diameter of at least 3, we show that, for every $\tau \in \{\mu, \mu_d, \mu_o, \mu_t\}$ and every constant $\varepsilon > 0$, it is not possible to design a polynomial-time algorithm that computes a τ-set whose size approximates the value of $\tau(G)$ within a factor of $n^{1/(\beta+1)-\varepsilon}$, where $\beta = 2$ if $\tau \in \{\mu, \mu_d\}$ and $\beta = 1$ if $\tau \in \{\mu_o, \mu_t\}$, unless P = NP.

Given a graph H of N vertices and a clique K_L of $L \geq 1$ vertices, consider the graph $G = K_L \square H$. We denote by H_i the layer consisting in the i-th copy of H in $K_L \square H$ and we denote by v_i the copy of $v \in V(H)$ that belongs to the layer H_i. We now prove some useful connections between τ-sets in $K_L \square H$ and independent sets in H.

Lemma 4. *Given a τ-set X for $K_L \square H$, with $\tau \in \{\mu, \mu_d, \mu_o, \mu_t\}$, where H is a graph with N vertices and $L \geq 1$, we can find in polynomial time a subset $X' \subseteq X$ that satisfies the following two conditions:*

- *for every $i \in [L]$, $X' \cap V(H_i)$ is an independent set of H_i;*
- *$|X'| \geq |X| - N^\beta$, where $\beta = 2$ if $\tau \in \{\mu, \mu_d\}$ and $\beta = 1$ if $\tau \in \{\mu_o, \mu_t\}$.*

Proof. We say that an edge $(u,v) \in E(H)$ *appears* k *times* in $K_L \square H$ w.r.t. X if there are k distinct copies $(u_{i_1}, v_{i_1}), \ldots, (u_{i_k}, v_{i_k})$ of the edge (u,v) such that $u_{i_j}, v_{i_j} \in X$ for every $j \in [k]$.

We show that no edge $(u,v) \in E(H)$ appears $k \geq 2$ times in $K_L \square H$ w.r.t. X. Indeed, if there were two distinct copies (u_i, v_i) and (u_j, v_j) of (u,v) in H such that $u_i, v_i, u_j, v_j \in X$, then u_i and v_j would not be X-visible as the only two shortest paths from u_i to v_j pass through vertices u_j and v_i, respectively. Furthermore, when $\tau \in \{\mu_o, \mu_t\}$, for each edge $(u,v) \in E(H)$ that appears $k = 1$ times in $K_L \square H$ w.r.t. X, i.e., such that $u_i, v_i \in X$ for some $i \in [L]$, we have that $u_j, v_j \notin X$ for every $j \in [L]$, with $i \neq j$. Indeed, if w.l.o.g. u_j were contained in X, u_i and v_j would not be X-visible.

We compute a set X'' in polynomial time as follows. We start with $X'' = \emptyset$. Next, for each edge $(u, v) \in E(H)$ that appears $k = 1$ times in $K_L \square H$ w.r.t. X, i.e., there is a exactly one copy (u_i, v_i) of (u, v) in H_i such that $u_i, v_i \in X$, we add both u_i and v_i to X''. By construction, X'' is a subset of X of size $|X''| \leq N^2$ as H contains at most $\binom{N}{2}$ edges, each of which contributes with at most 2 vertices in X''. The upper bound on the size of X'' can be refined to $|X''| \leq N$ when $\tau \in \{\mu_o, \mu_t\}$ for the following reason. Adjacent edges in H that appear $k = 1$ times in $K_L \square H$ w.r.t. X must all appear in the same copy of H in $K_L \square H$. This implies that each vertex u of H contributes with at most 1 vertex in X''.

Let $X' = X \setminus X''$. By construction, $|X'| \geq |X| - N^2$ when $\tau \in \{\mu, \mu_d\}$ and $|X'| \geq |X| - N$ when $\tau \in \{\mu_o, \mu_t\}$. Moreover, each edge of H does not appear in $K_L \square H$ w.r.t. X'. As a consequence, $X' \cap V(H_i)$ is an independent set of H_i, for every $i \in [L]$. $\qquad\square$

Lemma 5. *Let H be a graph with $N \geq 2$ vertices containing a vertex z of degree $N - 1$ and let $S \subseteq V(H)$ be an independent set of H that does not contain z. Let L be a positive integer. The graph $K_L \square H$ has diameter of at most 3. Moreover, the set $X = \cup_{i \in [L]} S_i$, where $S_i = \{v_i \mid v \in S\}$ is the copy of S in H_i, is a μ_t-set of $K_L \square H$.*

Proof. The graph H has diameter of at most 2 as z is adjacent to all other vertices of H. As a consequence, the graph $K_L \square H$ has diameter of at most 3 as all L copies of z in $K_L \square H$ form a clique.

By definition, the endvertices of any edge of $K_L \square H$ are always X-visible. As copies of the same vertex of H are connected via a clique in $K_L \square H$, we only need to consider the case of copies u_i and v_j of distinct vertices u and v of H such that (u_i, v_j) is not an edge in $K_L \square H$.

As $z_i \notin S_i$ and $z_j \notin S_j$, we have $z_i, z_j \notin X$. Therefore, if $i = j$, both u_i and $v_j = v_i$ are adjacent to z_i and thus they are X-visible. For the case $i \neq j$ we split the proof into two subcases, according to whether $(u, v) \in E(H)$ or not.

We consider the subcase $(u, v) \notin E(H)$. First of all, we observe that u_i and v_j are at a distance of 3 in $K_L \square H$. As $K_L \square H$ contains the three edges (u_i, z_i), (z_i, z_j), and (z_j, v_j) and $z_i, z_j \notin X$, it follows that u_i and v_j are X-visible.

We consider the subcase $(u, v) \in E(H)$. First of all, we observe that u_i and v_j are a distance of 2 in $K_L \square H$. In particular, there are two shortest paths from u_i to v_j: the one passing through u_j and the one passing through v_i. As both S_i and S_j are copies of S, they are both independent set of layers H_i and H_j, respectively. This implies that either $u_i, u_j \in X$ or $v_i, v_j \in X$. In either cases, u_i and v_j are X-visible. $\qquad\square$

Corollary 1. *Let H be a graph with $N \geq 2$ vertices containing a vertex z of degree $N - 1$. Let L be a positive integer. Then $\mu_t(K_L \square H) \geq L \cdot \alpha(H)$.*

We can show the inapproximability results on graphs of diameter 3.

Theorem 4. *For every $\tau \in \{\mu, \mu_d, \mu_o, \mu_t\}$ and for every constant $\varepsilon > 0$, the problem of computing a τ-set of maximum cardinality on graphs G with n vertices and diameter 3 cannot be approximated within a factor of $n^{1/3-\varepsilon}$ for $\tau \in \{\mu, \mu_d\}$ and a factor of $n^{1/2-\varepsilon}$ for $\tau \in \{\mu_o, \mu_t\}$, unless $\mathsf{P} = \mathsf{NP}$.*

Proof. First of all we observe that, for every constant $\delta > 0$, the problem of computing a maximum independent set of a graph H with $N \geq 2$ vertices and containing a vertex of degree $N - 1$ is not approximable within a factor of $O(N^{1-\delta})$, unless $\mathsf{P} = \mathsf{NP}$. This is because given a graph H' with $N - 1$ vertices as an input instance of the MAX-INDEPENDENT-SET problem, we can construct a new graph H by adding to H' a new vertex z of degree $N - 1$ that is adjacent to all the vertices of H'. As $\alpha(H) = \alpha(H')$, any independent set of H whose size approximates $\alpha(H)$ within a factor of $O(N^{1-\delta})$ can be used to compute an independent set of H' whose size approximates $\alpha(H')$ within a factor of $O(N^{1-\delta})$ in polynomial time.

Given a graph H with $N \geq 2$ vertices containing a vertex z of degree $N-1$ as our input instance of the MAX-INDEPENDENT-SET problem, we build the graph $K_L \square H$ having $n = NL = N^{1+\beta}$ vertices by setting $L = N^\beta$, where $\beta = 2$ if $\tau \in \{\mu, \mu_d\}$ and $\beta = 1$ if $\tau \in \{\mu_o, \mu_t\}$.

By Lemma 5, $K_L \square H$ is a graph of diameter 3. Let $\varepsilon = \delta/(\beta + 1)$. We show that the existence of any polynomial-time algorithm that computes a τ-set X for $K_L \square H$ whose size approximates $\tau(K_L \square H)$ within a factor of $n^{1/(\beta+1)-\varepsilon}$ would imply the existence of a polynomial-time algorithm that computes an independent set S of H whose size approximates $\alpha(H)$ within a factor of $O(N^{1-\delta})$, i.e., such that $|S| = \Omega\left(\frac{\alpha(H)}{N^{1-\delta}}\right)$.

For the sake of contradiction, assume that there is a polynomial-time algorithm that computes a τ-set X for $K_L \square H$ such that $|X| \geq \frac{\tau(K_L \square H)}{n^{1/(\beta+1)-\varepsilon}}$. By Corollary 1, we have that $\tau(K_L \square H) \geq \mu_t(K_L \square H) \geq L \cdot \alpha(H) = N^\beta \cdot \alpha(H)$. Therefore, $|X| \geq \frac{N^\beta \cdot \alpha(H)}{N^{1-\delta}}$.

By Lemma 4, given X and $K_L \square H$, we can compute in polynomial time a subset $X' \subseteq X$ of size $|X'| \geq |X| - N^\beta$ such that, for every $i \in [L]$, $S_i := X' \cap V(H_i)$ is an independent set of H_i. We now compute an independent set S of H as follows. If all S_i's are empty sets, then $S = \{v\}$ where v is any arbitrary vertex of H. If some S_i are not empty, then let $i^* \in \arg\max\{|S_i| \mid i \in [L]\}$ and let $S = \{v \mid v_{i^*} \in S_{i^*}\}$. As $|S_{i^*}| \geq |X'|/L$, we have that $|S| \geq |X'|/L$. By construction, we always return an independent set S of size of at least $\max\{1, |X'|/L\}$. As a consequence, when $\alpha(H) < 2N^{1-\delta}$, $|S|$ already approximates $\alpha(H)$ within a factor of $O(N^{1-\delta})$. When $\alpha(H) \geq 2N^{1-\delta}$, we have

$$|S| \geq \frac{|X'|}{L} = \frac{|X| - N^\beta}{N^\beta} \geq \frac{\frac{N^\beta \cdot \alpha(H)}{N^{1-\delta}} - N^\beta}{N^\beta} = \frac{\alpha(H)}{N^{1-\delta}} - 1 = \Omega\left(\frac{\alpha(H)}{N^{1-\delta}}\right).$$

$\square$

6 Open Problems

For the mutual-visibility number we presented a polynomial-time algorithm that finds a μ-set with size $\Omega\left(\sqrt{n/D}\right)$, where D is the average distance in G. It would be interesting to study better algorithms to close the gap with the inapproximability result shown in Sect. 4 for the same problem. Of course, providing good approximation algorithms for finding the maximum values of the other invariants is also a challenging problem.

Given the inapproximability results for the problem of finding a maximum τ-set, for $\tau \in \{\mu, \mu_d, \mu_o, \mu_t\}$, it would be relevant to introduce a relaxed version of the mutual visibility. Given a set of vertices X, we could say that two vertices are in visibility if there exists a path connecting each pair of vertices without a further vertex in X such that the length of the path is at most σ times the distance between the two vertices, where $\sigma \geq 1$ is a fixed constant. Then, it would be interesting studying the computational complexity of the corresponding visibility problems, as well as the properties of the relative invariants from a graph theoretical point of view.

In this setting, we observe that the inapproximability results of Theorem 2 and Theorem 3 carry over to the case in which $1 \leq \sigma < 3/2$ as they hold for the class of diameter-2 graphs. Similarly, the inapproximability results of Theorem 4 on graphs of diameter 3 also hold when $1 \leq \sigma < 4/3$. Moreover, we observe that the reduction provided in Theorem 2, that we slightly modify by removing the vertex y from G, results in a graph G of diameter 2 in which the problem of computing a μ-set of maximum size is still APX-hard for every $\sigma \geq 1$.

References

1. Alimonti, P., Kann, V.: Hardness of approximating problems on cubic graphs. In: Bongiovanni, G., Bovet, D.P., Di Battista, G. (eds.) CIAC 1997. LNCS, vol. 1203, pp. 288–298. Springer, Heidelberg (1997). https://doi.org/10.1007/3-540-62592-5_80
2. Anand, B.S., Ullas Chandran, S.V., Changat, M., Klavžar, S., Thomas, E.J.: Characterization of general position sets and its applications to cographs and bipartite graphs. Appl. Math. Comput. **359**, 84–89 (2019)
3. Badri, S., Cicerone, S., Di Fonso, A., Di Stefano, G.: An optimal algorithm for geodesic mutual visibility on hexagonal grids. In: Proceedings of SSS 2024. LNCS, vol. 14931, pp. 161–176. Springer, Cham (2024)
4. Bhagat, S.: Optimum algorithm for the mutual visibility problem. In: Proceedings of WALCOM: 2020. LNCS, vol. 12049, pp. 31–42. Springer, Cham (2020)
5. Bilò, D., Di Fonso, A., Di Stefano, G., Leucci, S.: On the approximability of graph visibility problems. CoRR, abs/2407.00409 (2024)
6. Brešar, B., Yero, I.G.: Lower (total) mutual-visibility number in graphs. Appl. Math. Comput. **465**, 128411 (2024)
7. Bujtás, C., Klavžar, S., Tian, J.: Total mutual-visibility in hamming graphs. arXiv (2023)
8. Caro, Y., Tuza, Z.: Improved lower bounds on k-independence. J. Graph Theory **15**(1), 99–107 (1991)

9. Chandran, U.S.V., Jaya Parthasarathy, G.: The geodesic irredundant sets in graphs. Int. J. Math. Comb. **4**, 135–143 (2016)
10. Cicerone, S., Di Fonso, A, Di Stefano, G., Navarra, A.: The geodesic mutual visibility problem for oblivious robots: the case of trees. In: Proceedings of ICDCN 2023, pp. 150–159. ACM (2023)
11. Cicerone, S., Di Fonso, A., Di Stefano, G., Navarra, A.: The geodesic mutual visibility problem: oblivious robots on grids and trees. Pervasive Mob. Comput. **95**, 101842 (2023)
12. Cicerone, S., Di Fonso, A., Di Stefano, G., Navarra, A.: Time-optimal geodesic mutual visibility of robots on grids within minimum area. In: Proceedings of SSS 2023. LNCS, vol. 14310, pp. 385–399. Springer (2023)
13. Cicerone, S., Di Fonso, A., Di Stefano, G., Navarra, A., Piselli, F.: Mutual visibility in hypercube-like graphs. In: Proceedings of SIROCCO 2024. LNCS, vol. 14662, pp. 192–207. Springer, Cham (2024)
14. Cicerone, S., Di Stefano, G.: Mutual-visibility in distance-hereditary graphs: a linear-time algorithm. In: Proceedings of LAGOS 2023, vol. 223, pp. 104–111. Elsevier (2023)
15. Cicerone, S., Di Stefano, G., Drožek, L., Hedžet, J., Klavžar, S., Yero, I.G.: Variety of mutual-visibility problems in graphs. Theor. Comput. Sci. **974**, 114096 (2023)
16. Cicerone, S., Di Stefano, G., Klavžar, S.: On the mutual visibility in Cartesian products and triangle-free graphs. Appl. Math. Comput. **438**, 127619 (2023)
17. Cicerone, S., Di Stefano, G., Klavžar, S., Yero, I.G.: Mutual-visibility in strong products of graphs via total mutual-visibility. Discret. Appl. Math. **358**, 136–146 (2024)
18. Cicerone, S., Di Stefano, G., Klavžar, S., Yero, I.G.: Mutual-visibility problems on graphs of diameter two. Eur. J. Comb. **120**, 103995 (2024)
19. Di Luna, G.A., Flocchini, P., Chaudhuri, S.G., Poloni, F., Santoro, N., Viglietta, G.: Mutual visibility by luminous robots without collisions. Inf. Comput. **254**, 392–418 (2017)
20. Di Stefano, G.: Mutual visibility in graphs. Appl. Math. Comput. **419**, 126850 (2022)
21. Dudeney, H.E.: Amusements in Mathematics. Nelson, Edinburgh (1917)
22. Klavžar, S., Patkós, B., Rus, G., Yero, I.G.: On general position sets in Cartesian products. RM **76**, 123 (2021)
23. Kuziak, D., Rodríguez-Velázquez, J.A.: Total mutual-visibility in graphs with emphasis on lexicographic and Cartesian products. Bull. Malays. Math. Sci. Soc. **46**, 197 (2023)
24. Manuel, P., Klavžar, S.: A general position problem in graph theory. Bull. Aust. Math. Soc. **98**(2), 177–187 (2018)
25. Manuel, P.D., Klavžar, S.: The graph theory general position problem on some interconnection networks. Fund. Inform. **163**(4), 339–350 (2018)
26. Poudel, P., Aljohani, A., Sharma, G.: Fault-tolerant complete visibility for asynchronous robots with lights under one-axis agreement. Theor. Comput. Sci. **850**, 116–134 (2021)
27. Shachnai, H., Srinivasan, A.: Finding large independent sets of hypergraphs in parallel. In: Rosenberg, A.L. (ed.) Proceedings of SPAA 2001, pp. 163–168. ACM (2001)
28. Sharma, G., Vaidyanathan, R., Trahan, J.L.: Optimal randomized complete visibility on a grid for asynchronous robots with lights. Int. J. Netw. Comput. **11**(1), 50–77 (2021)

29. Tian, J., Klavžar, S.: Graphs with total mutual-visibility number zero and total mutual-visibility in Cartesian products. Discussiones Mathematicae. Graph Theory **44**(4), 1277–1291 (2024)
30. Tian, J., Kexiang, X.: The general position number of Cartesian products involving a factor with small diameter. Appl. Math. Comput. **403**, 126206 (2021)
31. Zuckerman, D.: Linear degree extractors and the inapproximability of max clique and chromatic number. Theory Comput. **3**(1), 103–128 (2007)

Algorithms for the Collaborative Delivery Problem with Monitored Constraints

Lotte Blank[1], Kien C. Huynh[2], Kelin Luo[3(✉)], and Anurag Murty Naredla[1]

[1] Institute of Computer Science, University of Bonn, Bonn, Germany
{lblank,anuragmurty}@uni-bonn.de
[2] Communications and Transport Systems, ITN, Linköping University, Linköping, Sweden
chihu36@liu.se
[3] Department of Computer Science and Engineering, University at Buffalo, Getzville, NY, USA
kelinluo@buffalo.edu

Abstract. We provide efficient algorithms to solve package delivery problems in which a sequence of drones work together to 'optimally' deliver a package from a source s to a target t. The package may be transferred from one drone to another only on a given line on which secure package transfers are ensured (we refer to this as *monitored constraints*). We allow the source and target to lie outside of this line. The drones have different starting locations, speeds, and rates of fuel consumption. Two notions of optimality are studied—fuel-efficient delivery (minimize total fuel consumption in a successful delivery), and fastest delivery (minimize total time for delivery).

Keywords: Delivery · Combinatorial Optimization · Algorithms

1 Introduction

Recently, the field of drone delivery has attracted significant attention, particularly in collaborative operations where multiple drones work together to complete a single task. This growing interest stems from the varying speeds and efficiency of individual drones (or agents), necessitating collaborative solutions [1–3]. Optimization objectives that have been considered include minimizing delivery time, minimizing the energy consumption of the agents, or a combination of both. Most studies assume that handovers can occur at any point or along any edge freely throughout the entire domain. Furthermore, some collaborative delivery studies are motivated by introducing restrictions based on factors such as each agent's travel distance due to energy constraints [6,7], or drones handover constraints [8]. Considering the drones' limited coverage areas, Erlebach et al. [8] proposed a model where each drone has a defined operating area, and handovers between drones can only occur within the overlap of their respective areas. This generalized setting, which incorporates handover restriction constraints, presents

© The Author(s), under exclusive license to Springer Nature Singapore Pte Ltd. 2025
S. Nakano and M. Xiao (Eds.): WALCOM 2025, LNCS 15411, pp. 62–78, 2025.
https://doi.org/10.1007/978-981-96-2845-2_5

significant challenges for algorithm design compared to simpler models without such restrictions.

In this paper, we introduce for the first time a variation of the problem where handovers are restricted to occur only within specific regions of the plane. This serves as an intermediate stage between the models previously discussed, both with and without handover constraints. Specifically, we incorporate handover constraints that apply uniformly to all drones rather than just a subset. Our objective is to analyze how these handover restrictions impact the drone's collaborative delivery. Specifically, we consider the following setting.

Monitored Package Handovers. The package must be transferred between drones at specific nodes or areas along a predefined line L, representing safe or secure zones. This monitored handover setting offers greater operational control, allowing for optimizations that enhance efficiency while ensuring security throughout the package's journey. However, the inclusion of monitored handovers increases the problem's complexity.

In this paper, we study collaborative package delivery problems where a sequence of drones work together to optimally deliver a package from a source s to a target t in the Euclidean plane, with both s and t being points on the plane. A set of drones is provided, each with a specific starting location, velocity, and consumption rate. The delivery is collaborative, allowing the package to be handed over between drones during transit. We focus on a monitored package handover setting along a line L (referred to as the transfer line), where package transfers are restricted to points on this line (which can represent polyline road paths in real-world scenarios). The objective may be to minimize total fuel consumption or to optimize delivery time.

Problem 1 (Fuel-efficient delivery (Sect. 3)). We are given a set of drones along with each drone's starting point and rate of fuel consumption, a line L and a source point s and a target point t. Find the minimum total consumption needed to deliver a package from s to t using these drones, such that package exchanges between drones are restricted to the line L.

Problem 2 (Fastest Delivery (Sect. 4)). We are given a set of drones along with each drone's starting point and velocity, a line L and a source point s and a target point t. Find the minimum time needed to deliver a package from s to t using these drones, such that package exchanges between drones are restricted to the line L.

1.1 Related Work

In the early work of collaborative delivery studies, Bärtschi et al. [1] and Bärtschi et al. [2], introduced the concept of collaborative drone delivery within the framework of a moving network or graph. The objective was to schedule a set of drones, each with specified speeds or fuel consumption rates, to transport a package from a source to a target point while minimizing either delivery time or total fuel consumption. Polynomial-time algorithms were provided for this problem:

for minimizing the total delivery time, they provide an $O(k^2m + kn^2 + \text{APSP})$ time algorithm, where k is the number of drones, n is the number of nodes, m is the number of edges in the given graph and APSP represents the time required to compute all-pairs shortest paths in a graph with n nodes and m edges [2]; for minimizing the total consumption, the problem can be reduced to a shortest path problem and solved in time $O(n^3)$ [1]. Later, Carvalho et al. [3] presented a faster algorithm with a time complexity of $O(kn \log n + km)$, which remains the fastest known algorithm for the problem. They also demonstrated that the minimum travel time drone delivery problem becomes NP-hard when considering two packages. All of these studies assume that handovers can occur freely at any point or along any edge throughout the entire graph.

Some research introduces collaborative strategies due to factors like energy constraints, which limit each agent's travel distance [4,5] or specific handover limitations [8]. The energy constraint is primarily motivated by realistic scenarios in which each agent possesses limited working time or energy, thereby limiting the total distance they can travel. This consideration necessitates exploring a collaborative path that leverages multiple agents to efficiently deliver packages. Chalopin et al. [4] first demonstrated that the energy-constrained drone delivery problem is NP-hard in general graphs. Subsequently, Chalopin et al. [5] showed that this variant remains NP-hard even on a path graph. Considering the drones' coverage areas, Erlebach et al. [8] proposed a model where each drone has a defined operating area, and handovers between drones can only occur within the overlap of their respective areas. They proved that the problem of minimizing delivery time is NP-hard, even when all drones have the same speed or when the underlying graph is a simple path. This generalized setting, which incorporates movement constraints, presents significant challenges for algorithm design compared to simpler models without handover restrictions. Due to page limitations, additional related work is included in the extended version of the paper.

1.2   Our Results

We state our results in this section. Relevant notation and definitions are provided in Sect. 2. In all the theorems below, n is the number of drones. In Sect. 3, we present a polynomial-time algorithm for the fuel-efficient delivery problem.

Theorem 1. *The fuel-efficient delivery problem restricted to exchanges of the package at a line L can be solved exact in $\mathcal{O}(n^3)$.*

In Sect. 4, we present a polynomial-time algorithm for the fastest delivery problem. As we know, drones are utilized in order of increasing speed in an optimal schedule. We will execute a single loop to sequentially explore the next faster drone, aiming to extend its delivery range in the minimum possible time.

Theorem 2. *The fastest delivery problem restricted to exchanges of the package at a line L can be solved exact in $O(n^2)$.*

For the special case when source node and destination node are co-linear with L, we further improve the running time using binary search.

Lemma 1. *When s and t are co-linear with L, there is an algorithm with $\Theta(n \log n)$ running time.*

The algorithms we present are exact and deterministic. Our approach begins by designating a picker (i.e., the first drone to pick up the package) and utilizes an $O(n^2)$ algorithm to identify the most fuel-efficient delivery from source to destination. The process then involves a simple loop to select the picker and subsequently determines the minimum fuel consumption across all possible deliveries. Although replacing nested loops by some form of binary search might seem to be possible improvements, we cannot use local information (like derivatives) to guide our search for optimal transfers. This is because the functions we optimize are neither convex nor concave. In Sect. 3, there is added complexity since we try to minimize the sum of scaled distance functions. It is important to note that we consider the general case where the starting locations of all drones, the source point s, and the target point t are not constrained to the transfer line L. Restricting the locations of s or the drones to the line could significantly improve running time due to more favorable properties of the functions. However, we do not explore these restricted settings in detail due to space limitations and mostly focus on the scenario where s, t, and drone starting points can be anywhere in Euclidean space. All omitted proof details can be found in the extended version of the paper.

2   Preliminaries

We study collaborative package delivery problems in which a sequence of drones work together to 'optimally' deliver a package from a source s to a target t. The domain of interest is the Euclidean plane and both s and t are points on this plane. We are given a set of drones in the plane along with their individual starting points, and the delivery is *collaborative*, i.e., the package may be transferred from one drone to another. We might wish to optimize the delivery time (in which case the speeds of the drones are given to us), or minimize the total fuel consumption (the rate of fuel consumption for each drone is provided). We consider the monitored package handover setting on a line L (called the transfer line), where a package may be transferred from one drone to another only at points on the line L, see an example in Fig. 1.

2.1   Notation and Definitions

We begin with some notation and setup for the problems. There is a total of n drones with indices $1, 2, \ldots, n$. The drone with index i will be referred to as *drone i*. The package initially lies at the source point s and needs to be delivered to the target point t. The package can be transferred from one drone to another at any point on a given line L, called the transfer line. The transfer line L is incident to the x-axis. We will denote the non-negative part of the x-axis by L^+, and the non-positive part by L^-. The set of drones from 1 to i is denoted [i]. For all drones $i \in [n]$, its starting position is denoted $s(i)$, its rate of

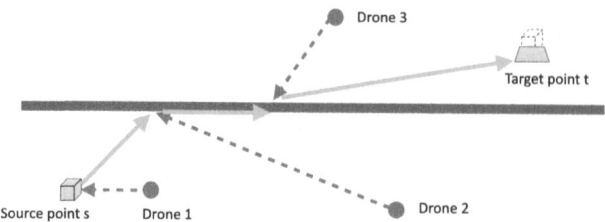

Fig. 1. Collaborative delivery example: Multiple drones work together to transport the package from the source point s to the target point t.

consumption is denoted $r(i)$, and its velocity is denoted by $v(i)$. The Euclidean distance between two points q, r on the plane is denoted $d(q, r)$. For any point q, the x and y coordinates are denoted by q_x and q_y respectively. For points on L, we will frequently not distinguish between the point itself and its x-coordinate.

The drone that picks up the package from s is called the *picker*, the drone that delivers the package to t the *deliverer* and every intermediate drone *carrier*. Note that it is possible that no exchanges happen on the line L and a single drone picks up the package from s and drops it to t. In this special case, the same drone is both the picker and the deliverer. If exchanges occur on the transfer line, the points on L where the package is transferred from one drone to another are called *exchange points*. The path taken by the various drones in order to move the package will be represented using a *drone sequence*—an ordered sequence of drones that participates in a package delivery from s to t. A drone sequence S may be denoted $[(i_1, e_1), (i_2, e_2), \ldots, (i_k, e_k)]$. Define $e_0 = s$. Then, the drone sequence S consists of tuples of the form $(i_j, e_j), 1 \leq j \leq k$ where i_j is the index of the drone that picks up the package from e_{j-1} and delivers it to e_j. The tuple preceding any tuple in a drone sequence is called its *parent tuple*.

Assumption 1. *The transfer line L is coincident with the x-axis, the x-coordinate of the source point s is 0.*

We begin with an initial observation.

Observation 1. *In a drone sequence with minimum consumption or delivery time, a drone either (i) does not move or (ii) moves in straight line segments: from its start to pick up a package at s or an exchange point on L, and then to t or the next exchange point on L.*

3   Fuel-Efficient Delivery with Exchanges on a Line

For the fuel-efficient delivery problem, we wish to use the minimum amount of fuel to deliver the package from s to t. Package handovers are permitted to occur only on the transfer line L. In any delivery sequence, the *consumption of drone i* is the product of its rate of consumption $r(i)$ and the total distance moved by i.

The problem is to minimize the total consumption—the sum of the consumption of all the drones. Fix a drone sequence $S = [(i_1, e_1), (i_2, e_2), \ldots, (i_k, e_k)]$. Observation 1 states that the path from the starting point $s(i_j)$ to the point e_{j-1} where a drone with index i_j picks up the package, and the path from there to the next exchange point e_j are both straight line segments. The sum of the consumption of all drones in the drone sequence is the *consumption of the sequence* to deliver the package from s to $e_k = t$, denoted $c(S)$. In other words:

$$c(S) = \sum_{j=1}^{k} r(i_j) \cdot [d(s(i_j), e_{j-1}) + d(e_{j-1}, e_j)]. \tag{1}$$

Throughout this section, we assume that $r(i) \geq r(i+1)$ for all $i \in [n-1]$. Using this assumption, we get the following basic observation, which can be proven using a simple exchange argument.

Observation 2. *In any optimal solution of the fuel-efficient delivery problem, the package is never transferred from a drone with a higher index to a drone with a lower index.*

The fuel-efficient delivery problem searches for a drone sequence of minimum consumption to carry a package from s to t. Such a sequence is called an *optimal sequence from s to t*. We will construct such an optimal sequence in an incremental manner. An optimal sequence from s to a target point x when we only consider drones in the set $\mathcal{D} \subseteq [n]$ is denoted by $S(\mathcal{D}, x)$. Later, we fix a picker P and then solve the fuel-efficient delivery problem with this fixed picker. A sequence of minimum consumption from s to a target point x with a fixed picker $P \in [n]$ is denoted $S_P(\mathcal{D}, x)$. Since the overall optimal sequence from $\mathcal{D}$ to deliver a package to x must use one of the drones as the picker, observe that

$$S(\mathcal{D}, x) \in \bigcup_{P \in \mathcal{D}} \{S_P(\mathcal{D}, x)\}. \tag{2}$$

The function $c : L^+ \to \mathbb{R}_{\geq 0}$ that plots the consumption of an optimal drone sequence to L^+ is called the *consumption function*, i.e., $c(x) = c(S([n], x))$. If we only allow drones in $\mathcal{D} \subset [n]$ to carry the package, we denote the consumption function as $c(\mathcal{D}, x) := c(S(\mathcal{D}, x))$. Finally, if we fix a picker $P \in \mathcal{D}$, the consumption of an optimal sequence will be denoted $c_P(\mathcal{D}, x) := c(S_P(\mathcal{D}, x))$. Our definitions imply

$$c(\mathcal{D}, x) = \min_{P \in \mathcal{D}} c_P(\mathcal{D}, x). \tag{3}$$

Let $\mathcal{D} \subseteq [n]$ be a set of drones and $P \in \mathcal{D}$ be a fixed picker. Then, we define for every drone i its *set of delivery points* $I_P(i, \mathcal{D})$ as a subset of L^+ in the following way. A point $x \in L^+$ is contained in $I_P(i, \mathcal{D})$ if and only if (i, x) is the last tuple in the optimal drone sequence $S_P(\mathcal{D}, x)$. In some cases, it may be possible to include x in the set of delivery points of more than one drone of $\mathcal{D}$. We avoid this situation by uniquely assigning x only to the set of delivery points of the highest drone index.

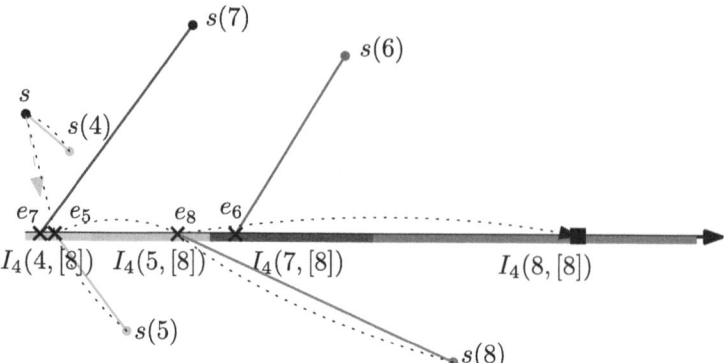

Fig. 2. Partition of L^+ into intervals that form the set of delivery points for the same drone (for details, see Algorithm 1). In this example, drone 4 is the picker which explains why we ignore drones of lower index. Drone 5 is the best delivery drone for interval $I_4(5, [8])$, and its optimal exchange point e_5 lies in $I_4(4, [8])$. Drone 6 is not the best delivery drone for any interval although an exchange point e_6 was determined before drones 7 and 8 were processed. Similarly, the sets of delivery points are computed for drone 7 and drone 8 as well. The dashed lines represent the paths that are taken by the drones to deliver the package to the black square in $I_4(8, [8])$.

3.1   Properties of Optimal Sequences with a Fixed Picker

In this section, we describe some properties of optimal drone sequences with a fixed picker P. We use Eqs. (2) and (3) to derive an algorithm for fuel-efficient delivery over all pickers. The lemmas and observations in this section guide the description of our algorithm and the proof of its correctness. See Fig. 2.

Observation 3. *Fix a picker P and consider a subset of drones $[k] \subset [n]$. Then $I_P(i, [k] \cup \{j\}) \subseteq I_P(i, [k])$ for all drones $i \in [k]$ and $k < j \leq n$.*

The following lemma allows us to treat exchange points on L^+ and L^- separately. We give the proof of it and the lemma afterwards in the extended version of the paper.

Lemma 2. *For every optimal sequence $S_P(\mathcal{D}, x) = [(i_1, e_1), \ldots, (i_k, e_k)]$ with a fixed picker $P \in \mathcal{D}$ (so, $i_1 = P$) and $e_k \in L^+$, we have $0 \leq e_1 < e_2 < \cdots < e_k$.*

Lemma 3 (Optimal Substructure). *If $S_P(\mathcal{D}, x) = [(i_1, e_1), (i_2, e_2), \ldots, (i_k, e_k)]$, then it holds that $S_P(\mathcal{D}, e_j) = [(i_1, e_1), (i_2, e_2), \ldots, (i_j, e_j)]$ for all $1 \leq j \leq k$.*

Lemma 4. *If $S_P(\mathcal{D}, x_1) = [(P, e_1)), (i, x_1)]$, then for all points $x_2 > x_1 > 0$, it holds that $c([(P, e_1), (i, x_2)]) < c([(P, x_2)]).*

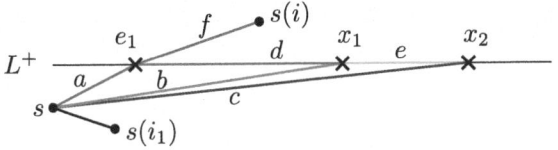

Fig. 3. Visualization of the notation in the proof of Lemma 4.

Proof. We use the following notation (see Fig. 3): $a = d(s, e_1)$, $b = d(s, x_1)$, $c = d(s, x_2)$, $d = d(e_1, x_1)$, $e = d(x_1, x_2)$, and $f = d(e_1, s(i))$. Since $x_1 \in I_P(i, \mathcal{D})$, it holds that $c([(P, e_1)), (i, x_1)]) \leq c([(P, x_1)])$. Hence,

$$r(P) \cdot a + r(i) \cdot d \leq r(P) \cdot a + r(i) \cdot (f + d) \leq r(P) \cdot b.$$

So, $\frac{r(i)}{r(P)} \leq \frac{b-a}{d}$. Further, by convexity it holds that $b < \frac{d}{d+e} \cdot c + \frac{e}{d+e} \cdot a$. Hence, $be - ae < cd - bd$ and $\frac{c-b}{e} > \frac{b-a}{d} \geq \frac{r(i)}{r(P)}$. Therefore, we get

$$r(P) \cdot c > r(P) \cdot b + r(i) \cdot e \geq r(P) \cdot a + r(i) \cdot (f + d + e)$$

and $c([(P, x_2)]) > c([(P, e_1), (i, x_2)])$.   □

Lemma 5. *Let $i, i' \in \mathcal{D}$ with $i < i'$ such that $I_P(i, \mathcal{D}) \neq \emptyset$ and $I_P(i', \mathcal{D}) \neq \emptyset$, i.e., there exist delivery points for both drones i and i'. Then, if $x_1 \in I_P(i, \mathcal{D})$ and $x_2 \in I_P(i', \mathcal{D})$, it holds that $0 \leq x_1 < x_2$.*

Proof. Let $x_2 \in I_P(i', \mathcal{D})$, $i < i'$ and $x_1 > x_2$, we show that then $x_1 \notin I_P(i, \mathcal{D})$. Let $S_P(\mathcal{D}, x_2)$ be an optimal sequence $[(i_1', e_1'), \ldots, (i_{k'}', e_{k'}')]$, where $i_1' = P$, $i_{k'}' = i'$ and $e_{k'}' = x_2$. Further, let $S = [(i_1, e_1), \ldots, (i_k, e_k)]$ be a sequence to $x_1 = e_k$, where $i_k = i$ and $i_1 = P$. By Observation 2, it holds that $r(i_l) > r(i')$ for all $l = 1 \ldots, k$. To conclude the proof, we show that S can not be an optimal sequence to x_1.

If $e_1 \leq x_2$, then $c(S) \geq c(S_P(\mathcal{D}, x_2)) + r(i')d(x_1, x_2) = c([(i_1', e_1'), \ldots, (i_{k'}', x_1)])$.

Otherwise, it holds that $e_1 > x_2$. If S is optimal, it must hold that $c([(i_1, e_1)]) = c(S_P(\mathcal{D}, e_1))$ by Lemma 3. However, this can not be the case by Lemma 4 as $x_2 \in I_P(i', \mathcal{D})$. Hence, in both cases S is not an optimal sequence to x_1 with picker P for the drone set $\mathcal{D}$.   □

Corollary 1. *The set of delivery points $I_P(i, \mathcal{D})$ for drone i is either empty, or an interval on L^+. For $i \in \mathcal{D}$, the intervals $I_P(i, \mathcal{D})$ partition L^+ and occur left-to-right in increasing order of i.*

3.2   The Algorithm

In this section, we give an algorithm that solves the fuel-efficient delivery problem. We first compute the consumption functions c_P for all (picker) drones

$P \in [n]$. Our discussion focuses on L^+ but the computation for the fixed picker function is symmetric on L^-, and we will omit mentioning it. The final algorithm performs computations for both rays separately and picks the best solution.

We compute the function c_P for a fixed picker P incrementally. So, given the consumption function $c_P([i], \cdot)$ for drones 1 to i, we construct the consumption function $c_P([i+1], \cdot)$ for drones 1 to $i+1$. We need some bookkeeping to ensure that the optimal sequence of drones can be retrieved for every point on L^+. By Corollary 1, $c_P([i], \cdot)$ is a piecewise-defined function over the intervals $I_P(j, [i])$ for the drones in $[i]$.

During the algorithm, we maintain a *block list* that stores a block for every drone j where $I_P(j, [i]) \neq \emptyset$. In this block list, the information needed to construct the consumption function c_P is stored. The *block* stored for a drone j is $[j, [L_j, R_j), (j^-, e(j)), f_j)]$, where

- $[L_j, R_j) = I_P(j, [i])$,
- $(j^-, e(j))$ is the parent tuple of (j, L_j) in the optimal drone sequence $S_P([i], L_j)$, and
- $f_j : [L_j, R_j) \rightarrow \mathbb{R}$ is the consumption function c_P restricted to the interval $[L_j, R_j)$. Note that this is expressible as a linear function of slope $r(j)$ for all drones $j \neq P$ by Theorem 3. For the picker P, f_P is a basically a scaled distance function defined for $[0, R_P)$ as $f_P(x) := r(P) \cdot (d(s(P), s) + d(s, x))$.

The block list is a list of those blocks ordered by the indices of the drones corresponding to the blocks. By Lemma 5, this is equivalent to sorting the blocks by the intervals $[L_j, R_j)$. Corollary 1 ensures that the block intervals form a partition of L^+.

Having a block list for drone set $[i]$ helps to retrieve the optimal sequence $S_P([i], x)$ to any point on $\mathbb{R}_{\geq 0}$—simply locate the block in which x lies, and follow the parent tuples until we reach a tuple $(l, e(l'))$, where $l = P$. We must eventually reach such a tuple since the index of the drone in a parent tuple always takes us to the block of a drone with lower index. The sequence of drones we encounter along with the exchange points (in reverse order) give us the optimal sequence $S_P([i], x)$.

We depict the set of delivery points with a fixed picker index in Fig. 2. After fixing a picker (index 4 in the figure), Algorithm 1 computes the set of delivery points for all drones until the last. The set of delivery points are color-coded according to the corresponding drone index. We ignore the relatively simpler step of determining the deliverer once we obtain this partition. The exact method used in Algorithm 1 is to ensure that the correct lower envelope for the total consumption function is computed once a picker is fixed. A graphical visualization of how the computation works is depicted in Fig. 4.

Theorem 3. *Algorithm 1 computes the block list for a fixed picker and the drone set $[n]$ in $O(n^2)$.*

Algorithm 1: Optimal BlockList: Fuel-efficient delivery with picker P

---

 1 **Function** ConstructBlockList(*picker P, drone set $[n]$, source point s*):

 2      $f_P(x) = r(P) \cdot (d(s(P), s) + d(s, x))$

 3      $BlockList = [(P, [0, \infty), null, f_P)]$

 4      $k = 1$ /*k denotes the length of the $BlockList$*/

 5   **for** $i = P + 1$ *to* n **do**

 6          $(i^-, e(i)) =$ ComputeParentTuple($BlockList, i$)

 7   UpdateBlockList($BlockList, i, k$)

 8   **return** $BlockList$

 9 **Function** ComputeParentTuple(*BlockList, i*):

10   **for** $(j, [L_j, R_j], (j', e(j)), f_j) \in BlockList$ **do**

11   Let e be such that $f'_j(e) + r(i)(d(s(i), e))' - r(i) = 0$

12          $B(x) = f_j(e) + r(i) \cdot (d(s(i), e) + x - e)$

13   **if** $f_i(0) > B(0)$ **then**

14   Set $f_i(x) = B(x)$ for all x

15   Set $e(i) = e$ and $i^- = j$

16   **return** $(i^-, e(i))$

17 **Function** UpdateBlockList(*BlockList, i, k*):

18      $(j, [L_j, R_j], (j^-, e(j)), f_j) = BlockList(k)$

19   **while** $f_i(L_j) \leq f_j(L_j)$ **do**

20   Remove $BlockList(k)$ from $BlockList$

21          $k = k - 1$

22          $(j, [L_j, R_j], (j^-, e(j)), f_j) = BlockList(k)$

23   Let $L_i \in [L_j, R_j]$ be such that $f_i(L_i) = f_j(L_i)$

24   Set $BlockList(k) = (j, [L_j, L_i], (j^-, e(j)), f_j)$ /*Update R_j*/

25      $BlockList(k + 1) = (i, [L_i, \infty), (i^-, e(i)), f_i)$ /*Add Block for i*/

26      $k = k + 1$ /*Update length of $BlockList$*/

Proof. We start to prove the running time. It takes $O(n)$ time for each call to ComputeParentTuple, which iterates over each earlier block to determine the parent tuple. Further, a simple charging argument shows that each call of UpdateBlockList takes constant amortized time, so all the calls together take total time $O(n)$. So, the total running time of ConstructBlockList for a picker P is in $O(n^2)$.

We prove the correctness via induction. First, note that $I_P(i, [n]) = \emptyset$ for $i < P$ by Observation 2. Initially, the block list consists just one block, which is the correct block for the drone set $[P]$. After iteration $i - 1$, we are given the correct block list for the drone set $[i - 1]$. Next, we prove that after iteration i, the updated block list is the block list for the drone set $[i]$. As drone i is the drone with the lowest consumption in $[i]$, its block must contain an interval $[L_i, \infty)$ (the drone with the lowest consumption rate must deliver to $+\infty$, and Corollary 1 tells us that the region $I_P(i, [i])$ is an interval).

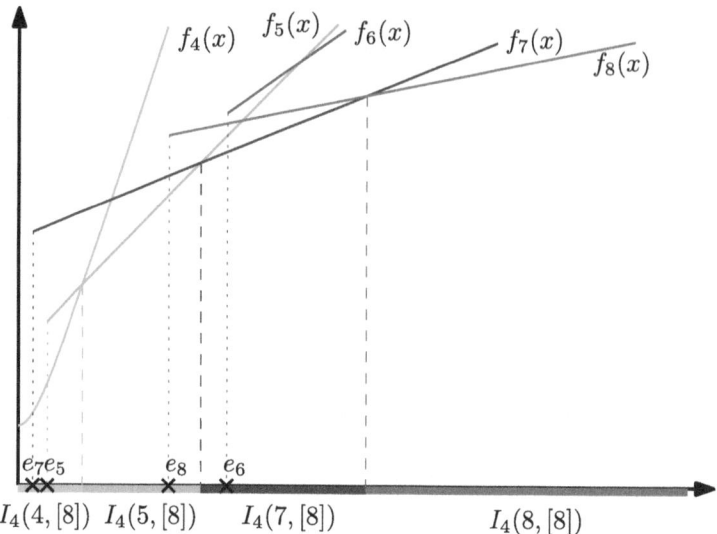

Fig. 4. The lower envelope is a consumption function $c_P([8], \cdot)$ for $P = 4$. This figure corresponds to the fixed picker solution obtained in Fig. 2. The rate of increase of the consumption decrease as we move from left to right. Apart from the picker function (which looks like a scaled distance function), all other functions in the graph are linear. Algorithm 1 basically performs the computations of the optimal exchange and crossover points with the total consumption graph.

For every point $x \in I_P(i, [i])$, we have to find an exchange point $e \le x$ where the package is handed over to drone i. This is the point e that minimizes $c_P([i - 1], e) + r(i) \cdot (d(s(i), e) + d(e, x))$. To find this minimum, ComputeParentTuple does a brute-force search over all drones j, where $I_P(j, [i - 1]) \ne \emptyset$. If $e \in I_P(j, [i - 1])$, then the package is handed over from drone j to drone i due to the optimal substructure property (Lemma 3). Further, it must hold that the derivative of $f_j(e) + r(i) \cdot (d(s(i), e) + (x - e))$ with respect to e is zero, i.e., $f_j'(e) + r(i) \cdot (d(s(i), e))' - r(i) = 0$. Because of how the functions f_j are constructed, there exists exactly one such point e for every j. Note, that this is independent of e. Hence, the point where the package is handed over from a drone j to i is the same for all points in $I_P(i, [i])$. ComputeParentTuple finds the minimum of all those values x and sets the parent tuple accordingly.

UpdateBlockList deletes all blocks from the block list where $L_j \ge L_i$ and updates the right end of the interval where $L_i \in [L_j, R_j)$. As L_i is computed correctly and ComputeParentTuple computes a parent tuple suitable for all points in $I_P(i, [i])$, the algorithm computes the correct block partition for drone set $[i]$ after iteration i by Corollary 1. □

It is straightforward to build upon the ideas described for the fixed picker in Sect. 3.1 to solve our original problem for fuel-efficient delivery. Suppose the sequence of minimum total consumption includes exchange points on L. Given the block list for a fixed picker $P \in [n]$, we compute for every deliverer $i \geq P$ and all blocks $[j, [L_j, R_j), (i^-, e(j)), f_j)]$ in the block list where $j < i$, the optimal point where the package can be handed over from j to i in $[L_j, R_j)$ such that the deliverer i brings the package from x to t. This is the point $x \in [L_j, R_j)$ that minimizes $C_{i,j} = f_j(x) + r(i) \cdot (d(s(i), x) + d(x, t))$. Since $f_j(x)$ is a linear function, x can be found in constant time for fixed i and j. The minimum over all $C_{i,j}$ is the most fuel efficient delivery from s to t, where the package was picked up by drone P. If there are no exchange points on L, the picker P is also the deliverer and the consumption of the delivery is $r(P) \cdot (d(s(P), s) + d(s, t))$. This computation takes constant time and we can assign the minimum of the two obtained values as the most fuel-efficient delivery with picker P. Hence, we get the following lemma.

Lemma 6. *For a fixed picker P, the most fuel efficient delivery from s to t can be computed in $O(n^2)$.*

The least fuel consumption among the most efficient deliveries for each picker $P \in [n]$ is the overall most fuel-efficient delivery. Therefore, we get:

Theorem 1. *The fuel-efficient delivery problem restricted to exchanges of the package at a line L can be solved exact in $\mathcal{O}(n^3)$.*

4   Fastest Delivery

This section deals with a problem similar to the previous section but with a different objective: we want to minimize the makespan of a delivery. In this setting, the handovers still need to occur on L and drones can have different speeds $v(i)$. We also assume that $\forall i : v(i) \leq v(i+1)$, and Assumption 1 is true for convenience. Subsection 4.1 provides the algorithm to solve this problem. Subsection 4.2 considers a special case where both s and t are co-linear with L which has a faster algorithm that runs in $\Theta(n \log n)$.

4.1   Algorithm for the General Case

Let $f_i^*(x) : \mathbb{R}_{\geq 0} \to \mathbb{R}_{\geq 0}$ be the function of the *shortest* time it takes to deliver the package to a point with coordinate x in L^+ using the drones from the set $[i]$. Note that some drones in this set might not be used at all. To prove some important properties of $f_i^*(x)$, the following observations which are similar to those from the previous section are needed:

Observation 4. *The package is never transferred from a faster drone (higher index) to a slower drone (lower index).*

Lemma 7 (Optimal Substructure). *Let $S(\mathcal{D}, x) = [(i_1, e_1), \ldots (i_k, e_k)]$ then it holds that $S(\mathcal{D}, e_j) = [(i_1, e_1), \ldots (i_j, e_j)], \forall 1 \leq j \leq k$.*

The proof of Lemma 7 is very similar to that of Lemma 3 so we will skip it for brevity.

The function $f_i^*(x)$ is a piecewise function where each piece is a part of a linear or a non-linear but convex function. In addition, each piece contains information about which drone is the latest one being used. Let I be the inverse image of a piece of drone j in $f_i^*(x)$ (I is an interval on L^+). If the drone visits some point on L^+ to pick up the package and moves to some other point x then its piece in $f_i^*(x)$ is $g(x) = \frac{1}{v(j)}x + c\big|_I$ where c is the time it takes for drone j to pick up the package. If drone j visits s and delivers it to some point on L^+, then its piece is $h(x) = \frac{1}{v(j)}(d(s(j),s) + d(s,x))\big|_I$. From this point onward, we will use $\eta_j(x)$ to denote $\frac{1}{v(j)}(d(s(j),s) + d(s,x))$ (so $h(x) = \eta_j(x)|_I$), and use $\gamma_j(x)$ to denote $\frac{1}{v(j)}x + c$ for some $c > 0$. Note that the same drone can have multiple pieces in $f_i^*(x)$.

The correctness and running time analysis of the main algorithm rely on the following property:

Lemma 8. $f_i^*(x)$ *is increasing over* $\mathbb{R}_{\geq 0}$.

Using the above lemma, given any point $x_1 \in L^+$ and a function $f_i^*(x)$ we can always infer the optimal sequence of drone exchanges that deliver the package to x_1. Assume that the piece $\rho(x)$, whose domain contains x_1, belongs to drone j. If $\rho(x)$ is a non-linear piece, drone j will deliver the package directly from s to x_1. Otherwise, consider the left breakpoint p_x of $\rho(x)$ and this point is shared with another piece $\rho'(x)$, whose domain contains p_x, belongs to drone l. This means that drone j will pick up the package from drone l at p_x at time $f_i^*(p_x)$. We redo the same step on $\rho'(x)$ and keep going over more pieces until either there is nothing left or the piece is non-linear. This will give us the optimal drone sequence with makespan $f_i^*(x_1)$.

The process to construct $f_n^*(x)$ is in Algorithm 2. We did not write the details of some basic procedures such as: SizeOf (return the number of pieces of $f_i^*(x)$), Dom (return the domain of a function), FindTangentPoint (in the case that a tangent point p_t between f_j and a line of slope $\frac{1}{v(i)}$ exists and $f_j(p_t) \geq d_i(p_t)$, return p_t otherwise return *null*). The algorithm has a main loop running from drone 1 to drone n. At loop i, we assume that we already have $f_{i-1}^*(x)$ from the previous loop and we will update $f_{i-1}^*(x)$ with the i-th drone to acquire $f_i^*(x)$. Two procedures will be run when adding drone i: AddNewPicker and AddNewCarrier.

AddNewPicker assumes that drone i will pick up the package at s and deliver it to some point on L^+. The function of time that it takes for drone i to do this is $\eta_i(x)$. The procedure will then compute a new function $\min(\eta_i(x), f_{i-1}^*(x))$. The other procedure, AddNewCarrier assumes that drone i will take the package from another drone on L^+ and move to the positive direction of L^+. Note that there could be multiple points on L^+ at which drone i could grab the package and perform better than the previous drone at that point. To compute all such pickup points for drone i, we will loop over all pieces of $f_{i-1}^*(x)$. Let $d_i(x) = \frac{1}{v(i)}d(s(i),x)$

Algorithm 2: FastDelivery

---

1 **Function** ConstructF(*drone set* $[n]$, *source point* s):

2     $f = [\eta_1(x)|_{\mathbb{R}\geq 0}]$ /*f is an array, the j-th element is accessed through f_j*/

3   **for** $i = 2$ *to* n **do**

4   AddNewPicker (f, i)

5   AddNewCarrier (f, i)

6   **return** $f_n^*(x) = f$ /*Return $f_n^*(x)$*/

7 **Function** AddNewPicker(f, i):

8   **if** $f_1(0) \geq \eta_i(0)$ **then**

9        $f = [\eta_i|_{\mathbb{R}\geq 0}]$ /*New drone beats all of the previous ones*/

10   **break**

11   **for** $j = 1$ *to* $SizeOf(f)$ **do**

12   Let p be a point s.t. $f_j(p) = \eta_i(p)$

13   **if** $p \neq null$ **then**

14   Remove f_{j+1} to $f_{SizeOf(f)}$ from f

15          $f_j = f_j|_{\mathrm{Dom}(f_j)\setminus[p_x,+\infty)}$

16          $f_{j+1} = \gamma_i|_{[p_x,+\infty)}$

17   Add f_j and f_{j+1} to f

18   **break**

19 **Function** AddNewCarrier(f, i):

20   **for** $j = 1$ *to* $SizeOf(f)$ **do**

21   Let p_i be a point s.t. $f_j(p_i) = d_i(p_i)$

22       $p_t = $ FindTangentPoint(f_j, $v(i)$)

23   **if** $p_t \neq null$ **then**

24   CarrierUpdate(f, p_t, j)

25   **else if** $p_i \neq null$ **then**

26   CarrierUpdate(f, p_i, j)

27 **Function** CarrierUpdate(f, p, j):

28   **for** $k = j + 1$ *to* $SizeOf(f)$ **do**

29   Let q be a point such that $f_k(q) = \gamma_i(q)$

30   **if** $q \neq null$ **then**

31   Remove f_{j+1} to f_k from f

32          $f_j = f_j|_{\mathrm{Dom}(f_j)\setminus[p_x,q_x]}$

33          $f_{j+1} = \gamma_i|_{p_x,q_x]}$

34   Add f_j and f_{j+1} to f

35          $j = k + 1$

36   **break**

---

be the function of time needed for drone i to fly directly to point $x \in L^+$. For each piece $\rho(x)$ of $f_{i-1}^*(x)$, we compute the best pickup point for i by finding these two points (if they exist) and pick the better one: the intersection (p_i)

between $\rho(x)$ and $d_i(x)$, or the tangent point (p_t) between $\rho(x)$ and a line of slope $\frac{1}{v(i)}$. Intuitively, the pickup point came from the intersection means that both drones (drone i and the drone of piece $\rho(x)$) arrive at p_i at the same time. If the pickup point is from a tangent point then drone i arrives at p_t first and then waits for the other drone to get there.

Let $\sigma(p,j)$ be the linear function that goes through $p = (x,y)$ and is restricted to $[x,+\infty)$ with slope $\frac{1}{v(j)}$ ($\sigma(p,j)$ is a ray). If the best pickup point x_1 exists for a piece using the above rules, we will compute the ray $\sigma(p,j)$ where $p = (x_1, \rho(x_1))$ and then compute $\min(\sigma(p,j), f_{i-1}^*(x))$ over $[x_1, +\infty)$. We will repeat this until we loop through all of the pieces in $f_{i-1}^*(x)$ which gives us $f_2(x)$. At the end of loop i, we compute $f_i^*(x) = \min(f_1(x), f_2(x))$. To save time, we can compute $f_1(x)$ first and then run AddNewCarrier on $f_1(x)$ instead. Once $f_n^*(x)$ is computed, we can find the best drone and pickup point to deliver the package to t by iterating over all drones and all pieces of $f_n^*(x)$. This entire process does not take care of the case where the best schedule may be using exactly one: naturally, we will also compute the best solution using only one drone and compare that with the schedule given the main procedure.

Lemma 9. *The above process returns the fastest delivery schedule from s to t.*

For the running time analysis, we need to bound the complexity of $f_i^*(x)$:

Lemma 10. $\forall i$, $f_i^*(x)$ *has at most $O(n)$ pieces.*

Next, we prove the running time of FastDelivery, confirms the Theorem 2:

Lemma 11. *The algorithm FastDelivery can be executed within $O(n^2)$.*

Proof. In each iteration of the main loop in FastDelivery, we run AddNewPicker and AddNewCarrier. For both AddNewPicker and AddNewCarrier, each has $O(n)$ loops since there are at most $O(n)$ pieces in $f_i^*(x), \forall i$ because of Lemma 10. Once the main algorithm returns the solution, we also need to compute the best solution using exactly one drone (who serves as both the picker and deliverer) and compare both of them. The time it takes to compute the best schedule with only one drone is $O(n)$ so it is dominated by the main algorithm. In total, the running time of the entire algorithm is $O(n^2)$. □

4.2 Algorithm for When s and t Are on L

We now consider the special case where s and t are co-linear with L (but the drones' starting locations can still be anywhere in the plane). Since it is a simpler problem, we can solve it more quickly than the general version:

Lemma 1. *When s and t are co-linear with L, there is an algorithm with $\Theta(n \log n)$ running time.*

To summarize, any function $f_{i-1}^*(x)$ is now a concave, piecewise linear function which allows us to use binary search to quickly look for the intersection point(s) with $d_i(x)$. This reduces the running time to $O(n \log n)$. The lower bound of the running time comes from reduction from sorting.

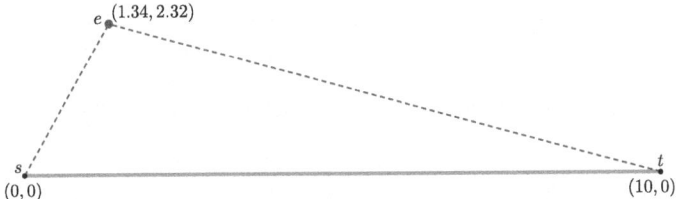

Fig. 5. If even one exchange point in addition to the line is permitted, we may have to revisit a line more than once. In the above example, a package needs to be delivered from s to t both on the same line. There is a drone of unit rate at the origin, and one of rate 0.794 at point e. Simple computations show that the best delivery schedule from s to t uses the first drone to take the package to point e and then the second drone delivers the package to t. We may similarly force multiple revisits to the same line if more exchange points/lines are added.

5   Conclusion

In this paper, we considered collaborative delivery problems with monitored package handover in the setting where transfers between two drones must happen on a transfer line. We study how to minimize total fuel consumption as well as delivery time. The following are interesting open problems:

- The algorithm for fuel-efficient delivery takes cubic time, while the algorithm for fastest delivery takes quadratic time. Can the running times be improved to quadratic and near-linear respectively?
- Future research could focus on developing efficient algorithms for scenarios where handovers occur on more complex arrangements, such as multiple lines or other configurations, instead of a single transfer line. While this would make the problem more practical, it also introduces significant complexity, requiring further investigation beyond our current approach. Our study addresses the simpler case of restricting handovers to a single line. Figure 5 illustrates complications, like the need to revisit the same line, that arise when this restriction is removed-challenges that could similarly affect the fastest delivery problem.

References

1. Bärtschi, A., et al.: Energy-efficient delivery by heterogeneous mobile agents. arXiv preprint arXiv:1610.02361 (2016)
2. Bärtschi, A., Graf, D., Mihalák, M.: Collective fast delivery by energy-efficient agents. In: Potapov, I., Spirakis, P., Worrell, J. (eds.) 43rd International Symposium on Mathematical Foundations of Computer Science (MFCS 2018). Leibniz International Proceedings in Informatics (LIPIcs), vol. 117, pp. 56:1–56:16. Schloss Dagstuhl – Leibniz-Zentrum für Informatik (2018)
3. Carvalho, I.A., Erlebach, T., Papadopoulos, K.: On the fast delivery problem with one or two packages. J. Comput. Syst. Sci. **115**, 246–263 (2021)

4. Chalopin, J., Das, S., Mihal'ák, M., Penna, P., Widmayer, P.: Data delivery by energy-constrained mobile agents. In: Flocchini, P., Gao, J., Kranakis, E., Meyer auf der Heide, F. (eds.) ALGOSENSORS 2013. LNCS, vol. 8243, pp. 111–122. Springer, Heidelberg (2014). https://doi.org/10.1007/978-3-642-45346-5_9

5. Chalopin, J., Jacob, R., Mihalák, M., Widmayer, P.: Data delivery by energy-constrained mobile agents on a line. In: Esparza, J., Fraigniaud, P., Husfeldt, T., Koutsoupias, E. (eds.) ICALP 2014. LNCS, vol. 8573, pp. 423–434. Springer, Heidelberg (2014). https://doi.org/10.1007/978-3-662-43951-7_36

6. Choi, Y., Schonfeld, P.M.: Optimization of multi-package drone deliveries considering battery capacity. In: Proceedings of the 96th Annual Meeting of the Transportation Research Board, Washington, DC, USA, pp. 8–12 (2017)

7. Coelho, B.N., et al.: A multi-objective green UAV routing problem. Comput. Oper. Res. **88**, 306–315 (2017)

8. Erlebach, T., Luo, K., Spieksma, F.C.: Package delivery using drones with restricted movement areas. In: Bae, S.W., Park, H. (eds.) 33rd International Symposium on Algorithms and Computation (ISAAC 2022). Leibniz International Proceedings in Informatics (LIPIcs), vol. 248, pp. 49:1–49:16. Schloss Dagstuhl – Leibniz-Zentrum für Informatik (2022)

A Piecewise Approach for the Analysis of Exact Algorithms

Katie Clinch[ID], Serge Gaspers[ID], Zixu He[ID], Abdallah Saffidine[ID], and Tiankuang Zhang[✉][ID]

School of Computer Science and Engineering, UNSW Sydney, Sydney, Australia
{k.clinch,serge.gaspers}@unsw.edu.au,
{tao.he,tiankuang.zhang}@student.unsw.edu.au,
abdallah.saffidine@gmail.com

Abstract. To analyze the worst-case running time of branching algorithms, the majority of work in exponential time algorithms focuses on designing complicated branching rules over developing better analysis methods for simple algorithms. In the mid-2000s, Fomin et al. introduced measure & conquer, an advanced general analysis method, sparking widespread adoption for obtaining tighter worst-case running time upper bounds for many fundamental NP-complete problems. Yet, much potential in this direction remains untapped, as most subsequent work applied it without further advancement. Motivated by this, we present piecewise analysis, a new general method that analyzes the running time of branching algorithms. Our approach is to define a similarity ratio that divides instances into groups and then analyze the running time within each group separately. The similarity ratio is a scale between two parameters of an instance. Instead of relying on a single measure and a single analysis for the whole instance space, our method allows us to take advantage of different intrinsic properties of instances with different similarity ratios.

To showcase its potential, we reanalyze two 17-year-old algorithms by Fomin et al. from 2007, solving 4-Coloring and #3-Coloring, respectively. The original analysis in their paper gave running times of $\mathcal{O}(1.7272^n)$ and $\mathcal{O}(1.6262^n)$, respectively, for these algorithms. Our analysis improves these running times to $\mathcal{O}(1.7207^n)$ and $\mathcal{O}(1.6225^n)$.

Keywords: Graph coloring · exact exponential time algorithms · algorithm analysis · c-Coloring · $\#c$-Coloring · 4-Coloring · #3-Coloring

1 Introduction

Branching algorithms have been widely used for more than 50 years to solve NP-complete problems. To achieve better worst-case running time upper bounds,

K. Clinch, S. Gaspers and T. Zhang—Research supported by the Australian Government through the Australian Research Council's Discovery Projects funding scheme (project DP210103849).

S. Nakano and M. Xiao (Eds.): WALCOM 2025, LNCS 15411, pp. 79–93, 2025.
https://doi.org/10.1007/978-981-96-2845-2_6

there are two main approaches. One is to design smarter algorithms that exploit the input's properties and make a large case distinction. The other approach is to improve the analysis of a simple (existing) algorithm. The first approach had been widely used in the past, but its effectiveness is often hamstrung by the design and the analysis of the algorithm being too intertwined. This means that such algorithms typically do not benefit from advances in algorithm analysis techniques. Even for simple algorithms, we typically do not know what the true worst-case running time is. Often, the known upper and lower bounds do not match. Therefore, a major issue in the field of exponential time algorithms is the development of better methods to analyze algorithms, in particular branching algorithms.

In the mid-2000s, Fomin et al. introduced measure & conquer (M&C) in [9], after some preliminary work had been done in the SAT community (see, e.g., [16]) and the quasiconvex analysis by Eppstein [5] in relation to graph coloring algorithms. Before this, in order to obtain a running time upper bound with respect to a variable n, one would track the progress of a branching algorithm in terms of n and solve a set of recurrences resulting from this analysis. M&C introduced the use of a potential function, a so-called measure, to track the progress of the algorithm. A measure is a weighted linear function of parameters of the instance; parameters such as the numbers of vertices of various degrees or the size of a solution. One then needs to minimize an upper bound on the measure with respect to n, subject to convex constraints arising from the analysis of the branching rules [13]. In [9], Fomin et al. used a linear function where all coefficients and weights are positive. Later, several authors expanded M&C to a wider set of measures: negative coefficients were used by [8], a small [19] or large [15] number of potentials were used that add terms to the measure, conditional on the properties of the instance or the algorithm, logarithmic terms were used by [14] and [12] and the use of compound piecewise linear measures was introduced by Wahlström [21]. We view all of these extensions under the umbrella of M&C, which has also found uses in the analysis of parameterized branching algorithms [10], whereas our piecewise analysis differs from M&C in significant detail. We use a measure that not only depends on the (sub-)instance being solved by the current node of the search tree, but also on the original instance.

The remainder of this paper is structured as follows. Section 2 provides all necessary definitions and lemmas required by other sections. Section 3 describes the piecewise analysis method and its relation with M&C. Following this, Sect. 4 presents our case study on c-Coloring and the application to 4-Coloring. The Sect. 5 presents the case study on $\#c$-Coloring and the application to $\#3$-Coloring. Finally, the conclusion summarizes our findings.

2   Preliminaries

Let $G = (V, E)$ be a simple undirected graph with vertex set V ($|V| = n$) and edge set E ($|E| = m$). Let $v \in V$ be a vertex in G. The *degree* of v in G is denoted by $d_G(v)$. The *maximum degree* of G is denoted by $\Delta(G)$. We denote the *open* *neighbourhood* of v by $N(v) = \{u \in V : uv \in E\}$ and the *closed neighbourhood* by $N[v] = N(v) \cup \{v\}$. For a vertex subset $V' \subseteq V$, we use $G[V']$ to denote the subgraph of G *induced* by V'. The set V' is an *independent set of G* if $G[V']$ has no edge. The set V' is a *vertex cover* of G if $V \setminus V'$ is an independent set. The graph $G[V \setminus \{v\}]$ is also denoted by $G - v$. $\mathcal{O}^*$-notation is similar to the usual $\mathcal{O}$-notation but allows hiding polynomial factors. We adopt the standard definitions of tree and path decompositions, along with established theorems, as presented in [3,7,18]. Proofs omitted due to space constraints can be found in the full version of the paper [2].

To prove the correctness of the algorithm for c-Coloring in later sections, the following lemma is required.

Lemma 1 ([1]). *The 3-Coloring problem can be solved in time $\mathcal{O}(1.3289^n)$ for a graph on n vertices.*

2.1 Measure and Conquer

The M&C method was first introduced by Fomin et al. [9]. They highlighted a disproportionate focus on developing sophisticated algorithms over analysis methods within the research on exact exponential time algorithms.

Their measures allow for a fine-grained tracking of the progress a branching algorithm makes when solving an instance and allow us to amortize "slow" branching rules with "fast" ones. A measure that is too simple can easily fail to reflect progress made by branching rules, resulting in loose running time upper bounds. Fomin et al. [9] observed that a number of parameters, beyond the input size, are good candidates for measure. A measure μ for a problem Π is a function $\mu : \mathcal{I} \to \mathbb{R}_{\geq 0}$ where $\mathcal{I}$ is the instance space (i.e., the set of inputs) of Π.

Lemma 2 ([11]). *Let $\mathcal{A}$ be a branching algorithm for a problem Π whose search trees have polynomial depth, and let $\mu(\cdot)$ be a measure for the instances of Π, such that for any input instance I, $\mathcal{A}$ reduces I to instances $I_1, \ldots, I_k$ in polynomial time, solves these recursively, and combines their solutions in polynomial time to solve I, and such that for any reduction done by Algorithm $\mathcal{A}$,*

$$(\forall I)\ 2^{\mu(I_1)} + \ldots + 2^{\mu(I_k)} \leq 2^{\mu(I)}. \tag{1}$$

Then $\mathcal{A}$ solves any instance I in $\mathcal{O}^*(2^{\mu(I)})$ time.

Based on Lemma 2, we can obtain an upper bound on the running time of the algorithm $\mathcal{A}$, provided that all constraints required by the lemma are satisfied. In other words, the running time analysis becomes an optimization problem that minimizes an upper bound on the measure by finding suitable weights (see [11,13]) without violating any constraint imposed by the lemma.

3 Introduction to Piecewise Analysis

In this section, we introduce the main contribution of this paper: *piecewise analysis*, a new tool for analyzing the worst-case running time of branching algorithms.

The intuition behind piecewise analysis is to group instances based on similarity and then analyze each group separately. To do this, we define a single similarity function, termed similarity ratio, that assigns a value to each instance. Then, instead of analyzing an algorithm $\mathcal{A}$ on the instance space $\mathcal{I}$ as a whole, we split $\mathcal{I}$ into *pieces* $P \subseteq \mathcal{I}$, each corresponding to a specific range of similarity ratio values. Then we analyze the worst-case running time of instances within each individual piece. Among all pieces, the one with the maximum value gives the upper bound of the running time of the algorithm $\mathcal{A}$ to $\mathcal{I}$.

This description naturally prompts the question: how do we construct a good similarity ratio? The short answer is to explore the relations between the existing parameters that are relevant to either the problem or the algorithm being analyzed.

The seemingly simple idea behind piecewise analysis has potential. To showcase its potency, we apply it to analyze two 17-year-old algorithms in [6] that solve c-Coloring and #c-Coloring respectively. By applying our new piecewise analysis method, we improve the running time analysis of the algorithms for 4-Coloring and #3-Coloring presented in [6]. The original analysis in their paper gave running times of $\mathcal{O}(1.7272^n)$ and $\mathcal{O}(1.6262^n)$ respectively for these algorithms, and our analysis improves these running times to $\mathcal{O}(1.7207^n)$ and $\mathcal{O}(1.6225^n)$. Notably, our result of $\mathcal{O}(1.7207^n)$ marks the first improvement in the running time for the 4-Coloring problem in 17 years. Meanwhile, a recently accepted ESA 2024 paper [22], which was published after our initial submission, reports a further improvement in the running time to $\mathcal{O}(1.7159^n)$ for 4-Coloring using a different approach.

With this context established, we now proceed to formally define piecewise analysis below.

Piecewise Analysis Framework

Input: Instance space $\mathcal{I}$ for an algorithm $\mathcal{A}$,

parameters $r : \mathcal{I} \to \mathbb{R}_{\geq 0}$, $q : \mathcal{I} \to \mathbb{R}_{\geq 0}$,

a finite cover $\mathcal{C} = \{P_0, P_1, \ldots, P_p\}$ of the instance space where $P_0 = \{I \in \mathcal{I} : r(I) = 0\}$ and p is a positive integer, and

lower and upper bounds $l, u : \mathcal{C} \setminus P_0 \to \mathbb{R}_{\geq 0}$ such that for all pieces $P \in (\mathcal{C} \setminus P_0)$ and any instance $I \in P$, we have $l(P) \leq \frac{q(I)}{r(I)} \leq u(P)$.

Requisite: If $r(I) = 0$, then I can be solved by $\mathcal{A}$ in polynomial time.

Oracle: For each piece $P \in (\mathcal{C} \setminus P_0)$, the oracle uses $l(P)$, $u(P)$ and parameter r to generate an upper bound on the running time $\mathcal{O}^*(\zeta_P^r)$ for all of the instances in the piece, where $\zeta_P \in \mathbb{R}_{>0}$ is a constant.

Output: Running time upper bound $max_{1 \leq i \leq p}\{\mathcal{O}^*(\zeta_{P_i}^r)\}$ for $\mathcal{A}$ over $\mathcal{I}$.

3.1   Details of the Piecewise Analysis Framework

The framework analyzes the running time for algorithm $\mathcal{A}$ on instance space $\mathcal{I}$ in terms of a parameter r. We assume that instances whose parameter r is 0, denoted by $P_0 = \{I \in \mathcal{I} : r(I) = 0\}$, are trivial and can be solved in polynomial time. For other instances in $\mathcal{I} \setminus P_0$, we divide them into pieces and select an auxiliary parameter q to form the similarity ratio $\frac{q}{r}$. For each instance, this similarity ratio must be a real number that falls within an interval $[l_{\mathcal{I}}, u_{\mathcal{I}}]$, where $l_{\mathcal{I}}, u_{\mathcal{I}} \in \mathbb{R}_{\geq 0}$. We then split this interval into segments, each of which is of the form $[l, u]$ such that $l \leq u$. For a segment $[l, u]$, a *piece* is defined as the set of all instances whose similarity ratio falls within $[l, u]$. Since the union of all segments covers the whole interval $[l_{\mathcal{I}}, u_{\mathcal{I}}]$, the union of $\{P_1, \ldots, P_p\}$ and P_0 form the instance cover $\mathcal{C}$ of the instance space $\mathcal{I}$.

The oracle is an external source that takes all the inputs mentioned and obtains a running time upper bound in terms of r for each piece $P \in (\mathcal{I} \setminus P_0)$. In practice, it can be any legitimate analysis that obtains a running time upper bound in terms of r.

The output of the framework is the maximum value among the running times of all pieces. Note that the piece P_0 does not affect this running time in the $\mathcal{O}^*$ notation.

3.2   Comparison with Measure and Conquer

When we apply M&C, we define a single measure μ for the entire instance space $\mathcal{I}$. By minimizing an upper bound for μ, we conduct a single analysis by computing one set of weights. This means that these weights are uniform across all instances, and this uniformity marks a key difference from our piecewise analysis.

The main impetus for this work is to allow different weights for different instances. Thus, in piecewise analysis, by defining a similarity ratio, we allow a different analysis for instances with different ratios. This results in multiple measures and multiple analyses. Furthermore, by dividing an instance space $\mathcal{I}$ into pieces, we can incorporate additional information provided by the pieces to better upper bound the measure.

M&C can be integrated into the piecewise analysis framework to analyze the running time of instances within each piece. In fact, if we apply piecewise analysis by setting the hyperparameter $p = 1$ and use the M&C analysis as the oracle, then this is equivalent to applying M&C on the entire instance space.

The M&C analysis within a piece P can benefit from additional information about instances provided by the piece P. For example, when the similarity ratio falls within a restricted interval, this may reflect some structural property of all instances within the piece P which we can exploit in our running time analysis.

3.3 Comparison with Walström's Compound Measure

Our work is not the first to explore different weights for different kinds of instances. Wahlström [20] uses a *compound measure* to allow measuring different kinds of instances in different ways.

His work using compound measures explores the idea that some intrinsic constraints are imposed by the properties of instances on various parameters. In [20], Wahlström gives the following example. If the average degree of G is greater than d, then there is at least one vertex of degree of at least $d + 1$. He describes this kind of connection as implicit states. To model its effect in branching applicability, he uses a compound piecewise linear measure. However, similar to M&C, this approach uses a single, though compound, measure and still conducts only a single analysis, whereas in piecewise analysis we conduct multiple analyses.

A limitation of the compound measure is the need to ensure that the measure of an instance transitions smoothly between the components of the measure, since an instance may transition from being measured by one component to another. In piecewise analysis, an instance does not transition from one piece to another, since the measure is determined by the original input instance.

Similar to how M&C can be integrated in piecewise analysis, we can also use compound measure to analyze running time within each piece of an instance cover.

4 The c-Coloring Problem – A Case Study

In this section, we explore how piecewise analysis can be used to analyze the running time of the c-Coloring algorithm in [6]. Our piecewise analysis of Fomin et al.'s algorithmic technique [6] improves the analysis of their 4-Coloring algorithm from $\mathcal{O}(1.7272^n)$ to $\mathcal{O}(1.7207^n)$, see Theorem 1.

In Subsect. 4.1, we elucidate the framework in [6] and describe the algorithm and its pseudocode for c-Coloring. In Subsect. 4.2, we combine piecewise analysis and measure & conquer to analyze the running time upper bounds of their algorithm. In Subsect. 4.3, we first demonstrate the faster running time achieved by applying piecewise analysis to 4-Coloring compared to the original, and then show how the number of pieces affects the running time analysis, using 4-Coloring as an example.

---

c-Coloring

Input: A graph $G = (V, E)$
Question: Can we color all the vertices of G with at most c colors so that no adjacent vertices share the same color?

4.1 The Algorithmic Technique and the Algorithm

The algorithmic technique in [6] solves c-Coloring by combining a pathwidth approach and an enumeration approach. Both approaches can be turned into

---

Algorithm 1: enumISPw(G, S, C)

Input : A graph G, an independent set S of G, and a set of vertices C such that $N(S) \subseteq C \subseteq V \setminus S$, integer $a \geq 3$, and $\alpha_i \in \mathbb{R}_{>0}$ for $i = 2, \ldots, a - 1$

Output: $\bigvee$enumIS(G, S', C') taken over all vertex covers $C' \supseteq C$ of G

1 **if** $(\Delta(G - (S \cup C)) \geq a)$

2      $\vee \big(\Delta(G - (S \cup C)) = a - 1 \ and \ |C| > (\alpha_{a-1}n + \gamma_{a-1}|S|)\big)$

3      $\vee \big(\Delta(G - (S \cup C)) = a - 2 \ and \ |C| > (\alpha_{a-2}n + \gamma_{a-2}|S|)\big)$

4      $\vee \cdots$

5      $\vee \big(\Delta(G - (S \cup C)) = 3 \ and \ |C| > (\alpha_3 n + \gamma_3 |S|)\big)$

6 **then**

7   choose a vertex $v \in V - (S \cup C)$ of maximum degree in $G - (S \cup C)$

8      $T1 \leftarrow$ enumISPw($G, S \cup \{v\}, C \cup N(v)$)

9      $T2 \leftarrow$ enumISPw($G, S, C \cup \{v\}$)

10   **return** $T1 \vee T2$

11 **else if** $\Delta(G - (S \cup C)) \leq 2 \ and \ |C| > (\alpha_2 n + \gamma_2 |S|)$ **then**

12   **return** enumIS(G, S, C)

13 **else**

14   Stop this algorithm and run Pw(G, S, C) instead

---

independent exact algorithms for c-Coloring, but they shine on different kinds of graphs.

The pathwidth approach provides a running time upper bound $\mathcal{O}^*(c^w)$ where w is the width of a path decomposition of the input graph G and c is the number of colors available. When G has a small pathwidth, then this approach runs fast. The enumeration approach, on the other hand, enumerates (maximal) independent sets and then checks if the corresponding (minimal) vertex covers are $(c - 1)$-colorable. Fomin et al. [6] demonstrated that the enumeration approach's worst-case instances have small pathwidth; moreover, it can detect these worst-case instances efficiently and hand over the computation to the pathwidth approach in these cases.

By exploiting this dichotomy and hybridizing these two approaches, graphs with nice graph-theoretic properties are differentiated from those with nice algorithmic properties. This distinction allows for the unique strength of each approach to be better utilized.

Algorithm 1 is adapted from [6] with minor modifications. Compared to [6], we pre-calculate an extra set of variables γ_i for $2 \leq i \leq a-1$ and use them for line 2 to 5 and line 11. These changes adjust the numeric thresholds for triggering the pathwidth approach without altering the application and combination of the two approaches. For clarity on whether the improvement comes from the modification or the application of piecewise analysis, see discussions in Subsect. 4.2 and Subsect. 4.3.

Details of the Algorithm. The key step of Algorithm 1 is branching on a vertex $v \in G - (S \cup C)$ of maximum degree in $G - (S \cup C)$, where S is an independent set and C covers all edges incident to S. The branching rule places v

---

Algorithm 2: enumIS(G, S, C)

Input : A graph G with independent set S and vertex set C, such that
$\Delta(G - (S \cup C)) \leq 2$ and $N(S) \subseteq C \subseteq V - S$
Output: $\bigvee$color(G, C') taken over all vertex covers $C' \supseteq C$

1 **if** $\Delta(G - (S \cup C)) > 0$ **then**
2   choose a vertex $v \in V - (S \cup C)$ of maximum degree in $G - (S \cup C)$
3     $T1 \leftarrow$ enumIS$(G, S \cup \{v\}, C \cup N(v))$
4     $T2 \leftarrow$ enumIS$(G, S, C \cup \{v\})$
5   **return** $T1 \vee T2$
6 **else**
7     $S \leftarrow V - C$
8   run color(G, C)

---

into either S or C, thus dividing the input (G, S, C) into two branching pathways: $(G, S \cup \{v\}, C \cup N(v))$ and $(G, S, C \cup \{v\})$ respectively (line 7 to line 10).

This branching continues unless we hit one of the precalculated thresholds, upon which we then run a pathwidth algorithm Pw (line 14). Details on how to estimate the upper bound of pathwidth w, obtain the path decomposition, and calculate values α_i and γ_i for $2 \leq i \leq a - 1$ can be found in [2].

If Pw is executed, we obtain a path decomposition of G efficiently and apply the pathwidth approach instead of continuing the enumeration approach. If the maximum degree of $G - (S \cup C)$ is at most 2 and Pw is not called, then enumIS(line 12) is called and continues to enumerate independent sets. It should be noted that the algorithm enumISis flexible, and we may not need to explore the entire search space.

Algorithm 1, by starting enumerating independent sets S, can determine a pathwidth upper bound w of the original input graph G. This enumeration has a twofold purpose: first, to apply the pathwidth approach if w is small enough (line 1 to 5 and line 11); and second, to solve c-Coloring by using branching if w is large.

Framework and Details of Subroutine enumIS. For enumIS, we employ a simplified version of the algorithm (see Algorithm 2) used in the paper.[1] It does the same two-way branching as enumISPw.

The input of enumISis a graph G, an independent set S and a vertex set C such that $N(S) \subseteq C \subseteq V - S$ and $\Delta(G - (S \cup C)) \leq 2$. The algorithm arbitrarily chooses a vertex v of maximum degree in $G - (S \cup C)$ and puts v into either S or C (line 2 to line 5) until the maximum degree of $G - (S \cup C)$ drops to 0 (see line 1). When this happens, any vertex left in $G - (S \cup C)$ is added to S (line 7) and then subroutine color is called (line 8). The algorithm color is an algorithm for $(c - 1)$-Coloring.

---

[1] In [4], there are two branching rules on degree 2 vertex: rules on non-triangle cycles and rules on chains. This is slightly more complex than our rules.

4.2   Running Time Analysis

In this subsection, we analyze the running time of the algorithm enumISPw (Algorithm 1) introduced in Subsect. 4.1. It is important to note that analyzing running time for the pathwidth approach becomes straightforward once we obtain a pathwidth upper bound w (see details in [2]). In contrast, the true challenge lies in analyzing the running time of the branching approach. In what follows, we will employ measure & conquer and piecewise analysis to analyze the running time of the branching approach in more detail.

Before moving forward, we introduce a new problem, d-colorable Vertex Cover, to benefit the running time analysis of enumISPw.

---

d-colorable Vertex Cover

Input: A graph $G = (V, E)$ and non-negative integer $k \le \frac{dn}{d+1}$
Question: Is there a vertex subset $X \subseteq V$ such that $G \setminus X$ is an empty graph, $|X| \le k$, and $G[X]$ is d-colorable?

Proposition 1. *A graph G is c-colorable if and only if it has a $(c-1)$-colorable vertex cover C with at most $\frac{(c-1)n}{c}$ vertices.*

Proposition 2. *c-Coloring can be solved in $\mathcal{O}^*(\tau^n)$ time if and only if $(c-1)$-colorable Vertex Cover can be solved in $\mathcal{O}^*(\tau^n)$ time.*

As such, for the remainder of this section, our analysis of enumISPwfocuses on d-colorable Vertex Cover.

Applying Measure and Conquer. In this subsection, we use measure & conquer to analyze the Algorithm 1. Note that [6] did not use measure & conquer, so we first need to define a measure that can capture the progress made in each branching step. We define the measure μ as follows:

$$w_1(n - |S \cup C|) + \sum_{i=2}^{a-1} w_{ki}b_i + w_s b_s + c'k, \tag{2}$$

where we set the parameters $b_i = \max(0, k - \alpha_i n - \gamma_i |S_i|)$ for all $2 \le i \le a-1$ and, in turn, $|S_i|$ denotes the number of vertices added into S when $\Delta(G-(S \cup C)) > i$ in Algorithm 1. The intuition behind b_i relates to the maximum number of vertices that can be added to C when $\Delta(G - (S \cup C)) \le i$. The pathwidth approach Pw is triggered the moment $\Delta(G - (S \cup C))$ decreases to i and b_i reaches 0, as the condition in (line 1 to 5) becomes false; if not, as referenced from line 1 to 5 and line 11 of the Algorithm 1, we have $|C| > (\alpha_i n + \gamma_i |S|)$ when $\Delta(G - (S \cup C)) \le i$. This limits the subsequent additions to C to a maximum of $b_i = k - \alpha_i n - \gamma_i |S|$ vertices. The parameter $b_s = n - k$ upper bounds the number of vertices added to the independent set S. We also include an extra term $c'k$, which corresponds only to the color subroutine. The constant c' is based solely on color (see line 8 of the Algorithm 2).[2] For each parameter, we also define a weight variable: w_1, w_s, w_{ki} for $2 \le i \le a - 1$.

---

[2] In fact, color can be any algorithm that validly solves the d-Coloring problem with a running time of $\mathcal{O}^*(2^{c'n'})$ where n' is the number of vertices of input graph.

The analysis now becomes a convex optimization problem (see Subsect. 2.1) to compute the values for the weights to minimize an upper bound on the objective function μ. The values assigned to the weights must satisfy all branching constraints, which can be found in [2].

When the algorithm finishes enumerating all minimal vertex covers C, the d-Coloring algorithm `color` is called on each induced graph $G[C]$ of size at most k. The running time for the subproblem corresponding to each of our vertex covers of size k is upper bounded by $\mathcal{O}^*(2^{c'k})$. Combining all of this, the overall running time for d-colorable Vertex Cover is $\mathcal{O}^*(2^\mu)$ based on Lemma 2.

To upper bound μ in terms of n, we first upper bound it in terms of both n and k by the following function:

$$w_1 n + \sum_{i=2}^{a-1} w_{ki}(k - \alpha_i n) + w_s(n - k) + c'k. \tag{3}$$

This is because $b_s = n - k$ by definition. Since the pathwidth approach is not triggered, the parameter $b_i \leq k - \alpha_i n$, because $b_i = k - \alpha_i n - \gamma_i |S_i|$ and $\gamma_i |S_i| \geq 0$ for $2 \leq i \leq a - 1$.

According to Proposition 1, if G is c-colorable, then $0 \leq k \leq \frac{(c-1)n}{c}$. This allows us to further upper bound the measure μ by

$$w_1 n + \sum_{i=2}^{a-1} w_{ki}\left(\frac{(c-1)n}{c} - \alpha_i n\right) + w_s n + \frac{(c-1)c'n}{c}. \tag{4}$$

The new upper bound is a running time in terms of only parameter n.

Example 1. Using the above objective function (4) along with the corresponding constraints outlined in [2] we can analyze 4-Coloring as follows. The subroutine `color` solves 3-Coloring in running time $\mathcal{O}(1.3289^n)$ (see Lemma 1), which means $c' = \log_2(1.3289)$ because $\mathcal{O}^*(2^{c'n}) = \mathcal{O}(1.3289^n)$. By setting the remainder of the variables to the values specified in [2] we obtain a running time upper bound $\mathcal{O}(1.7275^n)$.

The running time presented for 4-Coloring in [6] is $\mathcal{O}(1.7272^n)$, which is very similar to the result we obtained in Example 1. Note that the result $\mathcal{O}(1.7275^n)$ comes only from applying M&C to `enumISPw`. This suggests that our previous modification on thresholds and the application of M&C do not significantly improve the analysis. As discussed in Subsect. 3.2, applying M&C directly is equivalent to applying piecewise analysis seeing the instance space as one piece. The next section demonstrates that a larger number of pieces quickly improves the running time.

Applying Piecewise Analysis. To apply piecewise analysis, it is essential to specify each input as required by the definition in Sect. 3.

Piecewise Analysis on the algorithm enumISPw

Input: Instance space $\mathcal{I} = \{(G, k) : G \text{ is a graph}, k \in \mathbb{Z}_{\geq 0} \text{ and } k \leq \frac{dn}{d+1}\}$,
           parameters $r((G, k)) := n$, $q((G, k)) := k$,
           a cover $\mathcal{C} := \{P_0, P_1, \ldots, P_p\}$ where $P_0 = \{(G, k) : G \text{ has no}$
           vertices, $k \in \mathbb{Z}_{\geq 0} \text{ and } k \leq \frac{dn}{d+1}\}$ and $P_i = \{(G, k) : G \text{ is a graph}$
           with $n > 0$ and $\frac{(i-1)(\frac{d}{d+1})}{p} \leq \frac{k}{n} \leq \frac{i(\frac{d}{d+1})}{p}\}$ for $i = 1, \ldots, p$ for
           $p \in \mathbb{Z}_{>0}$,
           lower bound $l(P_i) = \frac{(i-1)(\frac{d}{d+1})}{p}$ for $i = 1, \ldots, p$, and
           upper bound $u(P_i) = \frac{i\frac{d}{d+1}}{p}$ for $i = 1, \ldots, p$.
Requisite: If $n((G, k)) = 0$, then the input instance is a yes-instance.
Oracle: For each piece $P \in \mathcal{C} \backslash P_0$, the oracle uses $l(P)$, $u(P)$ and parameter
           n to generate a running time upper bound $\mathcal{O}^*((\zeta_P)^n)$ for all of the
           instances in the piece, where $\zeta_P \in \mathbb{R}_{>0}$ is a constant.
Output: Running time upper bound $\max_{1 \leq i \leq p} \{\mathcal{O}^*((\zeta_{P_i})^n)\}$ for enumISPw
           on $\mathcal{I}$.

The first input is the instance space of d-colorable Vertex Cover, denoted by $\mathcal{I}$. By definition of the problem, $\mathcal{I} = \{(G, k) : G \text{ is a graph}, k \in \mathbb{Z}_{\geq 0} \text{ and } k \leq \frac{dn}{d+1}\}$. The next two inputs r and q are chosen to be n and k, respectively. Since we would like the final running time to be expressed in terms of n, we set $r := n$. If $n = 0$, then G is a trivial yes-instance, which satisfies the requirement. Accordingly, we define $P_0 = \{(G, k) : G \text{ has no vertices}, k \in \mathbb{Z}_{\geq 0} \text{ and } k \leq \frac{dn}{d+1}\}$. The choice of q is less obvious, but it is a natural choice since k is at most linear in n, and the value $\frac{k}{n}$ falls nicely into $[0, \frac{d}{d+1}]$. With the P_0 and similarity ratio, we construct the instance cover in the following way. We *evenly* split the interval $[0, \frac{d}{d+1}]$ into p segments $[l_i, u_i]$ for $i = 1, \ldots, p$, where $p \in \mathbb{Z}_{>0}$. For a segment $[l_i, u_i]$, we define a piece $P_i = \{(G, k) : G \text{ is a graph}, l(P_i) \leq \frac{k}{n} \leq u(P_i)\}$ where $l(P_i) = l_i = \frac{(i-1)(\frac{d}{d+1})}{p}$ and $u(P_i) = u_i = \frac{i(\frac{d}{d+1})}{p}$. As introduced in Sect. 3, since the union of all segments is the whole interval $[0, \frac{d}{d+1}]$, the union of pieces $\{P_1, \ldots, P_p\}$ and P_0 form the instance cover $\mathcal{C}$ of the instance space $\mathcal{I}$. If we conduct a running time analysis on each piece, we cover the whole instance space $\mathcal{I}$. We will ignore P_0 from this point forward, as instances in this piece are null graphs with varying values of $k \in \mathbb{Z}_{\geq 0}$ and hence are all yes-instance.

The intuition behind this construction is to group instances (G, k) according to the closeness of their similarity ratio. With the upper and lower bounds of similarity ratio in each piece, we can further upper bound the measure.

Upper Bound Measure within Each Piece. To upper bound the measure μ (see Eq. 2), it is sufficient to upper bound Eq. 3, recalled here: $w_1 n + \sum_{i=2}^{a-1} w_{ki}(k - \alpha_i n) + w_s(n - k) + c'k$.

Now we show how to upper bound Eq. 3 within a piece. Given a piece $P \in (\mathcal{C} \backslash P_0)$, let $[l, u]$ be an interval containing all similarity ratio values of instances in

Table 1. 4-Coloring running time on 1, 10, 50, 100 and 550 pieces

#Pieces	1	10	50	100	550
Running time	$\mathcal{O}(1.7275^n)$	$\mathcal{O}(1.7257^n)$	$\mathcal{O}(1.7217^n)$	$\mathcal{O}(1.7212^n)$	$\mathcal{O}(1.7207^n)$

P. Assuming $P \in (\mathcal{C} \setminus P_0)$ is the piece that (G, k) belongs to, we have $\frac{k}{n} \in [l, u]$. Thus, we can establish that $k \leq un$ and $-k \leq -ln$. Now we can upper bound Eq. 3 by

$$w_1 n + \sum_{i=2}^{a-1} w_{ki}(un - \alpha_i n) + w_s(n - ln) + c'un. \tag{5}$$

Based on Lemma 2, the running time upper bound for instances in piece P is $\mathcal{O}^*(2^\mu)$. Substituting μ by Eq. 5 thus gives P the below running time upper bound:

$$\mathcal{O}^*\left(2^{n\left(w_1 + \sum_{i=2}^{a-1} w_{ki}(u-\alpha_i) + w_s(1-l) + c'u\right)}\right). \tag{6}$$

4.3 Result for 4-Coloring

By setting p = 550 and a = 7, we obtain the following theorem.

Theorem 1. *4-Coloring can be solved in* $\mathcal{O}(1.7207^n)$ *time.*

When applying piecewise analysis, we can choose how many pieces to divide the instance space $\mathcal{I}$ into. In the 4-Coloring case study, dividing the space into more pieces yields tighter upper bounds. Table 1 presents the upper bounds for 4-Coloring with $p = 1, 10, 50, 100,$ and 550.[3] Note that all analyses converge, and increasing the number of pieces no longer reduces the running time beyond a certain point.

A new running time of $\mathcal{O}(1.3217^n)$ for 3-Coloring, claimed in [17], is not officially accepted. If accepted, their result will further improve our 4-Coloring result to $\mathcal{O}(1.7146^n)$ (see Theorem 2).

Theorem 2. *4-Coloring can be solved in running time* $\mathcal{O}(1.7146^n)$ *if the 3-Coloring subroutine algorithm we applied has running time* $\mathcal{O}(1.3217^n)$.

5 The #c-Coloring Problem – A Case Study

In contrast to the decision problem c-Coloring, the counting version #c-Coloring instead is to count the number of c-colorings of an input graph. Despite this difference, the [6] framework used in Sect. 4 also applies to #c-Coloring with some minor changes.

---

[3] Refer to [2] for the relevant variable values. The values are computed using AMPL. The code is available at https://github.com/AlgorithmsWorkingGroup/coloring_pwa.

Our running time for the algorithm of [6] to solve #c-Coloring is $\mathcal{O}(1.6225^n)$. This outperforms the running time $\mathcal{O}(1.6262^n)$ presented in the original paper. Note that our result for #3-Coloring is looser than the current record of $\mathcal{O}(1.5858^n)$ given in [23]. However, our findings highlight the improvement in the running time analysis due to the piecewise analysis method.

We first describe the adapted algorithm and then, in a similar manner to Sect. 4, we conduct an analysis with M&C and piecewise analysis. Finally, our results for #3-Coloring are presented in Subsect. 5.2.

5.1 The Algorithm and Analysis

The algorithms for #c-Coloring (see Algorithm enumISPwCounting in [2]) are very similar to those for c-Coloring (Algorithm, enumISPw). The detailed descriptions, comparisons, and pseudocode can be found in [2]. The analysis is also conducted in a similar manner to Sect. 4. We first define a new problem #d-colorable Vertex Cover and then prove that the Algorithm enumISPwCounting solves the #d-colorable Vertex Cover in Proposition 3.

#d-colorable Vertex Cover

Input: A graph $G = (V, E)$ and non-negative integer $k \leq \frac{dn}{d+1}$
Question: How many vertex subsets $X \subseteq V$ are there, such that $G \setminus X$ is an empty graph, $|X| \leq k$, and $G[X]$ is d-colorable?

Proposition 3. *#c-Coloring can be solved in $\mathcal{O}^*(\tau^n)$ time if and only if #$(c-1)$-colorable Vertex Cover can be solved in $\mathcal{O}^*(\tau^n)$ time.*

Next, we define a measure μ for enumISPwCountingand then upper bound its running time by applying piecewise analysis.

Here, the measure μ is defined as

$$\mu = w_1(n - |S \cup C|) + \sum_{i=2}^{a-1} w_{ki}b_i + w_s b_s + c'k, \tag{7}$$

where all variables and weights are defined the same way as in Sect. 4. For the branching constraints, the key difference is an extra constraint for branching on an isolated vertex v in $G - (S \cup C)$ (see [2]).

The application of piecewise analysis to this extended set of branching rules mirrors the analysis in Subsect. 4.2 and is therefore omitted for brevity.

5.2 Application to #3-Coloring

Theorem 3. *Let G be a graph on n vertices, then the #3-Coloring can be solved in running time $\mathcal{O}(1.6225^n)$.*

6   Conclusion

The main contribution of this paper is a new method for the running time analysis of branching algorithms called piecewise analysis. It allows us to differentiate instances by defining a similarity ratio, giving us the flexibility to decide which instances should be analyzed similarly and which differently. Using c-Coloring and $\#c$-Coloring as case studies, we showcased how piecewise analysis can be applied, obtaining faster running times for 4-Coloring and #3-Coloring without significant changes to two 17-year-old algorithms.

Additionally, piecewise analysis helps identify the set of instances that bottleneck the running time analysis by identifying the piece with the worst upper bound, thus improving the algorithm for these instances directly improves the overall running time.

References

1. Beigel, R., Eppstein, D.: 3-coloring in time $O(1.3289)^n$. J. Algorithms **54**(2), 168–204 (2005). https://doi.org/10.1016/j.jalgor.2004.06.008
2. Clinch, K., Gaspers, S., Saffidine, A., He, Z., Zhang, T.: A piecewise approach for the analysis of exact algorithms. CoRR abs/2402.10015 (2024). https://doi.org/10.48550/arXiv.2402.10015
3. Edwards, K., McDermid, E.: A general reduction theorem with applications to pathwidth and the complexity of MAX 2-CSP. Algorithmica **72**(4), 940–968 (2015). https://doi.org/10.1007/s00453-014-9883-7
4. Eppstein, D.: Small maximal independent sets and faster exact graph coloring. In: Dehne, F., Sack, J.-R., Tamassia, R. (eds.) WADS 2001. LNCS, vol. 2125, pp. 462–470. Springer, Heidelberg (2001). https://doi.org/10.1007/3-540-44634-6_42
5. Eppstein, D.: Quasiconvex analysis of multivariate recurrence equations for backtracking algorithms. ACM Trans. Algorithms **2**(4), 492–509 (2006). https://doi.org/10.1145/1198513.1198515
6. Fomin, F.V., Gaspers, S., Saurabh, S.: Improved exact algorithms for counting 3- and 4-colorings. In: Lin, G. (ed.) COCOON 2007. LNCS, vol. 4598, pp. 65–74. Springer, Heidelberg (2007). https://doi.org/10.1007/978-3-540-73545-8_9
7. Fomin, F.V., Gaspers, S., Saurabh, S., Stepanov, A.A.: On two techniques of combining branching and treewidth. Algorithmica **54**(2), 181–207 (2009)
8. Fomin, F.V., Golovach, P.A., Kratochvíl, J., Kratsch, D., Liedloff, M.: Branch and recharge: Exact algorithms for generalized domination. Algorithmica **61**(2), 252–273 (2011). https://doi.org/10.1007/S00453-010-9418-9
9. Fomin, F.V., Grandoni, F., Kratsch, D.: A measure & conquer approach for the analysis of exact algorithms. J. ACM **56**(5), 25:1–25:32 (2009). https://doi.org/10.1145/1552285.1552286
10. Gaspers, S.: Measure & conquer for parameterized branching algorithms. Parameterized Complexity News: Newsletter of the Parameterized Complexity Community (2009)
11. Gaspers, S.: Exponential Time Algorithms - Structures, Measures, and Bounds. VDM (2010)
12. Gaspers, S., Lee, E.J.: Faster graph coloring in polynomial space. Algorithmica **85**(2), 584–609 (2023). https://doi.org/10.1007/S00453-022-01034-7

13. Gaspers, S., Sorkin, G.B.: A universally fastest algorithm for Max 2-Sat, Max 2-CSP, and everything in between. J. Comput. Syst. Sci. **78**(1), 305–335 (2012). https://doi.org/10.1016/J.JCSS.2011.05.010

14. Gaspers, S., Sorkin, G.B.: Separate, measure and conquer: faster polynomial-space algorithms for Max 2-CSP and counting dominating sets. ACM Trans. Algorithms **13**(4), 44:1–44:36 (2017). https://doi.org/10.1145/3111499

15. Iwata, Y.: A faster algorithm for dominating set analyzed by the potential method. In: Marx, D., Rossmanith, P. (eds.) IPEC 2011. LNCS, vol. 7112, pp. 41–54. Springer, Heidelberg (2012). https://doi.org/10.1007/978-3-642-28050-4_4

16. Kullmann, O.: New methods for 3-SAT decision and worst-case analysis. Theor. Comput. Sci. **223**(1–2), 1–72 (1999). https://doi.org/10.1016/S0304-3975(98)00017-6

17. Meijer, L.: 3-coloring in time $\mathcal{O}(1.3217^n)$. CoRR abs/2302.13644 (2023). https://doi.org/10.48550/ARXIV.2302.13644

18. Robertson, N., Seymour, P.D.: Graph minors. III. Planar tree-width. J. Comb. Theory Ser. B **36**(1), 49–64 (1984). https://doi.org/10.1016/0095-8956(84)90013-3

19. Wahlström, M.: Exact algorithms for finding minimum transversals in rank-3 hypergraphs. J. Algorithms **51**(2), 107–121 (2004). https://doi.org/10.1016/J.JALGOR.2004.01.001

20. Wahlström, M.: Algorithms, measures and upper bounds for satisfiability and related problems. Ph.D. thesis, Linköping University, Sweden (2007)

21. Wahlström, M.: A tighter bound for counting max-weight solutions to 2SAT instances. In: Grohe, M., Niedermeier, R. (eds.) IWPEC 2008. LNCS, vol. 5018, pp. 202–213. Springer, Heidelberg (2008). https://doi.org/10.1007/978-3-540-79723-4_19

22. Wu, P., Gu, H., Jiang, H., Shao, Z., Xu, J.: A faster algorithm for the 4-coloring problem. In: Chan, T.M., Fischer, J., Iacono, J., Herman, G. (eds.) 32nd Annual European Symposium on Algorithms, ESA 2024, 2–4 September 2024, Royal Holloway, London, UK. LIPIcs, vol. 308, pp. 103:1–103:18. Schloss Dagstuhl - Leibniz-Zentrum für Informatik (2024). https://doi.org/10.4230/LIPICS.ESA.2024.103

23. Zhu, E., Wu, P., Shao, Z.: Exact algorithms for counting 3-colorings of graphs. Discret. Appl. Math. **322**, 74–93 (2022). https://doi.org/10.1016/j.dam.2022.08.002

Parameterized Complexity of (d, r)-Domination via Modular Decomposition

Gennaro Cordasco[1($\boxtimes$)], Luisa Gargano[2], and Adele A. Rescigno[2]

[1] Department of Psychology, University of Campania "L.Vanvitelli", Caserta, Italy
gennaro.cordasco@unicampania.it
[2] Department of Computer Science, University of Salerno, Fisciano, Italy
{lgargano,arescigno}@unisa.it

Abstract. With the rise of social media, misinformation has significant negative effects on the decision-making of individuals, organizations and communities within society. Identifying and mitigating the spread of fake information is a challenging issue. We consider a generalization of the Domination problem that can be used to detect a set of individuals who, through an awareness process, can prevent the spreading of fake narratives. The considered problem, named (d, r)-Domination generalizes both distance and multiple domination. We study the parameterized complexity of the problem according to standard and structural parameters. We give fixed-parameter algorithms as well as polynomial compressions/kernelizations for some variants of the problem and parameter combinations.

Keywords: Parameterized Complexity · Kernelization Algorithms · Domination · Contamination minimization

1 Introduction

Domination is a fundamental concept in graph theory, which deals with the idea of dominating sets within a graph. In this problem, you seek to find the smallest set of vertices in a graph in such a way that every vertex in the graph is either in the dominating set or adjacent to a vertex in the dominating set. Dominating sets are critical in various real-world applications across fields that involve networks, connections, and coverage [33,34].

The Domination Problem in graph theory has several important variants that focus on different aspects of the problem. We focus on generalizations to distance domination and multiple domination that provide ways to solve practical problems related to the physical or operational constraints of networks.

L. Gargano and A. A. Rescigno—Work partially supported by project SERICS (PE00000014) under the MUR National Recovery and Resilience Plan funded by the European Union- NextGenerationEU.

One application that inspires this paper is epidemiology. Graph-based information dissemination algorithms can be used to study the spread of real and fake information in a network. While these algorithms are not designed to intercept fakes, they can be used as part of a broader strategy to identify and mitigate the spread of fakes. Controlling the spread of fake news is an ongoing challenge. Strategies to reduce the size of propagation by immunizing/removing vertices have been investigated in several papers [2,45]. In this paper, we focus on the effective production of accurate and evidence-based information to combat misinformation. To this end, we study the (d,r)-domination problem [3], which includes both multiplicity and distance.

Apart from our inspiring application the (d,r)-domination problem, as the classical domination problem, has many applications. For instance it can be used, from the other side, as a mechanism to identify key influencers or individuals who can drive (mis)information diffusion within a network [7–12]. In wireless sensor networks, it helps minimize the number of sensors while ensuring full coverage and connectivity [15]. In facility location and placement, it helps determine the optimal locations for services, facilities, or resources to ensure that they are accessible to a maximum number of people while minimizing the number of locations needed [29].

The Problem. Let $G = (V, E)$ be a graph. For a set of vertices $X \subseteq V$, we denote by $G[X]$ the induced subgraph of G generated by X. Moreover, for $v \in V$, by[1] $N(v) = \{u \in V \mid (u, v) \in E\}$ the neighborhood of v. This notation can also be extended to a set $X \subseteq V$, we let $N(X) = \{y \in V \setminus X \mid \exists x \in X, (x, y) \in E\}$ be the set of vertices outside X that have a neighbor in X. For any two vertices $u, v \in V$, we denote by $\delta(u, v)$ the distance between u and v in G and by $N_r(v) = \{u \in V \mid u \neq v, \ \delta(u, v) \leq r\}$ the *neighborhood of radius r around v*. Clearly, $N_1(v) = N(v)$.

A *Dominating set* in a graph $G = (V, E)$ is a subset of V such that every vertex not in the set has at least one neighbor in the set. In this paper, we will consider the following generalization introduced in [3]: Given a *demand d* and a *radius around r*, a vertex $v \in V \setminus S$ is *dominated by S* if there exist at least d vertices in S that are at a distance at most r from v.

In an information diffusion setting, we consider a networked population of people who could be misled by a word-of-mouth dissemination approach. It can be assumed that an individual gets immune when exposed to enough debunking information. For further clarification, an individual is considered immunized when he/she receives the debunking information from a number of neighbors at least equal to its demand d. Furthermore, every individual has a circle of trust (level of trust) defined by a radius r around it. Debunking information from inside the circle of trust is the only source that can be trusted.

Specifically, we examine how to use (d, r)-dominating sets to identify a collection of individuals who can stop the spread of false narratives by initiating a debunking (or prebunking) immunization campaign. The vaccination operation

---

[1] We will use the subscript G whenever the graph is not clear from the context.

on a vertex prevents the contamination of the vertex itself as well as the spread
of such false narratives when there is a debunking immunization campaign in
place. Therefore, to prevent the spread of malicious items, we are looking for a
small subset S of vertices (also known as the immunizing set) which allows us to
minimize the propagation of false information by disseminating the debunking
information. Depending on the vertex demand d and the radius r, which char-
acterizes the vertices' circle of trust, the immunizing set should be able to cover
each vertex d times within a maximum distance of r.

Definition 1. *Given an undirected graph $G = (V, E)$ and integers $r, d > 0$, a
subset of vertices $S \subseteq V$ is a (d, r)-dominating set of G if $|N_r(v) \cap S| \geq d$, for
all $v \in V \setminus S$.*

In this paper, we study the (d, r)-DOMINATION problem that asks for a minimum (d, r)-dominating set.

(d, r)-DOMINATION:
Input: An undirected graph $G=(V, E)$ and two integers $r, d>0$
Output: A (d, r)-dominating set of minimum size.

Since the problem can be solved independently in each connected component
of the input graph, from now on, we assume that the input graph is connected.

1.1   Parameterized Complexity and Compression/Kernelization

We study the (d, r)-DOMINATION problem from a parameterized point of view.
Parameterized complexity is a refinement of classical complexity theory in which
one takes into account not only the input size but also other aspects of the
problem given by a parameter p. A problem Q with input size n and parameter
p is called *fixed-parameter tractable (FPT)* if it can be solved in time $f(p) \cdot n^c$,
where f is a computable function only depending on p and c is a constant.
Downey and Fellows introduced a hierarchy of parameterized complexity classes
$FPT \subseteq W[1] \subseteq W[2] \subseteq \ldots$ that is believed to be strict, see [19,25].
Once fixed-parameter tractability of a problem is established, we can ask
whether we can go even one step further by showing the existence of a polynomial
compression/kernelization algorithm.

Definition 2. *A compression for a parameterized problem Q is a polynomial-
time algorithm, which takes an instance (I, p) and produces an instance (I', p')
called a kernel, such that $(I, p) \in Q$ iff $(I', p') \in Q'$ for another problem Q' and
the size $|I'|+p'$ is bounded by $f(p)$ for some computable function f, called the size
of the kernel. If f is a polynomial function, the algorithm is called polynomial
compression, denoted PC.*

*If $Q = Q'$ then the algorithm is called kernelization. Moreover, if $Q = Q'$ and
f is a polynomial function then the algorithm is called polynomial kernelization,
denoted PK.*

Kernelization can be seen as a special case of compression when the algorithm is required to produce another instance of the same problem. Moreover by Theorem 1.6 in [27], we know that if there is a polynomial compression of a problem Q into another problem Q' and the decision versions of Q and Q' are both NPC, then Q also admits a polynomial kernel. The reason is that any two NPC problems have a polynomial-time reduction between them. So, for two NPC problems, polynomial compression and polynomial kernel are equivalent, if we do not consider kernel size. In this paper we will treat the two concepts separately in order to emphasize the fact that a kernelization is often a straightforward preprocessing strategy in which inputs to the algorithm are replaced by a smaller input, achieved by applying a set of reduction rules that cut away parts of the instance that are easy to handle. It is known that for decidable problems, fixed-parameter tractability is equivalent to the existence of a kernelization algorithm, but not necessarily a PK.

Related Results. The (d, r)-DOMINATION problem was proved to be NP-complete even when restricted to bipartite graphs and chordal graphs [3]. Bounds on the minimum size of a solution of (d, r)-DOMINATION are presented in [37]. Subsequent results for some special classes of graphs are given in [32, 42]. (d, r)-DOMINATION generalizes several well-known, and widely studied, generalized domination problems.

- When $d = r = 1$ the problem becomes the classic DOMINATING SET problem.
- When $r = 1$ and $d \geq 2$, the problem becomes the d-DOMINATION problem [23, 24]. For any d, this problem is known to remain NP-hard in the classes of split graphs. From a positive point of view, it is solvable in linear time in the class of graphs having the property that any block is a clique, a cycle or a complete bipartite graph (this includes trees, block graphs, cacti, and block-cactus graphs), and it is solvable in polynomial time in the class of chordal bipartite graphs [41]. The d-DOMINATION problem was also studied with respect to its (in)approximability properties, both in general [6] and in restricted graph classes [1].
- When $d = 1$ and $r \geq 2$, the problem becomes the r-DISTANCE DOMINATION problem [35]. This problem, for any r, remains NP-hard even if we restrict the graph to belong to certain special classes of graphs, including bipartite or chordal graphs [5]. Slater presented a linear time algorithm for the r-DISTANCE DOMINATION problem in forests [47]. An approximate algorithm for unit disk graphs was presented in [16].

From a parameterized complexity point of view, it is known that the (d, r)-DOMINATION problem, as well as the d-DOMINATION problem and the r-DISTANCE DOMINATION problem are W[2]-hard with respect to the size k of the solution since all these problems have the DOMINATING SET problem as special case [18]. Moreover, since $(d, 1)$-DOMINATION remains NP-hard for each $d \geq 2$ we have that (d, r)-DOMINATION is para-NP-hard with respect to d and W[2]-hard with respect to $d + k$, even when the radius r is equal to 1.

For graphs of bounded branchwidth, it was presented in [36] an algorithm for the d-DOMINATION problem parameterized by d and the branchwidth $\mathtt{bw}(G)$ of graph G. Due to the linear relation between the branchwidth $\mathtt{bw}(G)$ and the treewidth $\mathtt{tw}(G)$ of graph G, the above result allows to obtain an algorithm for the d-DOMINATION problem parameterized by d and the treewidth $\mathtt{tw}(G)$. An XP algorithm for the d-DOMINATION problem parameterized by cliquewidth $\mathtt{cw}(G)$ of graph G was presented in [7]. For the r-DISTANCE DOMINATION problem, an algorithm parameterized by r and the treewidth $\mathtt{tw}(G)$ was presented in [4].

In this paper, we are interested in the analysis of (d, r)-DOMINATION with respect to some structural parameters of the input graph G: The neighborhood diversity $\mathtt{nd}(G)$, the iterated type partition $\mathtt{itp}(G)$, and the modular-width $\mathtt{mw}(G)$ of G. FPT algorithms for DOMINATING SET for $\mathtt{nd}(G)$, $\mathtt{itp}(G)$, and $\mathtt{mw}(G)$ parameters were presented in [13,44], and [46], respectively. Recently, Lafond and Luo [38] presented an FPT algorithm for d-DOMINATION parameterized by $\mathtt{nd}(G)$ and proved that this problem is W[1]-hard with respect to $\mathtt{itp}(G)$.

The existence of PKs for the DOMINATING SET and r-DISTANCE DOMINATION problems with respect to the size k of the solution in special classes of graphs has also been studied [20,22,26]. No PK for the DOMINATING SET problem exists when parameterized by the solution size and the vertex cover number $\mathtt{vc}(G)$ [17]. A PK for the DOMINATING SET problem parameterized by $\mathtt{itp}$, and consequently by $\mathtt{nd}$, has been provided in [13].

Our Results. In this paper, we give some results with respect to some standard and structural parameters of G: the demand d, the modular-width $\mathtt{mw}(G)$, the iterated type partition number $\mathtt{itp}(G)$ and the neighborhood diversity $\mathtt{nd}(G)$.

We present:

- a FPT algorithm for (d, r)-DOMINATION parameterized by $\mathtt{mw}(G) + d$.
- a PC for $(1, r)$-DOMINATION parameterized by $\mathtt{mw}(G)$ and, consequently, by $\mathtt{itp}(G)$ and $\mathtt{nd}(G)$.
- a PC for (d, r)-DOMINATION parameterized by $\mathtt{itp}(G) + d$.
- a PK for (d, r)-DOMINATION parameterized by $\mathtt{nd}(G) + d$.

Table 1 gives a summary of known and new parameterized complexity results with respect to the considered parameters.

2   Modular Decomposition

The notion of modular decomposition of graphs was introduced by Gallai in [31], as a tool to define hierarchical decompositions of graphs.

Definition 3. *The following operations can be used to construct a graph:*

(O1) The creation of an isolated vertex.
(O2) The disjoint union of two graphs denoted by $G_1 \oplus G_2$, i.e., $G_1 \oplus G_2$ is the graph with vertex set $V(G_1) \cup V(G_2)$ and edge set $E(G_1) \cup E(G_2)$.

Table 1. Parameterized complexity with respect to demand (d), neighborhood diversity (`nd`), iterated type partition number (`itp`) and modular-width (`mw`).

Parameters	$(1, 1)$-DOMINATION	$(1, r)$-DOMINATION	$(d, 1)$-DOMINATION	(d, r)-DOMINATION
`nd`	PK [13]	**PC** [Th.4]	FPT [38]	**FPT** [Cor.1]
`itp`	PK [13]	**PC** [Theorem 4]	W[1]-hard [38]	W[1]-hard [38]
`mw`	FPT [46]		W[1]-hard [38]	W[1]-hard [38]
	PC [Theorem 3]	**PC** [Theorem 4]		
`nd` and d	–	–	**PK** [Theorem 7]	**PK** [Theorem 7]
`itp` and d	–	–	**PC** [Lemma 5]	**PC** [Theorem 6]
`mw` and d	–	–	**FPT** [Lemma 3]	**FPT** [Theorem 2]

(O3) The complete join of two graphs denoted by $G_1 \otimes G_2$, i.e., $G_1 \otimes G_2$ has vertex set $V(G_1) \cup V(G_2)$ and edge set $E(G_1) \cup E(G_2) \cup \{(u, w) \mid u \in V(G_1),\ w \in V(G_2)\}$.

(O4) The substitution of the vertices $1, \ldots, \ell$ of an outline graph H by the graphs $G_1, \ldots, G_\ell$, denoted by $H(G_1, \ldots, G_\ell)$, is the graph with vertices $\bigcup_{1 \leq i \leq \ell} V(G_i)$ and edge set $\bigcup_{1 \leq i \leq \ell} E(G_i) \cup \{(u, w) \mid u \in G_i,\ w \in G_j, \overline{(i, j)} \in E(H),\ 1 \leq i \neq j \leq \ell\}$.

Notice that (O2) and (O3) are special cases of (O4) with $\ell = 2$.

A *module* of a graph $G = (V, E)$ is a set $M \subseteq V$ such that for all $u, v \in M$, $N(u) \setminus M = N(v) \setminus M$. Two modules M and M' are adjacent if, in G, every vertex of M is adjacent to every vertex of M', and independent if no vertex of M is adjacent to a vertex of M'. The empty set, the vertex set V, and all singletons $\{v\}$ for each $v \in V$ are always modules and they are called *trivial modules*. A graph is called *prime* if all of its modules are trivial. A partition $\mathcal{M} = \{M_1, M_2, \ldots, M_\ell\}$ of V is called a *modular partition* of G if every element of $\mathcal{M}$ is a module of G. Hence, the modular partition of $G = G_1 \oplus G_2$ and $G = G_1 \otimes G_2$ in operations (O2)-(O3) is $\mathcal{M} = \{M_1, M_2\}$ where $G_1 = G[M_1]$ and $G_2 = G[M_2]$ and the modular partition of $G = H(G_1, \ldots, G_\ell)$ in operation (O4) is $\mathcal{M} = \{M_1, \ldots, M_\ell\}$ where for each[2] $i \in [\ell], G_i = G[M_i]$.

Operations (O1)-(O4) taken to construct a graph, form a parse-tree of the graph. A *parse tree* of a graph G is a tree $T(G)$ that captures the decomposition of G into modules. The leaves of $T(G)$ represent the vertices of G (operation (O1)). The internal vertices of $T(G)$ capture operations on modules: Disjoint union of its children (operation (O2)), complete join (operation (O3)) and substitution (operation (O4)). Figure 1 depicts a graph G and the corresponding parse tree.

Modular-Width. The Modular-width parameter was introduced in [30]. The *width* of a graph G is the maximum size of the vertex set of an outline graph H used in operation (O4) to construct G, if any, it is 2 otherwise; this number

---

[2] For a positive integer a, we use $[a]$ to denote the set of integers $[a] = \{1, 2, \ldots, a\}$.

corresponds to the maximum number of children of a vertex in the corresponding parse tree. The *modular-width* is the minimum width such that G can be obtained from some sequence of operations (O1)-(O4). Finding a parse-tree, of a given graph G, corresponding to the modular-width, can be done in linear-time [14].

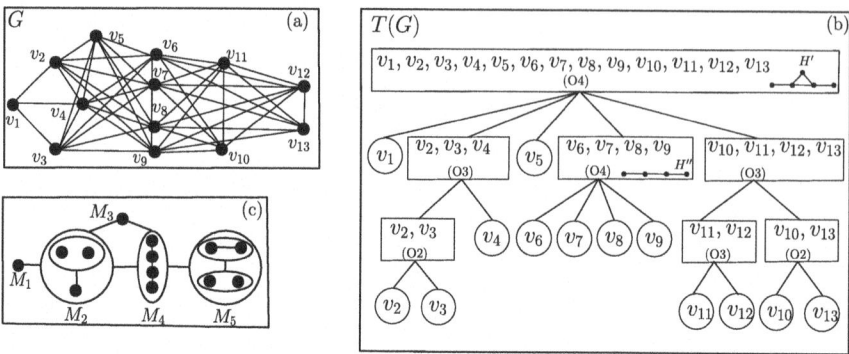

Fig. 1. (a) A graph G. (b) The parse tree $T(G)$ associated with a decomposition of G into modules. The width of the presented decomposition is 5. (c) A hierarchical representation of the decomposition of G into modules.

$\mathcal{G}$-Modular Cardinality. A variant of modular-width restricted to fixed graph classes has been proposed in [38]. A graph class $\mathcal{G}$ is a (possibly infinite) set of graphs containing at least one non-empty graph.

Definition 4 [38]. *Let $\mathcal{G}$ be a graph class. For a graph G (not necessarily in $\mathcal{G}$), a modular partition $\mathcal{M} = \{M_1, \ldots, M_\ell\}$ of G is called a $\mathcal{G}$-modular partition if each $G[M_i]$ belongs to $\mathcal{G}$. The $\mathcal{G}$-modular cardinality of G, denoted by $\mathcal{G}$-mc(G), is the minimum cardinality of a $\mathcal{G}$-modular partition of G.*

Therefore, a graph G of $\mathcal{G}$-modular cardinality $\ell \geq 1$ and modules $M_1, \ldots, M_\ell$, can be seen as $G = H(G_1, \ldots, G_\ell)$, according to substitution operation (O4) (including (O2) or (O3) when $\ell = 2$), in which $G_i = G[M_i] \in \mathcal{G}$, for each $i \in [\ell]$. If $\ell = |V(G)|$, we say that G is prime with respect to the class $\mathcal{G}$.

Neighborhood Diversity. The neighborhood diversity $\mathsf{nd}(G)$ of a graph G was introduced by Lampis in [40]. It can be seen as the $(\mathcal{K} \cup \mathcal{I})$-*modular cardinality* of G, where $\mathcal{K}$ and $\mathcal{I}$ denote the class of *clique* and of *independent set* graphs, respectively. Hence, a graph G of neighborhood diversity $\ell \geq 1$ and modules $M_1, \ldots, M_\ell$, can be seen as $G = H(G_1, \ldots, G_\ell)$, in which $G_i = G[M_i] \in \mathcal{K} \cup \mathcal{I}$, for each $i \in [\ell]$.

Iterated Type Partition. Given a graph G, the *iterated type partition* of G, introduced in [13], is defined by iteratively contracting *clique* and *independent set* modules until no more contractions are possible; that is, the obtained graph is prime with respect to the class $\mathcal{K} \cup \mathcal{I}$. The iterated type partition number, denoted $\mathtt{itp}(G)$, is the number of vertices of the obtained prime graph. An example of a graph G with $\mathtt{itp}(G) = 5$ and its iterative identification is given in Fig. 2.

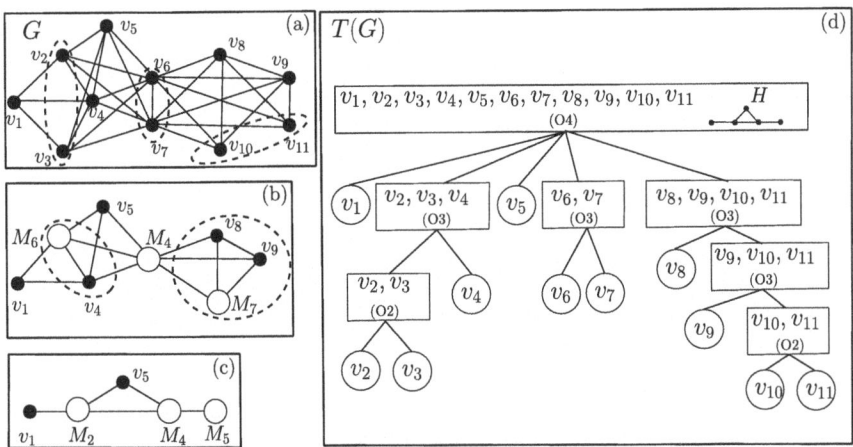

Fig. 2. (a)–(c) A graph G with *iterated type partition* number 5 and its iterative identification. Dashed circles describe the identified *clique* or *independent set* modules. (d) The parse tree $T(G)$ associated with a decomposition of G into modules. Apart from the root, all internal vertices use only operations (O2) and (O3).

It can be shown that the vertices of the obtained *prime graph* represent modules that are *cographs*. Therefore, the *iterated type partition* number of a graph is equivalent to its $\mathcal{C}$-*modular cardinality*, where $\mathcal{C}$ denotes the cograph class.

We know that for each graph G we have $\mathtt{mw}(G) \leq \mathtt{itp}(G) \leq \mathtt{nd}(G)$. See also [13] for a discussion on the relations between modular decomposition parameters and other parameters.

2.1 Modular Decomposition on Power Graphs

The *r-th graph power* of a graph $G = (V, E)$, denoted by G^r, is a graph that has the same set of vertices of G and an edge exists between two vertices if and only if their distance in G is at most r, that is, $V(G^r) = V$ and $E(G^r) = \{(u, v) \mid u, v \in V,\ \delta_G(u, v) \leq r\}$.

Lemma 1. *Let $G = (V, E)$ be any connected undirected graph. It holds*

(i) for every graph class $\mathcal{G}$ such that $\mathcal{K} \subseteq \mathcal{G}$, $\mathcal{G}$-mc$(G^r) \le \mathcal{G}$-mc(G);
(ii) mw$(G^r) \le$ mw(G).

Leveraging Lemma 1, we establish the following result.

Theorem 1. *For any $d, r > 0$, if the $(d, 1)$-DOMINATION problem is FPT parameterized by* mw *(resp. $\mathcal{G}$-mc(G) with $\mathcal{K} \subseteq \mathcal{G}$), then the (d, r)-DOMINATION problem is FPT parameterized by* mw *(resp. $\mathcal{G}$-mc(G) with $\mathcal{K} \subseteq \mathcal{G}$).*

Proof. We know that there exists an algorithm $\mathcal{A}_{mw}$ (resp. $\mathcal{A}_{mc}$) that solves the $(d, 1)$-DOMINATION problem within time $f(\text{mw}(G)) \cdot n^c$ (resp. $f(\mathcal{G}\text{-mc}(G)) \cdot n^c$) where c is a constant. Starting from G one can easily build in polynomial time G^r. Then we observe that a $(d, 1)$-*dominating set* of G corresponds to a $(d, 1)$-*dominating set* of G^r and vice-versa. Hence, using $\mathcal{A}_{mw}$ (resp. $\mathcal{A}_{mc}$) on G^r we are able to solve the (d, r)-DOMINATION problem within time $f(\text{mw}(G^r)) \cdot n^{c'}$ (resp. $f(\mathcal{G}\text{-mc}(G^r)) \cdot n^{c'}$) where c' is a constant. Moreover by Lemma 1 we know that mw$(G^r) \le$ mw(G) (resp. if $\mathcal{K} \subseteq \mathcal{G}$, $\mathcal{G}$-mc$(G^r) \le \mathcal{G}$-mc(G)) and the result holds.   □

Theorem 1 implies that, by applying the FPT algorithm for DOMINATING SET parameterized by mw(G) (presented in [46]), we immediately have that $(1, r)$-DOMINATION is FPT parameterized by mw(G) and, consequently, by itp(G) and nd(G). Moreover, recalling that both in case of nd(G) and itp(G) modules are in $\mathcal{K} \cup \mathcal{I}$, one can apply Theorem 1 on the faster algorithms for DOMINATING SET parameterized by nd(G) (presented in [44]) and itp(G) (presented in [13]).

Moreover, by Theorem 1, we have that, using the FPT algorithm [38] for the $(d, 1)$-DOMINATION problem on G^r, the following result holds.

Corollary 1. (d, r)-DOMINATION *problem is FPT parameterized by* nd.

3   A FPT Algorithm for (d, r)-DOMINATION Parameterized by mw Plus the Demand d

Given a graph $G = (V, E)$, and let $T(G)$ be the parse-tree of G where the number of children of each vertex in $T(G)$, representing the (O4) operation, is at most mw(G). Our strategy traverses the tree $T(G)$ in an opposite breadth-first fashion. For each vertex $w_M \in T(G)$, representing one of the operations (O1)-(O4) of Definition 3 and corresponding to a module M, in order to be able to reconstruct the solution recursively, we calculate optimal solutions under different hypotheses based on the following consideration.

Since all the vertices $v \in M$ share the same neighborhood $N(v) \setminus M$, we can solve the problem locally, considering that the solution, adopted by modules adjacent to M, dominates all the vertices in M with the same multiplicity (i.e., for each $u, v \in M$, $|N(v) \cap (S \setminus M)| = |N(u) \cap (S \setminus M)|$). Hence, we can partition the solution S into two sets $S' \subseteq V \setminus M$ and $S'' \subseteq M$. If we assume that $|N(v) \cap S'| = a$, then all the vertices in M are dominated a times by vertices in S' and we need to build a solution $S'' \subseteq M$, which dominates each vertex $v \in M$

Algorithm 1: MODULECOSTCOMPUTATION

Input: A graph $G[M] = H(G_1, \ldots, G_\ell)$, costs $c_{M_i}(t)$, for each $i \in [\ell]$ and $t \in [d]$.

Output: The costs $c_M(t)$, for each $t \in [d]$.

1 **for** $t = 1, 2, \ldots, d$ **do**

2     $c_M(t) = \infty$

3   **for each** $t_1, t_2, \ldots, t_\ell \in \{0, 1, \ldots, t\}$ **do** // here $c_{M_i}(0) = 0$, for each $i \in [\ell]$

4   **if** $\left(\text{for each } i \in [\ell],\ t_i \geq t - \sum_{(i,j) \in E(H)} c_{M_j}(t_j)\right)$ **then**

5   **if** $\left(\sum_{i=1}^{\ell} c_{M_i}(t_i) < c_M(t)\right)$ **then** $c_M(t) = \sum_{i=1}^{\ell} c_{M_i}(t_i)$

6 **return** c_M

---

at least $t = \max\{d - a, 0\}$ times, that is we need a $(t, 1)$-Dominating set of $G[M]$. Hence, we compute the optimal solution for $G[M]$ considering all the possible demands $t = 0, 1, \ldots, d$, where $t = 0$ corresponds to the case $a \geq d$ (i.e., all the vertices are already dominated by at least d neighbors) and, $t = d$ corresponds to the case $a = 0$ (i.e., adjacent modules do not contain any vertex in S').

Definition 5. *For each vertex* $w_M \in T(G)$, *representing a module* M *and for each* $t \in \{0, 1, \ldots, d\}$ *we define the cost of the module* M *with demand* t, *denoted* $c_M(t)$, *as the size of a minimum* $(t, 1)$-*Dominating set of* $G[M]$.

It is worth observing that:

- For each module M, we have $c_M(0) = 0$.
- For each module leaf $w_{\{v\}}$ of the parse tree $T(G)$, which corresponds to a vertex $v \in V$, we have $c_{\{v\}}(t) = 1$, for each $t \in [d]$.
- The size of the solution of our $(d, 1)$-DOMINATION problem on G corresponds to $c_V(d)$ (i.e., the cost of the root vertex for $t = d$).

Recalling that operations (O2) and (O3) of Definition 3 are special case of (O4), the following Algorithm 1 enables us to compute the costs of an internal vertex w_M representing the (O4) operation, that is $G[M] = H(G_1, \ldots, G_\ell)$, in which $\mathcal{M} = \{M_1, M_2, \ldots, M_\ell\}$ is a modular partition of $G[M]$, $G_i = G[M_i]$ and $\ell \leq \mathtt{mw}(G)$, assuming that the costs for the children modules $G_1, G_2, \ldots, G_\ell$ have already been computed.

Lemma 2. *Let* $G[M] = H(G_1, \ldots, G_\ell)$, *in which* $G_i = G[M_i]$ *and for each* $i \in [\ell], t \in [d]$ *let* $c_{M_i}(t)$ *be the size of a minimum* $(t, 1)$-*Dominating set of* G_i. *The Algorithm 1 computes, in time* $O(d(d + 1)^\ell)$, *for each* $t \in [d]$, *all the values* $c_M(t)$.

Lemma 3. $(d, 1)$-DOMINATION *is solvable in time* $O(|V(G)|d(d + 1)^{\mathtt{mw}})$.

Proof. By observing that every vertex in the parse tree $T(G)$ has at least two children and there are exactly $|V(G)|$ leaves, we have that $T(G)$ has at most $|V(G)| - 1$ internal vertices and the result in Lemma 2. Hence, using the Algorithm 1 for each internal vertex, we can easily build in time $O(|V(G)|d(d+1)^{\mathtt{mw}})$

the optimal cost $c_V(d)$ of the solution for the root vertex, which corresponds to the size of the solution of the problem. The optimal set S can be computed within the same time by a standard backtracking technique.   □

By Lemma 3 and by using the same arguments of the proof of Theorem 1, the following result holds.

Theorem 2. (d, r)-DOMINATION *is solvable in time* $O(|V(G)|d(d + 1)^{\mathtt{mw}})$.

The next section shows a stronger result in case $d = 1$.

3.1   A PC for $(1, r)$-DOMINATION Parameterized by mw

We shift our attention to the $(1, 1)$-DOMINATION problem, that is the classical DOMINATING SET problem on connected graphs.

Lafond and Luo [39] proved the nonexistence of a PC for the DOMINATING SET problem – without the restriction of a connected input graph – parameterized by mw, while Luo [43] showed that the DOMINATING SET problem – without the restriction of a connected input graph – has a polynomial Turing compression (PTC) parameterized by mw. A PTC is a relaxed version of a PC defined as follows.

Definition 6. *A Turing compression for a parameterized problem Q is a polynomial-time algorithm with the ability to access an oracle for another problem Q', that can decide whether an instance $(I, p) \in Q$ with queries, called Turing kernels, of size at most $f(p)$ for some computable function f. In particular, Q admits a polynomial Turing compression (PTC) if f is a polynomial function.*

In the following, we restrict our attention to connected graphs and prove the existence of a PC for $(1, 1)$-DOMINATION parameterized by mw, that is, if we constrain the DOMINATING SET problem to connected graphs, then it admits a PC parameterized by mw.

Let $G = (V, E)$ be a connected undirected graph and let $T(G)$ be the parse-tree of G where the number of children of each vertex in $T(G)$, representing the (O4) operation, is at most $\mathtt{mw}(G)$. Hence, there exists a modular partition $\mathcal{M} = \{M_1, M_2, \ldots, M_\ell\}$ where $\ell \leq \mathtt{mw}(G)$ and $G = H(G_1, \ldots, G_\ell)$, in which $G_i = G[M_i]$, for some outline graph H having ℓ vertices.

Lemma 4. *Let $G = H(G_1, \ldots, G_\ell)$, in which $G_i = G[M_i]$, be a connected undirected graph. There exists a solution S of the DOMINATING SET problem for G such that, for each $i \in [\ell]$.*

(a) $|S \cap M_i| \leq 1$;
(b) if $|S \cap M_i| = 0$, then there exists $j \in N_H(i)$ such that $|S \cap M_j| = 1$;
(c) if $|S \cap M_i|=1$ and for each $j \in N_H(i)$ holds $|S \cap M_j|=0$, then $\bigcap_{v \in M_i}(N_{G_i}(v) \cup \{v\}) \neq \emptyset$.

We now define a variation of DOMINATING SET that will be useful to present the Polynomial Compression.

COLORED DOMINATION:
Input: A graph $G = (V, E)$, a coloring function $c : V \to \{B, W\}$, and an integer k.
Output: A *colored dominating set*, that is, a set $S \subseteq V$ such that

- $|S| \leq k$,
- $N(v) \cap S \neq \emptyset$, for each $v \in V$ such that either $c(v) = B$ or $v \notin S$.

The following Theorem shows that an instance $\langle G, k \rangle$ of the decision version of DOMINATING SET on $G = H(G_1, \ldots, G_\ell)$ with parameter $\mathtt{mw}(G)$ can be reduced to an instance $\langle H, c, k \rangle$ of COLORED DOMINATION.

Theorem 3. DOMINATING SET *admits a PC with respect to* $\mathtt{mw}$, *for the class of connected graphs*.

By Lemma 1, we know that $\mathtt{mw}(G^r) \leq \mathtt{mw}(G)$. Observing that a $(1, r)$-*dominating set* of G corresponds to a *dominating set* of G^r and vice-versa, we have that the above strategy, applied on G^r, allows to devise a PC parameterized by $\mathtt{mw}$ for the $(1, r)$-DOMINATION problem on G. By Theorem 3, the following result holds.

Theorem 4. $(1, r)$-DOMINATION *admits a PC with respect to* $\mathtt{mw}$.

4   A PC for (d, r)-DOMINATION Parameterized by itp Plus the Demand d

This section is devoted to devising a PC for $(d, 1)$-DOMINATION problem parameterized by the iterated type partition number $\mathtt{itp}$ plus the demand d.

Let $I = \langle G, k \rangle$ be an input of the decision version of the $(d, 1)$-DOMINATION problem on $G = (V, E)$. As observed in Sect. 2, any graph $G = (V, E)$ is described by a modular partition $\{M_1, M_2, \ldots, M_{\mathtt{itp}}\}$ where $\mathtt{itp} = \mathtt{itp}(G)$ and can be seen as $G = H(G_1, \ldots, G_{\mathtt{itp}})$, for some outline graph H having $\mathtt{itp}$ vertices, and for each $i \in \{1, 2, \ldots, \mathtt{itp}\}$, $G_i = G[M_i]$ is a cograph.

We are going to define an instance I' of an Integer Linear Programming (ILP) problem on the variables $x_{i,t} \in \{0, 1\}$, for each $i \in \{1, 2, \ldots, \mathtt{itp}\}$ and $t \in [d]$, which we will prove to be equivalent to I.

The ILP uses the costs $c_{M_i}(t)$ associated with the modules M_i, for each $i \in \{1, 2, \ldots, \mathtt{itp}\}$ and each $t \in [d]$, which can be obtained exploiting Algorithm 1. The binary variable $x_{i,t} = 1$ means that we are choosing a solution $S_i \subseteq M_i$ such that S_i is a minimum $(t, 1)$-Dominating set of G_i. By Lemma 2 we know that $|S_i| = c_{M_i}(t)$. The constraints of the ILP are as follows:

(1) $\displaystyle\sum_{i\in\{1,2,\ldots,\mathtt{itp}\},\ t\in[d]} c_{M_i}(t)\cdot x_{i,t} \leq k$

(2) $\displaystyle\sum_{(i,j)\in E(H),\ t\in[d]} c_{M_j}(t)\cdot x_{j,t} + \sum_{t\in[d]} t\cdot x_{i,t} \geq d \qquad \forall i \in \{1,2,\ldots,\mathtt{itp}\}$

(3) $\displaystyle\sum_{t\in[d]} x_{i,t} \leq 1 \qquad\qquad\qquad\qquad\qquad \forall i \in \{1,2,\ldots,\mathtt{itp}\}$

(4) $x_{i,t} \in \{0,1\} \qquad\qquad\qquad\qquad\qquad \forall i \in \{1,2,\ldots,\mathtt{itp}\}, \forall t \in [d],$

where:

- the inequality (1) guarantees that the size of the considered solution is at most k;
- the inequalities (2) guarantee that the vertices of each module M_i are dominated d times. Specifically, the first sum corresponds to the number of times a vertex in M_i is dominated by vertices of the solution belonging to adjacent modules, and the second sum corresponds to the number of times a vertex in M_i is dominated by vertices of the solution belonging to M_i;
- the inequalities (3) guarantee that, for each module M_i at most one of the minimum $(t, 1)$-dominating set of G_i, for $t = 1,\ldots d$, is chosen (i.e., at most one among $c_{M_i}(1), c_{M_i}(2),\ldots,c_{M_i}(d)$ is chosen). We notice that when $\sum_{t\in[d]} x_{i,t} = 0$, then no vertex in the solution is a vertex in M_i and then, all the vertices in M_i are dominated by vertices of the solution belonging to adjacent modules (that are by (2) at least d).

To obtain the desired PC algorithm, we are going to use the Frank and Tardos' algorithm [28] that enables to compress linear inequalities to an encoding length that is polynomial in the number of variables.

Theorem 5 [28]. *There is an algorithm that, given a vector $w \in \mathbb{Q}^r$ and an integer N, in polynomial time finds a vector $\overline{w} \in \mathbb{Z}^r$ with $\|\overline{w}\|_\infty \leq 2^{4r^3} N^{r(r+2)}$ such that $sign(w\cdot b) = sign(\overline{w}\cdot b)$ for all vectors $b \in \mathbb{Z}^r$ with $\|b\|_1 \leq N - 1$.*

In particular, we will use the same approach as adopted in [21] (see Corollary. 2).

Corollary 2 [21]. *There is an algorithm that, given a vector $w \in \mathbb{Q}^r$ and a rational $W \in \mathbb{Q}$, in polynomial time finds a vector $\overline{w} \in \mathbb{Z}^r$ with $\|\overline{w}\|_\infty = 2^{O(r^3)}$ and an integer $\overline{W} \in \mathbb{Z}$ with total encoding length $O(r^4)$, such that $w\cdot x = W$ if and only if $\overline{w}\cdot x = \overline{W}$ for every vector $x \in \{0,1\}^r$.*

The following Lemma shows that $I = \langle G, k\rangle$ and the ILP I' are equivalent, that is, G admits a $(d, 1)$-Dominating set of size at most k if and only if I' admits a feasible solution. Moreover, we can compress I' to a new equivalent instance I'' such that $|I''|$ is bounded by $f(\mathtt{itp}, d)$, where f is a polynomial function.

Lemma 5. *The* $(d,1)$-DOMINATION *problem, parameterized by* itp *plus* d, *admits a PC.*

By Lemma 1, we know that $\text{itp}(G^r) \leq \text{itp}(G)$. Observing that a (d,r)-*dominating set* of G corresponds to a $(d,1)$-*dominating set* of G^r and vice-versa, we have that the above strategy, applied on G^r, allows to obtain a PC parameterized by itp plus d for the (d,r)-DOMINATION problem on G. By Lemma 5, the following result holds.

Theorem 6. *The* (d,r)-DOMINATION *problem, parameterized by* itp *plus* d, *admits a PC.*

5   A PK for (d,r)-DOMINATION Parameterized by nd Plus the Demand d

Since $\text{nd}(G) \geq \text{itp}(G)$, Theorem 6 applies also to neighborhood diversity. In the following, we show that in the latter case, we can go further by devising a PK using a straightforward pruning strategy.

Let $I = \langle G, k \rangle$ be an instance of the decision version of the $(d,1)$-DOMINATION problem on $G = (V, E)$, where $G = (V, E)$ is described by a modular partition $\{M_1, M_2, \ldots, M_{\text{nd}}\}$ where $\text{nd} = \text{nd}(G)$ and can be seen as $G = H(G_1, \ldots, G_{\text{nd}})$, for some outline graph H having nd vertices, and for each $i \in \{1, 2, \ldots, \text{nd}\}$, $G_i = G[M_i]$ is a clique or an independent set.

The following Lemma shows that there exists an optimal solution S such that for each module, the number of vertices belonging to the solution is bounded to $2d - 1$.

Lemma 6. *Let* $G = H(G_1, \ldots, G_{\text{nd}})$, *in which* $G_i = G[M_i]$ *is a clique or an independent set. There exists a solution S of the* (d,r)-DOMINATION *problem for G such that, for each* $i \in \{1, 2, \ldots, \text{nd}\}$, $|S_i = S \cap M_i| \leq 2d - 1$.

Theorem 7. *The* (d,r)-DOMINATION *problem, parameterized by* nd *plus* d, *admits a PK.*

Proof. Let $G = (V, E)$ be a graph. We will reduce an instance $I = \langle G, k \rangle$ of the decision version of the $(d,1)$-DOMINATION problem on $G = (V, E)$, to an instance $I' = \langle G', k \rangle$ of the same problem on a pruned graph G'. We recall that $G = (V, E)$ is described by a modular partition $\{M_1, M_2, \ldots, M_{\text{nd}}\}$ where $\text{nd} = \text{nd}(G)$ and can be seen as $G = H(G_1, \ldots, G_{\text{nd}})$, for some outline graph H having nd vertices. Specifically, G' is obtained from G pruning each module M_i, having more than $2d$ vertices, to any subset of M_i having $2d$ vertices.

We prove that G admits a (d,r)-*dominating set* of size at most k if and only if G' admits a (d,r)-*dominating set* of size at most k.

First, assume that S is a solution of the (d,r)-DOMINATION problem on G and $|S| \leq k$. By Lemma 6, we may assume that for each $i \in \{1, 2, \ldots, \text{nd}\}$, $|S_i = S \cap M_i| \leq 2d - 1$.

Let $S' = \bigcup_{i=1}^{\mathrm{nd}} S_i'$, where S_i' is any set of $|S_i|$ vertices in M_i. Clearly $|S'| = |S|$. Since S is a solution of the (d, r)-DOMINATION problem G, then any vertex in M_i is dominated by some vertices in S_i (if G_i is a clique) and by some vertices in S_j with $j \neq i$. In both cases, the choice of nodes in subset S_i, for each $i \in \{1, 2, \ldots, \mathrm{nd}\}$ does not matter. The only thing that matters is the cardinality of such sets. Since S' keeps these cardinalities unchanged we have that S' is a solution of the (d, r)-DOMINATION problem on G'.

Assume now that S' is a solution of the (d, r)-DOMINATION problem on G'. Let $S = S'$, since all the nodes in a module share the same neighborhood then it is not hard to verify that S is a solution of the (d, r)-DOMINATION problem on G.

Finally, we note that the number of vertices of G' is at most $2d \times \mathrm{nd}(G)$, hence $|I'|$ is bounded by $f(\mathrm{nd}, d)$. $\square$

References

1. Bakhshesh, D., Farshi, M., Hasheminezhad, M.: Complexity results for k-domination and α-domination problems and their variants. arXiv:1702.00533 [cs .CC] (2017)
2. Albert, R., Jeong, H., Barabási, A.-L.: Error and attack tolerance of complex networks. Nature **404**, 378–382 (2000)
3. Bean, J.T., Henning, M.A., Swart, H.C.: On the integrity of distance domination in graphs. Aust. J. Combin. **10**, 29–43 (1994)
4. Borradaile, G., Le, H.: Optimal dynamic program for r-domination problems over tree decompositions. In: Proceedings of IPEC 2016, pp. 8:1–8:23 (2016)
5. Chang, G.J., Nemhauser, G.L.: The k-domination and k-stability problem on graphs. Technical report 540, Cornell University (1982)
6. Cicalese, F., Milanic, M., Vaccaro, U.: On the approximability and exact algorithms for vector domination and related problems in graphs. Discrete Appl. Math. **161**(6), 750–767 (2013)
7. Cicalese, F., Cordasco, G., Gargano, L., Milanic, M., Vaccaro, U.: Latency-bounded target set selection in social networks. Theor. Comput. Sci. **535**, 1–15 (2014)
8. Cordasco, G., Gargano, L., Rescigno, A.A.: On finding small sets that influence large networks. Soc. Netw. Anal. Min. **6**(1), 1–20 (2016). https://doi.org/10.1007/s13278-016-0408-z
9. Cordasco, G., Gargano, L., Mecchia, M., Rescigno, A.A., Vaccaro, U.: Discovering small target sets in social networks: a fast and effective algorithm. Algorithmica **80**(6), 1804–1833 (2018)
10. Cordasco, G., Gargano, L., Rescigno, A.A., Vaccaro, U.: Evangelism in social networks: algorithms and complexity. Networks **71**(4), 346–357 (2018)
11. Cordasco, G., Gargano, L., Rescigno, A.A.: Active influence spreading in social networks. Theoret. Comput. Sci. **764**, 15–29 (2019)
12. Cordasco, G., et al.: Whom to befriend to influence people. Theoret. Comput. Sci. **810**, 26–42 (2020)
13. Cordasco, G., Gargano, L., Rescigno, A.A.: Parameterized complexity for iterated type partitions and modular-width. Discrete Appl. Math. **350** (2024)
14. Corneil, D., Habib, M., Paul, C., Tedder, M.: A recursive linear time modular decomposition algorithm via LexBFS. arXiv:0710.3901 [cs.DM] (2024)

15. Das, B., Bharghavan, V.: Routing in ad-hoc networks using minimum connected dominating sets. In: International Conference on Communications, Montreal, Canada (1997)
16. Datta, A., Larmore, L., Devismes, S., Heurtefeux, K., Rivierre, Y.: Competitive self-stabilizing k-clustering. In: Proceedings of ICDCS 2012, pp. 476–485 (2012)
17. Dom, M., Lokshtanov, D., Saurabh, S.: Incompressibility through colors and IDs. In: Proceedings of 36th ICALP. LNCS, vol. 5555, pp. 378–389. Springer (2009)
18. Downey, R.G., Fellows, M.R.: Fixed-parameter tractability and completeness. Congr. Numer. **87**, 161–187 (1992)
19. Downey, R.G., Fellows, M.R.: Parameterized Complexity. Springer, New York (1999)
20. Drange, P.G., et al.: Kernelization and sparseness: the case of dominating set. In: Proceedings of STACS16. LIPIcs, vol. 47, pp. 31:1–31:14 (2016)
21. Etscheid, M., Kratsch, S., Mnich, M., Röglin, H.: Polynomial kernels for weighted problems. J. Comput. Syst. Sci. **84**, 1–10 (2017)
22. Einarson, C., Reidl, F.: A general kernelization technique for domination and independence problems in sparse classes. In: Proceedings of IPEC 2020. LIPIcs, vol. 180, 11:1–11:15 (2020)
23. Fischer, N., Kunnemann, M., Redzic, M.: The effect of sparsity on k-dominating set and related first-order graph properties. CoRR abs/2312.14593 (2023)
24. Fink, J.F., Jacobson, M.S.: n-Domination in Graphs. Graph Theory with Applications to Algorithms and Computer Science, pp. 283–300. Wiley (1985)
25. Flum, J., Grohe, M.: Parameterized Complexity Theory. Texts in Theoretical Computer Science. An EATCS Series. Springer (2006)
26. Fomin, F.V., Lokshtanov, D., Saurabh, S., Thilikos, D.M.: Linear kernels for (connected) dominating set on graphs with excluded topological subgraphs. In: Proceedings of STACS. LIPIcs, vol. 20, pp. 92–103 (2013)
27. Fomin, F.V., Lokshtanov, D., Saurabh, S., Zehavi, M.: Kernelization: Theory of Parameterized Preprocessing. Cambridge University Press (2019)
28. Frank, A., Tardos, É.: An application of simultaneous Diophantine approximation in combinatorial optimization. Combinatorica **7**(1), 49–65 (1987)
29. Haynes, T.W., Hedetniemi, S.T., Slater, P.J.: Fundamentals of Domination in Graphs. Marcel Dekker Inc., New York (1998)
30. Gajarský, J., Lampis, M., Ordyniak, S.: Parameterized algorithms for modular-width. In: Proceedings of IPEC, vol. 8246, pp. 163–176 (2013)
31. Gallai, T.: Transitiv orientierbare Graphen. Acta Math. Acad. Scientiarum Hung. **18**, 26–66 (1967)
32. Harutyunyan, L.: On the total (k,r)-domination number of random graphs. arXiv:1511.07249 (2015)
33. Haynes, T.W., Hedetniemi, S.T., Slater, P.J.: Fundamentals of Domination in Graphs. Marcel Dekker. Inc., New York (1998)
34. Haynes, T.W., Hedetniemi, S.T., Slater, P.J.: Domination in Graphs - Advanced Topics. Marcel Dekker. Inc., New York (1998)
35. Henning, M.A.: Distance domination in graphs. In: Haynes, T.W., Hedetniemi, S.T., Henning, M.A. (eds.) Topics in Domination in Graphs. DM, vol. 64, pp. 205–250. Springer, Cham (2020). https://doi.org/10.1007/978-3-030-51117-3_7
36. Ishii, T., Ono, H., Uno, Y.: (Total) Vector domination for graphs with bounded branchwidth. Discret. Appl. Math. **207**, 80–89 (2016)
37. Jothilakshmi, G., Pushpalatha, A.P., Suganthi, S., Swaminathan, V.: (k, r)-domination in graphs. Int. J. Contemp. Math. Sci. **6**(29), 1439–1446 (2011)

38. Lafond, M., Luo, W.: Parameterized complexity of domination problems using restricted modular partitions. In: MFCS 2023, pp. 61:1–61:14 (2023)
39. Lafond, M., Luo, W.: Preprocessing complexity for some graph problems parameterized by structural parameters. In: Proceedings of LAGOS 2023, pp. 130–139 (2023)
40. Lampis, M.: Algorithmic meta-theorems for restrictions of treewidth. Algorithmica **64**, 19–37 (2012)
41. Lan, J.K., Chang, G.J.: Algorithmic aspects of the k-domination problem in graphs. Discrete Appl. Math. **161**(10–11), 1513–1520 (2013)
42. Lu, Y., Hou, X., Xu, J.-M.: On the (2,2)-domination number of trees. Discuss. Math. Graph Theory **30**(2), 185–199 (2010)
43. Luo, W.: Polynomial turing compressions for some graph problems parameterized by modular-width. In: Proceedings of Computing and Combinatorics (COCOON 2023). LNCS, vol. 14422. Springer (2023)
44. Koutecký, M.: Solving hard problems on neighborhood diversity. Master Thesis. Charles University in Prague (2013)
45. Newman, M.E.J., Forrest, S., Balthrop, J.: Email networks and the spread of computer viruses. Phys. Rev. E **66** (2002)
46. Románek, M.: Parameterized algorithms for modular-width. Bachelor's Thesis, Masaryk University (2013)
47. Slater, P.J.: R-domination in graphs. J. Assoc. Comp. Mach. **23**(3), 446–450 (1976)

Multipacking and Broadcast Domination on Cactus Graphs and Its Impact on Hyperbolic Graphs

Sandip Das and Sk Samim Islam[(✉)]

Indian Statistical Institute, Kolkata, Kolkata, India
sandipdas@isical.ac.in, samimislam08@gmail.com

Abstract. For a graph G, $\mathrm{mp}(G)$ is the multipacking number, and $\gamma_b(G)$ is the broadcast domination number. It is known that $\mathrm{mp}(G) \leq \gamma_b(G)$ and $\gamma_b(G) \leq 2\,\mathrm{mp}(G) + 3$ for any graph G, and it was shown that $\gamma_b(G) - \mathrm{mp}(G)$ can be arbitrarily large for connected graphs. It is conjectured that $\gamma_b(G) \leq 2\,\mathrm{mp}(G)$ for any general graph G. We show that, for any cactus graph G, $\gamma_b(G) \leq \frac{3}{2}\,\mathrm{mp}(G) + \frac{11}{2}$. Although cactus graphs form a narrow graph class, we used some non-trivial techniques to provide the bound. These techniques make an important step towards generating a tighter bound for general graphs. We also show that $\gamma_b(G) - \mathrm{mp}(G)$ can be arbitrarily large for cactus graphs and asteroidal triple-free graphs by constructing an infinite family of cactus graphs which are also asteroidal triple-free graphs such that the ratio $\gamma_b(G)/\mathrm{mp}(G) = 4/3$, with $\mathrm{mp}(G)$ arbitrarily large. Moreover, we provide an $O(n)$-time algorithm to construct a multipacking of cactus graph G of size at least $\frac{2}{3}\,\mathrm{mp}(G) - \frac{11}{3}$, where n is the number of vertices of the graph G. The hyperbolicity of the cactus graph class is unbounded. For 0-hyperbolic graphs, $\mathrm{mp}(G) = \gamma_b(G)$. Moreover, $\mathrm{mp}(G) = \gamma_b(G)$ holds for the strongly chordal graphs which is a subclass of $\frac{1}{2}$-hyperbolic graphs. Now it's a natural question: what is the minimum value of δ, for which we can say that the difference $\gamma_b(G) - \mathrm{mp}(G)$ can be arbitrarily large for δ-hyperbolic graphs? We show that the minimum value of δ is $\frac{1}{2}$ using a construction of an infinite family of cactus graphs with hyperbolicity $\frac{1}{2}$.

Keywords: Cactus graph · Hyperbolic graph · Multipacking · Dominating broadcast

1 Introduction

Covering and packing are fundamental problems in graph theory and algorithms [8]. In this paper, we study two dual covering and packing problems called *broadcast domination* and *multipacking*. The broadcast domination problem is motivated by telecommunication networks. Imagine a network with radio

---

Some of the proofs and figures are omitted in this conference version of the paper. The full version of the paper is available in arXiv [12].

towers that can transmit information within a certain radius r for a cost of r. The goal is to cover the entire network while minimizing the total cost. The multipacking problem is its natural packing counterpart and generalizes various other standard packing problems. Unlike many standard packing and covering problems, these two problems involve arbitrary distances in graphs, which makes them challenging. The goal of this paper is to study the relation between these two parameters in the class of cactus graphs.

For a graph $G = (V, E)$ with vertex set V, edge set E and the diameter $diam(G)$, a function $f : V \rightarrow \{0, 1, 2, ..., diam(G)\}$ is called a *broadcast* on G. Suppose G is a graph with a broadcast f. Let $d(u, v)$ be the length of a shortest path joining the vertices u and v in G. We say $v \in V$ is a *tower* of G if $f(v) > 0$. Suppose $u, v \in V$ (possibly, $u = v$) such that $f(v) > 0$ and $d(u, v) \leq f(v)$, then we say v *broadcasts* (or *dominates*) u, and u *hears* the broadcast from v.

For each vertex $u \in V$, if there exists a vertex v in G (possibly, $u = v$) such that $f(v) > 0$ and $d(u, v) \leq f(v)$, then f is called a *dominating broadcast* on G. The *cost* of the broadcast f is the quantity $\sigma(f)$, which is the sum of the weights of the broadcasts over all vertices in G. So, $\sigma(f) = \sum_{v \in V} f(v)$. The minimum cost of a dominating broadcast in G (taken over all dominating broadcasts) is the *broadcast domination number* of G, denoted by $\gamma_b(G)$. So,

$$\gamma_b(G) = \min_{f \in D(G)} \sigma(f) = \min_{f \in D(G)} \sum_{v \in V} f(v), \text{ where } D(G) \text{ is the set of all dominating}$$

broadcasts on G.

Suppose f is a dominating broadcast with $f(v) \in \{0, 1\}$ for each $v \in V(G)$, then $\{v \in V(G) : f(v) = 1\}$ is a *dominating set* on G. The minimum cardinality of a dominating set is the *domination number* which is denoted by $\gamma(G)$.

An *optimal broadcast* or *optimal dominating broadcast* on a graph G is a dominating broadcast with a cost equal to $\gamma_b(G)$. A dominating broadcast is *efficient* if no vertex hears a broadcast from two different vertices. Therefore, no tower can hear a broadcast from another tower in an efficient broadcast. There is a theorem that says, for every graph there is an optimal efficient dominating broadcast [13]. Define a ball of radius r around v by $N_r[v] = \{u \in V(G) : d(v, u) \leq r\}$. Suppose $V(G) = \{v_1, v_2, v_3, \ldots, v_n\}$. Let c and x be the vectors indexed by (i, k) where $v_i \in V(G)$ and $1 \leq k \leq diam(G)$, with the entries $c_{i,k} = k$ and $x_{i,k} = 1$ when $f(v_i) = k$ and $x_{i,k} = 0$ when $f(v_i) \neq k$. Let $A = [a_{j,(i,k)}]$ be a matrix with the entries

$$a_{j,(i,k)} = \begin{cases} 1 & \text{if } v_j \in N_k[v_i] \\ 0 & \text{otherwise.} \end{cases}$$

Hence, the broadcast domination number can be expressed as an integer linear program:

$$\gamma_b(G) = \min\{c \cdot x : Ax \geq \mathbf{1}, x_{i,k} \in \{0, 1\}\}.$$

The *maximum multipacking problem* is the dual integer program of the above problem. In 2013, Brewster, Mynhardt, and Teshima [6] obtained a generalization of 2-packings called multipackings. A *k-multipacking* is a set $M \subseteq V$ in a graph $G = (V, E)$ such that $|N_r[v] \cap M| \leq r$ for each vertex $v \in V(G)$ and for every integer $1 \leq r \leq k$. The k-*multipacking number* of G is the maximum cardinality of a k-multipacking of G and it is denoted by $\text{mp}_k(G)$. If M is a k-multipacking, where $k = diam(G)$, the diameter of G, then M is called a *multipacking* of G, and the k-multipacking number of G is called *multipacking number* of G, denoted by $\text{mp}(G)$. A *maximum multipacking* is a multipacking M of a graph G such that $|M| = \text{mp}(G)$. If M is a multipacking, we define a vector y with the entries $y_j = 1$ when $v_j \in M$ and $y_j = 0$ when $v_j \notin M$. So,

$$\text{mp}(G) = \max\{y \cdot \mathbf{1} : yA \leq c, y_j \in \{0, 1\}\}.$$

A *multipacking* of a subgraph H is a set $M' \subseteq V(H)$ in a graph G such that $|N_r[v] \cap M'| \leq r$ for each vertex $v \in V(H)$ and for every integer $r \geq 1$.

A k-packing is a set of vertices in G, such that the shortest path between each pair of the vertices from the set has at least $k + 1$ edges. When $k = 1$, the 1-packing set problem is called the independent set problem. The usual 2-packing is a 1-multipacking of a graph.

Brief Survey: Packing is a well-studied research topic in graph theory. In 1985, Hochbaum, and Shmoys [21] proved that finding a maximum k-packing set in an arbitrary graph is an NP-hard problem, for any k. Very few graph classes have polynomial time algorithms to solve this kind of problems (e.g., rings and trees) [24]. In 2018, Flores-Lamas, Fernández-Zepeda, and Trejo-Sánchez [16] provided a polynomial time algorithm to find a maximum 2-packing set in a cactus. The application areas of these problems include information retrieval, classification theory, computer vision, biomedical engineering, scheduling, experimental design, and financial markets. Butenko [7] discussed these applications in detail.

Multipacking was formally introduced in Teshima's Master's Thesis [25] in 2012 (also see [4,8,13,23]). However, until now, there is no known polynomial-time algorithm to find a maximum multipacking of general graphs, and the problem is also not known to be NP-hard. However, there are polynomial-time algorithms for trees and more generally, strongly chordal graphs [5] to solve the multipacking problem. See [9] for the geometric version of this problem.

Broadcast domination is a generalization of domination problems. Erwin [14, 15] introduced broadcast domination in his doctoral thesis in 2001. For general graphs, an optimal dominating broadcast can be found in polynomial-time $O(n^6)$ [20]. The same problem can be solved in linear time for trees [5]. See [17] for other references concerning algorithmic results on these problems.

It is known that $\text{mp}(G) \leq \gamma_b(G)$ [6]. In 2019, Beaudou, Brewster, and Foucaud [2] proved that $\gamma_b(G) \leq 2\,\text{mp}(G) + 3$ and they conjectured that $\gamma_b(G) \leq 2\,\text{mp}(G)$ for every graph. Moreover, they provided an approximation algorithm that constructs a multipacking of a general graph G of size at least $\frac{\text{mp}(G)-3}{2}$. Hartnell and Mynhardt [19] constructed a family of connected graphs such that the difference $\gamma_b(G) - \text{mp}(G)$ can be arbitrarily large and in fact, for which the ratio $\gamma_b(G)/\text{mp}(G) = 4/3$. Therefore, for general connected graphs,

$$\frac{4}{3} \leq \lim_{\mathrm{mp}(G) \to \infty} \sup \left\{ \frac{\gamma_b(G)}{\mathrm{mp}(G)} \right\} \leq 2.$$

A natural question comes to mind: What is the optimal bound on this ratio for other graph classes? It is known that, for any connected chordal graph G, $\gamma_b(G) \leq \lceil \frac{3}{2} \mathrm{mp}(G) \rceil$ [11]. It is also known that $\gamma_b(G) - \mathrm{mp}(G)$ can be arbitrarily large for connected chordal graphs [11].

Our Contribution: A *cactus* is a connected graph in which any two cycles have at most one vertex in common. Equivalently, it is a connected graph in which every edge belongs to at most one cycle. Cactus is a superclass of tree and a subclass of outerplanar graph. In this paper, we study the multipacking problem on the cactus graph. We establish a relation between multipacking and dominating broadcast on the same graph class. The proofs of the claims marked by (∗) are deferred to the arXiv paper [12]. We start by bounding the multipacking number of a cactus:

Theorem 1. *Let G be a cactus, then $\gamma_b(G) \leq \frac{3}{2} \mathrm{mp}(G) + \frac{11}{2}$.*

Note that, for δ-hyperbolic graphs (defined in this section later), $\gamma_b(G) \leq \lfloor \frac{3}{2} \mathrm{mp}(G) + 2\delta \rfloor$ [10]. Connected chordal graphs are 1-hyperbolic. Earlier we mentioned that for any connected chordal graph G, $\gamma_b(G) \leq \lceil \frac{3}{2} \mathrm{mp}(G) \rceil$ [11]. The cactus graphs are far from being chordal or hyperbolic since cactus graphs can have unbounded hyperbolicity and that shows the importance of Theorem 1. The proof of Theorem 1 is based on finding some paths and a cycle (if needed) in the graph whose total size is almost double the radius. The set of every third vertex on these paths and the cycle yields a multipacking under some conditions.

The hardness problem of multipacking has been repeatedly addressed by numerous authors, yet it has persisted as an unsolved challenge for the past decade. However, polynomial-time algorithms are known for trees and more generally, strongly chordal graphs [5]. Even for trees, the algorithm for finding a maximum multipacking is very non-trivial. The complexity of the multipacking problem for the planar graphs is also unknown till now. Moreover, there is no approximation algorithm for the planar graphs. Cactus is a subclass of the planar graph class. We have already mentioned that, there is a polynomial time algorithm to find a maximum 2-packing set in a cactus [16]. A multipacking is a 2-packing, but the reverse is not true. In this paper, we provide an approximation algorithm to find multipacking on cactus. Our proof technique is based on the basic structure of the cactus graphs, and the technique is completely different from the existing polynomial time solution for the 2-packing problem on the cactus.

Theorem 2. *If G is a cactus graph, there is an $O(n)$-time algorithm to construct a multipacking of G of size at least $\frac{2}{3} \mathrm{mp}(G) - \frac{11}{3}$ where $n = |V(G)|$.*

Hartnell and Mynhardt [19] constructed a family of connected graphs H_k such that $\mathrm{mp}(H_k) = 3k$ and $\gamma_b(H_k) = 4k$. There is no such construction for trees and more generally, strongly chordal graphs, since multipacking number and broadcast domination number are the same for these graph classes [5]. We provide a simpler family of connected graphs G_k such that $\mathrm{mp}(G_k) = 3k$ and $\gamma_b(G_k) = 4k$ (Fig. 6). Not only that, the family of graphs G_k covers many graph classes, including cactus (a subclass of outerplanar graphs), AT-free[1] (a subclass of C_n-free graphs, for $n \geq 6$) graphs, etc. We state that in the following theorem:

Theorem 3. *For each positive integer k, there is a cactus graph (and AT-free graph) G_k such that $\mathrm{mp}(G_k) = 3k$ and $\gamma_b(G_k) = 4k$.*

Theorem 3 directly establishes the following corollary.

Corollary 1. *The difference $\gamma_b(G) - \mathrm{mp}(G)$ can be arbitrarily large for cactus graphs (and for AT-free graphs).*

We also make a connection with the *fractional* versions of the two concepts dominating broadcast and multipacking, as introduced in [3].

As mentioned earlier, for general connected graphs, the range of the expression $\lim_{\mathrm{mp}(G) \to \infty} \sup\{\gamma_b(G)/\mathrm{mp}(G)\}$ is $[4/3, 2]$. It might be very difficult to find the exact value of this expression. For trees the value is 1 since $\gamma_b(G) = \mathrm{mp}(G)$ holds for trees, and more generally for strongly chordal graphs[2] [5]. We tried to find a more general graph class that has the range of the expression $\lim_{\mathrm{mp}(G) \to \infty} \sup\{\gamma_b(G)/\mathrm{mp}(G)\}$ tighter than the existing bound $[4/3, 2]$. We found that the cactus graph has a tighter bound $[4/3, 3/2]$ for the expression. This indicates that the bound for the general graph can be improved further. Theorem 1 and Theorem 3 yield the following:

Corollary 2. *For cactus graphs G,* $\dfrac{4}{3} \leq \displaystyle\lim_{\mathrm{mp}(G) \to \infty} \sup\left\{\dfrac{\gamma_b(G)}{\mathrm{mp}(G)}\right\} \leq \dfrac{3}{2}.$

Corollary 2 yields that, for cactus graphs, we cannot form a bound in the form $\gamma_b(G) \leq c_1 \cdot \mathrm{mp}(G) + c_2$, for any constant $c_1 < 4/3$ and c_2.

We also answer some questions of the relation between broadcast domination number and multipacking number on the hyperbolic graph classes. Gromov [18] introduced the δ-hyperbolicity measures to explore the geometric group theory through Cayley graphs. Let d be the shortest-path metric of a graph G. The graph G is called a δ-*hyperbolic graph* if for any four vertices $u, v, w, x \in V(G)$, the two larger of the three sums $d(u,v)+d(w,x)$, $d(u,w)+d(v,x)$, $d(u,x)+d(v,w)$ differ by at most 2δ. A graph class $\mathcal{G}$ is said to be hyperbolic if there exists a constant δ such that every graph $G \in \mathcal{G}$ is δ-hyperbolic. It is known that a graph is 0-hyperbolic iff it is a block graph[3] [22]. Trees are 0-hyperbolic graphs, since

---

[1] An independent set of three vertices such that each pair is joined by a path that avoids the neighborhood of the third is called an *asteroidal triple*. A graph is asteroidal triple-free or AT-free if it contains no asteroidal triples.

[2] A graph is *strongly chordal* if it is chordal and every cycle of even length (≥ 6) has an odd chord i.e., an edge that connects two vertices that are an odd distance (> 1) apart from each other in the cycle.

[3] A graph is a *block graph* if every block (maximal 2-connected component) is a clique.

trees are block graphs. Moreover, it is known that all strongly chordal graphs are $\frac{1}{2}$-hyperbolic, and also $\mathrm{mp}(G) = \gamma_b(G)$ holds for the strongly chordal graphs. Since block graphs are strongly chordal graphs, therefore $\mathrm{mp}(G) = \gamma_b(G)$ holds for the graphs with hyperbolicity 0. Furthermore, we know that $\gamma_b(G) - \mathrm{mp}(G)$ can be arbitrarily large for chordal graphs which is a subclass of 1-hyperbolic graphs. This leads to a natural question: what is the minimum value of δ, for which we can say that the difference $\gamma_b(G) - \mathrm{mp}(G)$ can be arbitrarily large for δ-hyperbolic graphs? We answer this question. First we show that the family of graphs G_k (Fig. 6) is $\frac{1}{2}$-hyperbolic. This yields the following theorem.

Theorem 4. *For each positive integer k, there is a $\frac{1}{2}$-hyperbolic graph G_k such that $\mathrm{mp}(G_k) = 3k$ and $\gamma_b(G_k) = 4k$.*

Theorem 4 directly establishes the following corollary.

Corollary 3. *The difference $\gamma_b(G) - \mathrm{mp}(G)$ can be arbitrarily large for $\frac{1}{2}$-hyperbolic graphs.*

We discussed earlier that for all 0-hyperbolic graphs, we have $\gamma_b(G) = \mathrm{mp}(G)$. Therefore, Corollary 3 yields the following.

Theorem 5. *For δ-hyperbolic graphs, the minimum value of δ is $\frac{1}{2}$ for which, the difference $\gamma_b(G) - \mathrm{mp}(G)$ can be arbitrarily large.*

It is known that, if G is a δ-hyperbolic graph, then $\gamma_b(G) \leq \lfloor \frac{3}{2}\mathrm{mp}(G) + 2\delta \rfloor$ [10]. This result, together with Theorem 4, yields the following.

Corollary 4. *For $\frac{1}{2}$-hyperbolic graphs G, $\dfrac{4}{3} \leq \lim\limits_{\mathrm{mp}(G)\to\infty} \sup \left\{ \dfrac{\gamma_b(G)}{\mathrm{mp}(G)} \right\} \leq \dfrac{3}{2}$.*

Organisation: In Sect. 2, we prove Theorem 1. In Sect. 3, we provide a $(\frac{3}{2}+o(1))$-factor approximation algorithm for finding multipacking on the cactus graphs. In Sect. 4, we prove that the difference $\gamma_b(G) - \mathrm{mp}(G)$ can be arbitrarily large for cactus graphs. In Sect. 5, we show that the difference $\gamma_b(G) - \mathrm{mp}(G)$ can be arbitrarily large for $\frac{1}{2}$-hyperbolic graphs also. We conclude in Sect. 6. We provide some basic definitions and notations used in Appendix.

2   Proof of Theorem 1

In this section, we prove a relation between the broadcast domination number and the radius of a cactus. Using this we establish our main result that relates the broadcast domination number and the multipacking number for cactus graphs.

We start with a lemma that is true for general graphs.

Lemma 1 (∗). *(Disjoint radial path lemma) Let G be a graph with radius r and center c, where $r \geq 1$. Let P be an isometric path in G such that $l(P) = r$ and c is one endpoint of P. Then there exists an isometric path Q in G such that $V(P) \cap V(Q) = \{c\}$, $r - 1 \leq l(Q) \leq r$ and c is one endpoint of Q.*

The main idea of proving Lemma 1 is the following: if the length of all the isometric paths (except P) with one endpoint c has length less than $r - 1$, then the radius of G will be less than r. In other case, if all the isometric paths (except P) of length at least $r - 1$ with one endpoint c share at least one common vertex except c with the path P, then also the radius of G will be less than r. In both cases, we arrive at a contradiction. The formal proof is written in the Appendix.

For this section, we assume that G is a cactus. Let c be a center and r be the radius of G where $r \geq 1$. We can find an isometric path P of length r whose one endpoint is c, since $\text{rad}(G) = r$. Let $P = (c, v_1, v_2, v_3, \ldots, v_r)$. Therefore, we can find an isometric path Q in G such that $V(P) \cap V(Q) = \{c\}$, $r - 1 \leq l(Q) \leq r$ and c is one endpoint of Q by Lemma 1. Let $Q = (c, w_1, w_2, w_3, \ldots, w_{r'})$ where $r - 1 \leq r' \leq r$. Now we define some expressions to study the subgraph $P \cup Q$ in G (See Fig. 1). For $v_i \in V(P) \setminus \{c\}$ and $w_j \in V(Q) \setminus \{c\}$, we define $X_{P,Q}(v_i, w_j) = \{P_1 : P_1$ is a path in G that joins v_i and w_j such that $V(P) \cap V(P_1) = \{v_i\}$, $V(Q) \cap V(P_1) = \{w_j\}$ and $c \notin V(P_1)\}$. Let $X_{P,Q} = \{(v_i, w_j) : X_{P,Q}(v_i, w_j) \neq \phi\}$. Since G is a cactus, it implies every edge belongs to at most one cycle. Therefore, $|X_{P,Q}(v_i, w_j)| \leq 1$ and also $|X_{P,Q}| \leq 1$. Therefore, there is at most one path (say, P_1) that does not pass through c and joins a vertex of $V(P) \setminus \{c\}$ with a vertex of $V(Q) \setminus \{c\}$ and P_1 intersects P and Q only at their joining points. So, the following observation is true.

Observation 1. *Let G be a cactus with $\text{rad}(G) = r$ and center c. Suppose P and Q are two isometric paths in G such that $V(P) \cap V(Q) = \{c\}$, $l(P) = r$, $r - 1 \leq l(Q) \leq r$ and both have one endpoint c. Then*

(i) $|X_{P,Q}| \leq 1$ and $|X_{P,Q}(v_i, w_j)| \leq 1$ for all (v_i, w_j).
(ii) $X_{P,Q} = \{(v_i, w_j)\}$ iff $|X_{P,Q}(v_i, w_j)| = 1$.

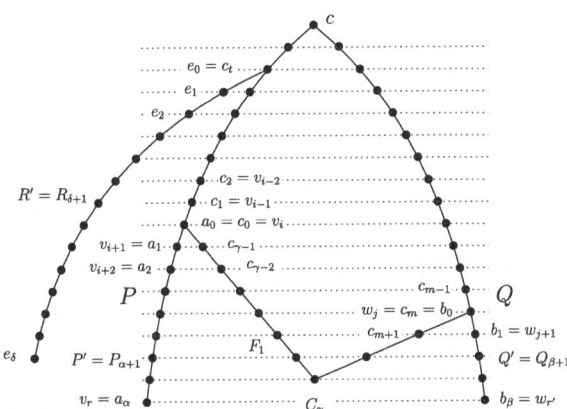

Fig. 1. The subgraph $P \cup Q \cup F_1 \cup R' = H_\gamma(c_0, \alpha, c_t, \delta, c_m, \beta)$

Finding a Special Subgraph $H_\gamma(c_0, \alpha, c_t, \delta, c_m, \beta)$ in Cactus G: From Observation 1, there are two possible cases: either $|X_{P,Q}| = 0$ or $|X_{P,Q}| = 1$. First, we want to study the case when $|X_{P,Q}| = 1$. In that case, suppose $X_{P,Q} = \{(v_i, w_j)\}$ and $X_{P,Q}(v_i, w_j) = \{F_1\}$ (See Fig. 1). We are interested in finding a multipacking set of G from the subgraph $P \cup Q \cup F_1$. Let $F_2 = (v_i, v_{i-1}, v_{i-2}, \ldots, v_1, c, w_1, w_2, \ldots, w_{j-1}, w_j)$, $P' = (v_i, v_{i+1}, \ldots, v_r)$ and $Q' = (w_j, w_{j+1}, \ldots, w_{r'})$. Here $F_1 \cup F_2$ is a cycle and P' and Q' are two isometric paths that are attached to the cycle. Suppose there is another isometric path R' which is disjoint with P' and Q' and whose one endpoint belongs to $V(F_2)$. Then $P \cup Q \cup F_1 \cup R' = F_1 \cup F_2 \cup P' \cup Q' \cup R'$ is a subgraph of G (See Fig. 1). Let $H = P \cup Q \cup F_1 \cup R'$. Note that, we can always find P and Q in G due to Lemma 1, but F_1 or R' might not be there in G, in that case, we can assume that $V(F_1)$ or $V(R')$ is empty in H. In the rest of this section, our main goal is to find a multipacking of G of size at least $\frac{2}{3} \text{rad}(G) - \frac{11}{3}$ from the subgraph H, so that we can prove Theorem 1. To give H a general structure, we want to rename the vertices. Suppose $F_1 \cup F_2$ is a cycle of length γ. Let $F_1 \cup F_2 = C_\gamma = (c_0, c_1, c_2, \ldots, c_{\gamma-2}, c_{\gamma-1}, c_0)$. Suppose $l(P') = \alpha$, $l(Q') = \beta$ and $l(R') = \delta$. We rename the paths P', Q' and R' as $P_{\alpha+1}$, $Q_{\beta+1}$ and $R_{\delta+1}$ respectively. Let $P_{\alpha+1} = (a_0, a_1, \ldots, a_\alpha)$, $Q_{\beta+1} = (b_0, b_1, \ldots, b_\beta)$ and $R_{\delta+1} = (e_0, e_1, \ldots, e_\delta)$ such that $c_0 = a_0$, $c_t = e_0$, $c_m = b_0$ (See Fig. 2). Here $P_{\alpha+1}$, $Q_{\beta+1}$ and $R_{\delta+1}$ are three isometric paths in G. According to the structure, we have $V(P_{\alpha+1}) \cap V(Q_{\beta+1}) = \phi$, $V(Q_{\beta+1}) \cap V(R_{\delta+1}) = \phi$, $V(R_{\delta+1}) \cap V(P_{\alpha+1}) = \phi$, $V(C_\gamma) \cap V(P_{\alpha+1}) = \{c_0\}$, $V(C_\gamma) \cap V(R_{\delta+1}) = \{c_t\}$, $V(C_\gamma) \cap V(Q_{\beta+1}) = \{c_m\}$. Now we can write H as a variable of α, β, γ and δ. Let $H = H_\gamma(c_0, \alpha, c_t, \delta, c_m, \beta)$.

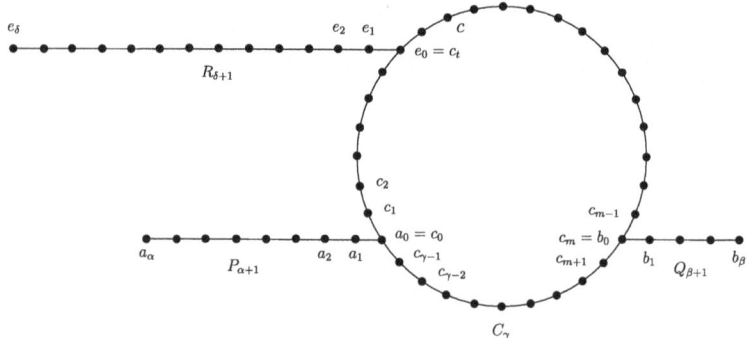

Fig. 2. The subgraph $H_\gamma(c_0, \alpha, c_t, \delta, c_m, \beta)$

In the rest of the section, we study how to find a multipacking of G from the subgraph $H_\gamma(c_0, \alpha, c_t, \delta, c_m, \beta)$. Using this, we prove Lemma 8 that yields Theorem 1 which is our main result in this section.

Lemma 2 ([2]). *Let G be a graph, k be a positive integer and $P = (v_0, v_1, \ldots, v_{k-1})$ be an isometric path in G with k vertices. Let $M = \{v_i : 0 \leq i \leq k, i \equiv$*

$0 \pmod 3\}$ be the set of every third vertex on this path. Then M is a multipacking in G of size $\left\lceil \frac{k}{3} \right\rceil$.

Note that, if $H_\gamma(c_0, \alpha, c_t, \delta, c_m, \beta)$ is a subgraph of G, then $P_{\alpha+1}$, $Q_{\beta+1}$ and $R_{\delta+1}$ are three isometric paths in G by the definition of $H_\gamma(c_0, \alpha, c_t, \delta, c_m, \beta)$. Moreover, since G is a cactus, two cycles cannot share an edge. Therefore, the following observation is true.

Observation 2. *If G is a cactus and $H_\gamma(c_0, \alpha, c_t, \delta, c_m, \beta)$ is a subgraph of G such that $\gamma \geq 3$ and c_0, c_t, c_m are distinct vertices of C_γ, then $H_\gamma(c_0, \alpha, c_t, \delta, c_m, \beta)$ is an isometric subgraph of G.*

Observation 3. *Let G be a cactus and $H_\gamma(c_0, \alpha, c_t, \delta, c_m, \beta)$ be a subgraph of G such that $\gamma \geq 3$ and c_0, c_t, c_m are distinct vertices of C_γ. Let F_1 and F_2 be two paths such that $F_1 = (c_m, c_{m+1}, \ldots, c_{\gamma-1}, c_0)$ and $F_2 = (c_0, c_1, \ldots, c_m)$. Then*

(i) If $l(F_1) > l(F_2)$, then $P_{\alpha+1} \cup F_2 \cup Q_{\beta+1}$ is an isometric path of G.
(ii) If $l(F_1) < l(F_2)$, then $P_{\alpha+1} \cup F_1 \cup Q_{\beta+1}$ is an isometric path of G.
(iii) If $l(F_1) = l(F_2)$, then both of $P_{\alpha+1} \cup F_1 \cup Q_{\beta+1}$ and $P_{\alpha+1} \cup F_2 \cup Q_{\beta+1}$ are isometric paths of G.

The following lemma tells why it is sufficient to find multipacking in $H_\gamma(c_0, \alpha, c_t, \delta, c_m, \beta)$ to provide a multipacking in G.

Lemma 3 (∗). *Let G be a cactus and $H_\gamma(c_0, \alpha, c_t, \delta, c_m, \beta)$ be a subgraph of G. If M is a multipacking of $H_\gamma(c_0, \alpha, c_t, \delta, c_m, \beta)$, then M is a multipacking of G.*

Now our goal is to find multipacking of $H_\gamma(c_0, \alpha, c_t, \delta, c_m, \beta)$. Whatever multipacking we find for the subgraph $H_\gamma(c_0, \alpha, c_t, \delta, c_m, \beta)$, that will be a multipacking for G by Lemma 3. There could be several ways to choose a multipacking set from the subgraph $H_\gamma(c_0, \alpha, c_t, \delta, c_m, \beta)$. But in order to prove the Lemma 8 which is the main lemma to prove Theorem 1, we have to ensure that the size of the multipacking is at least $\frac{2}{3} \text{rad}(G) - \frac{11}{3}$.

We discuss three choices to find a multipacking in $H_\gamma(c_0, \alpha, c_t, \delta, c_m, \beta)$ in the following subsections.

From now, we can write H in place of $H_\gamma(c_0, \alpha, c_t, \delta, c_m, \beta)$ for the simplicity.

Finding a Multipacking of $H_\gamma(c_0, \alpha, c_t, \delta, c_m, \beta)$ According to Choice-1:

Here we find two paths of H so that the set of every third vertex (with some exceptions) on those paths provide a multipacking of H. Let $\overline{P}_{\alpha_1+1} = (c_0, c_1, \ldots, c_{\alpha_1})$ and $\overline{Q}_{\beta_1+1} = (c_m, c_{m+1}, \ldots, c_{m+\beta_1})$ where $0 \leq \alpha_1 \leq m-1$ and $0 \leq \beta_1 \leq (\gamma-1) - m$. Now we choose every third vertex (with some exceptions) from the paths $P_{\alpha+1} \cup \overline{P}_{\alpha_1+1}$ and $Q_{\beta+1} \cup \overline{Q}_{\beta_1+1}$ to construct a set, called $M_\gamma(c_0, \alpha, \alpha_1, c_m, \beta, \beta_1)$ (See Fig. 3). Formally we can define, $M_\gamma(c_0, \alpha, \alpha_1, c_m, \beta, \beta_1) = \{a_i : 0 \leq i \leq \alpha, i \equiv 0 \pmod 3\} \cup \{c_i : 0 \leq i \leq \alpha_1, i \equiv 0 \pmod 3\} \cup \{b_i : 0 \leq i \leq \beta, i \equiv 0 \pmod 3\} \cup \{c_i : m \leq i \leq m+\beta_1, i \equiv m \pmod 3\} \setminus \{c_0, c_m\}$. This set will be a multipacking of G under the conditions stated in the following lemma.

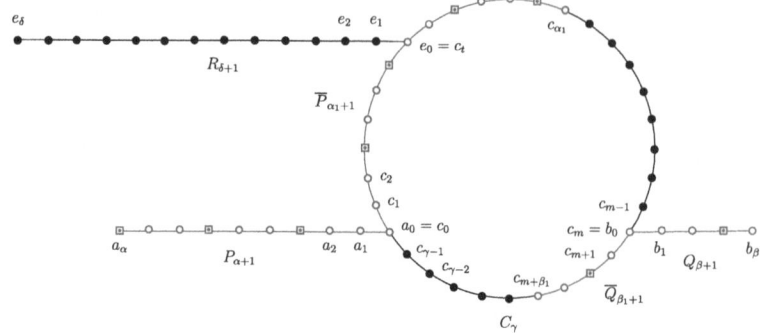

Fig. 3. The circles and squares represent the subgraph $P_{\alpha+1} \cup \overline{P}_{\alpha_1+1} \cup Q_{\beta+1} \cup \overline{Q}_{\beta_1+1}$ and the squares represent the set $M_\gamma(c_0, \alpha, \alpha_1, c_m, \beta, \beta_1)$ in this figure.

Lemma 4 ($*$). *Let G be a cactus and $H_\gamma(c_0, \alpha, c_t, \delta, c_m, \beta)$ be a subgraph of G. Let $\alpha_1, \alpha_2, \beta_1, \beta_2$ be non negative integers such that $\alpha_2 = (m-1) - \alpha_1$, $\beta_2 = (\gamma - 1) - (m + \beta_1)$. If $\alpha_1 \leq 3\beta_2 + \alpha_2 + \beta_1$ and $\beta_1 \leq 3\alpha_2 + \beta_2 + \alpha_1$, then $M_\gamma(c_0, \alpha, \alpha_1, c_m, \beta, \beta_1)$ is a multipacking of G of size at least $\lfloor \frac{\alpha+\alpha_1+1}{3} \rfloor + \lfloor \frac{\beta+\beta_1+1}{3} \rfloor - 2$.*

Substitute $\alpha_2 = (m-1) - \alpha_1$ and $\beta_2 = (\gamma - 1) - (m + \beta_1)$ in Lemma 4. This yields the following.

Lemma 5. *Let G be a cactus and $H_\gamma(c_0, \alpha, c_t, \delta, c_m, \beta)$ be a subgraph of G. Let α_1 and β_1 be non negative integers such that $\alpha_1 \leq m - 1$ and $\beta_1 \leq (\gamma - 1) - m$. If $\alpha_1 \leq \lfloor \frac{\gamma}{2} \rfloor - 1$ and $\beta_1 \leq \lfloor \frac{\gamma}{2} \rfloor - 1$, then $M_\gamma(c_0, \alpha, \alpha_1, c_m, \beta, \beta_1)$ is a multipacking of G of size at least $\lfloor \frac{\alpha+\alpha_1+1}{3} \rfloor + \lfloor \frac{\beta+\beta_1+1}{3} \rfloor - 2$.*

Finding a Multipacking of $H_\gamma(c_0, \alpha, c_t, \delta, c_m, \beta)$ According to Choice-2:

Here we find a path and a cycle of H so that the set of every third vertex (with some exceptions) on these subgraphs provide a multipacking of H. Consider the path $P_{\alpha+1}$ and the cycle C_γ. Now we choose every third vertex (with some exceptions) from these subgraphs to construct a set, called $M'_\gamma(c_0, \alpha)$ (See Fig. 4). Formally we can define, $M'_\gamma(c_0, \alpha) = \{a_i : 0 \leq i \leq \alpha, i \equiv 0 \pmod 3\} \cup \{c_i : 0 \leq i \leq \gamma - 1, i \equiv 0 \pmod 3\} \setminus \{c_0\}$.

Similarly, for the path $Q_{\beta+1}$ and the cycle C_γ, we define the set $M'_\gamma(c_m, \beta) = \{b_i : 0 \leq i \leq \beta, i \equiv 0 \pmod 3\} \cup \{c_i : 0 \leq i \leq \gamma - 1, i \equiv m \pmod 3\} \setminus \{c_m\}$. These sets will be a multipacking of G under the conditions stated in the following lemma.

Lemma 6. *Let G be a cactus and $H_\gamma(c_0, \alpha, c_t, \delta, c_m, \beta)$ be a subgraph of G. Then $M'_\gamma(c_0, \alpha)$ is a multipacking of G of size at least $\lfloor \frac{\gamma}{3} \rfloor + \lfloor \frac{\alpha}{3} \rfloor - 1$ and $M'_\gamma(c_m, \beta)$ is a multipacking of G of size at least $\lfloor \frac{\gamma}{3} \rfloor + \lfloor \frac{\beta}{3} \rfloor - 1$.*

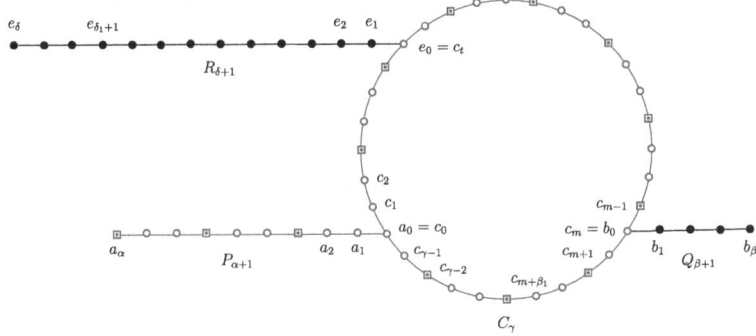

Fig. 4. The circles and squares represent the subgraph $P_{\alpha+1} \cup C_\gamma$ and the squares represent the set $M'_\gamma(c_0, \alpha)$ in this figure.

Proof. Let $H = H_\gamma(c_0, \alpha, c_t, \delta, c_m, \beta)$, $M' = M'_\gamma(c_0, \alpha)$. From the definition of M', $|N_r(v) \cap M'| \leq r$ for all $v \in V(H)$ and $r \geq 1$. Hence M' is a multipacking of H size at least $\lfloor \frac{\gamma}{3} \rfloor + \lfloor \frac{\alpha}{3} \rfloor - 1$. Therefore, M' is a multipacking of G by Lemma 3. By the similar reason $M'_\gamma(c_m, \beta)$ is a multipacking of G of size at least $\lfloor \frac{\gamma}{3} \rfloor + \lfloor \frac{\beta}{3} \rfloor - 1$.   □

Finding a Multipacking in $H_\gamma(c_0, \alpha, c_t, \delta, c_m, \beta)$ According to Choice-3:
Next, we discuss a more general choice, similar to the last one. In the last choice, we did not use the path $R_{\delta+1}$ to construct a multipacking. Here we are going to use $R_{\delta+1}$ also when it is there in G. Let $\overline{R}_{\delta,\delta_1} = (e_{\delta_1+1}, e_{\delta_1+2}, \ldots, e_\delta)$ which is a path and a part of the path $R_{\delta+1}$. Now we choose every third vertex (with some exceptions) from the subgraphs $P_{\alpha+1}$, $\overline{R}_{\delta,\delta_1}$ and C_γ to construct a set, called $M'_\gamma(c_0, \alpha, c_t, \delta, \delta_1)$ (See Fig. 5). Formally we can define, $M'_\gamma(c_0, \alpha, c_t, \delta, \delta_1) = \{a_i : 0 \leq i \leq \alpha, i \equiv 0 \pmod 3\} \cup \{e_i : \delta_1 + 2 \leq i \leq \delta, i \equiv \delta_1 + 1 \pmod 3\} \cup \{c_i : 0 \leq i \leq \gamma - 1, i \equiv 0 \pmod 3\} \setminus \{c_0\}$.

Similarly, if we consider the subgraphs $Q_{\beta+1}$, $\overline{R}_{\delta,\delta_1}$ and C_γ, we can define, $M'_\gamma(c_m, \beta, c_t, \delta, \delta_1) = \{b_i : 0 \leq i \leq \beta, i \equiv 0 \pmod 3\} \cup \{e_i : \delta_1 + 2 \leq i \leq \delta, i \equiv \delta_1 + 1 \pmod 3\} \cup \{c_i : 0 \leq i \leq \gamma - 1, i \equiv m \pmod 3\} \setminus \{c_m\}$.

These sets will be a multipacking of G under some conditions, that has been stated in the following lemma.

Lemma 7 (∗). *Let G be a cactus and $H_\gamma(c_0, \alpha, c_t, \delta, c_m, \beta)$ be a subgraph of G. If $\delta_1 = \lfloor \frac{\gamma}{2} \rfloor - d(c_0, c_t)$ and $\delta \geq \delta_1$, then $M'_\gamma(c_0, \alpha, c_t, \delta, \delta_1)$ is a multipacking of G of size at least $\lfloor \frac{\gamma}{3} \rfloor + \lfloor \frac{\alpha}{3} \rfloor + \lfloor \frac{\delta-\delta_1}{3} \rfloor - 1$. Moreover, if $\delta_2 = \lfloor \frac{\gamma}{2} \rfloor - d(c_m, c_t)$ and $\delta \geq \delta_2$, then $M'_\gamma(c_m, \beta, c_t, \delta, \delta_2)$ is a multipacking of G of size at least $\lfloor \frac{\gamma}{3} \rfloor + \lfloor \frac{\beta}{3} \rfloor + \lfloor \frac{\delta-\delta_2}{3} \rfloor - 1$.*

Now we are ready to prove Lemma 8. Here is a small observation before we start proving the lemma. We use this observation in the proof.

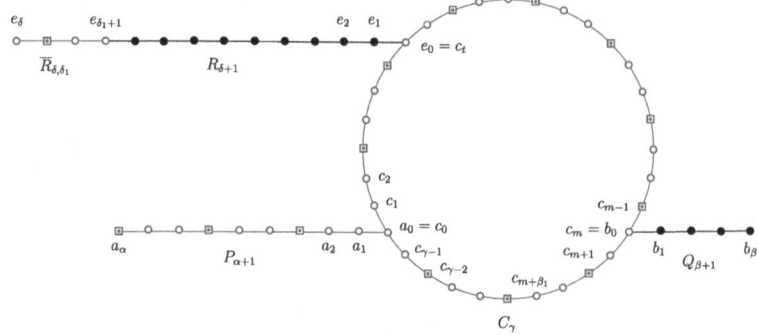

Fig. 5. The circles and squares represent the subgraph $P_{\alpha+1} \cup \overline{R}_{\delta,\delta_1} \cup C_\gamma$ and the squares represent the set $M'_\gamma(c_0, \alpha, c_t, \delta, \delta_1)$ in this figure.

Observation 4. *If r is a positive integer, then $\lfloor \frac{r}{3} \rfloor + \lfloor \frac{r-1}{3} \rfloor \geq \lfloor \frac{2r-1}{3} \rfloor - 1$, $\lfloor \frac{r}{3} \rfloor \geq \frac{r}{3} - \frac{2}{3}$, and $\lfloor \frac{r}{2} \rfloor + \lceil \frac{r}{2} \rceil = r$.*

Lemma 8 (∗). *Let G be a cactus with radius $\mathrm{rad}(G)$, then $\mathrm{mp}(G) \geq \frac{2}{3}\mathrm{rad}(G) - \frac{11}{3}$.*

Proof Sketch. Here, we present a very brief sketch of the proof for Lemma 8. Let $\mathrm{rad}(G) = r$ and c be a center of G. If $r = 0$ or 1, then $\mathrm{mp}(G) = 1$. Therefore, for $r \leq 1$, we have $\mathrm{mp}(G) \geq \frac{2}{3}\mathrm{rad}(G)$. Now assume $r \geq 2$. Since G has radius r, there is an isometric path P in G whose one endpoint is c and $l(P) = r$. Let Q be a largest isometric path in G whose one endpoint is c and $V(P) \cap V(Q) = \{c\}$. If $l(Q) = r'$, then $r - 1 \leq r' \leq r$ by Lemma 1. Let $P = (v_0, v_1, \ldots, v_r)$ and $Q = (w_0, w_1, \ldots, w_{r'})$ where $v_0 = w_0 = c$. From Observation 1, we know that $|X_{P,Q}| \leq 1$.

If $|X_{P,Q}| = 0$, we take the path $P \cup Q$ which is an isometric path of length $r + r'$ and choose every third vertex to the path to construct a multipacking of size at least $\mathrm{mp}(G) \geq \lceil \frac{2}{3}\mathrm{rad}(G) \rceil$ by Lemma 2.

Suppose $|X_{P,Q}| = 1$. We have already discussed in this section that we can find a subgraph $H_\gamma(c_0, \alpha, c_t, \delta, c_m, \beta)$ in G and we use Lemma 5, Lemma 6 and Lemma 7 under several cases to construct a multipacking of size at least $\frac{2}{3}\mathrm{rad}(G) - \frac{11}{3}$. □

Theorem 6 ([15,25]). *If G is a connected graph of order at least 2 having radius $\mathrm{rad}(G)$, multipacking number $\mathrm{mp}(G)$, broadcast domination number $\gamma_b(G)$ and domination number $\gamma(G)$, then $\mathrm{mp}(G) \leq \gamma_b(G) \leq \min\{\gamma(G), \mathrm{rad}(G)\}$.*

Proof of Theorem 1. We have $\gamma_b(G) \leq \mathrm{rad}(G)$ from Theorem 6. From Lemma 8, we get $\mathrm{mp}(G) \geq \frac{2}{3}\mathrm{rad}(G) - \frac{11}{3}$. Therefore, $\frac{2}{3}\cdot\gamma_b(G) - \frac{11}{3} \leq \mathrm{mp}(G) \implies \gamma_b(G) \leq \frac{3}{2}\mathrm{mp}(G) + \frac{11}{2}$. □

3   An Approximation Algorithm to Find Multipacking in Cactus Graphs

We provide the following algorithm to approximate multipacking of a cactus graph. This algorithm is a direct implementation of the proof of Theorem 1.

Approximation Algorithm: Since G is a cactus graph, we can find a center c and an isometric path P of length r whose one endpoint is c in $O(n)$-time, where n is the number of vertices of G. After that, we can find another isometric path Q having length $r-1$ or r whose one endpoint is c and $V(P) \cap V(Q) = \{c\}$. Then the flow of the proof of Lemma 8 provides an $O(n)$-time algorithm to construct a multipacking of size at least $\frac{2}{3}\operatorname{rad}(G) - \frac{11}{3}$. Thus we can find a multipacking of G of size at least $\frac{2}{3}\operatorname{rad}(G) - \frac{11}{3}$ in $O(n)$-time. Moreover, $\operatorname{mp}(G) \geq \frac{2}{3}\operatorname{rad}(G) - \frac{11}{3} \geq \frac{2}{3}\operatorname{mp}(G) - \frac{11}{3}$ by Theorem 6. This implies the existence of a $(\frac{3}{2} + o(1))$-factor approximation algorithm for the multipacking problem on cactus graph. Therefore, we have the following.

Theorem 2. *If G is a cactus graph, there is an $O(n)$-time algorithm to construct a multipacking of G of size at least $\frac{2}{3}\operatorname{mp}(G) - \frac{11}{3}$ where $n = |V(G)|$.*

4   Unboundedness of the Gap Between Broadcast Domination and Multipacking Number of Cactus and AT-Free Graphs

In this section, we prove that the difference between the broadcast domination number and the multipacking number of the cactus and AT-free graphs can be arbitrarily large. Moreover, we make a connection with the *fractional* versions of the two concepts dominating broadcast and multipacking for cactus and AT-free graphs.

To prove that the difference $\gamma_b(G) - \operatorname{mp}(G)$ can be arbitrarily large, we construct the graph G_k as follows. Let $A_i = (a_i, b_i, c_i, d_i, e_i, a_i)$ be a 5-cycle for each $i = 1, 2, \ldots, 3k$. We form G_k by joining b_i to e_{i+1} for each $i = 1, 2, \ldots, 3k-1$ (See Fig. 6). This graph is a cactus graph (and AT-free graph). We show that $\operatorname{mp}(G_k) = 3k$ and $\gamma_b(G_k) = 4k$.

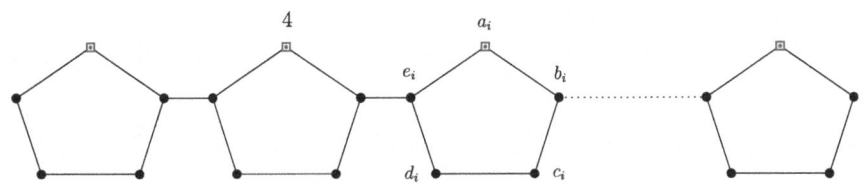

Fig. 6. The G_k graph with $\gamma_b(G_k) = 4k$ and $\operatorname{mp}(G_k) = 3k$. The set $\{a_i : 1 \leq i \leq 3k\}$ is a maximum multipacking of G_k.

Lemma 9 (∗). $\mathrm{mp}(G_k) = 3k$, *for each positive integer* k.

Fractional Multipacking: R. C. Brewster and L. Duchesne [3] introduced fractional multipacking in 2013 (also see [26]). Suppose G is a graph with $V(G) = \{v_1, v_2, v_3, \ldots, v_n\}$ and $w : V(G) \to [0, \infty)$ is a function. So, $w(v)$ is a weight on a vertex $v \in V(G)$. Let $w(S) = \sum_{u \in S} w(u)$ where $S \subseteq V(G)$. We say w is a *fractional multipacking* of G, if $w(N_r[v]) \leq r$ for each vertex $v \in V(G)$ and for every integer $r \geq 1$. The *fractional multipacking number* of G is the value $\max\limits_{w} w(V(G))$ where w is any fractional multipacking and it is denoted by $\mathrm{mp}_f(G)$. A *maximum fractional multipacking* is a fractional multipacking w of a graph G such that $w(V(G)) = \mathrm{mp}_f(G)$. If w is a fractional multipacking, we define a vector y with the entries $y_j = w(v_j)$. So,

$$\mathrm{mp}_f(G) = \max\{y.\mathbf{1} : yA \leq c, y_j \geq 0\}.$$

So, this is a linear program which is the dual of the linear program $\min\{c.x : Ax \geq 1, x_{i,k} \geq 0\}$. Let, $\gamma_{b,f}(G) = \min\{c.x : Ax \geq 1, x_{i,k} \geq 0\}$.

Using the strong duality theorem for linear programming, we can say that

$$\mathrm{mp}(G) \leq \mathrm{mp}_f(G) = \gamma_{b,f}(G) \leq \gamma_b(G).$$

Lemma 10 (∗). *If* k *is a positive integer, then* $\mathrm{mp}_f(G_k) = \gamma_b(G_k) = 4k$.

Lemma 9 and Lemma 10 imply the following results.

Theorem 3. *For each positive integer* k, *there is a cactus graph (and AT-free graph)* G_k *such that* $\mathrm{mp}(G_k) = 3k$ *and* $\gamma_b(G_k) = 4k$.

Corollary 1. *The difference* $\gamma_b(G) - \mathrm{mp}(G)$ *can be arbitrarily large for cactus graphs (and for AT-free graphs).*

Corollary 5. *The difference* $\mathrm{mp}_f(G) - \mathrm{mp}(G)$ *can be arbitrarily large for cactus graphs (and for AT-free graphs).*

5   1/2 Hyperbolic Graphs

In this section, we show that the family of graphs G_k (Fig. 6) is $\frac{1}{2}$-hyperbolic using a characterization of $\frac{1}{2}$-hyperbolic graphs. Using this fact, we show that the difference $\gamma_b(G) - \mathrm{mp}(G)$ can be arbitrarily large for $\frac{1}{2}$-hyperbolic graphs. If the length of a shortest path P between two vertices x and y of a cycle C of G is smaller than the distance between x and y measured along C, then P is called a *bridge* of C. In G, a cycle C is called *well-bridged* if for any vertex $x \in C$ there exists a bridge from x to some vertex of C or if the two neighbors of x from C are adjacent.

Theorem 7 ([1]). *A graph* G *is* $\frac{1}{2}$-*hyperbolic if and only if all cycles* C_n, $n \neq 5$, *of* G *are well-bridged and none of the graphs depicted in [1, 12] occur as isometric subgraphs of* G.

Theorem 7 yields the following.

Lemma 11. *The family of graphs G_k (Fig. 6) is $\frac{1}{2}$-hyperbolic.*

Lemma 9, Lemma 10 and Lemma 11 imply the following theorem.

Theorem 4. *For each positive integer k, there is a $\frac{1}{2}$-hyperbolic graph G_k such that $\mathrm{mp}(G_k) = 3k$ and $\gamma_b(G_k) = 4k$.*

Corollary 3. *The difference $\gamma_b(G) - \mathrm{mp}(G)$ can be arbitrarily large for $\frac{1}{2}$-hyperbolic graphs.*

From Theorem 4 and Lemma 10, we have the following.

Corollary 6. *The difference $\mathrm{mp}_f(G) - \mathrm{mp}(G)$ can be arbitrarily large for $\frac{1}{2}$-hyperbolic graphs.*

6   Conclusion

We have shown that the bound $\gamma_b(G) \leq 2\,\mathrm{mp}(G) + 3$ for general graphs G can be improved to $\gamma_b(G) \leq \frac{3}{2}\,\mathrm{mp}(G) + \frac{11}{2}$ for cactus graphs. Moreover, $\gamma_b(G) - \mathrm{mp}(G)$ can be arbitrarily large for cactus, AT-free and $\frac{1}{2}$-hyperbolic graphs.

It remains an interesting open problem to determine the best possible value of the expression $\lim_{\mathrm{mp}(G) \to \infty} \sup\{\gamma_b(G)/\mathrm{mp}(G)\}$ for general connected graphs and for cactus graphs. This problem could also be studied for other interesting graph classes.

References

1. Bandelt, H.J., Chepoi, V.: 1-hyperbolic graphs. SIAM J. Discret. Math. **16**(2), 323–334 (2003)
2. Beaudou, L., Brewster, R.C., Foucaud, F.: Broadcast domination and multipacking: bounds and the integrality gap. Australas. J. Combin. **74**(1), 86–97 (2019)
3. Brewster, R., Duchesne, L.: Broadcast domination and fractional multipackings. Manuscript (2013)
4. Brewster, R.C., Beaudou, L.: On the multipacking number of grid graphs. Discrete Math. Theor. Comput. Sci. **21** (2019)
5. Brewster, R.C., MacGillivray, G., Yang, F.: Broadcast domination and multipacking in strongly chordal graphs. Discret. Appl. Math. **261**, 108–118 (2019)
6. Brewster, R.C., Mynhardt, C.M., Teshima, L.E.: New bounds for the broadcast domination number of a graph. Cent. Eur. J. Math. **11**(7), 1334–1343 (2013)
7. Butenko, S.: Maximum independent set and related problems, with applications. University of Florida (2003)
8. Cornuéjols, G.: Combinatorial Optimization: Packing and Covering. SIAM (2001)
9. Das, A.K., Das, S., Islam, S.S., Mitra, R.M., Roy, B.: Multipacking in Euclidean plane. arXiv preprint arXiv:2411.12351 (2024)
10. Das, S., Foucaud, F., Islam, S.S., Mukherjee, J.: Relation between broadcast domination and multipacking numbers on chordal and other hyperbolic graphs. arXiv preprint arXiv:2312.10485 (2023)

11. Das, S., Foucaud, F., Islam, S.S., Mukherjee, J.: Relation between broadcast domination and multipacking numbers on chordal graphs. In: Conference on Algorithms and Discrete Applied Mathematics, pp. 297–308. Springer (2023)

12. Das, S., Islam, S.S.: Multipacking and broadcast domination on cactus graphs and its impact on hyperbolic graphs. arXiv preprint arXiv:2308.04882 (2023)

13. Dunbar, J.E., Erwin, D.J., Haynes, T.W., Hedetniemi, S.M., Hedetniemi, S.T.: Broadcasts in graphs. Discret. Appl. Math. **154**(1), 59–75 (2006)

14. Erwin, D.J.: Dominating broadcasts in graphs. Bull. Inst. Combin. Appl. **42**(89), 105 (2004)

15. Erwin, D.J.: Cost domination in graphs. Ph.D. Thesis, Western Michigan University (2001)

16. Flores-Lamas, A., Fernández-Zepeda, J.A., Trejo-Sánchez, J.A.: Algorithm to find a maximum 2-packing set in a cactus. Theoret. Comput. Sci. **725**, 31–51 (2018)

17. Foucaud, F., Gras, B., Perez, A., Sikora, F.: On the complexity of broadcast domination and multipacking in digraphs. Algorithmica **83**(9), 2651–2677 (2021)

18. Gromov, M.: Hyperbolic groups. In: Essays in Group Theory, pp. 75–263. Springer (1987)

19. Hartnell, B.L., Mynhardt, C.M.: On the difference between broadcast and multipacking numbers of graphs. Utilitas Math. **94**, 19–29 (2014)

20. Heggernes, P., Lokshtanov, D.: Optimal broadcast domination in polynomial time. Discret. Math. **306**(24), 3267–3280 (2006)

21. Hochbaum, D.S., Shmoys, D.B.: A best possible heuristic for the k-center problem. Math. Oper. Res. **10**(2), 180–184 (1985)

22. Howorka, E.: On metric properties of certain clique graphs. J. Combin. Theory Ser. B **27**(1), 67–74 (1979)

23. Meir, A., Moon, J.W.: Relations between packing and covering numbers of a tree. Pac. J. Math. **61**(1), 225–233 (1975)

24. Mjelde, M.: k-packing and k-domination on tree graphs. Master's thesis, The University of Bergen (2004)

25. Teshima, L.E.: Broadcasts and multipackings in graphs. Ph.D. thesis (2012)

26. Teshima, L.E.: Multipackings in graphs. arXiv preprint arXiv:1409.8057 (2014)

Evaluating Monotone Circuits on Surfaces

Samir Datta[1] and Chetan Gupta[2](✉)

[1] Chennai Mathematical Institute, Chennai, TN, India
sdatta@cmi.ac.in
[2] Indian Institute of Technology Roorkee, Roorkee, UP, India
chetan.gupta@cs.iitr.ac.in

Abstract. In this paper, we study the circuit value problem for monotone Boolean circuits (that is, circuits with $\wedge, \vee$ but no negation gates) that are embedded on a surface of bounded genus, and all inputs to the circuits lie on bounded number of input faces. We show that this problem belongs to complexity class **LogDCFL**. This along with the result of Cook [7], yields a simultaneously space-efficient ($O(\log^2 n)$-space) and polynomial time algorithm for the problem. It also gives a highly parallel algorithm (simultaneously $O(\log n)$-time with polynomially many processors).

This generalises the previous bound of **LogDCFL** for the problem on one input face monotone planar circuits [6]. More precisely, we show that if a monotone circuit is embedded on a surface of polylogarithmic genus g and has k faces on which all the inputs are present, then the circuit can be evaluated on a **CROW-PRAM** (concurrent read *owner* write parallel random access machine) in time $O(g \log (k + g) \log n)$ using $n^{O(1)}$ many processors.

We introduce a new distance metric in single sink DAGs that can be computed in deterministic logarithmic space (**L**) and is useful in partitioning the circuit into subcircuits such that each one is a one-input face monotone planar circuit. We also show that the partitioning procedure is done in deterministic logarithmic space. Thus we are able to side-step the barrier of computing the usual distance in bounded genus graphs, for which the best bound known is **UL** ∩ **coUL** [18,29] and therefore not known to be contained in **LogDCFL**. We also non trivially modify the parallel algorithm by Delcher-Kosaraju [9] for **MPCVP** to reduce the dependence of the parallel running time on the number of input faces from linear to logarithmic.

1 Introduction

The Circuit Value Problem **CVP** – "Given a Boolean circuit consisting of AND ($\wedge$), OR ($\vee$), NOT ($\neg$)-gates and a Boolean assignment to its input values, what is the value output by the circuit?" – occupies an important place in complexity theory as the archetypal **P**-complete problem. Various relaxations of the problem are known to be **P**-complete such as *Monotone* Circuit Value Problem **MCVP** where there are no $\neg$ gates and *Planar* Circuit Value Problem **PCVP** where the

S. Nakano and M. Xiao (Eds.): WALCOM 2025, LNCS 15411, pp. 127–142, 2025.
https://doi.org/10.1007/978-981-96-2845-2_9

circuit is itself embedded as a planar graph [14]. It is somewhat remarkable that a combination of the previous two problems *Monotone Planar* Circuit Value Problem MPCVP is parallelisable and therefore contained in NC [9,21,30] and hence is not expected to be P-complete. A series of papers have been devoted to refining the exact complexity of MPCVP and its restrictions [3,6,9,19,22]. Layering the circuit and restricting the outer face of the circuit to contain all the inputs are two specializations which dramatically improves the complexity bound on MPCVP.

MCVP can also be viewed as a generalization of reachability in a restricted class of graphs. This is so because reachability in a single sink DAG can be viewed as a circuit in which all gates are ∨-gates. Thus, MPCVP is a generalization of single sink planar DAG reachability. There has been considerable work on trying to pin down the exact complexity of planar reachability [4,28] and of other topologically restricted reachability instances such as those embedded on a surface of bounded genus [16,18]. Another thread of work has focused on DAG reachability with few sources/sinks in planar [1] and somewhat non-planar instances [1,25,26].

Overall there are three common restrictions on MCVP that make the problem efficiently parallelizable:

- Topological: The circuit is embedded on a plane or on a surface of low genus.
- Layering: The gates are partitioned into layers with edges between adjacent layers only.
- Single input face: There is a face containing all the inputs of the circuit.

The first set of papers [11,15] on the topic imposed all three kinds of restrictions and gave the LogCFL $\subseteq$ NC^2 bound on the problem. Subsequent work [21,22] gave parallel algorithms for the problems which were processor efficient (i.e. used linearly many processors) but had weaker bounds on the running time with and without the layering constraint. An independent thread of work [9] used the single input face constraint and used it to solve the (node bimodal) MPCVP problem in NC. Notice that they assumed that each gate has only two inputs by expanding gates with larger fan-in as a tree – this is directly possible only if the inputs and outputs of a gate do not intersperse i.e. as a directed graph, the each node is bimodal. The bound on upward planar, layered and single input face MPCVP was optimized to LogDCFL in [3]. Later, [19] removed the upward planar and layering restrictions but kept the single input face restriction to prove a bound using both the LogDCFL bound from [3] and planar longest path in DAGs which is known to be in UL $\subseteq$ NL. Also, in [19] the topological restrictions were relaxed from planar to toroidal yielding a L(LogCFL) = SAC^2 $\subseteq$ NC^3 bound for the monotone toroidal circuit value problem with no restriction on number of input faces.

In this work, we dispense completely with the layering restriction and show a smooth tradeoff between parameterized versions of the other two restrictions on the one hand and the complexity of the MCVP problem on the other. This provides a common generalization of the MPCVP results and of the DAG-reachability results mentioned above.

Overview of Our Results. The primary problem of interest for us is what we call kMgCVP – this stands for monotone circuit value problem for genus g circuits such that there are k faces of the embedded circuit that contain all the inputs. Thus the case with $k = 1, g = 0$ is the usual single input face MPCVP that was studied in [6] building on [3] and using insights from [19]. In this work, building on [3,6,19] we prove an upper bound on kMgCVP made precise in theorem 1 from which we derive a number of important corollaries.

In the following, CROW$[t(n)]$ is the class of languages accepted by a Concurrent Read Owner Write PRAM (or CROW PRAM) machine in (parallel) time $O(t(n))$ and with $n^{O(1)}$ processors. Notice that Dymond and Ruzzo [12] proved that CROW$[\log n]$ is precisely LogDCFL, i.e. recognizable by a deterministic logspace machine equipped additionally with a polynomial height stack. Our main results can be summarised as follows:

Theorem 1. kMgCVP *is in* CROW$[g\log(k + g)\log n]$.

From Theorem 1, we get the following as immediate corollaries.

Corollary 1. kMgCVP *problem when the number of input faces $k = O(1)$ and the genus $g = O(1)$ can be solved in* LogDCFL.

Corollary 2. *The* MGCVP *problem with no restriction on the number of input faces and where the genus $g = O(1)$ is in* CROW$[\log^2 n]$.

Corollary 3. *The* MGCVP *problem with no restriction on the number of input faces and with the genus $g = (\log n)^{O(1)}$ is in* NC.

It is worthwhile to point out that the bounds in Corollary 1 and Corollary 3 are completely new. While Corollary 2 improves on the AC^1(LogCFL) = SAC^2 bound proved in [19], since CROW$[\log^2 n] \subseteq$ AC^1(CROW$[\log n]$) = AC^1(LogDCFL). We summarise previous and currently reported work in Table 1.

Notice that the length of the longest path was used in [19] and previous papers to layer the graph so that it becomes possible to use the LogDCFL algorithm of [3]. Since layering is in UL ∩ coUL for planar DAGs they got a complexity bound larger than ours. In [6], layering was done by using a grid embedding and so the complexity of layering was reduced to L. Some cut-and-paste surgery was also required, which was the main technical content of [6]. However, generalizing this to larger genus seems difficult because it is hard to work with a grid embedding on a surface. Even if we could work with a grid-embedded fundamental polygon of the surface – we would need to do major surgery as in [18]. We are not sure how to do this while preserving the grid embedding. We take a different approach as follows.

An important idea behind our results is a new measure of the "distance" between two nodes in an acyclic digraph that is conceptually simple as well as computable in L. This allows us to chop up a planar circuit into annuli consisting of vertices spanning a range of distances to t such that the number of input faces in each annulus is at most half of those in the parent graph. Thus we can, in $O(\log k)$ number of steps (where k is the number of input faces in the

Table 1. Previously known and ⎡new⎤ results (∞ refers to unbounded number of input faces)

Restrictions			Bound
Topological	Layering	#input faces	
Upward Planar	✓	1	DSPACE[$\log^2 n$] [15]
Upward Planar	✓	1	LogCFL[11]
Upward Planar	✓	1	LogDCFL = CROW[$\log n$] [3]
Planar ($g = 0$)	✓	∞	EREW[$\log^2 n$] [21]
Planar ($g = 0$)	✗	∞	CRCW[$\log^4 n$] [9]
Planar ($g = 0$)	✗	∞	EREW[$\log^6 n$] [22]
Planar ($g = 0$)	✗	1	LogDCFL $\oplus$ (UL $\cap$ coUL) [19]
Planar ($g = 0$)	✗	1	LogDCFL [6]
Toroidal ($g = 1$)	✗	∞	AC^1(LogCFL) = SAC^2 [19]
$g = O(1)$	✗	$O(1)$	⎡LogDCFL⎤
$g = (\log n)^{O(1)}$	✗	∞	⎡NC⎤

original planar graph) reach the base case of one input face. Each of these one input face circuits can be solved via [6] in LogDCFL and by the equivalence of LogDCFL and CROW[$\log n$] [12] by an Owner PRAM in logarithmic time. Though we can reduce the k-input face circuit into circuits with one-input face in $O(\log k)$ steps, we cannot naively combine these one-input face circuits back (after evaluating them) to obtain the evaluation of the original k-input face circuit in $O(\log k)$ steps. Therefore, composing the computation of the individual PRAMs we get an $O(k \log n)$ parallel time algorithm for planar i.e. genus zero graphs. We further improve this to $O(\log k \log n)$ parallel time by aptly modifying a technique introduced by Delcher and Kosaraju [9] that allows indeterminates in the monotone circuit and partially evaluates a circuit with indeterminates multiple times using divide and conquer till all indeterminates are eliminated.

Moving on to circuits embedded on a genus g surface, we chop them into $O(g)$ subcircuits each of which is either planar with $O(g + k)$ number of input faces or of constant depth (even though it may be embedded on a high genus surface). Both can hence be evaluated in CROW[$\log (g + k) \log n$]. Composing the functions computed by the $O(g)$ pieces, we are able to get a bound of CROW($g \log (g + k) \log n$)).

Our notion of "distance" for a node v is based on a number N_v, which is the number of nodes that can reach node v. Thus, distance between two nodes u and v, $d_{\#}(u, v)$ can be defined as $N_v - N_u$ if u can reach v and ∞ (alternatively, undefined) otherwise. Notice that if reachability in a class of DAGs is in $\mathcal{C}$ then $d_{\#}(.,.)$ is computable in $\mathsf{L}^{\mathcal{C}}$. In our case $\mathcal{C}$ is almost invariably L ensuring that $d_{\#}(.,.)$ is also computable in L.

Organization of the Paper. The rest of the paper is organized as follows. In Sect. 2, we state some pertinent previous results, talk briefly about some relevant complexity classes and define some necessary notation that we use in the paper. In the first part of Sect. 3, we prove our result for planar circuits, in the second we improve the dependence on the number of faces and in the third part we generalize it to bounded genus circuits. Finally, in Sect. 4, we conclude and leave some open questions for future work.

2   Preliminaries

A Boolean circuit is a DAG (directed acyclic graph) which contains three types of nodes: *source nodes, a sink node* and *internal nodes*. Each internal node is labelled by an AND, OR or NOT gate. Edges in the graph represent the connection (wires) between two nodes (gates). In a Boolean circuit, source nodes are fed with an input binary string, and the circuit produces an output on the sink node by applying the sequence of Boolean operations represented by internal nodes. A circuit is called *n-input* circuit if the number of source nodes in the circuit is n. Given a *n-input* circuit along with a binary string of length n, problem of evaluating the output of the circuit (value at sink node) is called *circuit-value problem* (CVP). A Boolean circuit which does not have any NOT gate is called a *monotone* circuit. A circuit that can be embedded on a plane without crossing its edges is called a planar circuit and the problem of evaluating those monotone circuits is called monotone planar circuit value problem (MPCVP). Circuits that can be embedded on surfaces are a natural extension of planar circuits. For a circuit embedded on a surface such that all the source nodes of the circuit lie on k-faces, we call it k-input-face circuit. We denote the problem of evaluating a k-input faces monotone circuit that can be embedded on a genus g surface as kMgCVP. Which is the problem we are focusing on in this paper. We use the following result proved for one-input-face planar circuits.

Lemma 1 [6]. *One-input-face monotone planar circuit value problem can be solved in* LogDCFL.

As we mentioned that a circuit could be represented by a DAG; thus from now on we will identify a circuit as a DAG and write everything in terms of graphs. If a node in the graph does not have any incoming edges of the graph incident on it, we call it a source node or input node. Similarly, a node with no outgoing edges is called a sink node of the graph.

Graph Theory. Most of the directed graphs that we use are Directed Acyclic Graphs or DAGs. A DAG is said to be connected when the underlying undirected graph is connected. In general, when we refer to digraphs as graphs we are referring to the underlying undirected graph.

Graphs that can be embedded on surfaces form an important class for us and we proceed to introduce them. See [10, Appendix B] for a more detailed exposition. A g-genus surface is a sphere with g-many handles on it. A graph is called a g-genus graph if g is the minimum integer such that the graph can be embedded on a g-genus surface without intersecting its edges. A 2-cell embedding of a graph is an embedding in which every face of the graphs is homeomorphic to an open disk. A graph of genus g always has a 2-cell embedding on a surface of genus g. For a graph G and surface S, we will use $g(G)$ and $g(S)$ to denote the genus of G and S respectively. We know that if G is embedded on S then $g(G) \leq g(S)$. Cycles in a surface embedded graph can be divided into two categories, *surface separating* cycles and *surface non-separating* cycles. Surface separating cycles are those cycles such that cutting a surface along those cycles divides the surface into at least two disjoint surfaces. Surface non-separating cycles are those cycles such that cutting a surface along these cycles does not separate the surface but reduces the genus of the surface. We will use the following lemmas about surface separating and surface non-separating cycles in surface embedded graphs (Lemma B.4 and Lemma B.5 from [10]).

Lemma 2 ([10]). *Let C be a surface separating cycle in a surface S, and S' and S'' be the surfaces obtained from S by cutting along C and capping the holes. Then $g(S) = g(S') + g(S'')$.*

Lemma 3 ([10]). *Let C is a surface non-separating cycle in a surface S, and S' be the surface obtained from S by cutting along C and capping the holes. Then $g(S') = g(S) - 1$.*

We will also use the following lemmas about the deterministic logarithmic space (L) computable properties of graphs.

Lemma 4 ([2,8]). *Given a graph G, we can check if G is planar or not in L.*

Lemma 5 ([26]). *Given a directed acyclic graph embedded on a surface of genus $2^{O(\sqrt{\log n})}$ with $2^{O(\sqrt{\log n})}$ source nodes, we can check whether there is a directed path from a node u to another node v in L.*

Our algorithm requires an embedding that has simultaneously bounded genus and bounded number of input faces. While it is possible to test for the embeddability of a graph on a surface of bounded genus and even obtain the embedding in L [13] – this does not suffice for our purpose since we do not know how to concurrently ensure that the embedding has a bounded number of input faces.

$V(G), E(G)$ and $F(G)$ to represent the set of nodes, set of edges and set faces in a graph G, respectively. Although, when we write that $v \in G$ (or $e \in G$, $f \in G$), we mean $v \in V(G)$ (respectively $e \in E(G), f \in F(G)$). We refer to the number of vertices, faces and edges in a graph G by $\#v(G), \#f(G), \#e(G)$ respectively.

Complexity Classes. Classically, the common parallel computation model is the PRAM or Parallel Random Access Machine [17] where many processors communicate via shared memory. A problem is said to be parallelisable if it can be solved in the PRAM model using $\log^{O(1)} n$ time and polynomial number of processors. PRAM models can further be distinguished on the basis of how they resolve read and write conflicts. Thus the weakest model is the EREW PRAM model or the exclusive read exclusive write model that stipulates that there are no concurrent writes or reads for any memory location. On the other extreme is the CRCW PRAM where concurrent reads and writes are both permitted with several ways of resolving conflicts, most of which are shown to be equivalent (see [17]). Intermediate between these two types of PRAM models are the CREW PRAM models where the writes are exclusive but reads can be concurrent. While the fine-grained mapping between CREW, EREW PRAM models on the one hand and Turing machine models or circuit models on the other, is not precise, CRCW PRAM models correspond naturally to both alternating Turing machines [23] and unbounded fan-in circuits [24]. We put it in perspective below.

– LogDCFL: class of languages reducible to deterministic context-free languages using logspace reductions [27]. Alternatively, they are languages accepted by deterministic AuxPDAs with a pushdown stack in polynomial time (colloquially "logspace with polynomial stack") [27]. Cook proved that LogDCFL is contained in the class of problems that can be solved simultaneously in polylogarithmic space and polynomial time SC [5, 7]. They are also known to be contained in LogCFL—the class languages that are logspace reducible to context-free languages. Alternatively, LogCFLs are the uniform version of the circuit class SAC^1, which is a class of languages that are accepted by a circuit family of depth $O(\log n)$ and size polynomial in the number of inputs (like AC^1) but only OR-gates have polynomial fan-out while the AND-gates have bounded fan-in. LogCFL is known to be contained in NC^2 (by just replacing large fan-in OR-gates by trees of fan-in 2) thus, so is LogDCFL though LogCFL is not known to be contained in SC.
  Dymond and Ruzzo [12] showed a PRAM characterisation of LogDCFL in terms of Owner writes (any subset of processors can read from a memory location but only the designated owner of a memory location can write to it – a notion more restrictive than the CREW PRAM model) in a model known as CROW-PRAM. They showed that LogDCFL is precisely the class of languages that can be solved by polynomially many processors in $O(\log n)$ time which form a concurrent read owner write PRAM or equivalently CROW[$\log n$]. Further [20] shows other circuit based characterisations of LogDCFL, LogCFL.
– UL or unambiguous logspace is the class of languages accepted by a nondeterministic Turing machine that is unambiguous i.e. has at most one accepting path on any input. This class is clearly contained in $NL \subseteq LogCFL \subseteq NC^2$ but like LogCFL is not known to be contained in SC. This class has achieved some prominence because planar restrictions of important problems like reachability [4, 28] and distance [29] are known to be contained here (or indeed in the slightly smaller class $UL \cap coUL$).
  These unambiguous classes are the main villains for us and we show how to circumvent these by instead using the logspace algorithms for single sink planar DAG reachability from [1] and similar reachability for DAGs with a

single sink embedded on a surface of bounded genus [25, 26].

In Fig. 1, we summarise the relationship among the various complexities classes relevant to this paper.

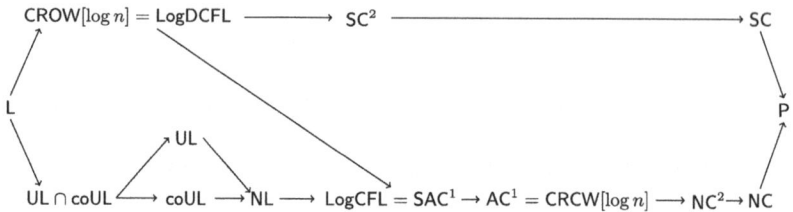

Fig. 1. Relevant complexity classes and relations among them.

CROW-Transducers: A CROW-PRAM accepts a language but we can as well use CROW-PRAMs to define functions that take a sequence of bits and output a polynomially bounded sequence of bits. In particular, we say that a function family $f_n : \{0,1\}^n \to \{0,1\}^m$, where $(m = m(n)$ is polynomially bounded in n) is computable by a CROW$[\log n]$ transducer if the map $f^{(i)} : x \mapsto (f(x))_i$ is[1] in CROW$[\log n]$ for every $i \in \{1, \ldots, m\}$.

We define functional composition of functional families with polynomially bounded outputs is the usual way: let, $f_n : \{0,1\}^n \to \{0,1\}^{m(n)}$ and and $f'_n : \{0,1\}^n \to \{0,1\}^{m'(n)}$ be two function families where $m(n), m'(n) = n^{O(1)}$. Define the function family $g = f' \circ f$ where, $g_n : \{0,1\}^n \to \{0,1\}^{m'(m(n))}$ maps $g_n : x \mapsto f'(f(x))$. The following lemma is what what makes the equivalence of LogDCFL and CROW$[\log n]$ useful for us:

Lemma 6. *If f, f' are functional families with polynomially bounded outputs computable in* CROW$[\log n]$ *then so is their composition $f' \circ f$.*

Proof. Given a string $x \in \{0,1\}^n$, each of the n^c bits of $f_n(x)$ can be computed in $e \log n$ time using n^d processors in via a CROW-PRAM. Here c, d, e are constants. Thus with n^{c+d} processors we can compute all bits of the outputof f Let, the corresponding constants for f' be c', d', e' Then the time to compute all the bits of $f' \circ f$ on input x is $(e + e') \log n$ and the number of processors required is $n^{c+d} + n^{(c+d)(c'+d')}$.

Moreover there are no concurrent writes in the composition since we can assign the same owners to memory locations as those in the computation of each bit of $f(x), f'(y)$ where $y = f(x)$.

---

[1] For a string $s \in \{0,1\}^n$, the i-th bit of s is represented by s_i for $i \in \{1, \ldots, n\}$.

3   Evaluating Monotone Circuits

Let us first define the notion of *distance* that we are using in this paper. If G is a directed acyclic graph and v is in a node in G then N_v represents the number of nodes in G which can reach v, i.e. nodes that have a directed path to v. We define distance $d_\#(u,v)$ between two nodes u and v as follows[2]:

$$d_\#(u,v) = \begin{cases} N_v - N_u, & \text{if there is directed path from } u \text{ to } v \\ \infty, & \text{otherwise.} \end{cases}$$

Remark 1. If there is a directed path from a node u to node v in G such that $u \neq v$ then $d_\#(u,w) > d_\#(v,w)$, for all nodes w reachable from both u and v.

Since G is a DAG and there is path from u to v, there will not be a path from v to u. Thus we know that $N_v > N_u$, which implies that $d_\#(u,w) > d_\#(v,w)$. If G is a DAG with only one sink node t then we define $G^{(i)}$ to be the subgraph of G induced by the vertices v such that $d_\#(v,t) \leq i$. This means $G^{(0)} = t$ (the sink node), and $G^{(n)} = G$.

Lemma 7. *For any connected DAG G, $G^{(i)}$ is a connected subgraph of G for all $i \in [n]$.*

Lemma 8. *In a bounded genus DAG with one sink, we can compute $d_\#(u,v)$ for any pair of nodes u and v in the graph in L.*

Above lemma follows simply from lemma 5. We can reverse the direction of all the edges of the single sink DAG so that the resulting graph becomes a single source DAG and then we can use Lemma 5 to check reachability.

In the following sections, we begin with a graph (planar or bounded genus) and divide it into multiple subgraphs. At any point, for any two nodes u and v of the graph, no matter whether u and v remain in the same subgraph or different after division, $d_\#(u,v)$ is always computed with respect to the initial graph.

3.1   Monotone Circuit Value Problem in Planar Graphs

In this section, we will prove that a monotone planar circuit such that all its inputs lie on at most k-faces, can be evaluated in $\mathsf{CROW}[\log k \log n]$. The high-level idea is as follows. We show that if G is the planar DAG that represents the given circuit such that it has at most k-input faces, we edge-partition G using a logspace procedure into three subgraphs L, M and R such that (i) L is a collection of DAGs each containing $(\frac{k}{2})$ input faces, (ii) R is a DAG that contains at most $(\frac{k}{2})$ input faces and, (iii) M is a 2-layered graph – a graph with two layers of nodes such that all the edges go from one layer to another. Then we recursively apply the same procedure on graphs L and R. We repeat this

---

[2] $d_\#(.,.)$ is a *quasimetric* that satisfies axioms of metric except symmetry. The fact that we can compute this in L may be of independent interest (see Lemma 8).

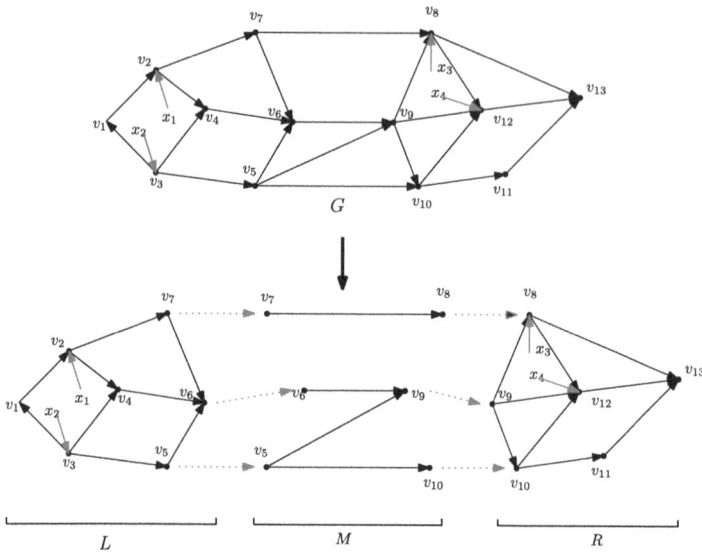

Fig. 2. Graph G with two input faces $f_1 = \{v_1, v_2, v_3, v_4\}$ and $f_2 = \{v_8, v_9, v_{12}\}$, edge partitioned into the graphs L, M and R such that f_1 is contained in L and f_2 is contained in R. Output from L (i.e. v_5, v_6, v_7) is input for M and output of M is input for R. After edge partition the outer face $\{v_8, v_9, v_{10}, v_{11}, v_{13}\}$ of R becomes a new input face in R, receiving inputs at v_8, v_9 and v_{10}.

procedure for $O(\log k)$ steps. In each recursive call, we decrease the number of input faces in each graph by half. Therefore, after $O(\log k)$ steps, we partition the graph into smaller graphs such that each graph that we obtain either has one-input face or is a 2-layered graph. A graph may be partitioned into multiple graphs in one recursive call, but the number of graphs that can be obtained is at most n (the number of nodes in G). A 2-layered CVP can be solved trivially in CROW[$\log n$] and we also know that one-input face MPCVP can be solved in CROW[$\log n$] [6]. Therefore, we can combine them and obtain CROW[$\log k \log n$] bound for evaluating G. Below we give a formal description of this idea.

First, we show how we edge-partition G into L, M and R. Note that when we partition G into many subgraphs, some faces of G may remain faces in one of the subgraphs, and some faces may get divided in this partition. More precisely, when we edge-partition the graph G, edges of a face f of G may appear in different subgraphs of G after the partition. Suppose that G is partitioned into subgraphs and H is one such subgraph. We say that a face of G is *contained* in H (or H *contains* a face of G), if H contains all the edges of the face. Similarly, we say that a face of G is incident on H if H contains some of the edges of that face. Note that a face of G that is contained in H is also incident on H, but a face of G that is incident on H might not be contained in H. Let F' be the set of input faces of G. Graph $G^{(n)}$ (which is nothing but G) contains all the input faces from F' and graph $G^{(0)}$ has a single node (sink t), i.e. contains no input faces of G. Thus, we can say that there exists a positive integer i such that $G^{(i)}$ contains at least $(\frac{k}{2} - 1)$ faces from F'. Let r be the largest integer such that $G^{(r)}$ contains at most $(\frac{k}{2} - 1)$ faces from F'. We define graphs L, M and R as follows.

- We define R to be the graph $G^{(r)}$.
- L is the graph induced by the nodes $V(L) := V(G) - V(G^{(r)})$.
- M is the graph induced by the edges: $E(M) := \{(u,v) \mid u \notin G^{(r)} \text{ and } v \in G^{(r)}\}$.

We also define two sets of vertices V_1 and V_2 as follows: $V_1 = \{u \mid (u,v) \in E(M)\}$ and $V_2 = \{v \mid (u,v) \in E(M)\}$. Note that $V_1 = V(L) \cap V(M)$, $V_2 = V(M) \cap V(R)$ and $V_1 \cup V_2 = V(M)$.

Lemma 9. *Edges of graph M form an edge-cut for graph G.*

From Lemma 9, we can say the edges of M form a cut for G, such that removing the edges of M from G divides it into graphs L and R. Faces in F' can now be divided into three classes: (i) faces which are contained in L, (ii) faces which are contained in R and (iii) faces which are incident to M. Faces of F', which are contained in L and R, remain input faces in L and R, respectively. However, some new input faces may appear on these graphs (for example, see Fig. 2). Let us first prove that R has a total of at most $(\frac{k}{2})$ input faces (note that this includes faces from F' as well as the new input faces that appear after partition). From Lemma 7, we know that R is a connected graph. We also know that R contains at most $(\frac{k}{2} - 1)$ faces from F'. Note that the edges of M are incident on the outer face of R. Thus sink nodes of M – nodes in the set V_2 are source nodes of R, and all these nodes lie on the outer face of R. Therefore the outer face of R will also become an input face in R. Therefore, we can conclude that R has at most $(\frac{k}{2})$ input faces. Let H be the graph defined as $H = M \cup R$. We will now prove that each graph in L also has at most $(\frac{k}{2})$ input faces. Before that, we prove the following lemma.

Lemma 10. *Graph $G^{(r+1)}$ is a subgraph of the graph H.*

From Lemma 10, we can say that H contains more than $(\frac{k}{2} - 1)$ input faces of G. Since L and $M \cup R$ are edge-disjoint subgraphs, there are at most $\frac{k}{2}$ faces from F' incident on L. In Lemma 9 we proved that there are no edges in G that go from R to L. Thus any source of L is already a source in G and so belongs to a face of F'. We can say that the (number of input faces in L) $\leq$ (number of faces from F' incident to L). We know that the number faces from F' incident to L are at most $\frac{k}{2}$. Therefore the number of input faces in L are at most $\frac{k}{2}$.

Now that we have divided the graph G into three subgraphs L, M and R, such that L and R are planar DAG with at most $\frac{k}{2}$ input faces and M is a 2-layered graph. We can recursively apply the same procedure in L and R. The only problem is that, unlike R, L may have many sink nodes and the algorithm that we have described works with graphs that have only one sink node. Assume that

$t_1, t_2, \ldots$ are sink nodes in L. Using Lemma 5 we can obtain graphs $L_1, L_2, \ldots$ such that L_i is the graph induced by the nodes that can reach t_i, in logspace. Then we can apply the same algorithm in each L_i recursively. At the end all the graphs that we obtain are one input face planar DAG or 2-layered. Since there are k input faces in the graph initially, the total number of graphs that we obtain are k. We know that we can evaluate 2-layered circuits in $\mathsf{AC}^0 (\subseteq \mathsf{LogDCFL})$ and one input face MPCVP in $\mathsf{LogDCFL} = \mathsf{CROW}[\log n]$. Notice that these circuits cannot be evaluated independently in parallel because output from one circuit may be an input for another. Therefore we will have to evaluate them sequentially. In order to evaluate the entire circuit represented by G, we need to sequentially compose the evaluations these k circuit. This can be done in $\mathsf{CROW}[k \log n]$ using Lemma 6.

3.2   Ensuring Logarithmic Dependence on the Number of Input Faces

We will use techniques from Delcher and Kosaraju [9] to reduce the parallel running time of our routines on k, the number of input faces in MPCVP from $O(k)$ to $O(\log k)$.

We need to establish some notation that follows [9] not in letter but in spirit. We start by extending the scope of monotone circuits from Boolean to trivalued inputs $\{0, 1, \boxed{?}\}$ where $\boxed{?}$ is supposed represent a variable. Thus we have $\boxed{?} \wedge 0 = 0 \wedge \boxed{?} = 0$, $\boxed{?} \wedge 1 = 1 \wedge \boxed{?} = \boxed{?}$; dually $\boxed{?} \vee 1 = 1 \vee \boxed{?} = 1$, $\boxed{?} \vee 0 = 0 \vee \boxed{?} = \boxed{?}$ and finally $\boxed{?} \wedge \boxed{?} = \boxed{?} = \boxed{?} \vee \boxed{?}$. In a sense $\boxed{?}$ can be thought to be a value intermediate between $0, 1$ and so that mononotonicity of $\wedge, \vee$ is preserved.

Delcher and Kosaraju [9] show how to reduce the evaluation of single input face MPCVP with trivalued inputs to the usual Boolean MPCVP.

Lemma 11 ([9]). *A single input face* MPCVP *with trivalued inputs* $\{0, 1, \boxed{?}\}$ *can be* L-*reduced to the evaluation of a single-face* MPCVP *instance with inputs from* $\{0, 1\}$.

From the above Lemma 11, the evaluation of a single input face MPCVP instance with trivalued inputs from $\{0, 1, \boxed{?}\}$ can be done in $\mathsf{CROW}[\log n]$ using [6] (which in turn uses [3] as the crucial subroutine) and [12] for the equivalence[3] between $\mathsf{LogDCFL}$ and $\mathsf{CROW}[\log n]$.

We generalise the notion of evaluation of monotone circuits with $\{0, 1\}$ inputs to what we call *distillation* of monotone circuits with trivalued inputs from $\{0, 1, \boxed{?}\}$. The definition is such that on Boolean inputs distillation is same a evaluation. We describe this in detail below. Then we show how to distill a trivalued k-input MPCVP instance in parallel time $O(\log k \log n)$ implying a similar time bound for evaluating a Boolean MPCVP circuit.

---

[3] Though of course, [9] used the bound of LogCFL from [11] for single input face monotone upward planar CVP leading to an inferior running time.

Let C be a monotone planar circuit where the $\boxed{?}$ inputs are constrained to lie on a single face while there exist at most k faces (including the face containing $\boxed{?}$-inputs) that contain all the Boolean inputs. We refer to these trivalued MPCVP instances C, x as $\boxed{?}$-SI-k-MPCVP (i.e. single-$\boxed{?}$-input k-input face MPCVP instances). If in addition, we also assume that the circuit outputs lie on a common face. then we call such associated $\boxed{?}$-SI-k-MPCVP instances as $\boxed{?}$-SISO-k-MPCVP (short for single-$\boxed{?}$-input face, k-input-face, single output face MPCVP instances. Note that the k-input faces include one $\boxed{?}$-input face.

Consider the evaluation of a $\boxed{?}$-SISO-k-MPCVP instance C, x to label all its gates with $\{0, 1, \boxed{?}\}$. We delete all wires incident on gates evaluating to $\{0, 1\}$ and identify such a gate with the Boolean value attained by it. Let the collection of the constant gates (with no incident wires) and the remaining gates that evaluate to $\boxed{?}$ and wires be C'. We first observe that C' is a subcircuit of C. All of its inputs that have outdegree more than zero are $\boxed{?}$-inputs of C; the single input face in C' contains only $\boxed{?}$-inputs; and a gate evaluating to $\boxed{?}$ may lose all gates that it feeds into, leading to the creation of new sources and thus new input faces. Let x' be the $\boxed{?}$-inputs in x.

Proposition 1. *If C, x is a $\boxed{?}$-SISO-k-MPCVP instance, C', x' is a $\boxed{?}$-SI-1-MPCVP instance and for every g' gate of C' there is a natural injective map from the gates and inputs of C', x' and the gates and inputs of C, x.*

Further, using the maps, $g' \mapsto g, x' \mapsto x$ the following holds:

Proposition 2. *Under the premise that C, x is a $\boxed{?}$-SISO-k-MPCVP instance, let b' be a Boolean assignment of values to x' and let b be the corresponding Boolean assignment to x. Under the assignment b' to x' each gate g' of C' gets the same value as g under b to x.*

To see this: let g be a $\vee$-gate (the case of $\wedge$-gate is dual) evaluating to $\boxed{?}$, with a gate h feeding into it where h evaluates to a value $\beta \in \{0, 1\}$. If $\beta = 1$ then g would evaluate to 1 also, contradicting that g evaluates to $\boxed{?}$. On the other hand if $\beta = 0$ then the value of g does not depend on h thus the deletion of the wire from h to g in C will not alter the value that g' gets under b' in C' as opposed to the value that g got under b in C.

The *distillate* of the instance C, x is an instance $\tilde{C}, \tilde{x}$. $\tilde{C}$ consists of those gates of C' that are either output gates of C' (including constant gates) or can reach such a gate in C'. $\tilde{x}$ is the set of those values of x' that are incident on gates in $\tilde{C}$. The following two are immediate because we remove all sink gates of C' that do not lie on the output face of C':

Proposition 3. *If C, x is a $\boxed{?}$-SISO-k-MPCVP instance, $\tilde{C}, \tilde{x}$ is a $\boxed{?}$-SISO-1-MPCVP instance with an injective map from the gates of $\tilde{C}$ to C and inputs $\tilde{x}$ to x.*

Proposition 4. *Under the premise that C, x is a $\boxed{?}$-SISO-k-MPCVP instance, let $\tilde{b}$ be a Boolean assignment of values to $\tilde{x}$ and let b be any corresponding Boolean assignment to x (where the $\boxed{?}$-values of x that are absent from $\tilde{x}$ are substituted arbitrarily). Under the assigment $\tilde{b}$ to $\tilde{x}$ each gate $\tilde{g}$ of $\tilde{C}$ gets the same value as gate g of C under b.*

The following is simple but crucial.

Lemma 12. *The distillate $\tilde{C}, \tilde{x}$ of a $\boxed{?}$-SISO-1-MPCVP instance C, x on n-gates can be obtained in $\mathsf{CROW}[\log n]$.*

We show how to obtain the distillate of a k-input face graph using the previous Lemma as the base case.

Lemma 13. *The distillate $\tilde{C}, \tilde{x}$ of a $\boxed{?}$-SISO-k-MPCVP instance C, x on n-gates can be obtained in $\mathsf{CROW}[\log k \log n]$.*

Observing that the distillate of a k-MPCVP instance (with no $\boxed{?}$-inputs) contains only gates with Boolean values associated with them and in particular the output gate has a value associated with it, we get:

Corollary 4. *A k-MPCVP instance can be evaluated in $\mathsf{CROW}[\log k \log n]$.*

3.3   Monotone Circuit Value Problem in Bounded Genus Graphs

In this section, we will prove that a monotone circuit embedded on a g genus surface such that all the inputs lie on at most k-faces, can be evaluated by CROW-PRAMS in $O(g \log (g + k) \log n)$ time using $(gkn)^{O(1)}$ many processors. The approach that we use in this section is similar to the one that we used in the previous section. Given a DAG G representing the monotone circuit embedded on a g-genus surface, we divide it into three subgraphs L, M and R such that L has genus at most $(g-1)$, M is a 2-layered graph, and R is a planar graph with at most $(g + \frac{3k-1}{2})$ input faces. Now we recursively apply the same procedure with L. In each recursive step, the genus of the resulting graphs decreases at least by one. Thus after g many iterations, we divide the graph G into subgraphs such that each of them is either a planar graph with at most $(g + \frac{3k-1}{2})$ input faces or a 2-layered graph. We know that we can solve the circuit value problem represented by a 2-layered graph trivially in $\mathsf{CROW}[\log n]$. From Sect. 3.1, we know that we can solve circuit value problem represented by a planar DAG with $(g + \frac{3k-1}{2})$-input faces in $\mathsf{CROW}[\log (g + k) \log n]$. Thus by combining them, we obtain a $\mathsf{CROW}[g \log (g + k) \log n]$ algorithm for evaluating G.

4   Conclusion and Open Questions

We show that the LogDCFL bound of [6] for single input face MCVP, that builds on [3,19], can be extended to monotone circuits with constantly many input faces and embedded on a surface of bounded genus. Further, for a monotone circuit with arbitrarily many input faces that is embedded on a surface of poly-logarithmic genus we show an NC bound.

We cannot hope to improve this bound significantly i.e. with polylogarithmic dependence on g instead of linear because of the following reasons. Suppose we are given a general monotone circuit as an input. It is trivial to obtain an embedding of such a circuit on a surface of $n^{O(1)}$ genus. Moreover, directed reachability in general graphs can also be solved in NL $\subseteq$ NC. Thus an algorithm with a run time depending polylogarithmic in g would result in an NC bound for MCVP. This would imply P $\subseteq$ NC because MCVP is P-complete [14].

Extending our results to circuits embedded on non-orientable surfaces is still open. It is known from [6] that single input face MPCVP is L-hard. Improving this lower bound to LogDCFL is another open question.

References

1. Allender, E., Barrington, D.A.M., Chakraborty, T., Datta, S., Roy, S.: Planar and grid graph reachability problems. Theory Comput. Syst. **45**(4), 675–723 (2009)
2. Allender, E., Mahajan, M.: The complexity of planarity testing. Inf. Comput. **189**(1), 117–134 (2004)
3. Barrington, D.A.M., Lu, C., Miltersen, P.B., Skyum, S.: On monotone planar circuits. In: Proceedings of the 14th Annual IEEE Conference on Computational Complexity, Atlanta, Georgia, USA, 4–6 May 1999, p. 24 (1999)
4. Bourke, C., Tewari, R., Vinodchandran, N.V.: Directed planar reachability is in unambiguous log-space. ACM Trans. Comput. Theory **1**(1), 4:1–4:17 (2009)
5. von Braunmühl, B., Cook, S.A., Mehlhorn, K., Verbeek, R.: The recognition of deterministic CFL's in small time and space. Inf. Control **56**(1/2), 34–51 (1983)
6. Chakraborty, T., Datta, S.: One-input-face MPCVP is hard for L, But in LogD-CFL. In: Arun-Kumar, S., Garg, N. (eds.) FSTTCS 2006. LNCS, vol. 4337, pp. 57–68. Springer, Heidelberg (2006). https://doi.org/10.1007/11944836_8
7. Cook, S.A.: Deterministic CFL's are accepted simultaneously in polynomial time and log squared space. In: STOC 1979, 30 April–2 May 1979, Atlanta, Georgia, USA, pp. 338–345 (1979)
8. Datta, S., Prakriya, G.: Planarity testing revisited. In: Ogihara, M., Tarui, J. (eds.) TAMC 2011. LNCS, vol. 6648, pp. 540–551. Springer, Heidelberg (2011). https://doi.org/10.1007/978-3-642-20877-5_52
9. Delcher, A.L., Kosaraju, S.R.: An NC algorithm for evaluating monotone planar circuits. SIAM J. Comput. **24**(2), 369–375 (1995)
10. Diestel, R.: Graph Theory, 5th edn. Springer, Heidelberg (2017). https://doi.org/10.1007/978-3-662-53622-3
11. Dymond, P.W., Cook, S.A.: Complexity theory of parallel time and hardware. Inf. Comput. **80**(3), 205–226 (1989)
12. Dymond, P.W., Ruzzo, W.L.: Parallel rams with owned global memory and deterministic context-free language recognition. J. ACM **47**(1), 16–45 (2000)
13. Elberfeld, M., Kawarabayashi, K.: Embedding and canonizing graphs of bounded genus in logspace. In: STOC 2014, New York, NY, USA, 31 May–03 June 2014, pp. 383–392 (2014)
14. Goldschlager, L.M.: The monotone and planar circuit value problems are log space complete for P. SIGACT News **9**(2), 25–29 (1977)

15. Goldschlager, L.M.: A space efficient algorithm for the monotone planar circuit value problem. Inf. Process. Lett. **10**(1), 25–27 (1980)
16. Gupta, C., Sharma, V.R., Tewari, R.: Reachability in $O(\log n)$ genus graphs is in unambiguous logspace. In: STACS, pp. 34:1–34:13 (2019)
17. Karp, R.M., Ramachandran, V.: Parallel algorithms for shared-memory machines. In: Handbook of Theoretical Computer Science, Volume A: Algorithms and Complexity, pp. 869–942. Elsevier and MIT Press (1990)
18. Kyncl, J., Vyskocil, T.: Logspace reduction of directed reachability for bounded genus graphs to the planar case. ACM Trans. Comput. Theory **1**(3), 8:1–8:11 (2010)
19. Limaye, N., Mahajan, M., Sarma, J.: Upper bounds for monotone planar circuit value and variants. Comput. Complex. **18**(3), 377–412 (2009)
20. McKenzie, P., Reinhardt, K., Vinay, V.: Circuits and context-free languages. In: Asano, T., Imai, H., Lee, D.T., Nakano, S., Tokuyama, T. (eds.) COCOON 1999. LNCS, vol. 1627, pp. 194–203. Springer, Heidelberg (1999). https://doi.org/10.1007/3-540-48686-0_19
21. Ramachandran, V., Yang, H.: An efficient parallel algorithm for the general planar monotone circuit value problem. SIAM J. Comput. **25**(2), 312–339 (1996)
22. Ramachandran, V., Yang, H.: An efficient parallel algorithm for the layered planar monotone circuit value problem. Algorithmica **18**(3), 384–404 (1997)
23. Ruzzo, W.L.: On uniform circuit complexity. J. Comput. Syst. Sci. **22**(3), 365–383 (1981)
24. Stockmeyer, L.J., Vishkin, U.: Simulation of parallel random access machines by circuits. SIAM J. Comput. **13**(2), 409–422 (1984)
25. Stolee, D., Bourke, C., Vinodchandran, N.V.: A log-space algorithm for reachability in planar acyclic digraphs with few sources. In: CCC 2010, Cambridge, Massachusetts, USA, 9–12 June 2010, pp. 131–138 (2010)
26. Stolee, D., Vinodchandran, N.V.: Space-efficient algorithms for reachability in surface-embedded graphs. In: CCC 2012, Porto, Portugal, 26–29 June 2012, pp. 326–333 (2012)
27. Sudborough, I.H.: On the tape complexity of deterministic context-free languages. J. ACM **25**(3), 405–414 (1978)
28. Tewari, R., Vinodchandran, N.V.: Green's theorem and isolation in planar graphs. Inf. Comput. **215**, 1–7 (2012)
29. Thierauf, T., Wagner, F.: The isomorphism problem for planar 3-connected graphs is in unambiguous logspace. Theory Comput. Syst. **47**(3), 655–673 (2010)
30. Yang, H.: An NC algorithm for the general planar monotone circuit value problem. In: SPDP 1991, 2–5 December 1991, Dallas, Texas, USA, pp. 196–203 (1991)

Ranking and Unranking of the Planar Embeddings of a Planar Graph

Giuseppe Di Battista[1] [ID], Fabrizio Grosso[1,2(✉)] [ID], Giulia Maragno[1],
and Maurizio Patrignani[1] [ID]

[1] University of Roma Tre, Rome, Italy
{giuseppe.dibattista,fabrizio.grosso,maurizio.patrignani}@uniroma3.it,
giu.maragno@stud.uniroma3.it
[2] CeDiPa - University of Perugia, Perugia, Italy
fabrizio.grosso@unipg.it

Abstract. Let $\mathcal{G}$ be the set of all the planar embeddings of a (not necessarily connected) n-vertex graph G. We present a bijection Φ from $\mathcal{G}$ to the natural numbers in the interval $[0 \dots |\mathcal{G}| - 1]$. Given a planar embedding $\mathcal{E}$ of G, we show that $\Phi(\mathcal{E})$ can be decomposed into a sequence of $O(n)$ natural numbers each describing a specific feature of $\mathcal{E}$. The function Φ, which is a ranking function for $\mathcal{G}$, can be computed in $O(n)$ time, while its inverse unranking function Φ^{-1} can be computed in $O(n\alpha(n))$ time. The results of this paper can be practically applied to uniformly generating the planar embeddings of a graph G at random or to enumerating such embeddings with an amortized constant delay. Additionally, they can be used for counting, enumerating, or uniformly generating constrained planar embeddings of G.

Keywords: Planarity · Graph Drawing · Ranking · Prüfer sequences

1 Introduction

The space of the planar embeddings of planar graphs is one of the favorite playgrounds of Graph Drawing and of Graph Algorithms. For instance, several algorithms (e.g., [2,6,17,20,26,27]) delve into this space to find graph drawings that are optimal according to some metrics, there are problems that are polynomially solvable (e.g., [7,18]) if a planar embedding is given and that become NP-complete when the choice of any planar embedding is allowed (e.g., [18,19]), and there are many algorithms that look for planar drawings whose underlying planar embedding must satisfy some constraints (e.g., [4,29,32]).

This space has been studied from various perspectives. As an example, all the planar embeddings of biconnected planar graphs can be represented by a data structure called SPQR-tree [15,16], and ranking and unranking functions were given for such graphs relying on PQ-trees [8,22] or segment graphs [23]. Also, the results of Cai [10] (see also [31]), which are extremely important in the perspective of our paper, measure the "extension" of such a space, providing the exact number of planar embeddings of a given planar graph.

S. Nakano and M. Xiao (Eds.): WALCOM 2025, LNCS 15411, pp. 143–159, 2025.
https://doi.org/10.1007/978-981-96-2845-2_10

Let G be a, not-necessarily connected, planar graph. In this paper we construct a bijective function Φ that maps a planar embedding of G to a natural number (*ranking* [24]) and such that its inverse maps a natural number to a planar embedding of G (*unranking* [24]). Both Φ and its inverse are efficiently computable. Also, given a planar embedding $\mathcal{E}$ of G, we show that $\Phi(\mathcal{E})$ can be decomposed into a sequence of $O(n)$ natural numbers each describing a specific feature of $\mathcal{E}$. These results can be practically applied to uniformly generating planar embeddings of G at random or enumerating such embeddings with an amortized constant delay. Additionally, they can be utilized for counting, enumerating, or uniformly generating constrained planar embeddings of G.

Formally, the paper is devoted to prove the following Theorem 1. In fact, Theorem 1 and the simple Lemma 1 that is shown in Sect. 2 imply the existence of a ranking and an unranking function that can be efficiently computed.

Let G be a planar graph. Consider a decomposition of G into its connected components and a decomposition of each connected component into its biconnected components and cut-vertices. For a cut-vertex v_ξ of G let δ_{v_ξ} be the degree of v_ξ in G, let $b(v_\xi)$ be the number of biconnected components containing v_ξ, and let $\delta_{v_\xi,\sigma}$ be the degree of v_ξ restricted to the biconnected component B_σ. Finally, consider the SPQR-trees representing the decompositions of the biconnected components into triconnected components. For a parallel P_ξ of triconnected components of G, let $\delta(P_\xi)$ be the number of its parallel components. Denote by $f(G)$ the number of faces of a planar embedding of G. Notice that all the planar embeddings of G have the same number of faces.

Theorem 1. *Let G be an n-vertex planar graph with connected components $G_1,\ldots,G_t$. Let $v_1,\ldots,v_w$ be the cut-vertices of G. Let $P_1,\ldots,P_y$ be the parallels of triconnected components of the SPQR-trees of all the biconnected components of G. Let $R_1,\ldots,R_z$ be the rigid triconnected components of the SPQR-trees of all the biconnected components of G. There exists a bijective function Φ whose domain is the set of the planar embeddings of G on the sphere and whose codomain is the set of tuples of natural numbers*

$$\langle a_1,\ldots,a_{t-1},b_1,\ldots,b_t,c_{1,1},\ldots,c_{1,b(v_1)},\ldots,c_{w,1},\ldots,c_{w,b(v_w)},$$
$$d_{1,1},\ldots,d_{1,b(v_1)-2},\ldots,d_{w,1},\ldots,d_{w,b(v_w)-2},p_1,\ldots,p_y,r_1,\ldots,r_z\rangle$$

where

- $a_\xi \in [0 \cdots \sum_{h=1}^{t}(f(G_h)-1)]$, *for* $\xi = 1,\ldots,t-1$;
- $b_\xi \in [0\ldots f(G_h)-1]$, *for* $\xi = 1,\ldots,t$;
- $c_{\xi,\sigma} \in [0\ldots \delta_{v_\xi,\sigma}-1]$, *for* $\xi = 1,\ldots,w$ *and* $\sigma = 1,\ldots,b(v_\xi)$;
- $d_{\xi,\sigma} \in [0\ldots \delta_{v_\xi}-\sigma-1]$, *for* $\xi = 1,\ldots,w$ *and* $\sigma = 1,\ldots,b(v_\xi)-2$;
- $p_\xi \in [0\ldots (\delta(P_\xi)-1)!-1]$, *for* $\xi = 1,\ldots,y$;
- $r_\xi \in [0,1]$, *for* $\xi = 1,\ldots,z$.

The number of elements of a tuple is $O(n)$. The function Φ can be computed in $O(n)$ time, while Φ^{-1} can be computed in $O(n\alpha(n))$ time, where α is the inverse Ackermann function.

The time bounds discussed in this paper assume the RAM model with uniform costs, as it is usual in the literature whenever the involved numbers can be represented with $O(\log n)$ bits [1]. This same model is sometimes assumed even when ranking permutations [28], a task that implies representing numbers up to $O(n!)$, hence using $O(\log(n!)) = O(n \log n)$ bits. Indeed, many libraries could be used to overcome this limitation with a little increase in cost per operation.

Theorem 1 is proved in three steps. First, in Theorem 2 we rank and unrank the planar embeddings of a biconnected component of G. This is simply achieved by exploiting the SPQR-tree associated with the biconnected component and by using a technique (see [28]) that allows to rank or unrank a permutation in linear time. More details are given in [14]. Second, in Sect. 3 we rank and unrank the arrangements of the planar embeddings of the biconnected components sharing a specific cut-vertex. To do that, we develop a new technique that associates a natural number to the position of a component around the cut-vertex. Finally, in Sect. 4, we rank and unrank the nesting of the connected components of G. To do that, we devise a variation of the Prüfer sequences, targeted to describe such a nesting. We believe this technique can have other applications. Observe that variations of the Prüfer sequences have already been studied for different problems. For instance, in [21] a variation is studied to represent the arrangements of the biconnected components into a block-cutvertex tree. Note that, although related to this paper, the cited results in [10] do not introduce efficient techniques that, given a planar embedding of G, can associate a natural number with the embedding and vice versa. We also observe that our techniques require identifying several components of G. The identification techniques that we adopt are partly introduced in Sect. 2 and partly shown where they are first used. An example that illustrates how Theorem 1 works is shown in Fig. 1. Some proofs have been omitted and can be found in the full version [14].

2   Preliminaries

We assume familiarity with the basic concepts of graphs and planarity and, therefore, we report here only the definitions that will be used extensively in this work. For further information, the reader could refer to [13,33]. We assume all the graphs to be undirected.

Connectivity. The structure of a graph $G = (V, E)$ can be analyzed in terms of connectivity of its vertices. G is *connected* if for each $u, v \in V$ with $u \neq v$ there exists a path in G connecting u and v. A vertex whose removal makes G non-connected is a *cut-vertex*. A non-connected graph can be subdivided into maximal connected subgraphs, called *connected components*. In general, a graph is k-connected if for each pair of distinct vertices $u, v \in V$ there exist k vertex-disjoint paths connecting u and v. 2-connected and 3-connected graphs are also called *biconnected* and *triconnected*, respectively. It is possible to subdivide a connected (biconnected) graph into maximal biconnected (triconnected) subgraphs, called biconnected (triconnected) components. The *block-cutvertex tree* of a connected graph shows the incidence relationships between its *blocks* (biconnected

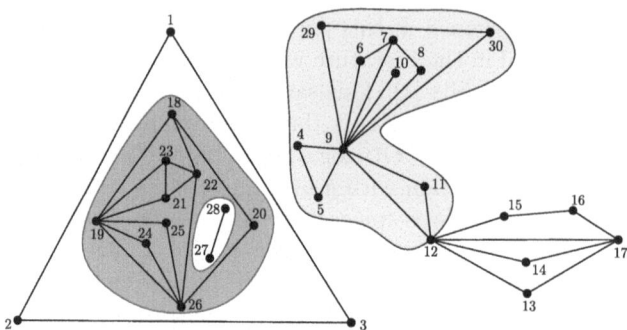

Fig. 1. A graph drawn according to its $754,705,812,645^{th}$ planar embedding on the sphere, over the 19,716,667,342,848 possible, corresponding to the tuple $\langle 0, 11, 1, 0, 1, 0, 0, 0, 1, 0, 1, 1, 0, 1, 2, 1, 6, 1, 4, 5, 1, 0 \rangle$, of Theorem 1. The numbers in the tuple correspond to the symbols used in the theorem as follows: $\langle a_1, a_2, a_3, b_1, b_2, b_3, b_4, c_{1,1}, c_{1,2}, c_{1,3}, c_{1,4}, c_{1,5}, c_{2,1}, c_{2,2}, d_{1,1}, d_{1,2}, d_{1,3}, p_1, p_2, p_3, \ r_1, r_2 \rangle$. In particular, each number of the tuple, as described in Lemma 1, is upper bounded by $\langle 17, 17, 17, 2, 9, 8, 1, 2, 3, 1, 2, 2, 2, 4, 9, 8, 7, 2, 6, 6, 2, 2 \rangle$. The numbers in red, blue, and green correspond to the embedding of the red biconnected component, of the biconnected components around the cut-vertex 9, and of the connected components, respectively, and are described in Fig. 2, Figs. 4 and 3, and Fig. 5. (Color figure online)

components) and its cut-vertices. A planar biconnected graph can be decomposed into its triconnected components. This decomposition is represented by its SPQR-tree. Details on the block-cutvertex tree and SPQR-tree decompositions can be found in [14].

Planarity and Embeddings. Given a graph $G = (V, E)$, a *drawing of G on the sphere* is a mapping of each vertex in V to a point of the sphere and of each edge (u, v) to a Jordan curve on the sphere from the point corresponding to u to the point corresponding to v. A drawing on the sphere is *non-intersecting* (or *planar*, with a slight bending of the terminology) if no two edges intersect except at common endpoints. A graph is *planar* if it admits a planar drawing on the sphere. A drawing Γ of a graph on the sphere determines a subdivision of the sphere into connected regions, called *faces*, and a circular ordering of the edges incident to each vertex, called *rotation system*. Visiting the (not necessarily connected) border of a face f of Γ in such a way to keep f to the left, determines a set of circular lists of vertices. Such a set is the *boundary* of f. Two drawings on the sphere are equivalent if they have the same rotation system and the same face boundaries. An *embedding* is an equivalence class of planar drawings on the sphere. Given a drawing Γ of G on the sphere and a point p internal to a face f, the *stereographic projection* [25] of Γ on the plane with p as center of projection produces a *drawing on the plane Γ'* of G, where f is the unbounded face, called *outer face*. Namely, let P be the plane tangent to the sphere at the point diametrically opposed to p. Each point $x \neq p$ of the sphere is mapped to the point x' of P that lies on the line through p and x. Observe that if no two edges of Γ cross, i.e., if Γ is planar, also no two edges of Γ' cross (we use the same term *planar* also for Γ'). All drawings on the plane obtained by projecting Γ preserve the same embedding as Γ, irrespective of the face chosen as the outer face. An *embedding on the plane* is the equivalence class of the drawings on the plane that share the same embedding and the same outer face. In the following, all the projections on the plane will be stereographic.

Identifiers. Given an n-vertex graph G we have *vertex identifiers* that uniquely associate each vertex with an integer in $[1 \ldots n]$. We also have *edge identifiers*: an edge (u, v) is identified by the pair of identifiers of u and of v in ascending order. Hence, often when referring to an edge (u, v) we assume that the identifier of u is smaller than the identifier of v. The edges of the skeletons of SPQR-trees are identified in the same way. All SPQR-trees that we consider are rooted at a reference edge, which is the edge with the minimum identifier.

Prüfer sequences. An example of ranking algorithm, that will be used later in this work, is given by the *Prüfer sequence*. In [30] Prüfer demonstrated a bijection between unrooted unordered trees of n nodes, with nodes labeled by distinct values in $[1 \ldots n]$, and tuples of length $n - 2$ of values in $[1 \ldots n]$. The result was used to demonstrate the Cayley's formula [12], which states that the number of unrooted labeled trees of n nodes is n^{n-2}. An explanation of the algorithm is given in [14].

The following lemma establishes a bijection between natural numbers and tuples of natural numbers whose elements have bounded values, generalizing to tuples a result that was already known for decision trees [5, Theorem 3.2]. The proof can be found in [14].

Lemma 1. *Let P be the set of all tuples $\langle b_1, b_2, \ldots, b_n \rangle$ such that $n \geq 1$, $0 \leq b_i < B_i$ ($i = 1 \ldots n$), and such that b_i and B_i are natural numbers. There exists a bijective function ψ with domain P and codomain the natural numbers in $[0 \ldots (\prod_{i=1}^{n} B_i) - 1]$. Both ψ and its inverse can be computed in $O(n)$ time.*

Biconnected Graphs. Let μ be a node of the SPQR-tree $\mathcal{T}^T$ of a biconnected planar graph G rooted at the edge with minimum identifier. Denote by $d(\mu)$ the depth of μ in $\mathcal{T}^T$ and by $e(\mu)$ the id of the edge with the minimum identifier in the pertinent graph of μ. We assign to μ the identifier $\langle d(\mu), e(\mu) \rangle$. We call *conventional order* the order of the nodes of $\mathcal{T}^T$ induced by their identifiers.

Theorem 2. *Let G be an n-vertex biconnected planar graph and denote by $\nu_1, \ldots, \nu_y$ and by $\mu_1, \ldots, \mu_z$ the P-nodes and the R-nodes of the SPQR-tree $\mathcal{T}^T$ of G in conventional order, respectively. There exists a bijective function χ whose domain is the set of the planar embeddings of G and whose codomain is the set of tuples of natural numbers $\langle p_1, \ldots, p_y, r_1, \ldots, r_z \rangle$, where $p_\xi \in [0 \ldots (\delta(\nu_\xi) - 1)! - 1]$ for $\xi = 1, \ldots, y$ and $r_\xi \in [0 \ldots 1]$ for $\xi = 1, \ldots, z$. Both the function χ and its inverse χ^{-1} can be computed in $O(n)$ time.*

Each number p_i (resp. r_j) in the tuple of Theorem 2, whose proof can be found in [14], represents a chosen permutation (resp. flip) in the possible embeddings of the corresponding P-node (resp. R-node). An example is given in Fig. 2.

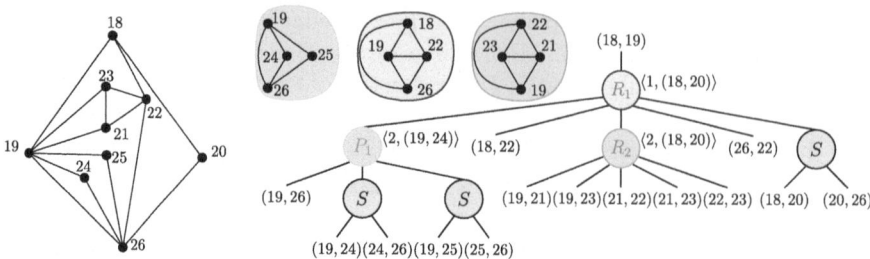

Fig. 2. The biconnected component depicted in red in Fig. 1, drawn according to its 21st embedding over the 23 possible. The embedding corresponds to the tuple $\langle 5, 1, 0 \rangle$ according to the variables $\langle p_1, r_1, r_2 \rangle$ of Theorem 2. In the SPQR-tree, each Q-node is represented by the corresponding edge. The skeletons of the R- and P-nodes are given in the bubbles with the corresponding colors. $p_1 = 5$ indicates that the children of the corresponding P-node are embedded according to the permutation number 5, among the 6 possible. $r_1 = 1$ and $r_2 = 0$ show the choices for the embeddings of the corresponding R-nodes. The P- and R-nodes of the SPQR-tree are equipped with their identifiers. In the context of the whole graph in Fig. 1 the values of the tuple refer to $\langle p_3, r_1, r_2 \rangle$ instead of $\langle p_1, r_1, r_2 \rangle$. (Color figure online)

3   Simply Connected Graphs

Let G be a connected graph and let $B_1, B_2, \ldots, B_x$ be the set of its biconnected components. Assume that each biconnected component B_j has a fixed planar embedding on the sphere $\mathcal{E}_j^\circ$. We describe a bijection from the embeddings of G on the sphere that preserve the embedding $\mathcal{E}_j^\circ$ of each block B_j to suitable sequences of natural numbers. Each sequence is composed of several sub-sequences, where each sub-sequence describes the arrangement choices around each cut-vertex of G independently. Namely, for each cut-vertex v, consider the $b(v)$ biconnected components of G that share v and rename them $B_1, B_2, \ldots, B_{b(v)}$. Cai proved in [10, Lemma 13] the following formula for the number E_v of the distinct planar arrangements on the sphere of the embeddings $\mathcal{E}_j^\circ$ around v, where δ_v is the degree of v and $\delta_{v,j}$ is the degree of v restricted to the biconnected component B_j.

$$E_v = \prod_{j=1}^{b(v)} \delta_{v,j} \prod_{j=1}^{b(v)-2} (\delta_v - j) \tag{1}$$

First, we state the following theorem, which is a counterpart of Lemma 13 of [10] and is the main result of this section. We then provide a procedure to transform a tuple into an embedding of the graph (Sect. 3.1) and vice versa (Sect. 3.2). The correctness of the procedures is discussed in [14].

Theorem 3. *Let G be a connected planar graph and let v be a cut-vertex of G. Let $B_1, B_2, \ldots, B_{b(v)}$ be the biconnected components of G containing v and let $\mathcal{E}_1^\circ, \mathcal{E}_2^\circ, \ldots, \mathcal{E}_{b(v)}^\circ$ be their planar embeddings on the sphere. There exists a bijection*

φ_v whose domain is the set of planar embeddings on the sphere of the subgraph $G(v) = B_1 \cup B_2 \cup \ldots B_{b(v)}$ preserving $\mathcal{E}_1^\circ, \mathcal{E}_2^\circ, \ldots, \mathcal{E}_{b(v)}^\circ$ and whose codomain is a sequence of natural numbers $\langle c_1, \ldots, c_{b(v)}, d_1, \ldots, d_{b(v)-2} \rangle$, where $c_j \in [0 \ldots \delta_{v,j} - 1]$, for $j = 1, \ldots, b(v)$ and $d_j \in [0 \ldots \delta_v - j - 1]$, for $j = 1, \ldots, b(v) - 2$. The function φ_v can be computed in $O(n)$ time, and its inverse φ_v^{-1} can be computed in $O(n\alpha(n))$ time, where α is the inverse of the Ackermann function.

Observe that the product of the ranges of the elements in the tuple of Theorem 3 $\prod_{i=1}^k \delta_{v,i} \prod_{h=1}^{k-2} \delta_v - h$ is exactly E_v and, therefore, the bijection ψ of Lemma 1 applied to $\langle c_1, \ldots, c_{b(v)}, d_1, \ldots, d_{b(v)-2} \rangle$ produces a number in the interval $[0 \ldots E_v - 1]$. In Fig. 1 is shown an example of the result of Theorem 3, where the rank of the embedding around the cut-vertex 9 (in blue) is 3653.

Cai also proved [10, Theorem 4] that the number E_G of distinct planar embeddings on the sphere of G, restricted to those that preserve the embedding $\mathcal{E}_j$ of each block B_j, is given by the product of the above numbers for all the w cut-vertices of G, i.e., $E_G = \prod_{i=1}^w E_{v_i}$. We have the following counterpart of Theorem 4 of [10] in terms of bijections.

Lemma 2. *Let G be a connected graph and let $\mathcal{E}_1, \mathcal{E}_2, \ldots, \mathcal{E}_x$ be the planar embeddings of its biconnected components $B_1, B_2, \ldots, B_x$. For each cut-vertex v_i of G, $i = 1, \ldots, w$, let φ_{v_i} be a bijection from the arrangements around v_i of the embeddings of those biconnected components that are incident to v_i to a number in the interval $[0 \ldots E_{v_i} - 1]$. There exists a bijection φ from the embeddings on the sphere of G whose restriction to B_j is $\mathcal{E}_j$ for $j = 1, \ldots, x$ to a sequence $\langle c_{1,1}, \ldots, c_{1,b(v_1)}, \ldots, c_{w,1}, \ldots, c_{w,b(v_w)}, d_{1,1}, \ldots, d_{1,b(v_1)-2}, \ldots, d_{w,1}, \ldots, d_{w,b(v_w)-2} \rangle$, where $c_{\xi,\sigma} \in [0 \ldots \delta_{v_\xi,\sigma} - 1]$, for $\xi = 1, \ldots, w$ and $\sigma = 1, \ldots, b(v_\xi)$; and $d_{\xi,\sigma} \in [0 \ldots \delta_{v_\xi} - \sigma - 1]$, for $\xi = 1, \ldots, w$ and $\sigma = 1, \ldots, b(v_\xi) - 2$.*

By Lemma 2 we have that the planar embeddings on the sphere of G, restricted to those that preserve the embedding $\mathcal{E}_i$ of each block B_i, can be ranked by combining the ranking function φ_v from Theorem 3 for the arrangement of the biconnected components around each cut-vertex v with Lemma 1.

3.1   Transforming a Tuple into an Embedding

Let G be a connected planar graph and let v be a cut-vertex of G. Let $B_1, B_2, \ldots, B_{b(v)}$ be the biconnected components of G containing v and let $\mathcal{E}_1^\circ, \mathcal{E}_2^\circ, \ldots, \mathcal{E}_{b(v)}^\circ$ be their planar embeddings on the sphere. We describe a transformation from a sequence of natural numbers $\langle c_1, \ldots, c_{b(v)}, d_1, \ldots, d_{b(v)-2} \rangle$, where $c_j \in [0 \ldots \delta_{v,j} - 1]$, for $j = 1, \ldots, b(v)$ and $d_j \in [0 \ldots \delta_v - j - 1]$, for $j = 1, \ldots, b(v) - 2$, to a planar embedding on the sphere of the subgraph $G(v) = B_1 \cup B_2 \cup \ldots B_{b(v)}$ preserving $\mathcal{E}_1^\circ, \mathcal{E}_2^\circ, \ldots, \mathcal{E}_{b(v)}^\circ$.

The first step is to use $c_1, c_2, \ldots c_{b(v)}$ to transform the embeddings on the sphere $\mathcal{E}_1^\circ, \mathcal{E}_2^\circ, \ldots, \mathcal{E}_{b(v)}^\circ$ into embeddings on the plane $\mathcal{E}_1^\square, \mathcal{E}_2^\square, \ldots, \mathcal{E}_{b(v)}^\square$ of $B_1, B_2, \ldots, B_{b(v)}$, respectively. Precisely, for each B_j, $j = 1, \ldots, b(v)$, let $e_0^j, \ldots, e_{\delta_{v,j}-1}^j$ be the edges of B_j incident to v ordered according to their identifiers. We select the c_j-th edge of this order as the *first edge* of $\mathcal{E}_j^\circ$ and denote it $first_j$. In the remainder of this section, we will assume that the edges $e_0^j, \ldots e_{\delta_{v,j}-1}^j$ are labeled according to a counter-clockwise visit of the adjacency list of v in $\mathcal{E}_j^\circ$ starting from $first_j = e_0^j$ and ending with $e_{\delta_{v,j}-1}^j$, which we also denote $last_j$. We project each $\mathcal{E}_j^\circ$ on the plane, obtaining $\mathcal{E}_j^\square$, so that the outer face of $\mathcal{E}_j^\square$ is the face of $\mathcal{E}_j^\circ$ that comes before $first_j$ and after $last_j$ in counter-clockwise order around v.

Now that all the biconnected components are independently embedded on the plane, we merge them by means of $b(v) - 1$ merging operations. We initialize $b(v)$ partial embeddings $\mathcal{P}_i^\square = \mathcal{E}_i^\square$, for $i = 1, \ldots, b(v)$. The j-th merging operation ($j = 2, \ldots, b(v)$) inserts, in the way that is specified below, $\mathcal{P}_j^\square$ (*source* of the merge) into a suitably chosen $\mathcal{P}_q^\square$ (*target* of the merge). After the merge, $\mathcal{P}_j^\square$ will not be used anymore and its edges are considered as edges of $\mathcal{P}_q^\square$. After $b(v) - 1$ such merge operations, we end up with a single partial embedding of $G(v)$ on the plane where all $\mathcal{E}_j^\square$ have been merged into. This embedding on the plane can then be regarded as an embedding on the sphere. The first merge operation merges source $\mathcal{P}_2^\square$ into target $\mathcal{P}_1^\square$. $\mathcal{E}_1^\square$ and $\mathcal{E}_2^\square$ are merged in such a way that in a counter-clockwise visit of the adjacency list of v starting from $first_1$ the edges of $\mathcal{E}_1^\square$ are not interleaved with the edges of $\mathcal{E}_2^\square$ and $first_2$ comes after $last_1$. We now show how to merge the remaining $b(v) - 2$ partial embeddings $\mathcal{P}_j^\square, j = 3, \ldots, b(v)$, guided by the values $d_j, j = 1, \ldots, b(v) - 2$. Namely, d_j identifies an edge incident to v of some component B^* different from B_{j+2}. In order to suitably select, for each of the $b(v) - 2$ iterations, the edge incident to v that determines the current merge operation, we create a sequence $\mathcal{S}$ of edges incident to v. We initialize $\mathcal{S}$ with the edges of $\mathcal{E}_1^\square$ in the linear ordering starting from $first_1$ and ending with $last_1$. Then we append to $\mathcal{S}$ the edges of $\mathcal{E}_2^\square$ in the linear ordering starting from $first_2$ and ending with $last_2$. Then, for $j = 3, \ldots, b(v)$, we append to $\mathcal{S}$ all the edges of $\mathcal{E}_j^\square$, with the exception of $first_j$, ordered according to their linear order in the embedding on the plane $\mathcal{E}_j^\square$. Finally, $\mathcal{S}$ is closed by the edges $first_j$ in reverse order, i.e., for $j = b(v), b(v) - 1, \ldots, 3$. We assume now that the edges of $\mathcal{S}$ are labeled by their position in $\mathcal{S}$, starting from 0. As all the edges adjacent to v are in $\mathcal{S}$, we have that $|\mathcal{S}| = \delta_v$. We also store a copy of $\mathcal{S}$, called $\mathcal{S}'$.

For $j = 3 \ldots b(v)$ we use d_{j-2} to perform a merge operation of $\mathcal{P}_j^\square$ into a suitably-chosen partial embedding. Let $e_{d_{j-2}}$ be the edge with label d_{j-2} in $\mathcal{S}$. Note that, since the upper-bound of the range $[0 \ldots \delta_v - k - 1]$ of the element d_k, for $k = 1, 2, \ldots, b(v) - 2$, is strictly decreasing as k increases, the chosen edge $e_{d_{j-2}}$ excludes the last $j - 2$ elements of $\mathcal{S}$, e.g., when choosing e_{d_2}, the last two cells, containing $first_3$ and $first_4$, are excluded. We distinguish two cases: (1) $e_{d_{j-2}}$ belongs to a partial embedding $\mathcal{P}_h^\square$ ($1 \le h \le b(v)$) distinct from $\mathcal{P}_j^\square$. (2) $e_{d_{j-2}}$ belongs to $\mathcal{P}_j^\square$.

If *case 1* applies, we merge the embeddings of $\mathcal{P}_j^\square$ into $\mathcal{P}_h^\square$ in such a way that in a counter-clockwise visit of the adjacency list of v starting from $first_h$, one encounters all the edges of $\mathcal{P}_h^\square$ up to $e_{d_{j-2}}$; then, right after $e_{d_{j-2}}$, one encounters $first_j$ and, consequently, all the edges of $\mathcal{P}_j^\square$; finally, after $last_j$ one encounters all the other edges of $\mathcal{P}_h^\square$. We replace in $\mathcal{S}$ $e_{d_{j-2}}$ with $first_j$, so that $e_{d_{j-2}}$ will

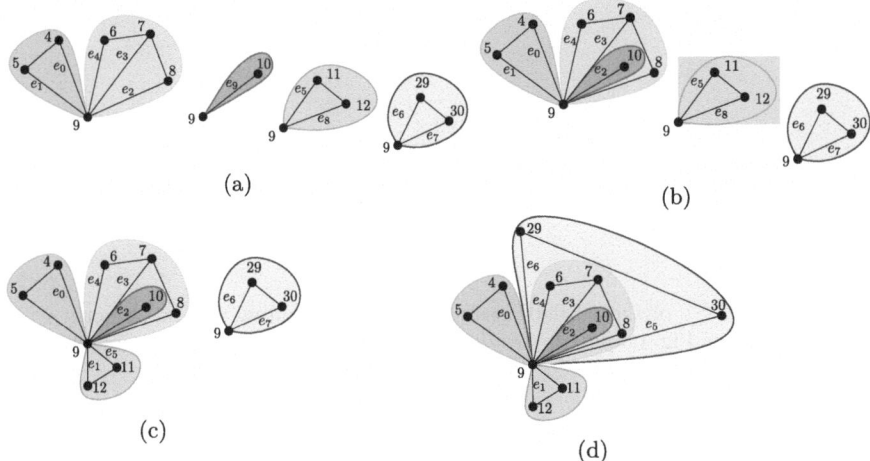

Fig. 3. The merge phase of the unranking of the embedding corresponding to the tuple $\langle 0, 1, 0, 1, 1, 2, 1, 6 \rangle$ around the cut-vertex 9 in Fig. 1. The edge labels change according to the sequence $\mathcal{S}$. The labels of the edges that have been substituted are not shown to avoid confusion. (a) The components after the unranking of the first edges using the values $\langle 0, 1, 0, 1, 1 \rangle$. (b) μ_3 (in red) must counter-clockwise follow the edge e_2, according to the tuple $\langle 2, 1, 6 \rangle$, so its first edge (e_9) is identified with e_2 (whose label is omitted to avoid confusion). (c) μ_4 (in green) is placed after e_1, and the first edge of μ_4 (e_8) is identified with e_1. (d) The edge e_6 belongs to μ_5 (in blue), therefore the edge e_6 is placed after $last_2$ (i.e. e_4) and $first_5$ (i.e. e_7) is placed after e_5 and identified with it. (Color figure online)

never be selected again and no other partial embedding will be inserted between $e_{d_{j-2}}$ and $first_j$. If $e_{d_{j-2}} = last_h$ then we update also $last_h = last_j$.

If *case 2* applies, let $e_{d_{j-2}}$ be the edge in position d_{j-2} in $\mathcal{S}'$. We use $\mathcal{S}'$ instead of $\mathcal{S}$ so that if $first_i$, for some $i < j$, has taken the place of $e_{d_{j-2}}$ in $\mathcal{S}$, we will not insert something between $e_{d_{j-2}}$ and $first_i$, but before $e_{d_{j-2}}$. We merge $\mathcal{P}_j^{\square}$ with $\mathcal{P}_1^{\square}$ by inserting $e_{d_{j-2}}$ and all the subsequent edges of $\mathcal{P}_j^{\square}$ right after $last_2$. We will insert $first_j$ right after the edge e^* that precedes $first_2$ in $\mathcal{P}_1^{\square}$ and all following edges of $\mathcal{P}_j^{\square}$ up to $e_{d_{j-2}}$ (not included) subsequently. Similarly to what has been done in *case 1*, we then replace in $\mathcal{S}$ the edge e^* with $first_j$, in such a way as to not insert any other edge between e^* and $first_j$. An example of the merge phase can be found in Fig. 3.

Regarding the time taken by this algorithm, the first phase, using the $c_1, \ldots,$ $c_{b(v)}$, requires $O(n)$ time, whereas the merging phase takes $O(n\alpha(n))$ time using a union-find data structure [34] to check which of the two cases applies.

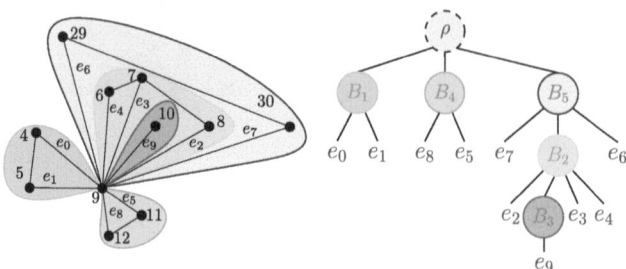

Fig. 4. The $3,653^{rd}$, of the $12,096$ possible, planar embedding around the cut-vertex 9 of the graph in Fig. 1, and its resulting nesting-tree. The corresponding tuple is $\langle 0, 1, 0, 1, 1, 2, 1, 6 \rangle$.

3.2 Transforming an Embedding into a Tuple

Let G be a connected planar graph and let v be a cut-vertex of G. Let $B_1, B_2, \ldots,$ $B_{b(v)}$ be the biconnected components of G containing v. Let $G(v) = B_1 \cup B_2 \cup \ldots B_{b(v)}$. Let $\mathcal{E}^\circ$ be an embedding on the sphere of $G(v)$ and let $\mathcal{E}_1^\circ, \mathcal{E}_2^\circ, \ldots, \mathcal{E}_{b(v)}^\circ$ be the restriction of $\mathcal{E}^\circ$ to the biconnected components $B_1, B_2, \ldots, B_{b(v)}$, respectively. We describe a transformation from $\mathcal{E}^\circ$ to a sequence of natural numbers $\langle c_1, \ldots, c_{b(v)}, d_1, \ldots, d_{b(v)-2} \rangle$, where $c_j \in [0 \ldots \delta_{v,j} - 1]$, for $j = 1, \ldots, b(v)$ and $d_j \in [0 \ldots \delta_v - j - 1]$, for $j = 1, \ldots, b(v) - 2$.

First we compute the values $c_1, \ldots, c_{b(v)}$. For each B_j, $j = 1, \ldots, b(v)$, we consider the counter-clockwise order σ_j of the edges incident to v of B_j and label the edges from 0 to $\delta_{v,j} - 1$ according to their position in σ_j starting from the edge with minimum index. We perform a visit of the adjacency list of v in counter-clockwise order starting from a random edge of $\mathcal{E}_1^\circ$. The first edge of $\mathcal{E}_1^\circ$ encountered after all the edges of $\mathcal{E}_2^\circ$ will be called $first_1$ and c_1 will be set to its label. We perform a second visit starting from $first_1$, the first edge encountered of each $\mathcal{E}_j^\circ$, for $j = 2, \ldots, b(v)$, will be $first_j$ and c_j will be set to its label.

We then label all the edges. We first put all the edges in a sequence $\mathcal{S}$ starting from $first_1$ and taking them, except for $first_i$ with $i = 3, \ldots, b(v)$, ordered primarily according to the increasing label of the component they belong to, and secondarily according to their counter-clockwise order around v starting from the $first_1$. Then we append the edges $first_i$ with $i = 3, \ldots, b(v)$ to $\mathcal{S}$ in decreasing order of their component labels. The label $\ell(e)$ of an edge e is its position in $\mathcal{S}$, starting from 0.

Next we compute an auxiliary ordered rooted tree $\mathcal{T}^B$ as follows. The nodes of $\mathcal{T}^B$ correspond either to biconnected components in $B_1, B_2, \ldots, B_{b(v)}$ (*component-nodes*) or to edges incident to v (*edge-nodes*). We call $n(B_j)$ the node of $\mathcal{T}^B$ corresponding to B_j. An edge-node can only be a leaf of $\mathcal{T}^B$. Initially $\mathcal{T}^B$ contains only a dummy root ρ and the *current node* γ of $\mathcal{T}^B$ is ρ. We visit the adjacency list of v counter-clockwise starting from $first_1$. When visiting an edge e we distinguish four cases: (i) If e is both the first and the last edge of a component B_j we add to $\mathcal{T}^B$ a component-node $n(B_j)$ as the last child of γ and an edge-node corresponding to e as child of $n(B_j)$ and set γ as the parent of γ. (ii) If e is the first, but it is not the last, edge of a component B_j we add to $\mathcal{T}^B$ a component-node $n(B_j)$ as the last child of γ and an edge-node corresponding to e as the first child of $n(B_j)$. We also set node $\gamma = n(B_j)$. (iii) If e is the last, but it is not the first, edge of a component B_j, we add to $\mathcal{T}^B$ an edge-node corresponding to e as the last child of γ and set γ as the parent of γ. (iv) If e is neither the first nor last edge of a component, we add to $\mathcal{T}^B$ an edge-node corresponding to e as the last child of γ. In addition, after the computation of $\mathcal{T}^B$, let π_2 be the path between the root ρ and $n(B_2)$, we save in all nodes in π_2 a Boolean value that indicates that those nodes belong to π_2. We then compute a pointer $jump_j$ for each node that belongs to π_2. First, we traverse the path π_2 top-down starting from the child of ρ and keeping the index i of the component with maximum index visited. For each node $n(B_j)$ we set $jump_j$ as the edge corresponding to the first edge-node that is a child of $n(B_j)$ on the right of π_2. Also, we set the value $max_j = i$ after having updated the value $i = j$ if $i < j$. Then we perform a post-order traversal of $\mathcal{T}^B$ starting from the child of ρ that belongs to π_2 and stopping at the parent of $n(B_2)$, looking at the children of a node in reverse order (right to left) and considering only the ones that are not on the right of π_2. During the visit, if a node B_j on π_2 has as left sibling a node B_i with $i > j$ we set $jump_i = jump_j$, we say that B_j does not belong to π_2 and B_i belongs to π_2. Otherwise, if a node B_j on π_2 has a parent B_i (different from ρ) and $max_j > j$ (i.e. there exists an ancestor of $n(B_j)$ with an index greater than j) we set $jump_i = jump_j$ and say that B_j does not belong to π_2.

Now we can exploit $\mathcal{T}^B$ to efficiently compute the values $d_1, \ldots, d_{b(v)-2}$. For $j = 1, \ldots, b(v) - 2$ consider node $n(B_{j+2})$ in $\mathcal{T}^B$. Two are the cases: Case 1: $n(B_{j+2})$ does not belong to π_2. Let β be the left sibling of $n(B_{j+2})$. If β is an edge-node, then let e be the edge corresponding to β, otherwise (β is a component-node labeled B_i) let e be $last_i$. We set $d_j = \ell(e)$. Case 2: $n(B_{j+2})$ belongs to π_2. We use the pre-computed pointers by setting $d_j = \ell(jump_{j+2})$. An example can be found in Fig. 4. After setting d_j, we update $\mathcal{S}$ as follows. Let β be the left sibling of $n(B_{j+2})$. If β is an edge-node, then let e be the edge corresponding to β, otherwise (β is a component-node B_i) let e be $last_i$. We replace e with $first_{j+2}$ in $\mathcal{S}$ and set $\ell(first_{j+2}) = \ell(e)$.

This algorithm takes $O(n)$ time as the notion of which component an edge belongs to does not change during the computation, and hence a union-find data structure is not required. Also, the computation of $first_i$, $\mathcal{T}^B$, and $jump_i$, for each i, involves a linear number of constant operations.

4   Non-connected Graphs

Let G be a planar graph with c connected components ($c \geq 1$). We sort the connected components of G in increasing order, based on the identifier of their vertex with the smallest identifier. The *identifier of a component* is its position in the sorted order. Hence we can call the components as $G_1, \ldots, G_c$.

Let $\mathcal{E}$ be an embedding of G and let $\mathcal{E}_1, \ldots, \mathcal{E}_c$ be the embeddings of $G_1, \ldots, G_c$, respectively, induced by $\mathcal{E}$. Consider the embedding $\mathcal{E}_i$ of G_i. The *label of a face* f of $\mathcal{E}_i$ is a triple $\langle i, (u, v), b \rangle$, where (u, v) is the edge of f with the smallest identifier and $b \in \{0, 1\}$ is 0 if traversing (u, v) from u to v face f is to the right of (u, v) and 1 otherwise. Observe that in some cases, when traversing (u, v) from u to v, face f is both to the right and to the left of (u, v). In this case f is the only face containing (u, v) and hence the value of b does not play any identification role. We denote by F_i the number of faces of $\mathcal{E}_i$. We can sort the faces of $\mathcal{E}_i$ according to their increasing label and the *identifier of a face* is the position of such a face in the order. Hence, we call them $f_1^i, \ldots, f_{F_i}^i$. We select as *reference face* f of $\mathcal{E}$ the face that contains the circular list f_1^1 of $\mathcal{E}_1$.

We associate with $\mathcal{E}$ a *nesting tree* $T_{\mathcal{E}}^N$, which is a rooted unordered tree labeled on the edges, and a *face tuple* $\langle o_1, \ldots, o_c \rangle$, where $o_i \in [1 \ldots F_i - 1]$ as follows. The nodes of $T_{\mathcal{E}}^N$ are the components of G plus a dummy node ρ which is the root of $T_{\mathcal{E}}^N$ and corresponds to the reference face f of $\mathcal{E}$. To build $T_{\mathcal{E}}^N$ we project $\mathcal{E}$ on the plane in such a way that f is the outer face. Note that this projection determines an *outer face* for each $\mathcal{E}_i$, which is the face of $\mathcal{E}_i$ that would be unbounded when all other components were removed. For each i, we set element o_i of the face tuple as the identifier of the outer face of $\mathcal{E}_i$. Hence, o_i is a number in $[0 \ldots F_i - 1]$. All the other faces of $\mathcal{E}_i$ are called *inner faces*. We consider the set of all the inner faces (each with its label) of $\mathcal{E}_1, \ldots, \mathcal{E}_c$. This set contains $\sum_{i=1}^c (F_i - 1)$ elements. We sort the set according to the increasing order of the labels. The sorting process uses the identifier of the component as the primary element, ensuring that the inner faces of a particular component are arranged consecutively in the ordering. Let $\pi(g)$ be the position of an inner face g in this ordering (starting from 1). There is an edge in $T_{\mathcal{E}}^N$ between ρ and a component G_i if a face f_i of $\mathcal{E}_i$ is a circular list of f. The label of edge (ρ, G_i) is set to zero. G_i is a parent of G_j if there exists a face of $\mathcal{E}$ that contains the circular lists of both the outer face of $\mathcal{E}_j$ and an inner face g of $\mathcal{E}_i$. The label of edge (G_i, G_j) is set to $\pi(g)$. Generalizing, given a planar graph G with c connected components $G_1, \ldots, G_c$ $(c \geq 1)$ with embeddings $\mathcal{E}_1, \ldots, \mathcal{E}_c$, respectively, we can define a nesting tree T_G^N and a face tuple $\langle o_1, \ldots, o_c \rangle$ as follows. The face tuple is any tuple such that $o_i \in [0 \ldots F_i - 1]$. T_G^N is any $c + 1$-nodes unordered tree with root ρ and such that the remaining nodes are the components of G.

Consider the set of integers $[1 \cdots \sum_{i=1}^c (F_i - 1)]$. We partition such a set into consecutive intervals $I_1 = [1 \cdots \sum_{i=1}^1 (F_i - 1)], I_2 = [1 + \sum_{i=1}^1 (F_i - 1) \cdots \sum_{i=1}^2 (F_i - 1)], \ldots, I_c = [1 + \sum_{i=1}^{c-1} (F_i - 1) \cdots \sum_{i=1}^c (F_i - 1)]$. The edges incident to ρ are labeled 0. The remaining edges of T_G^N are labeled in such a way that all the edges between a parent G_h $(h = 0, \ldots, c)$ and its children have a label in I_h.

As we have seen before, given an embedding $\mathcal{E}$ of G, it is possible to compute the corresponding nesting tree and face tuple. Also, this can be done in $O(n)$ time using standard data structures for the embeddings.

Conversely, given a set of embeddings $\mathcal{E}_1, \ldots, \mathcal{E}_c$, a nesting tree, and a face tuple, built with the above requirements, it is easy to construct a corresponding embedding $\mathcal{E}$ of G. This is done in linear time by constructing an embedding on the plane as follows. (1) The numbers specified in the face tuple are used to select an outer face for the embedding of each component. (2) The components that are children of ρ are embedded in the outer face. (3) The parent-child edges of the nesting tree are used to insert the embedding of the child component into an inner face of the parent component. (4) The face of the parent component in which to insert the child component is obtained from the label of the edge.

We summarize the above discussion with the following lemma.

Lemma 3. *Let G be a planar graph and let $G_1, \ldots, G_c$ be the connected components of G equipped with embeddings $\mathcal{E}_1, \ldots, \mathcal{E}_c$. Let F_i be the number of faces of $\mathcal{E}_i$. There exists a bijective function F whose domain is the set of embeddings of G and whose codomain is the set of pairs composed of a face tuple and a nesting tree. Both the function F and its inverse can be computed in $O(n)$ time.*

As we have seen so far, an embedding is fully specified by a pair consisting of a nesting tree and of a face tuple. Also, any of such pairs univocally determines an embedding. Unfortunately, this fact solves our encoding problem only partially. Namely, a face tuple is already a very simple encoding of some features of an embedding. Conversely, a nesting tree is far from being simply encoded in a set of numbers. Hence, in the reminder of this section we concentrate on how to efficiently encode a nesting tree. Namely we will show a bijection between nesting trees and tuples of $c-1$ numbers each chosen in the interval $[0 \cdots \sum_{i=1}^{c}(F_i - 1)]$.

Rooted unordered trees with $c + 1$ nodes and with a common root ρ are in one-to-one correspondence with tuples of c elements each in the interval $[0 \ldots c]$ via the Prüfer encoding described in Sect. 2. Unfortunately, this encoding does not provide information about the internal face of the parent component that contains the embedding of its child component, thereby lacking the specification of the parent-child relationship. On the other hand, we have established an order for the internal faces of the embedding of all the components that allows to retrieve, given an element in the order, which is the component it belongs to.

From the above discussion, given a nesting tree $\mathcal{T}_G^N$, we use the following variation of the Prüfer algorithm to determine a tuple of $c - 1$ numbers. The algorithm has $c - 1$ iterations. At each iteration i it selects the leaf ℓ (i.e. a component) of $\mathcal{T}_G^N$ with the smallest identifier, inserts into the i-th element of the tuple the label of the edge incident to ℓ and deletes ℓ from $\mathcal{T}_G^N$. Since the labels are in the interval $[0 \cdots \sum_{i=1}^{c}(F_i - 1)]$, each element of the tuple belongs to such an interval. In addition, the tuple can be computed in $O(n)$ time by implementing the algorithm as described in [9,11,35].

Conversely, given a tuple τ of $c - 1$ numbers each in the interval $[0 \cdots \sum_{i=1}^{c}(F_i - 1)]$, we construct a nesting tree $\mathcal{T}_G^N$ as follows. We perform a preprocessing step computing an additional tuple τ' with $c - 1$ elements as follows. We scan τ and for $i = 1, \ldots, c - 1$ we extract the i-th element τ_i of τ. If $\tau_i = 0$, we insert 0 in the i-th position of τ'. Otherwise, suppose that $\tau_i \in I_h$. We insert h in the i-th position of τ'. Furthermore, we associate with each component G_i of G the number δ_i of occurrences of i in τ' plus one. Intuitively, in this way we precompute the number of nodes that will be adjacent to G_i in $\mathcal{T}_G^N$. Also, we associate ρ with the number δ_0 of occurrences of 0 in τ plus two. Finally, we insert into $\mathcal{T}_G^N$ a node for each component G_i plus a node ρ. After the preprocessing step, the algorithm proceeds with $c - 1$ iterations. In each iteration i, we select the component G_h with $\delta_h = 1$ and the smallest h. We insert an edge in $\mathcal{T}_G^N$ from child G_h to the parent component G_k such that k is the i-th element of τ', if $k = 0$ the parent will be ρ. We label such an edge with the i-th element of τ. We finally decrement δ_h and δ_k. After the last iteration, there remains one component G_l such that $\delta_l = 1$, while $\delta_0 = 2$. We insert an edge in $\mathcal{T}_G^N$ from child G_l to the parent component ρ with label 0.

Observe that appending a 0 to τ' results exactly in the Prüfer encoding of a rooted tree with $c + 1$ nodes having the sought adjacencies between components and ρ as its root. This tree is exactly $\mathcal{T}_G^N$ without the edge labels. Having a set of labels for the edges that is dependent only on the parent node allows us to specify a label instead of a parent node in the encoding. We summarize the above discussion in the following lemma.

Lemma 4. *Let G be a planar graph and let $G_1, \ldots, G_c$ be the connected components of G equipped with embeddings $\mathcal{E}_1, \ldots, \mathcal{E}_c$. Let F_i be the number of faces of $\mathcal{E}_i$. There exists a bijective function ψ whose domain is the set of nesting trees of G and whose codomain are the tuples of length $c - 1$ whose elements are values in the interval $[0 \cdots \sum_{i=1}^{c}(F_i - 1)]$. Both the function ψ and its inverse can be computed in $O(n)$ time.*

Now we are ready to prove the following theorem.

Theorem 4. *Let G be a planar graph with connected components $G_1, \ldots, G_c$ ($c > 1$). Let $\mathcal{E}_1, \ldots, \mathcal{E}_c$ be planar embeddings for $G_1, \ldots, G_c$ such that embedding $\mathcal{E}_i$ has F_i faces. There exists a bijective function φ whose domain is the set of planar embeddings of G such that each G_i has embedding $\mathcal{E}_i$ and whose codomain is the set of tuples of natural numbers $\langle a_1, \ldots, a_{c-1}, b_1, \ldots, b_c \rangle$ where $a_i \in \{0, \ldots, \sum_{j=1}^{c}(F_j - 1)\}$ and $b_i \in \{0, \ldots, F_i - 1\}$. Both the function φ and its inverse can be computed in $O(n)$ time.*

Proof. The function φ is the combination of the function F from Lemma 3 and the function ψ from Lemma 4.

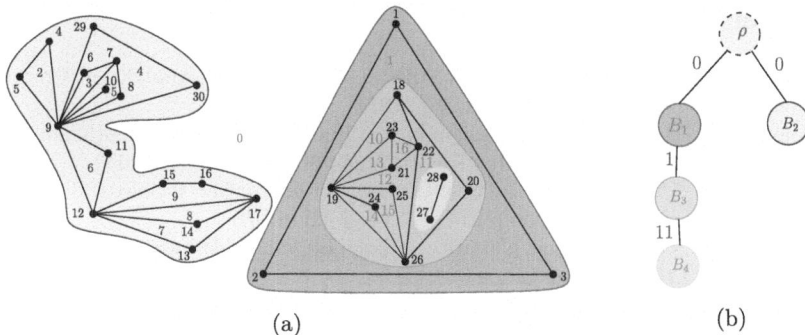

(a)   (b)

Fig. 5. (a) The non-connected components of the graph in Fig. 1, with the same embedding. The internal faces of each component are labeled with colored numbers, and the reference face is labeled with 0. The outer face of each component is defined by the numbers $\langle 0, 1, 0, 0 \rangle$. (b) The corresponding nesting tree, whose structure is encoded in the tuple $\langle 0, 11, 1 \rangle$. This is the 188th nesting over the $78,896$ possible. (Color figure online)

5   Conclusions and Future Work

In this paper, we addressed the problem of ranking and unranking all planar embeddings of an n-vertex graph G, which may not be connected. In particular, we produced a ranking function Φ that can be computed in $O(n)$ time, and its inverse unranking function Φ^{-1} that can be computed in $O(n\alpha(n))$ time. In addition, we showed that the natural number associated with a planar embedding can be decomposed into a sequence of a linear number of values, each associated with a specific feature of the embedding. This property has practical implications, allowing us to generate embeddings uniformly at random by independently generating each value in the sequence. It also facilitates the counting, enumeration, and generation uniformly at random of constrained embeddings.

As future work we point out the problem of devising a linear time algorithm for the unranking of simply connected graphs, whose current $O(n\alpha(n))$-time bound slightly impacts the unranking of the whole graph. Further, we would like to extend the results of this paper to beyond planar embeddings of graphs. Analogously to the topological morphing between two embeddings of a biconnected graph [3], the concepts introduced in this paper could lead to topological morphing algorithms for non-connected and simply connected graphs.

Acknowledgement. This research was supported, in part, by MUR of Italy (PRIN Project no. 2022ME9Z78 – NextGRAAL and PRIN Project no. 2022TS4Y3N – EXPAND). The second author was supported by Ce.Di.Pa. - PNC Programma unitario di interventi per le aree del terremoto del 2009–2016 - Linea di intervento 1 sub-misura B4 - "Centri di ricerca per l'innovazione" CUP J37G22000140001.

References

1. Aho, A.V., Hopcroft, J.E., Ullman, J.D.: The Design and Analysis of Computer Algorithms. Addison-Wesley, Reading (1974)
2. Angelini, P., Di Battista, G., Patrignani, M.: Finding a minimum-depth embedding of a planar graph in $O(n^4)$ time. Algorithmica **60**(4), 890–937 (2011)
3. Angelini, P., Cortese, P.F., Di Battista, G., Patrignani, M.: Topological morphing of planar graphs. Theor. Comput. Sci. **514**, 2–20 (2013)
4. Angelini, P., et al.: Testing planarity of partially embedded graphs. ACM Trans. Algorithms **11**(4) (2015). Article No. 32
5. Bender, E.A., Williamson, S.G.: Foundations of Combinatorics with Applications. Courier Corporation (2005)
6. Bertolazzi, P., Di Battista, G., Didimo, W.: Computing orthogonal drawings with the minimum number of bends. IEEE Trans. Comput. **49**, 826–840 (2000)
7. Bertolazzi, P., Di Battista, G., Liotta, G., Mannino, C.: Upward drawings of triconnected digraphs. Algorithmica **12**(6), 476–497 (1994)
8. Booth, K.S., Lueker, G.S.: Testing for the consecutive ones property, interval graphs, and graph planarity using PQ-tree algorithms. J. Comput. Syst. Sci. **13**(3), 335–379 (1976)
9. Chen, H.C., Wang, Y.L.: An efficient algorithm for generating prüfer codes from labelled trees. Theory Comput. Syst. **33**, 97–105 (2000)
10. Cai, J.: Counting embeddings of planar graphs using DFS trees. SIAM J. Discret. Math. **6**(3), 335–352 (1993)
11. Caminiti, S., Finocchi, I., Petreschi, R.: On coding labeled trees. Theor. Comput. Sci. **382**(2), 97–108 (2007). Latin American Theoretical Informatics
12. Cayley, A.: A theorem on trees. Quart. J. Math. **23**, 376–378 (1878)
13. Di Battista, G., Eades, P., Tamassia, R., Tollis, I.G.: Graph Drawing: Algorithms for the Visualization of Graphs. Prentice-Hall (1999)
14. Di Battista, G., Grosso, F., Maragno, G., Patrignani, M.: Ranking and unranking of the planar embeddings of a planar graph (2024). https://arxiv.org/abs/2411.10319
15. Di Battista, G., Tamassia, R.: On-line maintenance of triconnected components with SPQR-trees. Algorithmica **15**(4), 302–318 (1996)
16. Di Battista, G., Tamassia, R.: On-line planarity testing. SIAM J. Comput. **25**(5), 956–997 (1996)
17. Didimo, W., Liotta, G., Ortali, G., Patrignani, M.: Optimal orthogonal drawings of planar 3-graphs in linear time. In: Chawla, S. (ed.) Proceedings of the ACM-SIAM Symposium on Discrete Algorithms (SODA 2020), pp. 806–825. ACM-SIAM (2020)
18. Didimo, W., Liotta, G., Patrignani, M.: HV-planarity: algorithms and complexity. J. Comput. Syst. Sci. **99**, 72–90 (2019)
19. Garg, A., Tamassia, R.: On the computational complexity of upward and rectilinear planarity testing. SIAM J. Comput. **31**(2), 601–625 (2001)
20. Gutwenger, C., Mutzel, P., Weiskircher, R.: Inserting an edge into a planar graph. In: 12th ACM-SIAM Symposium on Discrete Algorithms, SODA 2001, pp. 246–255. Society for Industrial and Applied Mathematics, Philadelphia (2001)
21. Kajimoto, H.: An extension of the Prüfer code and assembly of connected graphs from their blocks. Graphs Combin. **19**, 231–239 (2003)
22. Karabeg, A.: Ranking planar embeddings using PQ-trees. In: Gimbel, J., Kennedy, J.W., Quintas, L.V. (eds.) Quo Vadis, Graph Theory? Annals of Discrete Mathematics, vol. 55, pp. 249–260. Elsevier (1993)

23. Kiem-Phong Vo, W.E.D., Williamson, S.G.: Ranking and unranking planar embeddings. Linear Multilinear Algebra **18**(1), 35–65 (1985)
24. Kreher, D.L., Stinson, D.R.: Combinatorial Algorithms: Generation, Enumeration, and Search. CRC Press (1999)
25. Leyshon, P.R., Lisle, R.J.: Stereographic Projection Techniques. Elsevier Science (1996). Original from the University of California
26. Mutzel, P., Weiskircher, R.: Optimizing over all combinatorial embeddings of a planar graph. In: 7th International IPCO Conference on Integer Programming and Combinatorial Optimization, pp. 361–376. Springer, London (1999)
27. Mutzel, P., Weiskircher, R.: Computing optimal embeddings for planar graphs. In: Proceedings of the 6th Annual International Conference on Computing and Combinatorics, COCOON 2000, pp. 95–104. Springer, London (2000)
28. Myrvold, W.J., Ruskey, F.: Ranking and unranking permutations in linear time. Inf. Process. Lett. **79**(6), 281–284 (2001)
29. Pizzonia, M., Tamassia, R.: Minimum depth graph embedding. In: Proceedings of the ESA 2000. LNCS, vol. 1879, pp. 356–367. Springer (2000)
30. Prufer, H.: Neuer bewis eines satzes uber permutationnen. Arch. Math. Phys. **27**, 742–744 (1918)
31. Stallmann, M.F.M.: On counting planar embeddings. Discret. Math. **122**(1–3), 385–392 (1993)
32. Tamassia, R.: Constraints in graph drawing algorithms. Constraints An Int. J. **3**(1), 87–120 (1998)
33. Tamassia, R. (ed.): Handbook on Graph Drawing and Visualization. Chapman and Hall/CRC (2013)
34. Tarjan, R.E., van Leeuwen, J.: Worst-case analysis of set union algorithms. J. ACM **31**(2), 245–281 (1984)
35. Wang, X., Wang, L., Wu, Y.: An optimal algorithm for prufer codes. J. Softw. Eng. Appl. **2**(2), 111–115 (2009)

Optimal Uniform Shortest Path Sampling

Simon Dreyer[ORCID], Antoine Genitrini[ORCID], and Mehdi Naima[✉][ORCID]

Sorbonne Université, CNRS, LIP6, 75005 Paris, France
{Simon.Dreyer,Antoine.Genitrini,Mehdi.Naima}@lip6.fr

Abstract. Random generation of shortest paths in graphs is utilized across various domains, including traffic-flow simulation and network topology exploration. In this paper, we address the challenge of uniform shortest path sampling in graphs from an algorithmic perspective. We introduce a new uniform shortest path sampling algorithm that uses a biased random walk operating in two stages. We demonstrate that our algorithm, when combined with a new variant of the Alias method is optimal in terms of worst-case running time and number of random bits needed, among all algorithms in its class. Furthermore, we present an efficient implementation of our algorithm in a low-level programming language and evaluate it on both real-world and synthetic datasets. We compare our theoretically optimal algorithm with other variants to assess its practical performance.

Keywords: Graph Algorithm · Random Generation · Uniform Sampling · Shortest Paths · Alias Sampling

1 Introduction

Graphs play a crucial role in understanding and addressing problems associated with networks and interconnected systems. They provide a theoretical basis for analyzing various structures, including social networks, communication systems, and transportation networks.

Sampling shortest paths is essential in numerous contexts, such as simulating traffic flow, studying the topology of large networks (like the internet and social networks), and assessing network damage. For instance, the authors of [4] employ this method to evaluate network damage, while [7] analyzes shortest path sampling as an approximation of traceroute paths. The practice of approximating graphs through shortest path sampling has been explored in [1,14,17,20], where the authors investigate the biases introduced by exploring a graph via random shortest paths. In [11], the authors discuss which graph properties can be well approximated using shortest path sampling and the influence of the number of samples on these properties. In the realm of traffic flow, the authors of [24] apply shortest path sampling to enhance network transmission capacities.

Research partly supported by the ANR-FWF project PANDAG ANR-23-CE48-0014-01 and by the ANR project FiT-ANR-19-LCV1-0005.

Despite extensive research, there is a noticeable gap in the explicit study of the algorithmic complexity of shortest path sampling. Typically, this problem involves fixed source and target nodes, aiming to sample a shortest path between them. Some studies, as mentioned in [4,6,7,15,18,24], suggest randomly selecting one shortest path from all possible paths without detailing the implementation. This approach faces challenges due to the potentially exponential number of shortest paths between two nodes in some graph families, such as grid graphs, making it impractical for simulating traffic in city graphs, which often resemble grid graphs.

Other studies recommend assigning random weights to edges to resolve ties among multiple shortest paths, then returning the unique shortest path left, as detailed in [5,9,14,23]. Each change of weights generates a new shortest path. However, this method introduces a bias, meaning that some shortest paths are more likely to be selected than others. To our knowledge, no theoretical or experimental studies have addressed this bias, which could affect the outcomes of experimental studies relying on biased sampling methods.

Main Contributions: We propose several algorithms for the uniform shortest path sampling from fixed source and target nodes and analyze their theoretical complexities as well as their behaviors in practice. We also propose a new variant of the Alias sampling (method to sample from a discrete probability distribution) that uses only integer arithmetic avoiding bias due to floating arithmetic approximations. This variant is then used in an unranking scheme to ensure random bits optimality as well as optimal running time.

The paper is organized as follows. In Sect. 2 after presenting our formalism we give the problem statement and review the state of the art. Then, in Sect. 3 we present a generic two-stage random walk algorithm, and demonstrate that, with appropriate weight settings, this algorithm uniformly samples shortest paths. Next, in Sect. 4 we discuss four different implementations of our proposed algorithm, conducting a detailed analysis to identify the optimal version. In Sect. 5 we present an unranking algorithm that uses the integer variant of the Alias method. Finally, in Sect. 6, we assess the performance of our algorithms using both real-world and synthetic graphs. Missing proofs in the main paper can be found in Appendix A.

2   Context of the Problem and State of the Art

In this article, we focus on graphs represented by $G = (V, E)$, where V is a finite set of nodes and $E = \{(u, v) \mid u, v \in V, u \neq v\}$ denotes the set of directed edges. Therefore, the graphs under consideration are directed and simple (no self-loops or multiple edges are allowed). We denote $n = |V|$ as the number of nodes and $m = |E|$ as the number of edges. An undirected graph can be viewed as a directed graph where if $(u, v) \in E$ then $(v, u) \in E$ as well. An edge $(u, v) \in E$ will be denoted as $u \to v$, where u is referred to as the starting point and v as the end point. A *walk* W in a graph is a sequence of edges such that the end point of one edge is the starting point of the next edge. The *length of a walk* W, denoted $|W|$, corresponds to the number of edges it contains. For a given walk, $s \in V$ usually denotes *the source node* and $t \in V$ denotes *the target node*. The *distance* from s to t, denoted $d(s,t)$, represents the minimal length among all walks from s to t. A *shortest path* from s to t is a walk W of minimal length, that is $|W| = d(s,t)$. We also denote by $\mathbf{W}_{st}$ the set of all shortest paths from s to t, note that $\mathbf{W}_{ss} = \{\epsilon\}$, with ϵ being the empty walk. Then we denote by $\sigma_s(t) = |\mathbf{W}_{st}|$ the number of shortest paths of $\mathbf{W}_{st}$. The set of all shortest paths starting from s is written as $\mathbf{W}_{s\bullet}$. Therefore, $\mathbf{W}_{s\bullet} = \cup_{t \in V} \mathbf{W}_{st}$ and $\sigma_{s\bullet} = |\mathbf{W}_{s\bullet}|$. Finally, we denote by $\mathbf{W}$ the set of all shortest paths in the graph: thus $\mathbf{W} = \cup_{v \in V} \mathbf{W}_{v\bullet}$ and $\sigma = |\mathbf{W}|$. Given a graph and fixed source s and target t, we are interested in:

Problem: Source-Target Uniform Shortest Path: Nodes s, t being fixed, give a random generation algorithm satisfying $\forall W \in \mathbf{W}_{st}, \mathbb{P}(W) = 1/\sigma_s(t)$ and for all $W \notin \mathbf{W}_{st}, \mathbb{P}(W) = 0$.

The simplest approach addressing this problem, involves precomputing all shortest paths from s to t and storing them in a list during a preprocessing stage. Queries can then be answered by returning a randomly selected element from this list. This approach is implicitly suggested by [4,6,7,15,18,24]. The preprocessing stage running time depends on the number of paths in $\mathbf{W}_{st}$, which can be exponential for certain graph families. See for example Fig. 4 in Appendix A.1. Therefore, we aim to design polynomial-time algorithms that still ensure uniform sampling.

To overcome this issue, one idea used in works such as [5,9,14,23] is to assign small random real weights to the edges of the graph. This ensures that only one shortest path has the minimum weight, which can then be found and returned using Dijkstra's Algorithm with backtracking. Therefore, redistributing random weights on the edges of the graph ensures that every shortest path from s to t has some probability of being sampled. However, the probabilities of the different shortest paths are not equal see Appendix A.1 for a counterexample.

3   A Random Walk Approach for Uniform Generation of Shortest Paths

In this section, source s and target t nodes are fixed and we design algorithms to tackle the source-target uniform shortest path problem. An overview of the complexity results of our approaches is presented in Table 1.

3.1   Two Stages Algorithm

We present a generic two-stage random walk algorithm to ensure the sampling of all shortest paths. The two phases are:

- *Preprocessing*: Compute the distributions, among the edges of each node, for the random walk. This step is done only once.
- *Queries*: For each query, generate a random walk in the graph.

Table 1. Overview of the discussed random sampling methods: running times correspond to the worst-case scenario given a graph G with n nodes and m edges. The value ℓ is the length of the sampled shortest path and Bit opt. corresponds to whether the algorithm is optimal in terms of random bits usage.

Algorithm	Bit opt.	Preprocessing	Space	Query	Section
BRW-linear	no	$O(m+n)$	$O(m+n)$	$O(n)$	4.1
BRW-Ordered	no	$O(m+n\log n)$	$O(m+n)$	$O(n)$	4.2
BRW-Binary	no	$O(m+n)$	$O(m+n)$	$O(\ell\log(\frac{n}{\ell}))$	4.3
BRW-Alias	no	$O(m+n)$	$O(m+n)$	$O(\ell)$	4.4
Unrank-Alias	yes	$O(m+n)$	$O(m+n)$	$O(\ell)$	5

We will discuss two random walk schemes: an *unbiased* one and a *biased* one. We first need to ensure that we never take edges in the graph that are not on a shortest path from s to t.

This type of two stages problems is usually called repetitive-mode [19] as opposed to single-shot. There are many problems on shortest paths that operate with this two stages procedure. We refer to [21] for a review of these methods.

Definition 1 (Predecessor Set [3]). *Let $G = (V, E)$ be a given graph and fix a node $s \in V$. For $v \in V$, the predecessor set of v is defined as:*

$$pre_s(v) = \{w \in V \mid s \rightarrow \cdots \rightarrow w \rightarrow v \in W_{sv}\}.$$

Said differently, the set $pre_s(v)$ contains the nodes w such that the edge $w \rightarrow v$ is the last one of a shortest path from s to v.

Definition 2 (Successor Graph). *Let s be a node. The successor graph related to s is a directed graph $G_s = (V_s, E_s)$ defined as follows:*

$$E_s = \{(w, v) \mid \forall v \in V, w \in pre_s(v)\}.$$

The set V_s corresponds to the nodes belonging to an edge from E_s which are the nodes accessible from s.

Said differently, G_s is the subgraph of G containing all the edges belonging to a shortest path starting from s. Given a successor graph $G_s = (V_s, E_s)$ and $v \in V_s$, we denote by $\mathbf{N}_v^-$ the list of incoming edges to node v. This list of elements is arbitrarily ordered, thus $\text{set}(\mathbf{N}_v^-) = pre_s(v)$.

Remark 1. The successor graph G_s is the union of all the BFS-trees rooted in s. It is acyclic. It contains exactly one source s and several sinks. Moreover, edges (u, v) of the successor graph go from a node at distance $k = d(s, u)$ to a node at distance $k + 1 = d(s, v)$.

The successor graph from s is computed in linear time in the size of the graph $O(n + m)$ by using a BFS that keeps all the optimal predecessors of a

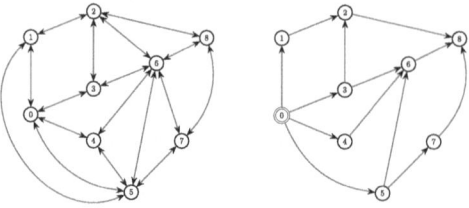

Fig. 1. (left) A graph and (right) the associated successor graph for $s = 0$.

node. Figure 1 introduces a graph-example and its associated successor graph G_0.

The successor graph from s ensures that starting a random walk with target t and taking the edges of G_s in a reversed way, the source s will be reached after exactly $k = d(s, t)$ transitions. The *generic random walk* approach is presented in Algorithm 1, once the preprocessing stage has been computed. During the latter preprocessing step, a *weight distribution* for edges of G_s is computed. We will discuss the weight distribution and the function rand_pred in Sect. 4. The second stage computes a *random walk* that starts from t and goes backwards until reaching s. Different random walks could be considered on $G_s = (V_s, E_s)$ by using different *weight functions* $\mathcal{W}$ that assign weights to the edges E_s. Then, the random walk standing on node $v \in V_s$ calculates a predecessor w given by the list $\mathbf{N}_v^- = [w_0, \ldots, w_k]$ according to the weight function $\mathcal{W}$.

Proposition 1 (Generic random walk complexity). *Any implementation of the preprocessing stage and Algorithm 1 needs $\Omega(m+n)$ operations and $\Omega(m+n)$ space for the* preprocessing step *and $\Omega(\ell)$ operations for the* random walk step *where ℓ is the length of the walk.*

Algorithm 1. Generic Random Walk

Input: s: source node, t: target node, $G_s = (V_s, E_s)$: Successor Graph from s, and
      $\mathcal{W}$: a weight function assigning weights to E_s
Output: path: A shortest path from s to t
 1: **function** RANDOM_WALK($s, t, G_s, \mathcal{W}$)
 2:     path $= [\,]$, $v = t$
 3:     **while** $v \neq s$ **do**
 4:         path $= [v] +$ path
 5:         $v =$ rand_pred($\mathbf{N}_v^-, \mathcal{W}$)   ▷ choosing a random predecessor according to $\mathcal{W}$
 6:     **end while**
 7:     **return** $[s] +$ path
 8: **end function**

The most straightforward random walk that we consider is the *Unbiased Random Walk* where the *weight function* $\mathcal{W}$ assigns weight 1 to all the edges in the successor graph. The random walk is thus unbiased. We can then show that the probability of sampling a shortest path $W \in \mathbf{W}_{st}$ is: $\mathbb{P}(W) = \prod_{v \in W \setminus \{s\}} \frac{1}{|\mathbf{N}_v^-|}$, which is not the uniform probability in general. Such a random walk is called *isotropic* in the literature (cf. e.g. [10]). The unbiased random walk (*URW*) necessitates $O(m + n)$ for the preprocessing step and $O(\ell)$ operations for the random generation step (ℓ being the length of the sampled walk).

3.2   Biased Random Walk: *BRW*

Now we present how to set the weights of $\mathscr{W}$ such that the uniformity of the sampling is guaranteed. The idea is to assign weights to the edges $(u, v) \in E_s$ of G_s based on the number of shortest paths arriving at u. Then, *Biased Random Walk (BRW)* is defined by assigning the weights of $\mathscr{W}$ as follows:

$$\forall (u, v) \in E_s, \mathscr{W}(u \to v) = \sigma_s(u). \tag{1}$$

Proposition 2 (BRW is uniform). *The biased random walk* BRW *solves the source-target uniform shortest path problem.*

Proof. Let W be the sampled random walk. Since it is computed on the successor graph G_s, $W \in \mathbf{W}_{st}$ is a shortest path. Now, let $W = v_0(:= s) \to v_1 \to \cdots \to v_k(:= t)$. The probability of sampling W can be computed by induction.

$$\mathbb{P}(W) = \prod_{i=1}^{k} \frac{\sigma_s(v_{i-1})}{\sigma_s(v_i)} = \frac{1}{\sigma_s(v_k)} = \frac{1}{\sigma_s(t)}.$$

The probability of going from v_k to v_{k-1} is $\sigma_s(v_{k-1})/\sigma_s(v_k)$. The same reasoning applies by induction (obviously $\sigma_s(s) = 1$), and then terms are telescoping. □

4   Implementations of BRW

Now we discuss the algorithmic complexity of computing *BRW*. The functions `compute_weights` (inside the preprocessing stage) and `rand_pred` (in Algorithm 1) are implemented in different ways depending on the amount of pre-processing done and the computations necessary in `rand_pred` afterward. The function `rand_pred` generates a random predecessor following a given discrete probability distribution. We propose four alternative implementations (linear, ordered, binary, alias) and prove that alias sampling provides the optimal algorithm for both stages. The *preprocessing phase* consists of computing the successor graph G_s and the weights of $\mathscr{W}$. This phase is done only once and stored in memory. The differences between implementations lie in the chosen ordering for $\mathbf{N}_v^-$ and in the computation of weights. The *queries* involve performing random walks on G_s following the distribution $\mathscr{W}$. The differences appear in the function `rand_pred`, which selects a predecessor w of v, where $w \in \mathbf{N}_v^-$.

For all implementations, we compute the values of $\sigma_s(v)$ for $v \in V$. These values are then used for the weight function $\mathscr{W}$. From the successor graph, it is

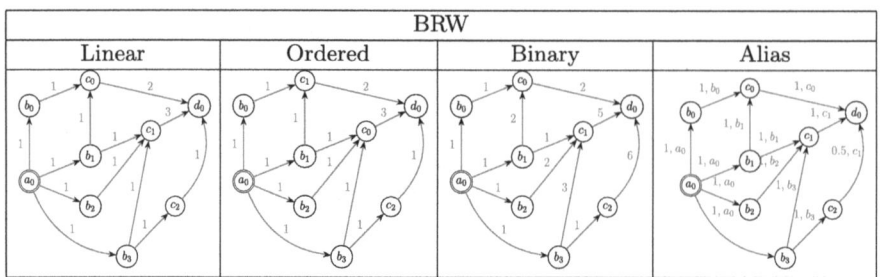

Fig. 2. Successor graph from $s = 0$ for the left graph of Fig. 1 with the different weight functions $\mathscr{W}$.

possible to dynamically compute the values of $\sigma_s(z)$ for all $z \in V$ by recurrence, noting that a shortest path from s to v is the concatenation of a shortest path from s to w with the edge (w, v), where w is a predecessor of v (i.e., $w \in \text{pre}_s(v)$). Thus, we have:

$$\sigma_s(s) = 1 \text{ and } \forall v \neq s, \ \sigma_s(v) = \sum_{w \in pre_s(v)} \sigma_s(w). \tag{2}$$

This recurrence can be found in [3, Lemma 3]. The values of σ_s will be used in `rand_pred` implementations of *BRW* as we shall see.

4.1   Linear Implementation

In the following implementation, the *preprocessing step* consists in computing the successor graph G_s from node s, and then computing the values of $\sigma_s(v)$ for all $v \in V$ using Eq. (2). An illustration of the successor graph with weights is given in second from left graph of Fig. 2. The *query step* then defines a predecessor by sampling a value $r \in [\![\sigma_s(v)]\!]$ (where $[\![n]\!] = \{1, \ldots, n-1\}$) and iterating through each predecessor of v to determine where r falls within the predecessor values. To run this procedure we implement the function `rand_pred`$(\mathbf{N}_v^-, \mathscr{W})$ that returns a predecessor w of v according the number of shortest paths to each $w \in \mathbf{N}_v^-$ with `rand_pred_it` that is given in Appendix A.2.

Proposition 3 (linear Preprocessing). *The preprocessing stage of* linear *to build the successor graph from* $s \in V$ *and to compute the values of* $\sigma_s(z)$ *for all* $z \in V$ *is done in time* $O(n + m)$.

Proof. A BFS is done, starting from the node s. This costs $O(n + m)$ operations. Then the values of $\sigma_s(v)$ are computed by the Recurrence (2) in linear time of the successor graph size.   □

The complexity of one query step corresponds to the biased random walk *BRW* and the random choice of predecessors.

Proposition 4 (linear Query). *The worst-case time complexity of the generation phase is* $O(n)$ *and this bound is tight.*

4.2   Ordered Implementation

In the *linear* implementation, we assumed an implicit order on the predecessor set $pres_s(v)$ of node v given by $\mathbf{N}^-$. The ordering can be optimized by minimizing the number of predecessor checks made by the function rand_pred_it. To achieve this, we order the list $\mathbf{N}_v^- = [w_0, \ldots, w_k]$ in decreasing order of the values of $\mathscr{W}(w_i \to v)$, so that $\mathscr{W}(w_0 \to v)$ is the largest value. For instance, in the successor graph given in Fig. 2, the node c_1 is renamed as c_0 in the ordered version. This minimizes the number of checks made in Line 5 of the function rand_pred_it. Let π be the ordering where the list $\mathbf{N}_v^-$ is ordered by decreasing values of $\sigma_s(w)$ where $w = \mathbf{N}_v^-[i]$ and $i \in [\![\mathbf{N}_v^-]\!]$. This classical ordering is called *boustrophedonic order* int the context of recursive generation [8].

Proposition 5 (Optimal order). *Ensuring uniform sampling of shortest paths, the ordering π is the order minimizing the number of predecessor checks for each possible values $r \in [\![\sigma_s(t)]\!]$.*

Therefore, the ordering π is better than any other ordering and improves the average running time. However, the worst-case time complexity is unchanged.

Proposition 6 (Ordered Preprocessing and Query). *The worst-case time of preprocessing is $O(m + n \log n)$ and the worst-case time complexity of the generation phase is $O(n)$ and this bound is tight.*

4.3   Binary Implementation

The procedure rand_pred used in Algorithm 1 corresponds to finding the position of an element r in a list. Thus, rather than using an iterative approach as it is the case now, it is possible to use binary search instead by adding few operations in the preprocessing stage. As we shall see this addition does not affect the theoretical running time of the preprocessing stage but obviously improves the theoretical bounds of the queries afterward.

In function rand_pred_it (Appendix A.2), we draw a value $r \in [\![\sigma_s(v)]\!]$ and if $[w_0, \ldots, w_{k-1}] = \mathbf{N}_v^-$, we want to return the node w_i such that

$$\sum_{j=0}^{i-1} \sigma_s(w_j) \leq r < \sum_{j=0}^{i} \sigma_s(w_j).$$

To optimize this, it is natural to work with partial sums of the weights of the incoming edges and perform a binary search on it. Thus, we assign new weights to the edges. In this scheme, the weights are defined as follows: for $v \in V_s$, let $[w_0, \ldots, w_{k-1}] = \mathbf{N}_v^-$. Then for each $w_i \to v \in E_s$, we assign the weight $\mathscr{W}(w_i \to v) = \sum_{j=0}^{i} \sigma_s(w_j)$. An illustration is given in the second from right graph of Fig 2. Thus, the number of shortest paths $\sigma_s(t)$ from s to t is just the largest label of the incoming edges of t. We will denote this edge by t^*. Now we can determine the edge taken by the random walk more efficiently. Given a node v and a value r we execute a binary search on $\mathbf{N}_v^-$ to find the index i such that

$$\mathscr{W}(w_{i-1} \to v) \leq r < \mathscr{W}(w_i \to v). \tag{3}$$

The predecessor of v we look for is then w_i. Thus, we use Algorithm 4 (cf. Appendix A.2) to implement rand_pred and modify the weights $\mathscr{W}$ computed during the preprocessing.

Proposition 7 (Binary Preprocessing). *The preprocessing time complexity is $O(n + m)$.*

Proof. Additionally to $O(n + m)$ for the successor graph construction G_s and the values of σ_s. The weights on G_s is done in $O(m)$ operations using an array containing the values of $\sigma_s(t)$ for all $t \in V$. We compute the partial sum step by step writing the weight of $(w_i \to v)$ at the step when we add $\sigma_s(w_i)$ to the partial sum. □

Proposition 8 (Binary Query). *Let W be the sampled shortest path, let $\ell = |W|$ then the generator's worst-case time complexity is $O\left(\ell \log \left(\frac{n}{\ell}\right)\right)$.*

Thus, the complexity of a query of the generator depends on the length ℓ of the generated shortest path. The binary implementation has a better worst-case running time than the linear and ordered. For instance, if the size of the sampled path $\ell = \log n$ (which is typical in many random graph models), the generation time in worst case is $O(\log^2 n)$ with binary implementation and $O(n)$ for the linear and ordered one.

4.4   Alias Implementation

We want to further improve the generation of the random walk. Note that for each node $v \in V_s$, we compute the distribution of the predecessors $\mathbf{N}_v^-$ already during the preprocessing stage. This distribution is fixed, and not dynamical. We can use this constraint to design a constant-time random generator that selects a predecessor by applying the classical *Alias method* defined by Walker [22].

Let us recall the foundations of the method. To each edge $(w_i, v) \in E$, we now associate a pair $\mathscr{W}(w_i \to v) = (t_i, al_i) \in [0, 1] \times \mathbf{N}_v^-$ where t_i is a real number in $[0, 1]$ called *threshold* and al_i is a predecessor in $\mathbf{N}_v^-$ called *alias*. We compute these weights (t, al) such that the following condition holds

$$\forall w_j \in \mathbf{N}_v^-, \; t_j + \sum_{\substack{i \in [\![|\mathbf{N}_v^-|]\!] \\ al_i = w_j}} (1 - t_i) = \frac{\sigma_s(w_j)}{\sigma_s(v)} \cdot |\mathbf{N}_v^-|. \tag{4}$$

Then, to choose a predecessor $w \in \mathbf{N}_v^-$ using the alias method, we draw an index $i \in [\![|\mathbf{N}_v^-|]\!]$ uniformly at random. Let $(t_i, al_i) = \mathscr{W}(w_i \to v)$, we choose the predecessor w_i with probability t_i and the predecessor al_i with probability $1 - t_i$. Thus, the determination of the predecessor is done in constant time. We then use the alias version rand_pred_al given in Algorithm 6 (Appendix A.2) to implement rand_pred of Algorithm 1. As an example, in the rightmost graph of Fig. 2, we have $\mathcal{W}(c_2 \to d_0) = (0.5, c_1)$ which means that in the case where the edge $(c_2 \to d_0)$ is drawn (case $i = 2$), we walk to c_2 with probability 0.5 and to c_1 otherwise.

Proposition 9. *For $v \in V$ and for all $w \in \mathbf{N}_v^-$:*

$$\mathbb{P}(rand\_pred\_al(\mathbf{N}_v^-, \mathcal{W}) = w) = \frac{\sigma_s(w)}{\sigma_s(v)}.$$

Proof. The proposition is a consequence of the Condition (4). Let $w_j \in \mathbf{N}_v^-$. Denote by X the uniformly drawn index in $[\![|\mathbf{N}_v^-|]\!]$. The probability law gives: $\mathbb{P}(w_j) = \frac{1}{|\mathbf{N}_v^-|}(\mathbb{P}(w_j|X = j) + \sum_{i \neq j} \mathbb{P}(w_j|X = i))$. For $i \neq j$ if $al_i \neq w_j$ then the probability $\mathbb{P}(w_j|X = i)$ is 0 else it is $1 - t_i$. Thus we have $\mathbb{P}(w_j) = \frac{1}{|\mathbf{N}_v^-|}(t_j + \sum_{\substack{i \neq j \\ al_i = w_j}} (1 - t_i)) = \frac{\sigma_s(w_j)}{\sigma_s(v)}$. □

Proposition 10 (Alias Preprocessing). *The preprocessing stage to compute the successor graph from s the associated weight function $\mathcal{W}$ for the Alias method is computed in time $O(n + m)$.*

Proof. The computation of the threshold and alias with respect to Condition (4) is done in linear time in the size of $\mathbf{N}_v^-$. To use the alias method at node $v \in V$, we compute $\mathcal{W}(w \to v)$ for all $w \in \mathbf{N}_v^-$, which costs $O(|\mathbf{N}_v^-|)$. Doing this for each node v in the successor graph costs $O(\sum_{v \in V} |\mathbf{N}_v^-|) = O(m)$ time. Since the rest of the preprocessing remains unchanged, the running time is then $O(n + m)$. □

Proposition 11 (Alias Query). *Let W be the sampled shortest path, let $\ell = |W|$ then the generator's worst-case time complexity is $O(\ell)$.*

Proof. To build the path W, we have to call ℓ times the function `rand_pred_al`. This function executes in constant time, since we only draw two uniform random variables. Thus the total time complexity is $O(\ell)$. □

Theorem 1. *The alias implementation is an optimal implementation in terms of the asymptotic worst case complexity for both preprocessing and sampling.*

Proof. It is a consequence of the Proposition 1 with the results from Propositions 10 and 11. □

5   Optimal Random Bit Complexity and Unranking

The Alias method is a powerful technique to obtain an efficient generation algorithm. However the main drawback concerning the method is that it relies on floating arithmetic which raises a number of questions related to the rounding approximations, the accuracy of the generation, the bias induced and the quantity of required random bits. We propose in the next section what seems a new approach by adapting the Alias method in the context of arbitrary-size integer arithmetic. We thus rely on the exact uniform distribution (that is central for us) and furthermore we are able to prove the optimality of our method for the number of random bits that are necessary during the sampling. As far as we know, we never encountered this adaptation in the literature.

5.1   A New Alias Method with Integers

We focus on a set $\mathcal{V} = \{v_0, v_1, \ldots, v_{n-1}\}$ of objects, each v_i has a positive integer weight w_i. We aim at sampling each v_i with probability w_i/W where $W = \sum_{j=0}^{n-1} w_j$. We first run a preprocessing step building two data structures of cumulated space complexity that is linear in n and that will allow during the random sampling step to get a random object with time complexity $O(1)$. Our approach is directly related to the Euclidean division of W by n, the number of objects. Let us introduce the quotient q and the remainder r satisfying

$$W = q \cdot n + r \quad \text{with} \quad 0 \leq r < n. \tag{5}$$

We then dispatch the total weight W among two tables: a first one of size n corresponding to the Alias table T. Cells of T are indexed between 0 to $n-1$ and each one contains a triplet. And a second table R of size r containing the remaining weight of the distribution. Let us devise how to distribute the objects inside the two tables. We first define two stacks S_0 and S_1 containing respectively the objects whose weight is larger than or equal to q and those whose weight is (strictly) smaller than q. While T is not full, we iteratively consider

- the first elements of each stack (if both exist), x_0 and x_1 with respective weight $y_0 \geq q$ and $y_1 < q$. We then distribute the weight y_0 as $q - y_1$ and $y_0 - (q - y_1)$. We take the first empty cell of the table T, and assign to it the triplet (y_1, x_1, x_0). There are two possibilities for the remaining part of x_0 with remaining weight $y_0 - (q - y_1)$. Either the later is still greater than or equal to q, and we keep it in S_0 with its remaining weight $y_0 - (q - y_1)$, or the remaining weight $y_0 - (q - y_1)$ is smaller than q and we move the object x_0 with its remaining weight $y_0 - (q - y_1)$ in the second stack S_1.
- the second stack S_1 is empty, and x_0 with weight y_0 is the first element from S_0. Then, we fill the next cell from T with the triplet $(q, x_0, \emptyset)$, and we deal with the remaining weight $y_0 - q$ for x_0 as in the previous case.

When the Alias table T is full, we fill the table R with the remaining weight of the distribution, whose value is r, in fact we put in each cell one of the remaining object (with weight 1). Eventually if an object has a remaining weight greater than 1 it will appear several times in R. The key idea proving the correctness of the building of T and R is based on the fact that S_0 is never empty (until T is full). This is due to the fact that whenever S_1 is not empty we remove one of its element at each loop.

We are now ready to state the sampling step algorithm satisfying the claimed constant time complexity.

Proposition 12. *The preprocessing requires $O(n)$ operations and is made only once. Then, in order to sample an object according to the weight distribution, one samples an integer $i \in [\![W]\!]$ uniformly at random, and then computes its Euclidean division by q: $i = q_i \cdot q + r_i$. If $q_i < n$, then one extracts the triplet (x, v, w) in $T[q_i]$ and returns the object v if $r_i < x$, otherwise, one returns the object w. If $q_i = n$, then one returns the object $R[r_i]$. The sampling process is done in $O(1)$ operations.*

A proof sketch is provided in Appendix A.3.

5.2   Unranking Algorithms

The problem of unranking objects emerges as one of the most fundamental challenges in combinatorial generation, as seen in Kreher and Stinson's book [12]. Usually, the approach for constructing structures is based on a recursive decomposition [16]. The schema involves leveraging this decomposition to build a larger object from smaller ones. The term *unranking* comes from the fact that we totally order the objects under consideration, thus each one gets a rank between 0 and the number of objects minus 1. Then we build the object of rank r from scratch. Our three algorithms BRW-linear, Ordered and Binary can easily be adapted to the context of unranking, without modifying their time-complexity. However, some care is needed to keep the time complexity of Alias method especially during the update of the rank after each predecessor determination. In our Alias method with integers, we derive in a way an unranking algorithm based on a multiset of objects where each object appears a number of times equal to its weight and the total order is related to the constructions of both tables T and R.

We are now ready to derive an unranking algorithm to reconstruct the shortest path of a given rank, as follows: the preprocessing stage consists of computing the predecessor graph G_s and $\forall v \in V, \sigma_s(t)$ using Eq. (2) and for each list $\mathbf{N}_v^-$ it computes augmented versions of the tables T and R as described in Sect. 5.1 keeping additionally partial sums (see Appendix A.3). The preprocessing still necessitates $O(n + m)$ operations. The query stage does the following: (1) Sample uniformly at random an integer $\rho \in [\![\sigma_s(t)]\!]$. (2) Alias method with integers is used to choose the predecessor x of t and update the rank as $\tilde{\rho}$ which requires $O(1)$ time. (3) Iterate the process with the new rank $\tilde{\rho}$ and new node x until reaching s. The query stage then requires $O(\ell)$ operations. We only use a single call for random bits in the rank sampling ($\rho \in [\![\sigma_s(t)]\!]$) which is clearly optimal. Since we need at least to distinguish between all shortest paths. The process is detailed in Appendix A.3. From the preceding, it follows that:

Theorem 2. *The unranking algorithm is an optimal implementation in terms of the asymptotic worst case complexity for both the preprocessing and the sampling stages and is also optimal in terms of random bits complexity.*

6   Experimental Evaluation and Application

Our algorithms are implemented in C to enhance efficiency. The code is open-source and is available here[1]. We conducted our experiments on an Intel(R) Xeon(R) Silver 4210R CPU at 2.40 GHz without parallel processes. In fact, the preprocessing can be fully parallelized since constructing successor graphs from each node is an independent task.

---

[1] https://github.com/simon-dreyer/Shortest_path_sampling.

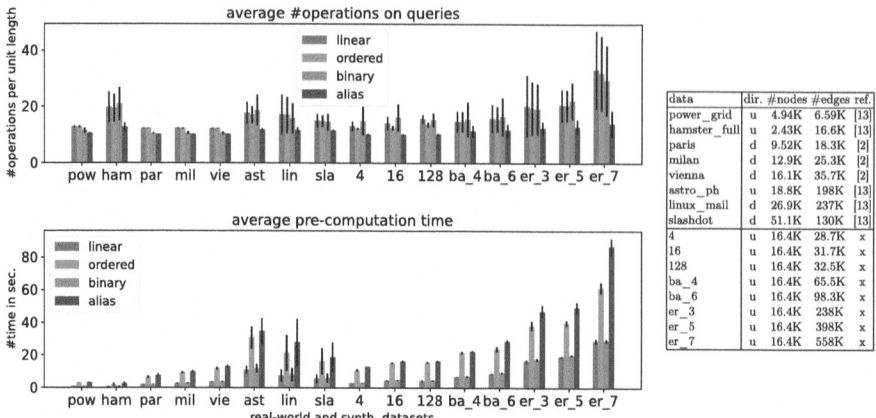

Fig. 3. (Up Left): Query complexity on the different graphs. The y-axis corresponds to the average number of operations (arithmetic, tests, affectation) made by 50 randomly selected source-target (s, t) pairs. For each (s, t) we generated 50 000 shortest paths and summed the number of operations made and divided them by $(50\ 000 \times d(s, t))$ the black bars gives the standard deviation. (Down Left): Preprocessing complexity on the different graphs. The y-axis corresponds to the average running time per node in seconds with the black bars giving the standard deviation. (Right): Real-world and synthetic datasets. Upper part corresponds to the real-world dataset and the bottom part corresponds to the synthetic one. Columns from left to right indicate whether the graph is directed 'd' or undirected 'u', the number of nodes and the number of edges. Our synthetic dataset contains from top to bottom: bi-dimensional grid graphs of 16 384 nodes with 4, 16 and 128 rows, Barabási-Albert graphs with parameter m equals to 4 and 6, Erdős-Rényi graphs with $p = \frac{i\log(n)}{n}$ for $i = 3.5$ and 7.

In our experiments, we compare *BRW* and its various implementations to assess their performance on both real-world and synthetic datasets. We used several real-world datasets from different domains, including scientific collaboration networks, city networks, social networks, and power grids. For synthetic networks, we used Erdős-Rényi and Barabási-Albert random graphs, and 2-d grids.

Figure 3 summarizes our experiments. On the preprocessing stage results show that the ordered implementation takes more time to compute than the linear and binary implementations due to the additional sorting step required. The Alias implementation also requires more computations since we need to associate a pair of values with each edge. Although the theoretical worst-case running times are similar for all variants, in practice, the Alias and ordered implementations are slower by a factor of at least 2. On the query stage, random walks run instantly once the preprocessing is complete. We therefore count the average number of operations performed by each query (sampling), normalized by the distance between pairs of sampled nodes to avoid having large differences between the dataset results. The Alias implementation consistently outperforms the others, confirming our theoretical analysis. The binary implementation performs well when the average distance is short. Therefore, for 2-dimensional grids where the average shortest path is linear, the binary implementation performs the worst. Finally, the ordered implementation is always at least as efficient as the linear one and sometimes better.

A   Appendix

A.1   Appendix Related to Section 2

Figure 4 shows that the number of shortest paths between two nodes can be exponential. Therefore the naive algorithm presented in Sect. 2 consisting of listing all shortest paths in the graph should be avoided.

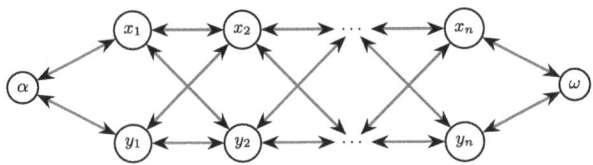

Fig. 4. Graph with $2n + 2$ nodes and 2^n shortest paths between α and ω.

In the following we show that the probability of each shortest path between α and ω of the graph G_k see Fig. 5 when we use the algorithm random weights to sample a path. This will show that random weights leads to a biased distribution on the sampled shortest paths.

We denote $a, b_1, \ldots, b_k, c_1, \ldots, c_k, d, e, f$ the random weights on the edges as in Fig. 5. The weights follows a continuous uniform distribution on $[1 - \frac{1}{n}, 1 + \frac{1}{n}]$. For the sake of simplicity, we recenter and reduce the variables. This does not change the selected shortest path (we just subtract by $1 - \frac{1}{n}$ every weights and then multiply them all by $\frac{n}{2}$). From now on, we suppose that all the weights follow the standard uniform distribution $U(0, 1)$.

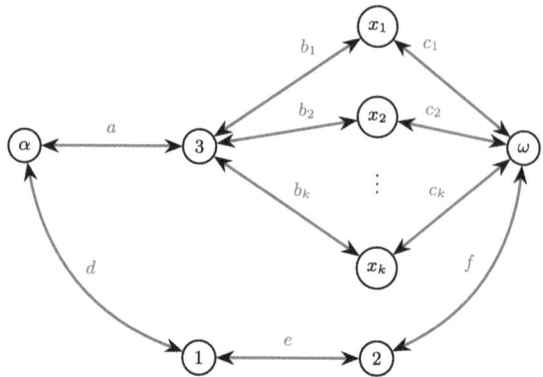

Fig. 5. The family of graphs G_k with the random variables defining the weights of the edges.

We define the following random variables:

- $Y_i = b_i + c_i$ for all $1 \leq i \leq k$
- $Z = d + e + f$: total weight of the bot path.

Let $W_0 = \alpha \to 1 \to 2 \to \omega$ and $W_i = \alpha \to 3 \to x_i \to \omega$ for $1 \leq i \leq k$. We want to compute the probability $\mathbb{P}(W_0) = 1 - \sum_{i=1}^{k} \mathbb{P}(W_i)$. All the weights follow the same distribution, by symmetry $\mathbb{P}(W_0) = 1 - k\mathbb{P}(W_1)$. Note that

$$\mathbb{P}(W_1) = \mathbb{P}\left((a + Y_1 < Z) \cap \left(\bigcap_{i=2}^{k}(a + Y_1 < a + Y_i)\right)\right)$$

$$= \mathbb{P}\left((a + Y_1 < Z) \cap \left(\bigcap_{i=2}^{k}(Y_1 < Y_i)\right)\right).$$

We condition on $t \in [0, 2]$ the value of Y_1. Denote by f_Y the probability density function of Y_1.

$$\mathbb{P}(W_1) = \int_0^2 \mathbb{P}\left((a + t < Z) \cap \left(\bigcap_{i=2}^{k}(t < Y_i)\right) \Big| Y_1 = t\right) f_Y(t)dt.$$

The weights are drawn independently thus the variables $(Y_i)_{1 \leq i \leq k}$ and $Z - a$ are independent. It follows:

$$\mathbb{P}\left((a + t < Z) \cap \left(\bigcap_{i=2}^{k}(t < Y_i)\right) \Big| Y_1 = t\right) = \mathbb{P}(t < Z - a).\prod_{i=2}^{k}\mathbb{P}(t < Y_i).$$

All the variables Y_i follow the same distribution (sum of two independent standard uniform variables). We denote F_Y the cumulative density function of the Y_i and F_{Z-a} that of $Z - a$. Then

$$\mathbb{P}(W_1) = \int_0^2 (1 - F_{Z-a}(t))(1 - F_Y(t))^{k-1} f_Y(t) dt.$$

Because the density function is the derivative of the cumulative density function we can do an integration by parts.

$$k\mathbb{P}(W_1) = 1 - F_{Z-a}(0) - \int_0^2 (1 - F_Y(t))^k f_{Z-a}(t) dt,$$

where f_{Z-a} is the density function of $Z - a$. Finally

$$\mathbb{P}(W_0) = F_{Z-a}(0) + \int_0^2 (1 - F_Y(t))^k f_{Z-a}(t) dt.$$

a and Z are independent. Y and Z follow the *Irwin-Hall distribution* for $n = 2$ and $n = 3$ respectively then

$$F_{Z-a}(0) = \mathbb{P}(Z < a) = \int_0^1 F_Z(t) dt = \frac{1}{4!} = \frac{1}{24}.$$

and the density of $Z - a$ is the convolution product of the density of Z and $-a$

$$f_{Z-a}(t) = \int_{-1}^0 f_Z(t - u) du = \int_0^1 f_Z(t + u) du.$$

The final formula for $\mathbb{P}(W_0)$ is then

$$\mathbb{P}(W) = \frac{1}{24} + \int_0^2 (1 - F_Y(t))^k \int_0^1 f_Z(t + u) du \, dt. \tag{6}$$

For $k = 2$ the Eq. (6) gives $\mathbb{P}(W_0) = \frac{737}{2016}$. As k grows to infinity the term $(1 - F_Y(t))^k$ converge to 0 for $t \in]0, 2]$. Moreover the function $t \mapsto (1 - F_Y(t))^k f_{Z-a}(t)$ is continuous on $[0, 2]$ then it is dominated by a constant. Then the dominated convergence theorem ensures that the integral converge towards 0. Therefore Eq. 6 gives $\mathbb{P}(W_0) \xrightarrow{k \to +\infty} \frac{1}{24}$.

A.2   Appendix Related to Section 3

Algorithm 2. Preprocessing step

Input: G: a graph, s: source node, t: target node
Output: G_s: Successor Graph from s and the values of $\mathscr{W}$ computed
 1: **function** PREPROCESSING(s, t)
 2:     $G_s = \texttt{BFS\_with\_predecessors}(G, s)$
 3:     $\mathscr{W} = \texttt{compute\_weights}(G_s)$
 4:     **return** $(G_s, \mathscr{W})$
 5: **end function**

Algorithm 3. Linear implementation of `rand_pred`

Input: $\mathbf{N}_v^-$: List of predecessors of each node v, $\mathscr{W} : E_s \to \mathbb{R}$ a weight function and
      σ_s: an array with the number of sh. paths from s to each node v
Output: $w \in \mathbf{N}_v^-$ according to the weights $\mathscr{W}$
 1: **function** RAND_PRED_IT$(\mathbf{N}_v^-, \mathscr{W}, \sigma_s)$
 2:     $r = \text{uniform}(\llbracket \sigma_s[v] \rrbracket)$
 3:     $i = 0$, $w = \mathbf{N}_v^-[i]$
 4:     $r = r - \mathscr{W}(w \to v)$
 5:     **while** $r \geq 0$ **do**
 6:         $i = i + 1$, $w = \mathbf{N}_v^-[i]$
 7:         $r = r - \mathscr{W}(w \to v)$
 8:     **end while**
 9:     **return** w
10: **end function**

The following proof stresses out that the order that minimizes the number of predecessors checks is the order where the nodes with the largest $\sigma_s(v)$ are those with the smallest index.

Proof (of Proposition 5). Fix a node $v \in V$. Let $\mathbf{N}_v^- = [w_0, \ldots, w_k]$. Note that `rand_pred_it`$(\mathbf{N}_v^-, \mathscr{W})$ returns the predecessor w_i for exactly $\sigma_s(w_i)$ different values of $r \in \llbracket \sigma_s(v) \rrbracket$. When the function returns w_i, it has iteratively checked all the i first predecessors. Then the average number of predecessors checked before finding the right one in `rand_pred_it` is $(1/\sigma_s(v)) \sum_{i=1}^{k} i \cdot \sigma_s(w_i)$. To minimize the sum the largest values of $\sigma_s(w_i)$ should be matched with the smallest indexes i. That is the ordering π.

We give the proof of the preprocessing complexity of the ordered algorithm.

Proof (of Proposition 6). We need to ensure that the function `rand_pred_it` traverses the values of $\mathbf{N}_v^- = [w_0, \ldots, w_k]$ in decreasing order of the values of $\sigma_s(w_i)$. To achieve this, we rewrite $\mathbf{N}_v^-$ after computing all the values $\sigma_s(v)$. We start by ordering the nodes $w \in V$ by decreasing values of $\sigma_s(w)$, which costs $O(n \log n)$. For each $w \in V$ we store its successors in $\mathbf{N}_w^+$. That is $v \in \mathbf{N}_w^+$ if and only if $w \in \mathbf{N}_v^-$. Then, for every $w \in V_s$ in decreasing order of $\sigma_s(w)$, and for every successor node $v \in \mathbf{N}_w^+$, we write w in $\mathbf{N}_v^-$. After this operation, all the $\mathbf{N}_v^-$ are rewritten in decreasing order of $\sigma_s(w)$. The operation of rewriting costs $O(m)$ as $\sum_{v \in V} |\mathbf{N}_v^+| \leq m$. Thus, the overall preprocessing stage cost becomes $O(m + n \log n)$.

Algorithm 4. Binary implementation of `rand_pred`

Input: $\mathbf{N}_v^-$: List of predecessors of each node v, $\mathscr{W} : E_s \to \mathbb{R}$ a weight function and
   σ_s: an array with the number of sh. paths from s to each node v
Output: $w \in \mathbf{N}_v^-$ according to the weights $\mathscr{W}$
1: **function** RAND_PRED_BIN($\mathbf{N}_v^-, \mathscr{W}, \sigma_s$)
2:     $r = \mathrm{uniform}(\llbracket \sigma_s[v] \rrbracket)$
3:     $i = \mathtt{bin\_search\_index}(\mathbf{N}_v^-, \mathscr{W}, r)$
4:     **return** $w = \mathbf{N}_v^-[i]$
5: **end function**

Algorithm 5. Binary implementation of `rand_pred`

1: **function** BIN_SEARCH_INDEX($\mathbf{N}_v^-, \mathscr{W}, r$)
2:     $i = -1, j = |\mathbf{N}_v^-| - 1$
3:     **while** $j - i > 1$ **do**
4:         $x = \lfloor \frac{i+j}{2} \rfloor, w = \mathbf{N}_v^-[x]$
5:         **if** $\mathscr{W}(w \to v) > r$ **then** $j = x$ **else** $i = x$
6:     **end while**
7:     **return** j
8: **end function**

Proof (of Proposition 8). Suppose that $W = x_0 \to \cdots \to x_\ell$. To find the predecessor of x_i we do a binary search on all the incoming edges to find the edge satisfying Eq. (3). This costs $O\left(\log(|\mathbf{N}_{x_i}^-|)\right)$ operations. So the total complexity of the random walk is of order of

$$\sum_{i=1}^{\ell} \log(|\mathbf{N}_{x_i}^-|) = \log\left(\prod_{i=1}^{\ell} |\mathbf{N}_{x_i}^-|\right).$$

Let n_k be the number of nodes at distance k from x_0. With the Remark 1, we obtain $|\mathbf{N}_{x_i}^-| \leq n_{i-1}$. Then, it follows

$$\prod_{i=1}^{\ell} |\mathbf{N}_{x_i}^-| \leq \prod_{i=1}^{\ell} n_{i-1} \leq \left(\frac{n_0 + \cdots + n_{\ell-1}}{\ell}\right)^{\ell},$$

by using the comparison between arithmetic and geometric means in the last step. We get: $\sum_{i=1}^{\ell} \log(|\mathbf{N}_{x_i}^-|) \leq \ell \log\left(\frac{n}{\ell}\right)$, by noting that $n_0 + \cdots + n_{\ell-1} \leq n.\square$

Algorithm 6. Alias implementation of `rand_pred`

Input: $\mathbf{N}_v^-$: predecessors of each v, $\mathscr{W}$: weight function
Output: $w \in \mathbf{N}_v^-$ using the alias method
 1: **function** RAND_PRED_AL($\mathbf{N}_v^-$, $\mathscr{W}$)
 2:     $i = \mathrm{uniform}(\llbracket|\mathbf{N}_v^-|\rrbracket)$, $t = \mathrm{uniform}([0,1])$
 3:     $w = \mathbf{N}_v^-[i]$, $(t', al) = \mathscr{W}(w \to v)$
 4:     **if** $t \leq t'$ **then return** w **else return** al
 5: **end function**

A.3   Appendix Related to Section 5

Proof (Proof Sketch of Proposition 12). We use the notations introduced in Eq. (5) and for each stack S_i $(i = 0, 1)$ we define its weight $\|S_i\|$ to be the cumulated weight of the elements it contains. Furthermore if a cell from T is not empty then it contains a weight q.

The algorithm is based on the following loop invariant:
At each iteration i, we remove a weight q from the total weight $\|S_0\| + \|S_1\|$ that is added to the total weight contained in T and the total weight of each element $v_i \in \mathscr{V}$ is unchanged (it is distributed over T, R, S_0 and S_1).

The algorithm terminates after n iterations. Finally, the sampling an integer $i \in \llbracket W \rrbracket$ and returning the result allows to sample an object according to the distribution of w_j for $j \in \llbracket n \rrbracket$. □

References

1. Achlioptas, D., Clauset, A., Kempe, D., Moore, C.: On the bias of traceroute sampling: or, power-law degree distributions in regular graphs. J. ACM **56**(4), 1–28 (2009)
2. Boeing, G.: OSMnx: new methods for acquiring, constructing, analyzing, and visualizing complex street networks. Comput. Environ. Urban Syst. **65**, 126–139 (2017)
3. Brandes, U.: A faster algorithm for betweenness centrality. J. Math. Sociol. **25**(2), 163–177 (2001)
4. Ciulla, F., Perra, N., Baronchelli, A., Vespignani, A.: Damage detection via shortest-path network sampling. Phys. Rev. E **89**(5), 052816 (2014)
5. Clauset, A., Moore, C.: Traceroute sampling makes random graphs appear to have power law degree distributions. arXiv preprint cond-mat/0312674 (2003)
6. Crespelle, C., Tarissan, F.: Evaluation of a new method for measuring the internet degree distribution: simulation results. Comput. Commun. **34**(5), 635–648 (2011)
7. Dall'Asta, L., Alvarez-Hamelin, I., Barrat, A., Vázquez, A., Vespignani, A.: Exploring networks with traceroute-like probes: theory and simulations. Theoret. Comput. Sci. **355**(1), 6–24 (2006)
8. Flajolet, P., Zimmermann, P., Van Cutsem, B.: A calculus for the random generation of labelled combinatorial structures. Theoret. Comput. Sci. **132**(1–2), 1–35 (1994)

9. Flaxman, A.D., Vera, J.: Bias reduction in traceroute sampling–towards a more accurate map of the internet. In: International Workshop on Algorithms and Models for the Web-Graph, pp. 1–15. Springer (2007)
10. Genitrini, A., Pépin, M., Peschanski, F.: A quantitative study of fork-join processes with non-deterministic choice: application to the statistical exploration of the state-space. Theor. Comput. Sci. **912**, 1–36 (2022)
11. Guillaume, J.L., Latapy, M.: Relevance of massively distributed explorations of the internet topology: Simulation results. In: Proceedings IEEE 24th Annual Joint Conference of the IEEE Computer and Communications Societies, vol. 2, pp. 1084–1094. IEEE (2005)
12. Kreher, D.L., Stinson, D.R.: Combinatorial Algorithms: Generation, Enumeration, and Search. CRC Press (1999)
13. Kunegis, J.: Konect: the Koblenz network collection. In: Proceedings of the 22nd International Conference on World Wide Web, pp. 1343–1350 (2013)
14. Lakhina, A., Byers, J.W., Crovella, M., Xie, P.: Sampling biases in IP topology measurements. In: IEEE INFOCOM 2003. Twenty-second Annual Joint Conference of the IEEE Computer and Communications Societies, vol. 1, pp. 332–341. IEEE (2003)
15. Leguay, J., Latapy, M., Friedman, T., Salamatian, K.: Describing and simulating internet routes. Comput. Netw. **51**(8), 2067–2085 (2007)
16. Nijenhuis, A., Wilf, H.S.: Combinatorial Algorithms. Computer Science and Applied Mathematics. Academic Press, New York (1975)
17. Petermann, T., De Los Rios, P.: Exploration of scale-free networks: do we measure the real exponents? Eur. Phys. J. B **38**, 201–204 (2004)
18. Pósfai, M., Fekete, A., Vattay, G.: Shortest-path sampling of dense homogeneous networks. Europhys. Lett. **89**(1), 18007 (2010)
19. Preparata, F.P., Shamos, M.I.: Computational Geometry: An Introduction. Springer (2012)
20. Rezvanian, A., Meybodi, M.R.: Sampling social networks using shortest paths. Phys. A: Stat. Mech. Appl. **424**, 254–268 (2015)
21. Sommer, C.: Shortest-path queries in static networks. ACM Comput. Surv. (CSUR) **46**(4), 1–31 (2014)
22. Walker, A.J.: New fast method for generating discrete random numbers with arbitrary frequency distributions. Electron. Lett. **8**(10), 127–128 (1974)
23. Wang, H., Van Mieghem, P.: Sampling networks by the union of m shortest path trees. Comput. Netw. **54**(6), 1042–1053 (2010)
24. Zhang, G.Q., Zhou, S., Wang, D., Yan, G., Zhang, G.Q.: Enhancing network transmission capacity by efficiently allocating node capability. Phys. A: Stat. Mech. Appl. **390**(2), 387–391 (2011)

An Efficient Implementation of Cosine Distance on Minimal Absent Word Sets Using Suffix Automata

Mohammad Tamimul Ehsan$^{(\boxtimes)}$ ⬤, Sk. Sabit Bin Mosaddek⬤,
and M Saifur Rahman⬤

Bangladesh University of Engineering and Technology, Dhaka, Bangladesh
`tamimehsan99@gmail.com, mrahman@cse.buet.ac.bd`

Abstract. The Cosine Distance on Minimal Absent Word Sets, CD-MAWS in short, was recently introduced for alignment-free phylogeny estimation. Here we introduce a refined CD-MAWS method, significantly reducing computational complexity from $\mathcal{O}(km^3n + km^2n\log n)$ to $\mathcal{O}(knm + nm^2)$ while maintaining tree quality. Here, m is the number of species, n is the size of the whole genome of a species, and k is the maximum length of a minimal absent word (MAW). This advancement is achieved through a revised cosine distance calculation method, binary encoding of MAWs, and the adoption of suffix automata for MAW generation, addressing the main computational bottleneck and setting a better runtime for alignment-free phylogenetic analysis.

Keywords: Phylogeny · CD-MAWS · Suffix Automata

1 Introduction

The study of phylogeny is crucial for understanding evolutionary relationships among species. The Cosine Distance on Minimal Absent Word Sets (CD-MAWS) [1] offers an alignment-free approach for phylogeny reconstruction from whole genome analysis. However, CD-MAWS is hampered by its computational inefficiency. Our improvements aim to streamline this process, offering a more practical solution for genomic analysis. Building on the foundational work of the CD-MAWS method, our research extends the application of Minimal Absent Words (MAWs) in phylogenetic analysis by addressing computational bottlenecks and enhancing methodological efficiency.

We introduce suffix automata for MAW generation in lexicographical order and leverage this ordered generation for determining the intersection set of MAWs between species pairs in an efficient way. Furthermore, we propose to encode each nucleotide in two bits so as to process any MAW of up to a certain length in a very efficient way in various operations. We thus reduce computational complexity while maintaining accuracy. Our work opens up new avenues for large-scale genomic studies, offering a faster and more scalable tool for unraveling the complexities of evolutionary histories.

2   Background and Related Works

2.1   Background

Phylogenetic analysis is essential for understanding evolutionary relationships among species. Traditional methods, such as multiple sequence alignment (MSA), have been widely used but come with significant computational challenges and limitations, particularly when dealing with large datasets or highly divergent sequences. To address these issues, alignment-free methods have gained attention in recent years [2–5]. These methods bypass the need for sequence alignment, thus reducing computational complexity.

Minimal Absent Words (MAWs) is a novel concept in alignment-free phylogenetic analysis. MAWs are sequences that are conspicuously absent from a given genome but present in related genomes. These absent sequences provide unique phylogenetic signals that help in constructing evolutionary relationships. The use of MAWs, combined with mathematical measures such as cosine distance, offers a promising approach for phylogeny estimation, balancing computational efficiency and accuracy.

2.2   Related Works

Several alignment-free methods have been proposed and benchmarked in recent studies. Zielezinski et al. (2019) [9] conducted a comprehensive benchmarking of alignment-free sequence comparison methods, highlighting the strengths and weaknesses of various approaches. Their study provides a foundation for evaluating new methods such as the one proposed in this paper.

One notable alignment-free method is the use of k-mers [10], which involves counting the frequency of fixed-length substrings within sequences. This method has been widely adopted due to its simplicity and effectiveness. However, it may not fully capture the phylogenetic signals present in the sequences, especially for more complex datasets. Another significant contribution to the field is the work of Liu et al. (2009) [8], who developed a rapid and accurate method for large-scale coestimation of sequence alignments and phylogenetic trees. Although not alignment-free, their approach sets a high standard for accuracy and computational efficiency.

In addition to k-mers and large-scale coestimation methods, techniques like spectral methods [11] and information-theoretic measures [12] have been explored for phylogenetic analysis. These methods offer alternative ways to capture evolutionary signals in sequence data without relying on traditional alignment procedures.

Minimal absent words (MAWs) are increasingly important in alignment-free sequence analysis [13–17]. They efficiently distill essential genomic sequence information, enabling comparisons without traditional alignment methods. MAWs support distance-based techniques such as Length Weighted Index [16] and Jaccard Distance [14], which aid in reconstructing phylogenies. Additionally, they facilitate the creation of composition vectors used in cosine similarity computation [18]. The CD-MAWS approach [1] specifically demonstrates their effectiveness through experiments on biological and simulated datasets, showcasing their potential in reconstructing species phylogeny. However, as the number of MAWs increases, memory constraints can prevent the method from generating vectors needed for cosine distance calculations, and the running time can also become prohibitively expensive. In this study, we integrated suffix automata and encoding techniques to address these limitations, achieving significantly improved runtime and memory usage.

3   Preliminary Concepts

3.1   Minimal Absent Words

Minimal Absent Words (MAWs) are words that do not appear in a string, but all their proper substrings do. For example, acg, gac, and gct are examples of MAWs for the string $x = actgcga$. However, $acta$ is an absent word but not a MAW of x. Notably, proper substrings are derived by deleting one or more characters from the beginning or end of the word. Proper prefixes and suffixes are similarly defined. The number of MAWs of a string of size n is $O(n)$ [19].

3.2   Cosine Distance for MAW Vectors

Cosine distance measures the cosine of the angle between two vectors, representing the presence or absence of MAWs in genomic sequences. It ranges from 0 (identical orientation, implying high similarity) to 1 (orthogonal orientation, implying low similarity). This metric is advantageous for its normalization, accounting for differences in sequence lengths and compositions. The distance can be shown as:

$$CD\text{-}MAWS(x, y) = 1 - \frac{V^x . V^y}{|V^x||V^y|}$$

Here V^x (V^y) is a vector representation of the sequence x (y), such that the value assigned to vector V^x (V^y) along dimension w identifies whether w is a MAW of x (y) or not. These vectors are also known as composition vectors.

3.3   Length of MAWs

Aurell et al. [6] derived lower and upper limit on the length within which the bulk of MAWs of a string of length N lie, assuming a random model:

$$l_{min} = \frac{\ln N - \ln \ln N}{\ln 4}$$

$$l_{max} = \frac{2 \ln N + \ln 9}{\ln 4}$$

Therefore, we use MAWs of length in this range to analyze a whole genome.

3.4 Suffix Automata

Suffix Automata. A suffix automaton is an efficient data structure used to represent the substring index of a given string. It enables the storage, processing, and retrieval of compressed information about all the string's substrings. The suffix automaton for a string S is the smallest directed acyclic graph with a designated initial vertex and a set of "final" vertices, such that every path from the initial vertex to any final vertex corresponds to a suffix of the string [22,23].

Here is an example of a tree of suffix links in the suffix automaton build for the string "abcbc". The nodes are labeled with the longest substring from the corresponding equivalence class (Fig. 1).

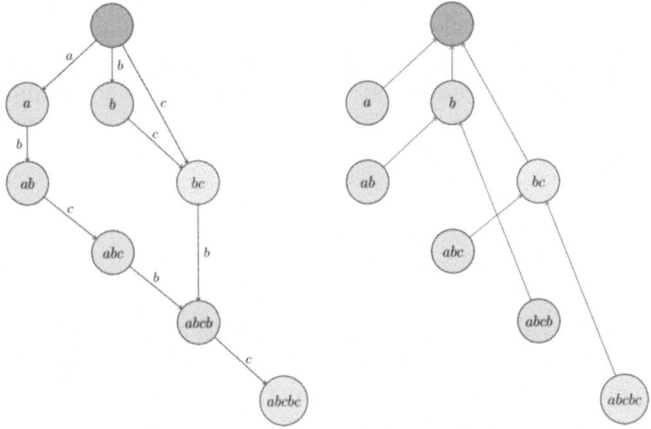

Fig. 1. A tree of suffix links in the suffix automaton build for the string "ABCBC". The nodes are labeled with the longest substring from the corresponding equivalence class. Example is taking from (https://cp-algorithms.com/string/suffix-automaton.html).

End Position Equivalency. A substring can occur in multiple positions of a string. For each substring, we can find the set of positions where the corresponding substring has ended. This set does not uniquely define a substring. In fact, multiple substrings can have the same set of end positions. If two substrings have the same set of end positions, they are considered to be in an equivalence class. For example, in string "abcbc", the substring "abcb" and "bcb" belong to the same equivalence class as both of them have the same set of end positions. Each node in Suffix automaton represents an equivalence class. It can easily be derived from the definition of suffix automaton. As the suffix automaton accepts all the suffixes, from each node there will be some path to the terminal node representing a set of suffixes. Thus, if two strings reach a node, they will have the same path to the end which means wherever those two strings have ended, they end at the exact same positions. Otherwise, there would have been different sets of suffixes starting from their endpoint.

Suffix Link. For two different substrings, if their set of end positions does not exactly match, then either one of them has to be a subset of the other or there will be no common end position among them. This is because, if two substrings have at least one common end position, it clearly signifies that one of the substring is a suffix of the other. Thus, the shorter substring will have a larger set of end positions. To keep track of the largest suffix that does not belong in the equivalence class, we introduce a suffix link between them. Formally, there will be a suffix link from an equivalence class to another if and only if the former class has a proper subset of end positions of the later one and the later class has the smallest set of end positions among all possible other sets. For example, in string "abcbc", the class which represents "abc" and class which represent "abcbc" has a suffix link toward the class with "bc".

4   Proposed Methodology

We break down our work into two primary segments. The first part is the distance calculation i.e. CD-MAWs. This segment assumes that the MAW set is already calculated and works on that. The second part concerns the calculation of the MAW set of a given string using suffix automata. The description of the proposed methodology and comparison are done seperately for the two segments.

4.1   Generating MAW Using Suffix Automata

Before we proceed, we would like to remind readers that linear-time algorithms for generating Minimal Absent Words (MAWs) already exist, as described by Barton et al. [19]. In this work, we propose a methodology for generating MAWs in lexicographically sorted order. Although our method is slower than the linear-time algorithm when generating MAWs alone, it achieves significantly better runtime when the output is required to be sorted in the downstream processing.

We will use the concept of suffix links and suffix tree to compute MAWs of a string. For instance, we are trying to find if a string $t = xabcdy$ is a MAW of another string s. We at first move to the node that contains the string $xabcd$. If we do not find this substring of t in s, then t is not a MAW of s. If the substring is present, then we attempt to proceed using edge y. If we find a node following edge y then t itself is present in s and thus is not a MAW. If we do not find it, then we move back using the suffix link. The property of suffix link dictates that we are in a node which is the longest possible suffix of previous node which is present in the string. If the string size in the node following the suffix link is less than $|t| - 2$, then t is not a MAW because it is smaller than the longest proper suffix of t. Otherwise, we again try to proceed using y from that node. If we can, it means there is a substring $abcdy$ present in s. Thus, by the definition, t is a MAW of s.

For example, consider the word $x = ACTGGA$. Its suffix automaton is shown in Fig. 2. We check if two strings $s = GAC$ and $t = CTGA$ are MAWs or not. For GAC, we traverse the trie up to GA to reach node 7. The longest prefix of s is thus a substring of the original word. But from there we cannot move forward with edge C. Therefore, s is not included in x. We then move backward using the suffix link to reach node 1. The length of the word in that node is one, thus having the potential of constructing the longest proper suffix of s. Node 1 has an outgoing edge C, so AC is included in x. Thus, both the longest proper prefix and longest proper suffix of s are included in x but not s itself, making s a MAW of x.

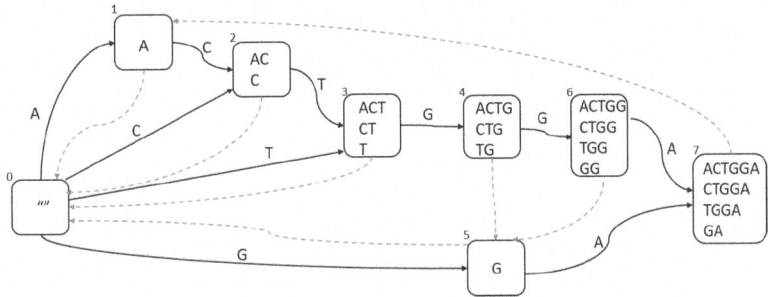

Fig. 2. Suffix automaton of 'ACTGGA'

Similarly, for t, we move to node 4 to find CTG by traversing through node 0, 2, 3 using edges C, T, G respectively. Next, we try to move forward using edge A but it is absent. So, $CTGA$ is not present in x but CTG is, which is the longest proper prefix of t. Now we move back using suffix link to node 5. However, this node only holds a length 1 string. Although it has a forward edge of A to construct a suffix of t, it is not going to be the longest proper suffix of t. Therefore, t is not a MAW of x.

4.2   Encoding MAWs

To compare two MAWs, which are character arrays or strings, we need to iterate over their length, resulting in a complexity of $O(k)$. However, as there are 4 characters in the DNA set: A, C, G, and T, we can encode them in binary as A (00), C (01), G (10), and T (11). Thus, each MAW can be encoded in a binary string. In many programming languages, including C++, a MAW of length 32 can thus fit in a 'long long' integer. On a 64-bit computer, the comparison can be done in constant time, and on a 32-bit computer, it takes only twice the time, which is also constant.

Following the equations in Sect. 3.3, the encoding solution can be implemented in analyzing genomes as large as ≈1.43 billion base pairs. While this is insufficient to cover for human genome (roughly 3.2 billion base pairs [20]), it can be used to analyze many other species with much shorter genome length. In case of human genome, or even the largest sequenced genome thus far (*Tmesipteris truncata*, aka *Tmesipteris oblanceolata*, a fern ally endemic to eastern Australia, with ≈160 billion base pairs [21]), the encoding can be done with two variables or a 128 bit integer.

4.3   Two Pointer Method

In the original CD-MAWS method, the proposed approach was to create a composition vector using MAWs for each species, and then calculate the pairwise cosine distances. However, by examining the structure of the cosine similarity, we observe that the numerator of the ratio implies the number of common MAWs, while the denominator is the product of the number of MAWs in each sequence. Without composing the actual vectors, we can directly find the denominator value as the size of the MAW set for each species. For the numerator, we can use a two-pointer method to find the common MAWs between two species. For this, the MAWs need to be sorted. After sorting, we iterate over them using the two-pointer technique to find the number of common MAWs.

5   Algorithm

The algorithm for constructing a suffix automaton is well-known and, therefore, is not detailed in this paper. The pseudo-code for generating MAWs is provided in Algorithm 1. Additionally, the authors assume that the two-pointer method is straightforward and have thus opted to omit it as well.

6   Runtime Analysis

At first we define the variables required for further analysis

$m = $ the total number of taxa in the dataset
$n = $ the size of DNA (the number of base pair) of a taxa
$k = $ the maximum size of a MAW (the number of base pair)

The overall process is divided into 3 stages, i) building the suffix automata, ii) generating MAWs, iii) calculating cosine distance

The time complexity of building suffix automaton is $\mathcal{O}(n)$. In order to generate m suffix automaton the time complexity will be $\mathcal{O}(nm)$.

Then to generate the MAWs, we run a dfs with max depth k. To find a MAW of length l we find if its longest prefix of length $l - 1$ is present in the string

---

Algorithm 1. Generating MAW from Suffix Automaton

```
1: procedure GENERATEMAW(node, word, len)
2:     if len ≥ maxK then
3:         return
4:     end if
5:     for c ∈ {A, C, G, T} do
6:         u ← node.ForwardLink(c)
7:         if u is null then
8:             if len + 1 < minK then
9:                 continue
10:            end if
11:            v ← node.BackwardLink
12:            if v is not null and v.len + 1 ≥ len then
13:                Write word + c to output
14:            end if
15:        else
16:            GENERATEMAW(u, word + c, len + 1)
17:        end if
18:    end for
19: end procedure
```

---

as substring and check if it can be extended any further. The total number of substring of length l in a string of length n is $n - l + 1$. So, the upper bound on the number of substring of size less than k is

$$\sum_{l=1}^{k} n - l + 1$$
$$= n + (n - 1) + (n - 2) + \ldots + (n - (k - 1))$$
$$= n * k - k * (k - 1)/2$$

So, the total number of search space to generate MAW is bounded by $\mathcal{O}(kn)$. As we have to generate the MAWs for m taxa, so the total time complexity to generate all the sets is $\mathcal{O}(knm)$.

In the final stage, we find the cosine distance between the taxa. To find the cosine distance between two taxa, we apply the method mentioned in Sect. 4.3. As we run a linear search on the set of MAW of the two taxa, the runtime is equal to the size of the MAW set of a taxa. As we have mentioned earlier, the number of MAWs of a string is $\mathcal{O}(n)$ and equality check of two MAWs can be done in $\mathcal{O}(k)$ (i.e., order of their length), Therefore the time complexity of finding the cosine distance between two taxa is $\mathcal{O}(kn)$. As we have a total m taxa in the dataset, so there will $m * (m - 1)/2$ pair of taxa. So, the time complexity of this stage is $\mathcal{O}(knm^2)$. We improve this step further by encoding the MAWs. Thus the comparision will be done in constant time. So, the time complexity becomes $\mathcal{O}(nm^2)$.

So, in total the time complexity of the whole process is

$$\mathcal{O}(nm) + \mathcal{O}(knm) + \mathcal{O}(nm^2)$$
$$= \mathcal{O}(knm + nm^2)$$

7   Comparative Analysis of MAW Generation

In this section we compare the execution time and memory usage of our proposed method of MAW generation based on Suffix Automata with the MAW generation algorithm proposed by Barton et al. [19] All the comparisons were conducted in a system with Microsoft Windows 10 Pro x64-based PC with WSL Kernel Version: 5.15.133.1-microsoft-standard-WSL2. The Linux Distribution in WSL was Ubuntu 22.04.2 LTS. The processor of the machine is AMD Ryzen 5, 3600 Mhz, 6 Core(s), 12 Logical Processor(s), and RAM was 8 GB. The programming language used is C++ with GNU GCC compiler.

7.1   Execution Time

Table 1 and Fig. 3 compares the execution times of the MAW generation algorithm proposed in this paper with the one in Barton et al., with an extra sorting step, for varying input sizes. Our proposed algorithm shows asymptotic improvements, particularly for large datasets.

Table 1. Execution time comparison (in seconds) between the MAW generation algorithms of Barton et al. [19] and our proposed one with Suffix Automata

Input Size (n)	Barton et al.	Sorting	Total	Proposed
10^2	0.0190	0.0068	0.0258	0.3695
10^3	0.0200	0.0090	0.0290	0.3485
10^4	0.0295	0.0218	0.0513	0.3553
10^5	0.1298	0.1133	0.2430	0.4573
10^6	1.2658	1.0448	2.3105	1.9138
10^7	14.5448	11.1833	25.7280	18.0760

The original CD-MAWS algorithm did not have any constraint on the order of the MAWs required in the MAW set. The algorithm of Barton et al. [19] also does not produce the MAW set in any particular order. However, given that our proposed algorithm processes a sorted list, the MAW set is to be sorted to facilitate the distance calculation step. As the suffix automata already outputs the MAW set in sorted order, it is asymptotically faster for this particular use case. For an input size of 10^7, our method shows a clear asymptotic advantage.

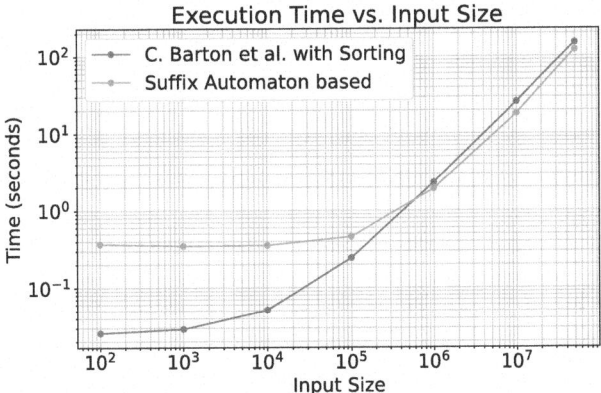

Fig. 3. Execution time comparison between the MAW generation algorithms of Barton et al. [19] and our proposed one with Suffix Automata.

Table 2. Peak memory usage (in MB) comparison between the MAW generation algorithms of Barton et al. [19] and our proposed one with Suffix Automata.

Input Size (n)	Barton et al.	Proposed
10^1	3.77	3.33
10^2	3.82	3.36
10^3	3.84	3.43
10^4	4.46	3.76
10^5	16.13	12.36
10^6	253.26	78.89
10^7	2479.24	620.97

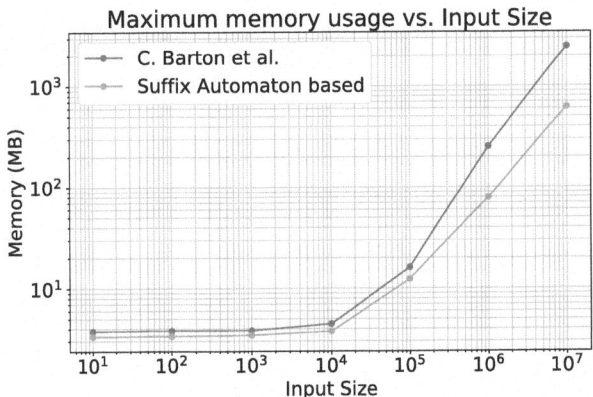

Fig. 4. Peak memory usage (in MB) comparison between the MAW generation algorithms of Barton et al. [19] and our proposed one with Suffix Automata.

7.2   Memory Usage

Table 2 and Fig. 4 summarize the peak memory consumption, showing significant reductions in memory usage with the modified algorithm.

8   Comparative Analysis of Cosine Distance Calculation

Previous CD-MAWs implementation introduced a method of composition vector where the length of the vector will be the number of MAW across all DNA. This vector can easily get very large in size. Thus it took more time to create those vectors and also to work with them. As shown in this section, the proposed two-pointer method and encoding techniques significantly reduce the runtime and memory footprint compared to the original CD-MAWS method. All the experiments in this section were conducted in the same computing system as mentioned in Sect. 7.

Since the original paper focused solely on finding the cosine distance without considering the MAW generation step, we also chose to skip the MAW generation step in order to make a fair comparison with the original approach.

8.1   Data Generation

We generated a simulated dataset with varying numbers of species using the Seq-gen Monte Carlo simulation tool [7]. The data comprised 1000 DNA sequences, each having 10000 nucleotides. We then prepared various sub-samples and ran our experiments 5 times on each of them and averaged the execution time and peak memory usage.

Table 3. Execution Time comparison (in seconds) for different versions of CD-MAWS for varying number of Taxa.

Taxa Count	CD-MAWS	CD-MAWS-SA	CD-MAWS-SA-Encoded
100	2.34	0.37	0.31
200	6.83	1.52	0.91
300	13.55	3.33	1.79
400	23.18	5.89	3.06
500	33.97	9.15	4.63
600	47.72	13.29	6.46
700	63.41	17.87	8.60
800	82.62	23.52	11.10
900	104.12	29.71	14.03
1000	129.76	36.53	17.01

8.2   Execution Time

A comparative runtime analysis of the original CD-MAWS method and our proposed method using suffix automata can be found in Table 3. We show two variants of our method: CD-MAWS using suffix automata (CD-MAWS-SA) and another one which also applies bit encoding for nucleotides (CD-MAWS-SA-Encoded). The corresponding graph of this data is presented in Fig. 5.

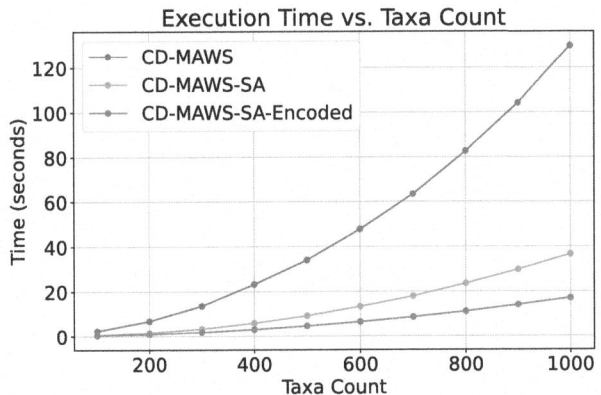

Fig. 5. Execution Time comparison (in seconds) for different versions of CD-MAWS for varying number of Taxa.

8.3   Peak Memory Usage

The composition vector in CD-MAWS generated heavy memory usage in the earlier implementations. By only working with the MAWs (and avoiding the composition vectors), we have significantly decreased the memory footprint. Additionally, encoding the MAWs further reduces memory usage. The memory usage comparison is provided in Table 4, with the corresponding graph presented in Fig. 6.

9   Results on Biological Dataset

We utilized datasets obtained from the AF project [9] to evaluate the performance of our improved method. Specifically, we tested our method on five datasets and compared its performance against the method described in CD-MAWS [1]. Detailed information about the datasets is provided in Table 5, while the time taken to execute both methods on these datasets is presented in Tables 6 and 7.

Table 4. Peak memory consumption (in MB) for different versions of CD-MAWS for varying number of Taxa

Taxa Count	CD-MAWS	CD-MAWS-SA	CD-MAWS-SA-Encoded
100	90.07	14.81	8.79
200	199.08	25.99	15.00
300	371.25	37.54	20.96
400	449.67	48.80	26.85
500	526.91	59.90	32.89
600	914.80	71.50	38.66
700	1008.31	82.65	45.53
800	1102.43	98.06	52.14
900	1198.46	111.13	57.99
1000	1286.33	116.76	62.69

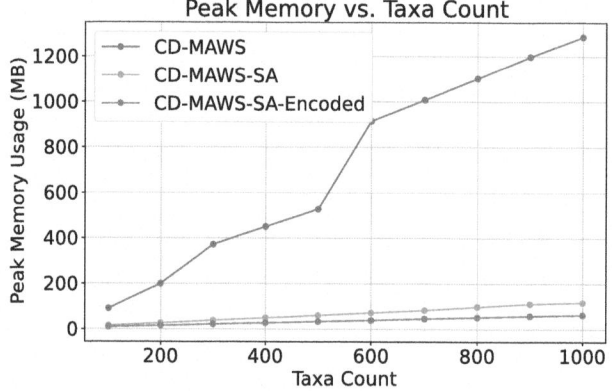

Fig. 6. Peak memory consumption (in MB) for different versions of CD-MAWS for varying number of Taxa

10   Discussion

The enhanced CD-MAWS method presented in this study significantly improves computational efficiency as well as memory consumption. By employing a two-pointer method on the sorted MAW set, encoding MAWs in binary, and utilizing suffix automata for MAW generation, we have reduced the computational complexity from $\mathcal{O}(km^3n + km^2n\log n)$ to $\mathcal{O}(knm + nm^2)$. These improvements address the primary computational bottlenecks of the original CD-MAWS method, making it more suitable for large-scale genomic studies.

Compared to previous alignment-free methods, such as k-mers and spectral methods, our approach provides a more nuanced understanding of evolutionary relationships through the use of MAWs. This method not only retains the

Table 5. Summary of the AFproject [9] datasets used in this study.

Dataset	Number of Species	Total Sequence Size (MB)
Fish mtDNA	25	0.42
E.coli/Shigella	29	144.3
E.coli/Shigella (HGT)	27	134.70
Yersinia (HGT)	8	37.50
Simulated (HGT)	33	74.40

Table 6. Execution Time comparison (in seconds) between CD-MAWS [1] and CD-MAWS-SA proposed in this paper, for AFproject [9] dataset

Dataset	CD-MAWS	CD-MAWS-SA
Fish mtDNA	1.86	0.24
E.coli/Shigella	5.60	1.00
E.coli/Shigella (HGT)	4.66	0.91
Yersinia (HGT)	1.11	0.126
Simulated (HGT)	15.54	2.51

advantages of being alignment-free but also improves the scalability of phylogenetic estimations. However, our study has limitations. Our implementation of MAW encoding currently support analyzing genomes no larger than ≈1.43 billion base pairs. However, this limitation can trivially be overcome by using two 64 bit integers to encode each MAW. With this implemented, experiments should be conducted with larger genomes. Additionally, while the binary encoding of MAWs enhances comparison speed, it may require further optimization for different hardware architectures.

Future research could focus on extending the CD-MAWS method to other types of genomic data, such as RNA or protein sequences, and integrating it with other phylogenetic tools to provide a more comprehensive analysis frame-

Table 7. Execution time comparison (in seconds) between the MAW generation algorithms of Barton et al. [19] followed by sorting in lexicographical order and our proposed one with Suffix Automata for AFproject [9] dataset

Dataset	Barton et al.	CD-MAWS-SA
Fish mtDNA	0.67	0.293
E.coli/Shigella	179.54	103.921
E.coli/Shigella (HGT)	168.84	99.99
Yersinia (HGT)	48.25	27.77
Simulated (HGT)	96.16	57.50

work. Exploring the use of machine learning techniques to predict evolutionary relationships based on MAWs could also be a promising direction.

11   Conclusion

In this paper, we presented an enhanced version of the CD-MAWS method for phylogenetic estimation, achieving a significant reduction in computational complexity while maintaining high accuracy. Our improvements, including refined cosine distance calculations, binary encoding of MAWs, and the use of suffix automata for MAW generation, have addressed the major computational challenges of the original method.

These advancements make the CD-MAWS method a more practical and scalable tool for large-scale genomic analyses, capable of efficiently handling modern datasets. The impact of our work lies in its potential to facilitate more accurate and faster phylogenetic estimations, contributing to a deeper understanding of evolutionary relationships. We have made our implementation of CD-MAWS open-source and freely available at https://github.com/TamimEhsan/cd-maws-sa. Looking forward, we anticipate that our method will be integrated into broader phylogenetic analysis pipelines and adapted for various types of genomic data. The continued development and refinement of alignment-free methods like CD-MAWS will be crucial for advancing the field of phylogenetic analysis and unlocking new insights into the evolutionary history of life.

Disclosure of Interests. The authors state that there is no competing interest for this study.

References

1. Anjum, N., Nabil, R.L., Rafi, R.I., Bayzid, M.S., Rahman, M.S.: CD-MAWS: an alignment-free phylogeny estimation method using cosine distance on minimal absent word sets. IEEE/ACM Trans. Comput. Biol. Bioinform. **20**(1), 196–205 (2023). https://doi.org/10.1109/TCBB.2021.3136792
2. Yi, H., Jin, L.: Co-phylog: an assembly-free phylogenomic approach for closely related organisms. Nucleic Acids Res. **41**, e75 (2013)
3. Klotzl, F., Haubold, B.: Phylonium: fast estimation of evolutionary distances from large samples of similar genomes. Bioinformatics **36**(7), 2040–2046 (2020)
4. Ondov, B.D., et al.: Mash screen: high-throughput sequence containment estimation for genome discovery. Genome Biol. **20**(1), 232 (2019)
5. Sarmashghi, S., Bohmann, K., Gilbert, M.T.P., Bafna, V., Mirarab, S.: Skmer: assembly-free and alignment-free sample identification using genome skims. Genome Biol. **20**(1), 1–20 (2019)
6. Aurell, E., Innocenti, N., Zhou, H.-J.: The bulk and the tail of minimal absent words in genome sequences. Phys. Biol. **13**(2), 026004 (2016)
7. Rambaut, A., Grass, N.C.: Seq-gen: an application for the Monte Carlo simulation of DNA sequence evolution along phylogenetic trees. Bioinformatics **13**(3), 235–238 (1997)

8. Liu, K., Raghavan, S., Nelesen, S., Linder, C.R., Warnow, T.: Rapid and accurate largescale coestimation of sequence alignments and phylogenetic trees. Science **324**(5934), 1561–1564 (2009)

9. Zielezinski, A., Girgis, H.Z., Bernard, G., et al.: Benchmarking of alignment-free sequence comparison methods. Genome Biol. **20**, 144 (2019). https://doi.org/10.1186/s13059-019-1755-7

10. Ondov, B.D., et al.: Mash: fast genome and metagenome distance estimation using minhash. Genome Biol. **17**(1), 132 (2016)

11. Abeysundera, M., Field, C., Gu, H.: Phylogenetic analysis based on spectral methods. Mol. Biol. Evol. **29**(2), 579–597 (2012). https://doi.org/10.1093/molbev/msr205

12. Liu, Z., Meng, J., Sun, X.: A novel feature-based method for whole genome phylogenetic analysis without alignment: application to HEV genotyping and subtyping. Biochem. Biophys. Res. Commun. **368**(2), 223–230 (2008). https://doi.org/10.1016/j.bbrc.2008.01.070. ISSN 0006-291X

13. Akon, M., Akon, M., Kabir, M., Rahman, M.S., Rahman, M.S.: ADACT: a tool for analysing (dis)similarity among nucleotide and protein sequences using minimal and relative absent words. Bioinformatics (2020). In Press

14. Rahman, M.S., Alatabbi, A., Athar, T., Crochemore, M., Rahman, M.S.: Absent words and the (dis)similarity analysis of DNA sequences: an experimental study. BMC. Res. Notes **9**, 186 (2016)

15. Silva, R.M., Pratas, D., Castro, L., Pinho, A.J., Ferreira, P.J.: Three minimal sequences found in Ebola virus genomes and absent from human DNA. Bioinformatics **31**(15), 2421–2425 (2015)

16. Chairungsee, S., Crochemore, M.: Using minimal absent words to build phylogeny. Theor. Comput. Sci. **450**, 109–116 (2012)

17. Garcia, S.P., Pinho, A.J.: Minimal absent words in four human genome assemblies. PLoS ONE **6**, e29344 (2011)

18. Belazzougui, D., Cunial, F.: A framework for space efficient string kernels. Algorithmica **79**(3), 857–883 (2017)

19. Barton, C., Heliou, A., Mouchard, L., et al.: Linear-time computation of minimal absent words using suffix array. BMC Bioinform. **15**, 388 (2014). https://doi.org/10.1186/s12859-014-0388-9

20. International Human Genome Sequencing Consortium: Initial sequencing and analysis of the human genome. Nature **409**, 860–921 (2001). https://doi.org/10.1038/35057062

21. Fernández, P., et al.: A 160 Gbp fork fern genome shatters size record for eukaryotes. iScience **27**(6), 109889 (2024)

22. Blumer, A., Blumer, J., Ehrenfeucht, A., Haussler, D., McConnell, R.: Building the minimal DFA for the set of all subwords of a word on-line in linear time. In: Paredaens, J. (ed.) ICALP 1984. LNCS, vol. 172, pp. 109–118. Springer, Heidelberg (1984). https://doi.org/10.1007/3-540-13345-3_9

23. Crochemore, M., Hancart, C.: Automata for matching patterns. In: Rozenberg G., Salomaa A. (eds.) Handbook of Formal Languages. 2, Linear Modeling: Background and Application, pp. 399–462. Springer (1997). ffhal-00620792f

Popularity on the 3D-Euclidean Stable Roommates

Steven Ge$^{(\boxtimes)}$ and Toshiya Itoh

Tokyo Institute of Technology, Meguro, Japan
ge.s.aa@m.titech.ac.jp, titoh@c.titech.ac.jp

Abstract. We study the 3D-Euclidean Multidimensional Stable Roommates problem with (strict) popularity. An agent's preference depends solely on the distance to its roommates. It prefers to be in a room where the sum of the distances to its roommates is minimal. We show that determining the existence of a strictly popular outcome in a 3D-Euclidean Multidimensional Stable Roommates game with room size 3 is co-NP-hard.

Keywords: Stable marriage problem · Stable roommates problem · Stable matching · Popularity · Coalition formation · Co-NP-hard

1 Introduction

The formation of stable coalitions in multi-agent systems is a computational problem with several variations, each imposing different conditions on coalition size. Arkin et al. [2] introduced a restricted variant of the stable roommates problem called the geometric stable roommates: each agent is associated with a point in d-dimensional metric space $\mathbb{R}^d$ and an agent prefers to be matched with an agent located close to it over an agent that is located far from it.

Chen and Roy [7] investigated a restricted variant of the geometric stable roommates, called the Euclidean d-Dimensional Stable Roommates (Euclid-d-SR) problem, where every agent is associated with a point in 2-dimensional Euclidean space. The agents are partitioned into rooms of size d. Chen and Roy [7] found that determining the existence of a core stable outcome for $d \geq 3$ is NP-complete for the Euclid-d-SR problem.

The Euclidean d-Dimensional model captures important aspects of real-world coalition formation scenarios, such as swarm robotics. A robot is a physical object, thus it has coordinates in space associated with it. When forming coalitions of robots to perform a task, it is desirable for the robots to be located close to each other as they will be able to group faster and start the task earlier.

Various notions that define the stability or optimality of a partitioning of agents exist. The notion of popularity was introduced by Gärdenfors [14] in 1975. Popular matchings have been an exciting area of research [16]. Cseh [9] has recently provided a survey on popular matchings. The notion of popularity

S. Nakano and M. Xiao (Eds.): WALCOM 2025, LNCS 15411, pp. 196–214, 2025.
https://doi.org/10.1007/978-981-96-2845-2_13

has the elements of both stability and optimality, whereas the aforementioned core stability studied by Chen and Roy [7] is only a notion of stability.

Computing a popular partition in a stable roommates game is NP-hard, even if the preferences are strict [11,13,15]. One method of obtaining tractability results from intractable coalition formation problems is to put restrictions on the agents' preferences for which the associated computational problems become tractable. This approach has been successfully used in the stable roommates problem [1,3,4,6,8,10]. Euclidean preferences follows the same type of approach.

We study an adapted version of the Euclid-d-SR problem named the 3D-Euclidean Multidimensional Stable Roommates problem, where we consider 3-dimensional space instead of 2-dimensional, with the notion of popularity. We show that determining the existence of a strictly popular outcome in a 3D-Euclidean Multidimensional Stable Roommates game with room size 3 is co-NP-hard. Our hardness proofs are inspired by the work by Chen and Roy [7] in combination with the work by Brandt and Bullinger [5].

2 Preliminaries

We use the notation and definitions related to the Euclidean d-Dimensional Stable Roommates problem described in the work of Chen and Roy [7]. Additionally, we use the notation and definitions related to popularity described in the work of Brandt and Bullinger [5]. For our hardness results, we construct polynomial-time reductions from the Planar and Cubic Exact Cover by 3-Sets (PC-X3C) problem. To ensure that this paper is self-contained, we include the relevant existing definitions in this section.

For $m \in \mathbb{N}^+$, let $[m] = \{1, \ldots, m\}$. Let ϵ denote a small fractional value with $0 < \epsilon < 0.001$, and let $d = \sqrt{1 - \left(\frac{\epsilon}{2}\right)^2}$.

A 3D-Euclidean Multidimensional Stable Roommates (3D-EuclidSR) game G is a triple (V, E, s), where the set $V = [s \cdot n]$ represents the set of agents, the embedding $E : V \to \mathbb{R}^3$ denotes the position of an agent in 3-dimensional Euclidean space, and $s \in \mathbb{N}^+$ is the room size. For an agent $a \in V$, let $E(a)_x$, $E(a)_y$, and $E(a)_z$ denote the x, y, and z coordinates of agent a in E, respectively.

An s-sized subset of V is called a room. An outcome $\pi = \{C_1, \ldots, C_k\}$ of G is a partitioning of the agents V into k rooms. Let $\pi(a)$ denote the room in π that contains agent $a \in V$.

For $p, q \in V$, let $\delta(p, q) = \sqrt{\sum_{i \in \{x,y,z\}} (E(p)_i - E(q)_i)^2}$ denote the Euclidean distance between agents p and q. Let $a \in V$ be an agent and $R \subseteq V$ be a room such that $a \in R$. We overload the notation by defining $\delta(a, R) = \sum_{a' \in R} \delta(a, a')$.

Let $a \in V$ be an agent and $R, T \subseteq V$ be two rooms such that $a \in R$ and $a \in T$. We write $R \succsim_a T$ to say that agent a weakly prefers R over T if a likes being in room R at least as much as being in room T. Additionally, we write $R \succ_a T$ and say that agent a strictly prefers room R over room T if $R \succsim_a T$ and

$T \not\succsim_a R$. If agent a weakly prefers R over T and T over R, we write $R \sim_a T$ and say that a is indifferent between R and T.

The preference of an agent solely depends on its distance to its roommates. Specifically, an agent prefers being in a room with roommates that are positioned close to themselves. Formally, let $a \in V$ be an agent and $R, T \subseteq V$ be two rooms such that $a \in R$ and $a \in T$. We have that $R \succsim_a T \iff \delta(a, R) \leq \delta(a, T)$.

Let $X = [m]$, where $m \in \mathbb{N}^+$ and $m \mod 3 = 0$, and let $C = \{A_1, \ldots, A_q\}$ be a collection of 3-element subsets of X. An instance of the X3C problem is a tuple $I = (X, C)$, which asks: does there exist $S \subseteq C$ such that S partitions X? We call such an S a solution of I.

The associated graph of I, denoted by $G(I)$, is a bipartite graph $(U \cup W, E)$ where $U = \{u_i | i \in X\}$, $W = \{w_j | A_j \in C\}$, and an edge $\{u_i, w_j\} \in E$ if and only if $i \in A_j$. We call vertices in U and W the element-vertices and set-vertices, respectively.

Instance I is an instance of the PC-X3C problem if each element $x \in X$ occurs in exactly three sets of C and the associated graph is planar. Note that for a valid instance of the PC-X3C problem, we have that $|X| \geq 6$. Additionally, we have that $|X| = |C|$.

Let π and π' be arbitrary outcomes. Let $N(\pi, \pi') = \{a \in N | \pi(a) \succ_a \pi'(a)\}$ and $\phi(\pi, \pi') = |N(\pi, \pi')| - |N(\pi', \pi)|$. We call $\phi(\pi, \pi')$ the popularity margin of π and π'.

An outcome π is strictly more popular than outcome π' if $\phi(\pi, \pi') > 0$. An outcome π is popular if for any outcome π' we have $\phi(\pi, \pi') \geq 0$. An outcome π is strictly popular if for any other outcome $\pi' \neq \pi$ we have $\phi(\pi, \pi') > 0$.

3   Strict Popularity co-NP-hard

In this section, we show that determining the existence of a strictly popular outcome is co-NP-hard when the room size is fixed to 3. We construct a 3D-EuclidSR game $G = (V, E, 3)$ from a PC-X3C instance $I = (X, C)$, such that there exists a solution $S \subseteq C$ that partitions X if and only if no strictly popular outcome exists for G. The 3D-EuclidSR game G consists of three parts: the bottom layer, the top layer, and the ascending layer. We construct separate sets of agents V_b, V_t, V_a, and embeddings E^b, E^t, E^a for the bottom layer, the top layer, and the ascending layer, respectively. Note that $V = V_b \cup V_t \cup V_a$, and for each $a \in V$, we have

$$E = \begin{cases} E^b(a) & \text{if } a \in V_b; \\ E^t(a) & \text{if } a \in V_t; \\ E^a(a) & \text{if } a \in V_a. \end{cases}$$

The idea of the construction is that the resulting 3D-EuclidSR game G has exactly one popular outcome π_S for each solution S of I, and exactly one additional popular outcome π_{pp}. The outcome π_S is called a reduced outcome, and π_{pp} is called the permanent popular outcome. Thus, the permanent popular outcome π_{pp} is strictly popular if and only if I has no solutions.

The bottom layer has a one-to-one correspondence to the PC-X3C instance I, where we have one agent associated with each element of the universe X and a triplet of agents associated with each 3-set in C. In a reduced outcome π_S, the rooms of each agent in the bottom layer consist exclusively of other agents from the bottom layer. In a permanent popular outcome π_{pp}, only the rooms of the agents in the bottom layer that are not associated with the universe X consist exclusively of other agents from the bottom layer. The room of an agent a in the bottom layer, associated with the universe X, in the permanent popular outcome π_{pp}, consists of a and two agents from the ascending layer.

The top and ascending layers connect the agents that are associated with the universe X of the bottom layer in such a way that π_{pp} and π_S, where S is a solution of I, are the only popular outcomes.

3.1 Construction

Bottom Layer. In the bottom layer, we leverage the fact that the associated graph $G(I)$ is planar and cubic. As noted by Valiant [17], $G(I)$ admits a specific planar embedding in $\mathbb{Z}^2$, called an orthogonal drawing. This embedding maps each vertex $v \in U \cup W$ to an integer grid point and each edge to a chain of non-overlapping horizontal and vertical segments along the grid (except at the endpoints). A horizontal segment is a collection of consecutive horizontal edges of the grid $\mathbb{Z}^2$, and a vertical segment is defined analogously. A bending point is a point where a horizontal segment intersects a vertical segment.

We use the following specific restricted orthogonal drawing $E^{G(I)}$ for $G(I)$:

Proposition 1. ([12]) *A planar graph with a maximum vertex degree of three can be embedded, in polynomial time, in the grid $\mathbb{Z}^2$ such that its vertices lie on the integer grid points and its edges are drawn using at most one horizontal and one vertical segment in the grid.*

The planar graph in Proposition 1 can be constructed without any empty vertical nor empty horizontal grid lines that intersect the graph. Here, "empty" refers to a grid line that does not intersect any vertex. If any vertical or horizontal grid line is empty, the graph can be compressed by shifting vertices to remove the empty grid line. An illustration of this shifting can be found in the full version of the paper.

In the orthogonal drawing $E^{G(I)}$, the edges of the grid $\mathbb{Z}^2$ are set to a length of 10. This length is chosen semi-arbitrarily to ensure sufficient distances between agents in subsequent constructions. Since there are no empty vertical nor empty horizontal grid lines that intersect the graph in $E^{G(I)}$, the height and width of the graph in the embedding are at most $10 \cdot (|X|+|C|)$. See Fig. 1 for an example orthogonal drawing.

For each element-vertex $u_i \in U$, we create an agent u_i for V_b, called an element-agent, with the same xy-coordinates as u_i and a z-coordinate of 0 in embedding E^b.

Additionally, for each edge $\{u_i, w_j\}$ in $G(I)$, if $\{u_i, w_j\}$ has a bending point in $E^{G(I)}$, we create an agent b_j^i for V_b, called a bending point agent, with the same

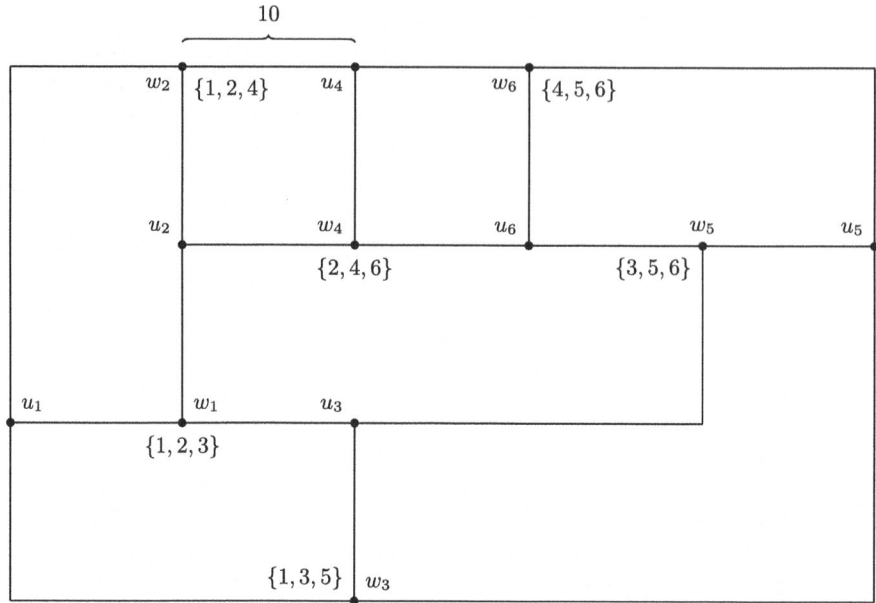

Fig. 1. Example orthogonal drawing $E^{G(I)}$, where $I = (X, C)$, $X = \{1, 2, 3, 4, 5, 6\}$, and $C = \{\{1, 2, 3\}, \{1, 2, 4\}, \{1, 3, 5\}, \{2, 4, 6\}, \{3, 5, 6\}, \{4, 5, 6\}\}$.

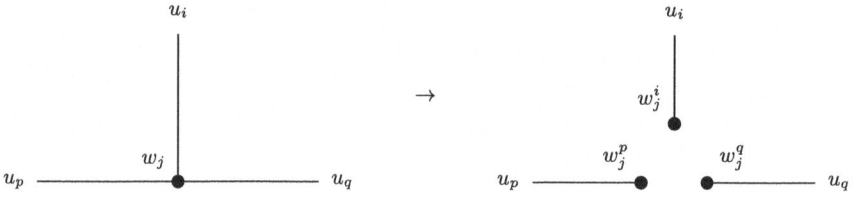

Fig. 2. Gadget for set-vertex w_j, where $C_j = \{i, p, q\}$ and w_j^i, w_j^p, w_j^q form equilateral triangle with edge length 1.

xy-coordinates as the bending point and a z-coordinate of 0 in the embedding E^b. Let $B = \{b_j^i | G(I)$ contains the edge $\{u_i, w_j\}$ with a bending point in $E^{G(I)}\}$ denote the set of all bending point agents.

Finally, for each set-vertex $w_j \in W$, we create three agents w_j^i for V_b, where $i \in C_j$, which form an equilateral triangle with edge length 1. Specifically, for a set-vertex $w_j \in W$, we can assume w.l.o.g. that it has three connecting edges going upwards, leftward, and rightward connecting to element-vertices u_i, u_p, and u_q, respectively. If this is not the case, we can rotate the coordinate system and apply the same reasoning. We create three set-agents $w_j^i, w_j^p, w_j^q \in V_b$ to replace w_j. The set-agents w_j^i, w_j^p, w_j^q are embedded such that they are on the segment of the upward, leftward, and rightward edge, respectively, and are equidistant, each separated by a distance of 1. As previous agents, each of these set-agents has z-coordinate 0. Let $W' = \{w_j^i, w_j^p, w_j^q | C_j = \{i, p, q\} \in C\}$ be the set of set-agents. See Fig. 2 for an illustration.

Now, modify $G(I)$ to create the graph $G'(I) = (V', E')$. Let V' be the vertices consisting of all element-agents, set-agents, and bending point agents, i.e., $V' = U \cup B \cup W'$. For each edge $\{u_i, w_j\}$ in $G(I)$, if $\{u_i, w_j\}$ has a bending point in $E^{G(I)}$, then we create the edges $\{u_i, b_j^i\}, \{b_j^i, w_j^i\}$ and place them in E'. Otherwise we create the edge $\{u_i, w_j^i\}$ and place it in E'. Let the embedding $E^{G'(I)} : V' \rightarrow \mathbb{Z}^2$ map the vertices in V' to the same x, y-coordinates that embedding E^b assigns. That is, for each $v \in V'$ we have $E^{G'(I)}(v) = (E_x^b(v), E_y^b(v))$. An illustration of this modification can be found in the full version of the paper.

Note that for each edge in E' its segment in embedding $E^{G'(I)}$ is either horizontal or vertical. We replace the segment in $E^{G'(I)}$ of each edge in $G'(I)$ with a chain of copies of three agents. This chain ensures that either all three agents w_j^i for $i \in C_j$ are matched in the same triple (indicating that the corresponding set is in the solution) or none of them is matched in the same triple (indicating that the corresponding set is not in the solution).

For an edge $e \in E'$, let s_e denote its corresponding segment in $E^{G'(I)}$. Let $S_{G'(I)} = \{s_e | e \in E'\}$ denote the set of all segments of the edges in $E^{G'(I)}$ and for a segment $s \in S_{G'(I)}$ let $l(s)$ denote the length of segment s.

For each segment $s \in S_{G'(I)}$, we create $\hat{n} = \left\lceil \frac{l(s)}{d} \right\rceil$ copies of the triple $A_s[z] = \{\alpha_s[z], \beta_s[z], \gamma_s[z]\}$, where $z \in [\hat{n}]$, for V_b. We embed the agents $A_s[z]$, where $z \in [\hat{n}]$, around segment s to connect the two vertices at the endpoints of s in embedding $E^{G'(I)}$. For convenience, when using $\hat{n}$, we refer to the constant corresponding to segment s which should be clear from the context. Let $f, t \in V'$ denote the vertices at the endpoints of segment s. We merge vertex t and $\gamma_s[\hat{n}]$ into one vertex, while referring to it by either name. For convenience, we shall use $\gamma_s[0]$ to refer to f. The agents will be embedded such that the following hold for each $z \in [\hat{n}]$:

- Each agent pair $\alpha_s[z]$ and $\beta_s[z]$ is separated by a distance of ϵ.
- The distance between agents $\alpha_s[z]$ (resp. $\beta_s[z]$) and $\gamma_s[z]$ is 1.
- The distance between agents $\alpha_s[z]$ (resp. $\beta_s[z]$) and $\gamma_s[z-1]$ is 1.

This embedding ensures that for a strictly popular outcome, either all $A_s[z]$, where $z \in [\hat{n} - 1]$, or all $\gamma_s[z-1], \alpha_s[z], \beta_s[z]$ must be matched together.

As the length of the chain $A_s[z]$, where $z \in [\hat{n}]$, is slightly longer than $l(s)$, the chain $A_s[z]$ is "bent" into the xy-plane in the negative z-direction in embedding E^b. The agents in $A_s[0]$ and $A_s[\hat{n}]$ have a z-coordinate closest to 0. The agents in $A_s[\lfloor \hat{n}/2 \rfloor]$ and $A_s[\lceil \hat{n}/2 \rceil]$ have the furthest z-coordinate from 0. For each $z \in [\hat{n}]$, the vertices $\alpha_s[z]$ and $\beta_s[z]$ have the same z-coordinate. See Figs. 3, 4 and 5 for an illustration.

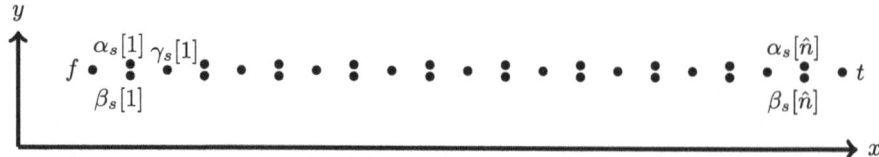

Fig. 3. Projection of example chain $A_s[z]$, where $z \in [\hat{n}]$, of segment s, where t, f are the endpoints of s, on the xy-plane.

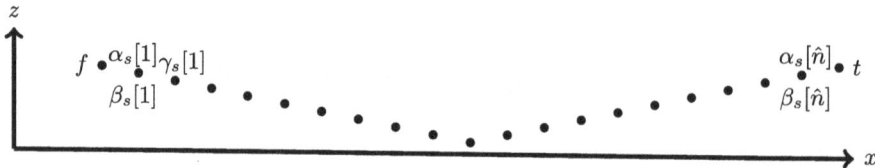

Fig. 4. Projection of example chain $A_s[z]$, where $z \in [\hat{n}]$, of segment s, where t, f are the endpoints of s, on the xz-plane.

In conclusion, for the bottom layer, $V_b = U \cup B \cup W' \cup \bigcup\limits_{s \in S_{G'(I)}, z \in [\hat{n}]} A_s[z]$.

These agents in $\bigcup\limits_{s \in S_{G'(I)}, z \in [\hat{n}]} A_s[z]$ are embedded in E^b as defined above and for each agent $a \in U \cup B \cup W'$, we have that $E^b(a) = (E_x^{G'(I)}, E_y^{G'(I)}, 0)$.

Top Layer. In the top layer, the agents are embedded in a structure resembling a snowflake. It consists of a triple of agents $d_{1,0}^1, d_{2,0}^1, d_{3,0}^1$, called the center agents, which are placed at the center of embedding E^t. The center agents form an equilateral triangle with an edge length of 1. From each of these center agents, a binary tree-like structure extends outward. In this structure, each parent and its two children also form an equilateral triangle. The edge length of the equilateral triangle at depth 0 is large, but it decreases exponentially as the depth increases. The total depth depends on the size of X. Finally, the edges are replaced by chains similar to those in the bottom layer. The constructed top layer will be placed on the xy-plane with a z-coordinate of $10|X|+10$, directly above the bottom layer to create sufficient separation.

We will construct an initial snowflake graph $G_{snow} = (V_{snow}, E_{snow})$ with embedding $E^{G_{snow}} : V_{snow} \to \mathbb{R}^2$. Let $k \in \mathbb{R}$ be such that $|X| = 3 \cdot 2^k$. The binary tree structure, attached to center vertices $d_{1,0}^1, d_{2,0}^1, d_{3,0}^1 \in V_{snow}$, will have a depth of $\lfloor k \rfloor$. We label the vertices in the binary tree structure attached to center vertex $d_{i,0}^1$ at depth j as $d_{i,j}^l \in V_{snow}$, where $l \in [j]$ denotes the vertex index. The edge length of the equilateral triangle formed by a vertex at depth j and its children is $10(|X|+|C|) + 4^{\lfloor k \rfloor - j + 2}$. See Fig. 6 for an example of the initial snowflake graph and its embedding.

We construct a balanced-binary-tree gadget with depth 2. A vertex at depth j forms an equilateral triangle with edge length $10(|X|+|C|) + 4^{2-j}$. The vertices

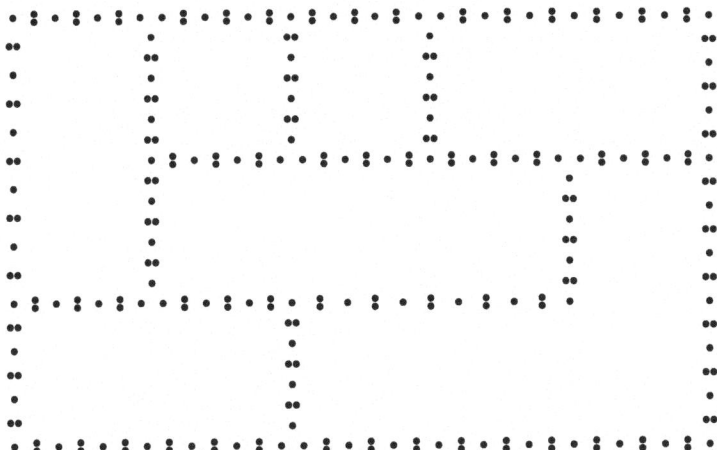

Fig. 5. Segments of embedding $E^{G'(I)}$ replaced with chains. The number of agents in the chains are reduced for clarity.

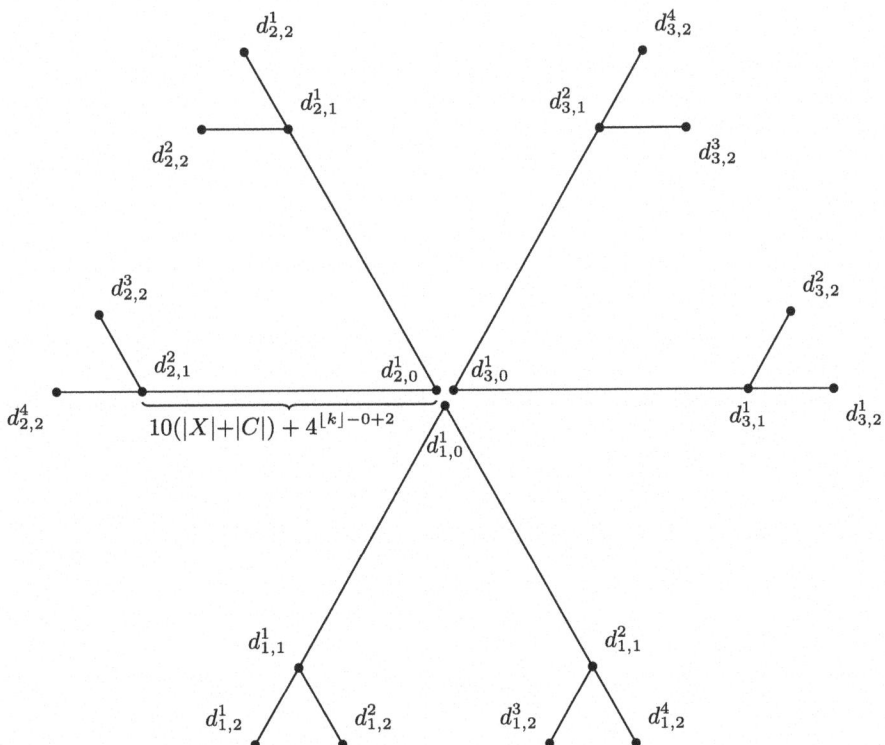

Fig. 6. Initial snowflake graph G_{snow} and its embedding $E^{G_{snow}}$ where $|X| = 12$. Note that $\lfloor k \rfloor = 2$.

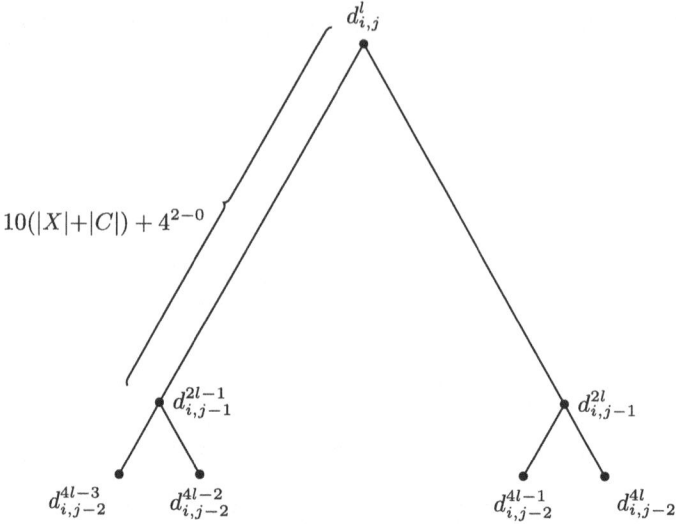

Fig. 7. balanced-binary-tree gadget.

in the balanced-binary-tree gadget will follow the same naming convention as the aforementioned vertices in the binary tree structure in the initial snowflake graph. See Fig. 7 for a depiction.

If $k \notin \mathbb{N}$, the number of leaves in G_{snow} is smaller than $|X|$. We replace $\frac{|X|-3 \cdot 2^{\lfloor k \rfloor}}{3}$ leaves with the balanced-binary-tree gadget to create the unbalanced snowflake graph $G'_{snow} = (V'_{snow}, E'_{snow})$. Let $d^l_{i,j}$ be a leaf that is replaced by the balanced-binary-tree gadget. We label the root vertex of the corresponding balanced-binary-tree gadget as $d^l_{i,j}$.

For the embedding $E^{G'_{snow}} : V'_{snow} \to \mathbb{R}^2$, the balanced-binary-tree gadgets are embedded in the same manner as the binary tree structures in $E^{G_{snow}}$. This replacement ensures that the number of leaves in G'_{snow} is equivalent to $|X|$. Note that it is always possible to replace (a part of) the leaves of G_{snow} with the balanced-binary-tree gadget to achieve the desired number of leaves in G'_{snow}. An example graph and embedding can be found in the full version of the paper.

We modify the snowflake graph by replacing each internal vertex with three vertices, which form an equilateral triangle with an edge length of 1. One of these vertices will be placed at the exact same location as the original internal vertex. The other two vertices will be placed on the edges between the internal vertex and its children. The vertices placed on the edges are connected to the child at the end of its corresponding edge. See Figs. 8 and 9 for an illustration of the gadget.

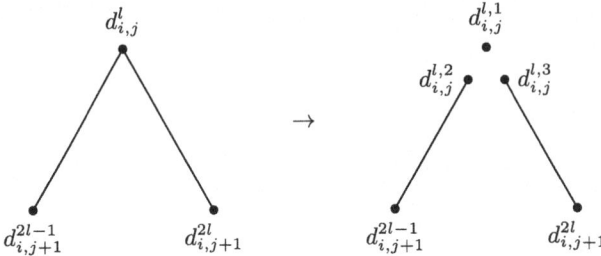

Fig. 8. internal-vertex gadget.

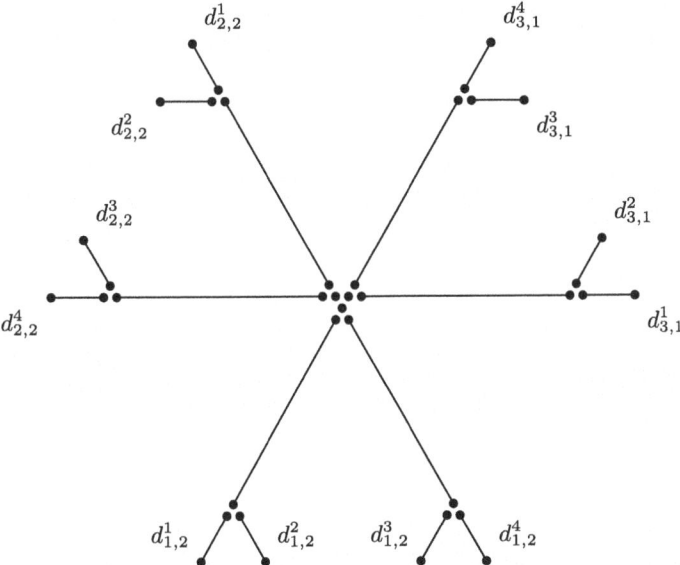

Fig. 9. Graph G_{triple}, which is internal vertex gadget applied to G_{snow} of Fig. 7.

Let $G_{triple} = (V_{triple}, E_{triple})$ denote the graph obtained after applying the internal-vertex gadget to graph G'_{snow}, and let embedding $E^{G_{triple}} : V_{triple} \to \mathbb{R}^2$ be as described above.

Let $d^l_{i,j}$ be a leaf of one of the trees in G'_{snow}. For convenience, we will also use the label $d^{l,1}_{i,j}$ to refer to the same leaf.

Let $l(e)$, where $e \in E_{triple}$, denote the length of e in embedding $E^{G_{triple}}$. The final step for the top layer is to replace each edge e in G_{triple} with $\hat{n} = \left\lceil \frac{l(e)}{d} \right\rceil$ copies of the triple $A_e[z] = \{\alpha_e[z], \beta_e[z], \gamma_e[z]\}$, where $z \in [\hat{n}]$. These triples are embedded in such a way that they maintain the same distance properties as those in the bottom layer. These properties are:

- The distance between agents $\alpha_s[z]$ and $\beta_s[z]$ is ϵ.
- The distance between agents $\alpha_s[z]$ (resp. $\beta_s[z]$) and $\gamma_s[z]$ is 1.
- The distance between agents $\alpha_s[z]$ (resp. $\beta_s[z]$) and $\gamma_s[z-1]$ is 1.

The embedding of the chain $A_s[z]$ is nearly identical to that in the bottom layer. Instead of "bending" the chain $A_s[z]$ in the negative direction, the chain is "bent" in the positive z-direction in embedding E^t in the top layer. See Figs. 10, 11, and 12 for an example.

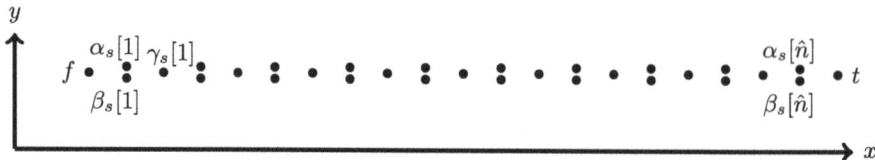

Fig. 10. Projection of example chain $A_s[z]$, where $z \in [\hat{n}]$, of edge e, where t, f are the endpoints of e, on the xy-plane.

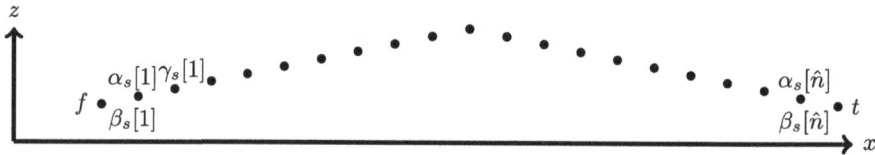

Fig. 11. Projection of example chain $A_s[z]$, where $z \in [\hat{n}]$, of edge e, where t, f are the endpoints of e, on the xz-plane.

In conclusion, for the top layer, we have $V_t = V_{triple} \cup \bigcup_{e \in E_{triple}, z \in [\hat{n}]} A_s[z]$.
As mentioned in the beginning of this section, the top layer is placed on the xy-plane with z-coordinate $10|X|+10$. Thus, for each agent $a \in V_{triple}$, we have that $E^t(a) = (E_x^{G_{triple}}, E_y^{G_{triple}}, 10|X|+10)$. The agents in $\bigcup_{e \in E_{triple}, z \in [\hat{n}]} A_s[z]$ are embedded in E^t as defined above.

Ascending Layer. The purpose of the ascending layer is to connect the bottom layer with the top layer. In the construction of the ascending layer, the embedding of the edges will not have any bending points.

Let us arbitrarily label the leaves in the top layer as $l_1, \ldots, l_m$. We shall connect vertex agent u_i from the bottom layer to the leaf l_i of the top layer, where $i \in [m]$.

We shall create a graph $G_{asc} = (V_{asc}, E_{asc})$ and embedding $E^{G_{asc}} : V_{asc} \to \mathbb{R}^3$. For each $u_i \in U$, let us create two auxiliary vertices u_i', l_i'. The vertex set V_{asc} consists of all element vertices from the bottom layer, the leaves from the top layer, and the aforementioned auxiliary vertices. That is, $V_{asc} = U \cup \{l_i, u_i', l_i' | i \in [m]\}$. In embedding $E^{G_{asc}}$, the element vertices and leaves are embedded in the same way as in embedding E^b and E^t, respectively. That is, for $i \in [m]$, $E^{G_{asc}}(u_i) = E^b(u_i)$ and $E^{G_{asc}}(l_i) = E^t(l_i)$.

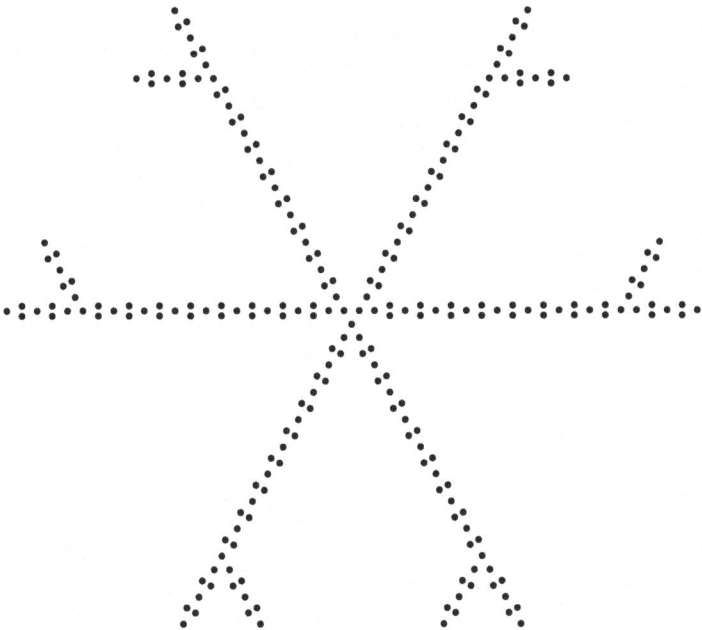

Fig. 12. The edges of Fig. 9 replaced with the chains. The number of agents and the distances between the agents are exaggerated for clarity.

The auxiliary vertex u_i' is placed directly above vertex u_i at height $10i$ in embedding $E^{G_{asc}}$. That is, $E^{G_{asc}}(u_i') = (E_x^b(u_i), E_y^b(u_i), 10i)$. The auxiliary vertex l_i' is placed directly below vertex l_i also at height $10i$ in embedding $E^{G_{asc}}$. That is, $E^{G_{asc}}(l_i') = (E_x^t(l_i), E_y^t(l_i), 10i)$.

The edges in E_{asc} consist of $\{u_i, u_i'\}, \{u_i', l_i'\}$, and $\{l_i', l_i\}$, for each $i \in [m]$. See Fig. 13 for an illustration.

Fig. 13. Initial ascending graph with element vertex u_i and leaf l_i.

Let $l(e)$, where $e \in E_{asc}$, denote the length of e in embedding $E^{G_{asc}}$. The final step for the ascending layer is to replace edge e in G_{asc} with $\hat{n} = \left\lceil \frac{l(e)}{d} \right\rceil$ copies of the triple $A_e[z] = \{\alpha_e[z], \beta_e[z], \gamma_e[z]\}$, where $z \in [\hat{n}]$. These triples are embedded such that they have the same distance properties as in the bottom and top layer.

As in the bottom layer and the top layer, the chain A_e is longer than edge e. We define 3 distinct bending directions for the edges $\{u_i, u_i'\}$, $\{u_i', l_i'\}$, and $\{l_i', l_i\}$ in embedding E^a. The chain $A_{\{u_i, u_i'\}}[z]$ will be bent in the u_i'-l_i' direction. The chain $A_{\{u_i', l_i'\}}[z]$ will be bent in the negative z-direction. The chain $A_{\{l_i', l_i\}}[z]$ will be bent in the l_i'-u_i' direction. See Fig. 14 for an illustration.

In conclusion, for the ascending layer we have that $V_a = V_{asc} \cup \bigcup_{e \in E_{asc}, z \in [\hat{n}]} A_e[z]$. These agents are embedded as described above.

3.2   Instance Size

We demonstrate that the size of the constructed 3D-EuclidSR game G is polynomial in the size of the X3C instance I. This can be shown by determining the order of the number of agents in each layer of G separately. The total number of agents is of the order of $\mathcal{O}(|X|^3)$. The derivation can be found in the full version of the paper.

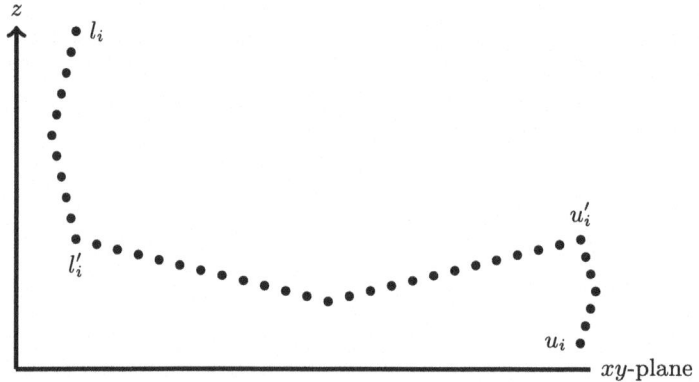

Fig. 14. Replacing the edges in Fig. 13 with the chains $A_e[z]$, where $e \in \{\{u_i, u_i'\}, \{u_i', l_i'\}, \{l_i', l_i\}\}$.

3.3   Permanent Popular Outcome

We define the permanent popular outcome π_{pp} in this section. In this outcome, every agent is assigned a room with its closest neighbors. That is, for any outcome π and any arbitrary agent a, we have that $\delta(a, \pi_{pp}(a)) \leq \delta(a, \pi(a))$.

Additionally, $\{d_{1,0}^{1,1}, d_{2,0}^{1,1}, d_{3,0}^{1,1}\} \in \pi_{pp}$ if and only if $\lfloor k \rfloor$ is even. The permanent popular outcome is unique and always exists, independent of whether I has a solution.

The description of how the agents are assigned to the rooms in each layer to construct the permanent popular outcome π_{pp} can be found in the full version of the paper. Example illustrations of the rooms can be found in Figs. 15, 16 and 17.

3.4   Reduced Outcome

In this section, we define the reduced outcome π_S, where $S \subseteq C$ is a solution of I. Similar to the permanent popular outcome, the reduced outcome will assign every agent to a room with its closest neighbors.

Additionally, $\{d_{1,0}^{1,1}, d_{2,0}^{1,1}, d_{3,0}^{1,1}\} \in \pi_S$ if and only if $\lfloor k \rfloor$ is odd. A reduced outcome exists if and only if I has a solution.

The procedure for assigning agents to rooms in each layer to construct the reduced outcome π_S is provided in the full version of the paper. Example illustrations of the rooms can be found in Figs. 18, 19 and 20.

3.5   Hardness

We show that determining the existence of a strictly popular outcome for a 3D-EuclidSR game G is co-NP-hard. The co-NP-hardness is demonstrated by proving that if the original PC-X3C instance $I = (X, C)$ has no solution, then

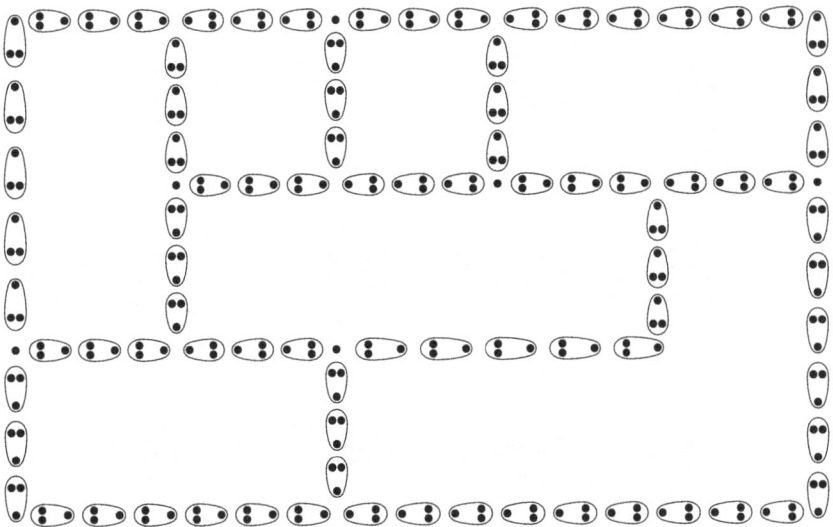

Fig. 15. Illustration of the rooms in the bottom layer from Fig. 5 in the permanent popular outcome.

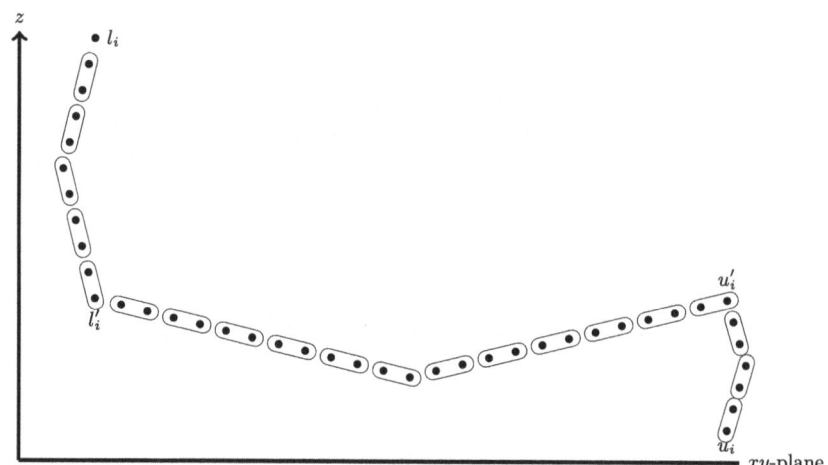

Fig. 16. Illustration of the rooms in the ascending layer in the permanent popular outcome.

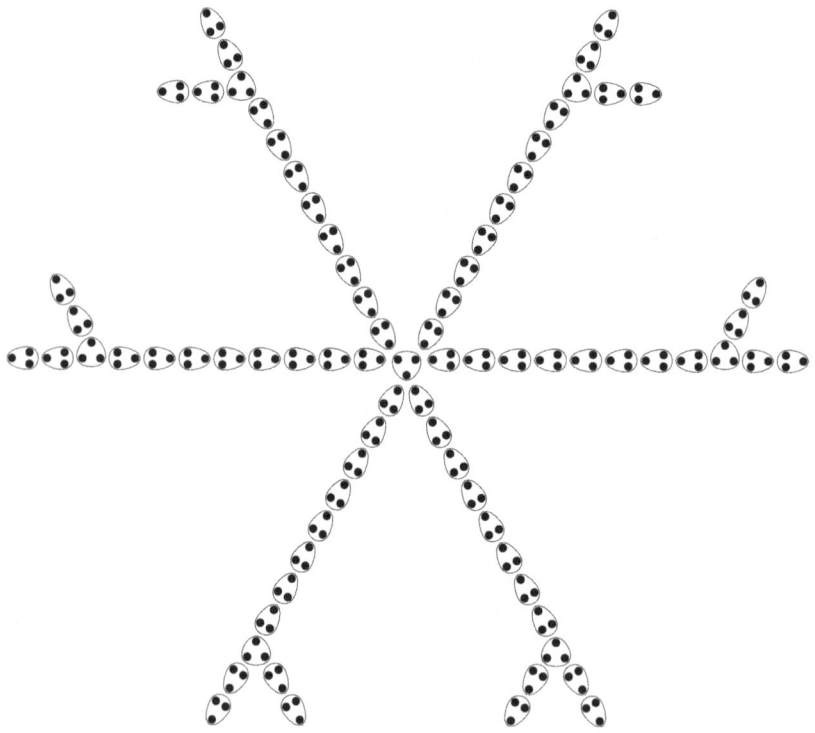

Fig. 17. Illustration of the rooms in the top layer in the permanent popular outcome.

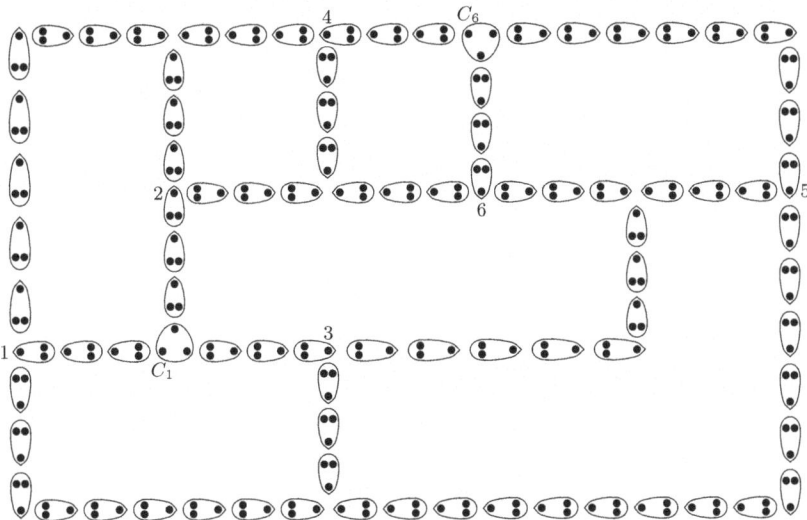

Fig. 18. Illustration of the rooms in the bottom layer from Fig. 5 in the reduced outcome, where the solution is $S = \{\{1, 2, 3\}, \{4, 5, 6\}\} = \{C_1, C_6\}$.

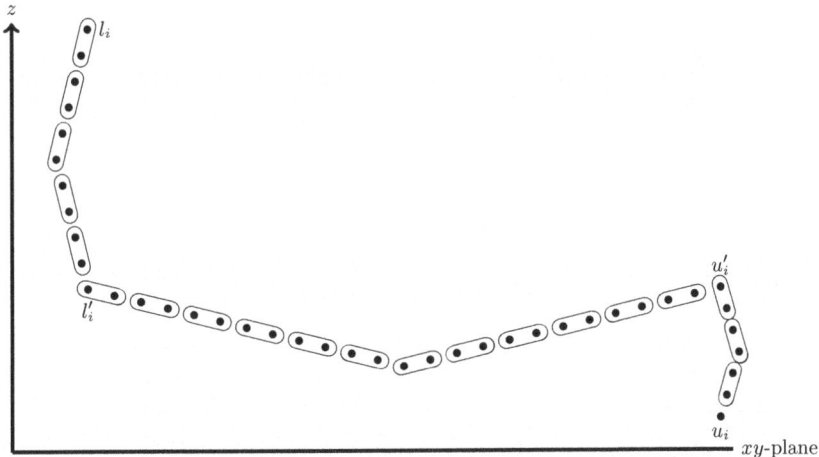

Fig. 19. Illustration of the rooms in the ascending layer in the reduced outcome.

π_{pp} is the only popular outcome. Therefore, π_{pp} would be the strictly popular outcome.

If instance I has a solution $S \subseteq C$, then G has multiple popular outcomes, including π_{pp} and π_S. Hence, G would not have a strictly popular outcome. The proof can be found in the full version of the paper.

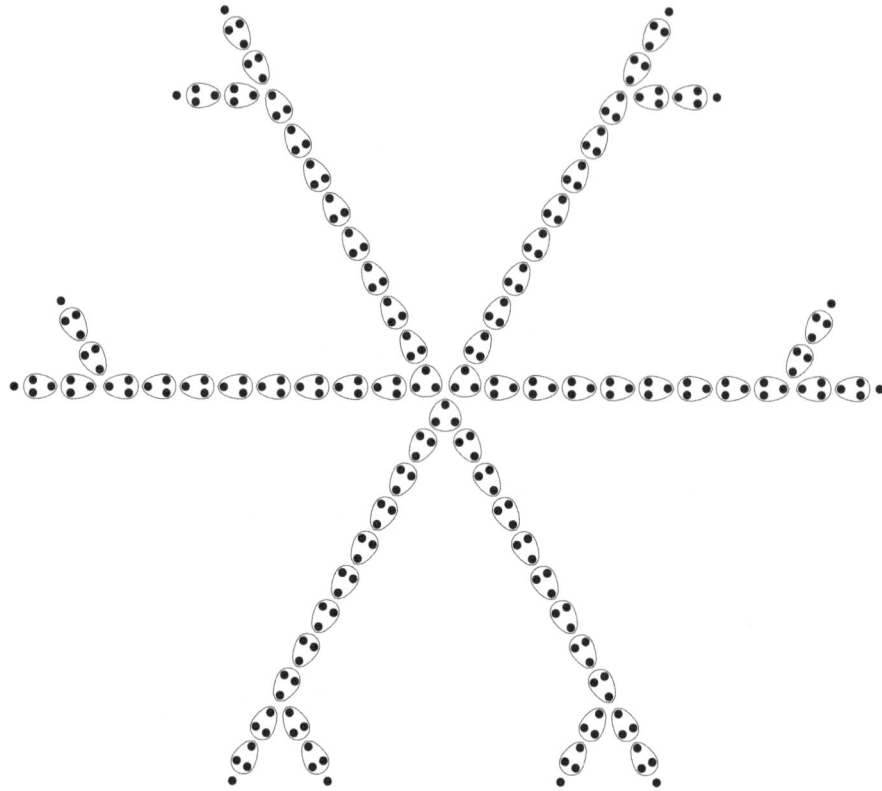

Fig. 20. Illustration of the rooms in the top layer in the reduced outcomes.

4 Conclusion

We have demonstrated that determining the existence of a strictly popular outcome for a 3D-EuclidSR game is co-NP-hard when the room size is fixed to 3.

We believe that our construction can be modified to show that the problem
remains hard, when the room size is greater than 3. Additionally, we believe that
our construction can be modified to show that determining the existence of a
popular outcome is co-NP-hard and that, under the assumption that P≠NP, a
mixed popular outcome cannot be computed in polynomial time. These are left
for future research.

The complexity of computing and determining the existence of a popular
outcome, when the room size is fixed to 2, is also yet unknown. Let us construct
a weighted graph $G' = (V, E', w)$, where the set of vertices V is equivalent to the
set of agents V in game G, the set of edges $E' = \{\{i, j\} | i \neq j \wedge i, j \in V\}$ forms
a complete graph, and the weight of edge $\{i, j\} \in E'$ is equal to the distance
between agents i and j in embedding E, i.e., $w(i, j) = \delta(i, j)$. We conjecture that

a minimum weight perfect matching M of G' corresponds to a popular outcome in G. However, we encountered difficulties while trying to prove our conjecture.

Another avenue for future research is demonstrating completeness for a certain complexity class. We conjecture that the problem is Π_2^p-complete.

Finally, it would be interesting to investigate whether a stronger result can be obtained by showing that determining the existence of a strictly popular outcome remains co-NP-hard, when the agents are assigned a point in 2-dimensional Euclidean space, i.e., popularity on the Euclid-d-SR problem. It is also possible that the problem becomes tractable in this case.

References

1. Abraham, D.J., Levavi, A., Manlove, D.F., O'Malley, G.: The stable roommates problem with globally-ranked pairs. In: Deng, X., Graham, F.C. (eds.) WINE 2007. LNCS, vol. 4858, pp. 431–444. Springer, Heidelberg (2007). https://doi.org/10.1007/978-3-540-77105-0_48

2. Arkin, E.M., Bae, S.W., Efrat, A., Okamoto, K., Mitchell, J.S., Polishchuk, V.: Geometric stable roommates. Inf. Process. Lett. **109**(4), 219–224 (2009). https://doi.org/10.1016/j.ipl.2008.10.003, https://www.sciencedirect.com/science/article/pii/S0020019008003098

3. Bartholdi, J., Trick, M.A.: Stable matching with preferences derived from a psychological model. Oper. Res. Lett. **5**(4), 165–169 (1986). https://doi.org/10.1016/0167-6377(86)90072-6, https://www.sciencedirect.com/science/article/pii/0167637786900726

4. Biró, P., Irving, R.W., Manlove, D.F.: Popular matchings in the marriage and roommates problems. In: Calamoneri, T., Diaz, J. (eds.) CIAC 2010. LNCS, vol. 6078, pp. 97–108. Springer, Heidelberg (2010). https://doi.org/10.1007/978-3-642-13073-1_10

5. Brandt, F., Bullinger, M.: Finding and recognizing popular coalition structures. J. Artif. Int. Res. **74** (2022). https://doi.org/10.1613/jair.1.13470, https://jair.org/index.php/jair/article/view/13470

6. Bredereck, R., Chen, J., Finnendahl, U.P., Niedermeier, R.: Stable roommates with narcissistic, single-peaked, and single-crossing preferences. Auton. Agent. Multi-Agent Syst. **34**(2), 53 (2020). https://doi.org/10.1007/s10458-020-09470-x, https://link.springer.com/article/10.1007/s10458-020-09470-x

7. Chen, J., Roy, S.: Multi-dimensional stable roommates in 2-dimensional euclidean space. In: Chechik, S., Navarro, G., Rotenberg, E., Herman, G. (eds.) 30th Annual European Symposium on Algorithms (ESA 2022). Leibniz International Proceedings in Informatics (LIPIcs), vol. 244, pp. 36:1–36:16. Schloss Dagstuhl – Leibniz-Zentrum für Informatik, Dagstuhl, Germany (2022). https://doi.org/10.4230/LIPIcs.ESA.2022.36, https://drops.dagstuhl.de/entities/document/10.4230/LIPIcs.ESA.2022.36

8. Chung, K.S.: On the existence of stable roommate matchings. Games Econom. Behav. **33**(2), 206–230 (2000). https://doi.org/10.1006/game.1999.0779, https://www.sciencedirect.com/science/article/pii/S0899825699907790

9. Cseh, Á.: Popular matchings, chap. 6, p. 105-122. Lulu. com (2017). https://archive.illc.uva.nl/COST-IC1205/Book/

10. Cseh, A., Juhos, A.: Pairwise preferences in the stable marriage problem. ACM Trans. Econ. Comput. **9**(1) (2021). https://doi.org/10.1145/3434427

11. Cseh, Á., Kavitha, T.: Popular matchings in complete graphs. Algorithmica **83**(5), 1493–1523 (2021). https://doi.org/10.1007/s00453-020-00791-7, https://link.springer.com/article/10.1007/s00453-020-00791-7

12. Di Battista, G., Liotta, G., Vargiu, F.: Spirality and optimal orthogonal drawings. SIAM J. Comput. **27**(6), 1764–1811 (1998). https://doi.org/10.1137/S0097539794262847

13. Faenza, Y., Kavitha, T., Powers, V., Zhang, X.: Popular matchings and limits to tractability, pp. 2790–2809. https://doi.org/10.1137/1.9781611975482.173, https://epubs.siam.org/doi/abs/10.1137/1.9781611975482.173

14. Gärdenfors, P.: Match making: assignments based on bilateral preferences. Syst. Res. Behav. Sci. **20**, 166–173 (1975). https://doi.org/10.1002/bs.3830200304, https://onlinelibrary.wiley.com/doi/10.1002/bs.3830200304

15. Gupta, S., Misra, P., Saurabh, S., Zehavi, M.: Popular matching in roommates setting is NP-hard. ACM Trans. Comput. Theory **13**(2) (2021). https://doi.org/10.1145/3442354

16. Manlove, D.: Algorithmics of Matching Under Preferences. In: Bull. EATCS (2013)

17. Valiant, L.G.: Universality considerations in VLSI circuits. IEEE Trans. Comput. **C-30**(2), 135–140 (1981). https://doi.org/10.1109/TC.1981.6312176

Independent Set Reconfiguration Under Bounded-Hop Token Jumping

Hiroki Hatano[1], Naoki Kitamura[1(✉)], Taisuke Izumi[1], Takehiro Ito[2], and Toshimitsu Masuzawa[1]

[1] Graduate School of Information Science and Technology Department of Computer Science, Osaka University, Suita, Osaka, Japan
{h-hatano,n-kitamura,t-izumi,masuzawa}@ist.osaka-u.ac.jp
[2] Graduate School of Information Sciences, Tohoku University, Sendai, Japan
takehiro@tohoku.ac.jp

Abstract. The independent set reconfiguration problem (ISReconf) is the problem of determining, for two given independent sets of a graph, whether one can be transformed into the other by repeatedly applying a prescribed reconfiguration rule. There are two well-studied reconfiguration rules, called the Token Sliding (TS) rule and the Token Jumping (TJ) rule, and it is known that the complexity status of ISReconf differs between the TS and TJ rules for some graph classes. In this paper, we analyze how changes in reconfiguration rules affect the computational complexity of ISReconf. To this end, we generalize the TS and TJ rules to a unified reconfiguration rule, called the k-Jump rule, which removes one vertex from a current independent set and adds a vertex within distance k from the removed vertex to obtain another independent set having the same cardinality. We give the following three results: First, we show that the computational complexity of ISReconf under the k-Jump rule for general connected graphs is identical for all $k \geq 3$. Second, we present a polynomial-time algorithm to solve ISReconf under the 2-Jump rule for split graphs. Third, we consider the shortest variant of ISReconf, which determines whether there is a transformation of at most ℓ steps, for a given integer $\ell \geq 0$. We prove that this shortest variant under the k-Jump rule is NP-complete for chordal graphs of diameter at most $2k+1$, for any $k \geq 3$.

1 Introduction

Combinatorial reconfiguration [4,5,8] has received much attention in the field of discrete algorithms and the computational complexity theory. A typical *reconfiguration problem* requires us to determine whether there is a step-by-step transformation between two given feasible solutions of a combinatorial (search) problem such that all intermediate solutions are also feasible and each step respects a prescribed reconfiguration rule. This type of reconfiguration problems have been studied actively for several well-known feasible solutions on graphs, such as independent sets, cliques, vertex covers, colorings, matchings, etc. (See surveys [4,8].)

S. Nakano and M. Xiao (Eds.): WALCOM 2025, LNCS 15411, pp. 215–228, 2025.
https://doi.org/10.1007/978-981-96-2845-2_14

While reconfiguration problems have been considered for a wide range of feasible solutions, there are no clear rules to define reconfiguration rules; the smallest change to a current solution is often adopted as the reconfiguration rule unless there is a motivation from the application side. Indeed, even the most well-used reconfiguration rules, called the Token Jumping and Token Sliding rules, have no clear motivation for their rules. In this paper, we study and analyze how changes in reconfiguration rules affect the computational complexity of reconfiguration problems.

1.1   Reconfiguration Rules and Related Known Results

In this paper, we consider the reconfiguration problem for independent sets of a graph [2,6], which is one of the most well-studied reconfiguration problems. A vertex subset of a graph $G = (V, E)$ is an *independent set of G* if it contains no vertices adjacent to each other. In the context of reconfiguration problems, an independent set is often interpreted to the placement of a set of *tokens*, i.e., we regard an independent set $I \subseteq V$ as the locations of $|I|$ tokens in the graph. Then, one reconfiguration step of a independent set corresponds to the movement of a single token from some vertex to another vertex (on which no token is placed), such that the token locations after the movement also forms an independent set. Notice that the size of the independent set before and after the token movement remains unchanged. Reconfiguration rules define the allowed movements of tokens, and there are two well-used rules, called *Token Sliding* and *Token Jumping* [6]:

– Token Sliding (TS) rule: a token is allowed to move only to a vertex adjacent to the current vertex; and
– Token Jumping (TJ) rule: a token is allowed to move to an arbitrary vertex (on which no token is placed) in the graph.

The independent set reconfiguration problem (ISReconf) is now the problem to determine whether a given initial independent set I_s of a graph G can be transformed into a given target independent set I_t of G (having the same size as I_s) by moving tokens one by one under the prescribed reconfiguration rule while preserving the independence of the token placements during the transformation. The optimization problem, the shortest independent set reconfiguration problem (Shortest-ISReconf), of the above decision problem is also derived naturally: Given independent sets I_s and I_t ($|I_s| = |I_t|$) of a graph G, find the smallest number of token movements required to transform I_s into I_t under the prescribed reconfiguration rule, if exists.

Under both the TS and TJ rules, ISReconf is known to be PSPACE-complete even for planar graphs of maximum degree three and bounded bandwidth [9]. Therefore, algorithmic developments have been obtained for several restricted graph classes. (See the survey [2] about ISReconf.) In particular, some known results show interesting contrasts of the complexity status between the TS and TJ rules, as follows: For split graphs, ISReconf is PSPACE-complete under the

TS rule [1], while it is solvable in polynomial time under the TJ rule [6].[1] For bipartite graphs, ISReconf is PSPACE-complete under the TS rule [7], while it is NP-complete under the TJ rule [7]. The latter contrast on bipartite graphs implies that there is a yes-instance on bipartite graphs such that even a shortest transformation requires a super-polynomial number of steps under the TS rule, with the assumption of NP $\neq$ PSPACE; on the other hand, any shortest transformation for bipartite graphs needs only a polynomial number of steps under the TJ rules.

1.2 Unified Reconfiguration Rule and Our Contributions

The main purpose of our paper is to analyze how changes in reconfiguration rules affect the computational complexity of reconfiguration problems. We regard the difference between the TS and TJ rules as "movable distance" of each token in one step: the TS rule allows a token to move to a vertex of distance one, while the TJ rule allows a token to move to a vertex of distance at most $D(G)$, where $D(G)$ is the diameter of G. From this perspective, we generalize the TS and TJ rules to a unified reconfiguration rule, called the k-Jump rule, which allows a token to move to a vertex within distance k from the current vertex, for an integer k, $1 \leq k \leq D(G)$. Then, the TS rule is the 1-Jump rule, and the TJ rule is the $D(G)$-Jump rule for a connected graph G.

This generalization yields a natural question of what complexity landscape lies in the case of $1 < k < D(G)$, particularly for the graph classes exhibiting different computational complexity between the TJ and TS rules. In this paper, we present three results addressing this question, which provides precise and interesting contrasts to the complexity status of (Shortest-)ISReconf. Throughout this paper, let $G = (V, E)$ be an input graph, $I_s \subseteq V$ be an initial independent set, and $I_t \subseteq V$ be a target independent set. We denote by a triple (G, I_s, I_t) an instance of ISReconf under the k-Jump rule. We say that (G, I_s, I_t) is *reconfigurable*, if I_s can be transformed into I_t under the k-Jumping rule.

The first result shows that the reconfigurability of an instance (G, I_s, I_t) does not change for any $k \geq 3$. Note that the following theorem holds for any connected graph G.

Theorem 1. *Let G be a connected graph, and $k \geq 3$ be an arbitrary integer. An instance (G, I_s, I_t) is reconfigurable under the k-Jump rule if and only if (G, I_s, I_t) is reconfigurable under the $D(G)$-Jump rule.*

While Theorem 1 shows that the reconfigurability of an instance does not change for all $k \geq 3$, it may differ between $k \leq 2$ and $k \geq 3$. For example, see Fig. 1, where G is a split graph. For split graphs, any instance with $|I_s| = |I_t|$ is reconfigurable under the k-Jump rule, $k \geq 3$ [6]. On the other hand, as we have seen in the example in Fig. 1, there exist instances for split graphs which

---

[1] Kamiński et al. [6] indeed gave a polynomial-time algorithm to solve Shortest-ISReconf under the TJ rule for even-hole-free graphs, which form a super graph class of split graphs.

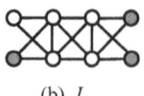

(a) I_s (b) I_t

Fig. 1. No instance for split graphs under the 2-Jump rule. Note that this is a yes-instance under the k-Jump rule, $k \geq 3$.

are not reconfigurable under the 2-Jump rule. Nonetheless, we give the following theorem, as our second result.

Theorem 2. *There exists a polynomial-time algorithm for* ISReconf *under the 2-Jump rule for split graphs.*

Recall that ISReconf under the 1-Jump (i.e., TS) rule is PSPACE-complete for split graphs [1]. Thus, the complexity status of ISReconf differs between $k = 1$ and $k = 2$.

Theorem 1 says that the complexity status of ISReconf is identical for all $k \geq 3$. Our third result shows that this does not hold for the optimization variant, Shortest-ISReconf. We note that Shortest-ISReconf under the $D(G)$-Jump rule is solvable in polynomial time for even-hole-free graphs [6], which include chordal graphs.

Theorem 3. *Let $k \geq 3$ be any integer. Then, there exists a graph class $\mathcal{G}_k$ such that $\mathcal{G}_k$ is a subclass of chordal graphs of diameter at most $2(k + 1)$ and* Shortest-ISReconf *under the k-Jump rule is NP-hard for $\mathcal{G}_k$.*

Tables 1 and 2 summarize our results and the comparison with known results. Recall that split graphs form a subclass of chordal graphs and chordal graphs form a subclass of even-hole-free graphs. We here give two remarks to Table 2. First, the complexity of Shortest-ISReconf under the k-Jump rule ($k \geq 3$) for split graphs is trivially obtained from the complexity of the problem under the $D(G)$-Jumping rule for split graphs, because the diameter of any split graph is at most three. Second, although Shortest-ISReconf is an optimization problem, we write the complexity as NP-complete, not NP-hard, in Table 2 to emphasize that there always exists a transformation of polynomial number of steps under the k-Jump rule ($k \geq 3$) for even-hole-free graphs. This fact can be obtained from the proof of Theorem 1: we will prove that any transformation with a polynomial number of steps under the $D(G)$-Jump rule can be converted into a transformation with a polynomial number of steps under the 3-Jump rule. It is known that there always exists a transformation with a polynomial number of steps under the $D(G)$-Jump rule if G is a connected even-hole-free graph [6].

1.3 Roadmap

In Sect. 2, we introduce fundamental definitions, terminologies, and notations. Sections 3, 4, and 5 respectively provide the results of Theorems 1, 2, and 3. Finally, we conclude the paper in Sect. 6 with several open problems related to

Table 1. Computational complexity of ISReconf for connected graphs G, where the term "PSPACE-c" is the abbreviation of PSPACE-complete.

Reconfiguration Rules	Even-hole-free	Chordal	Split
TS (1-Jump)	PSPACE-c [1]	PSPACE-c [1]	PSPACE-c [1]
2-Jump	open	open	P (Theorem 2)
k-Jump ($k \geq 3$)	P (Theorem 1)	P (Theorem 1)	P (Theorem 1)
TJ ($D(G)$-Jump)	P [6]	P [6]	P [6]

Table 2. Computational complexity of Shortest-ISReconf for connected graphs G, where the term "NP-c" means NP-complete.

Reconfiguration Rules	Even-hole-free	Chordal	Split
TS (1-Jump)	PSPACE-c [1]	PSPACE-c [1]	PSPACE-c [1]
k-Jump ($3 \leq k \leq \frac{D(G)-1}{2}$)	NP-c (Theorem 3)	NP-c (Theorem 3)	P (trivial)
TJ ($D(G)$-Jump)	P [6]	P [6]	P [6]

our new rule. Due to space limitation, we omit all the detailed proofs, which are described in the full version [3].

2   Preliminaries

In this section, we explain the notation used in this paper. For sets X and Y, their symmetric difference is defined as $X \triangle Y = (X \cup Y) \backslash (X \cap Y)$.

We consider only undirected graphs that are simple and connected[2]. For a graph G, we sometimes denote by $V(G)$ and $E(G)$ the vertex set and the edge set of G, respectively. The set of the vertices adjacent to v in G is denoted by $N_G(v)$, that is, $N_G(v) = \{w \in V(G) \mid \{v, w\} \in E(G)\}$. Let $\mathsf{dist}_G(u, v)$ denote the distance in G between vertices $u, v \in V(G)$. A subset $S \subseteq V(G)$ is called an *independent set of* G if no two vertices in S are adjacent in G. Let $\mathcal{I}(G)$ be the set of all independent sets of G.

Let k be a positive integer. For two independent sets $I, I' \in \mathcal{I}(G)$, we write $I \overset{k}{\leftrightarrow} I'$ if and only if $|I \backslash I'| = |I' \backslash I| = 1$ and $\mathsf{dist}_G(u, v) \leq k$ for $u \in I \backslash I'$, $v \in I' \backslash I$. Let $\overset{k}{\rightleftharpoons}$ be the transitive closure of $\overset{k}{\leftrightarrow}$. Note that, from the definitions, $\overset{k}{\leftrightarrow}$ and $\overset{k}{\rightleftharpoons}$ satisfy the symmetry. For two independent sets $I, J \in \mathcal{I}(G)$, a sequence $\langle I_0, I_1, \ldots, I_\ell \rangle$ of independent sets of G is called a *reconfiguration sequence* between I and J (under the k-Jump rule) if $I_0 = I$, $I_\ell = J$, and $I_i \overset{k}{\leftrightarrow} I_{i+1}$ for all i, $0 \leq i < \ell$. The *length* of a reconfiguration sequence

---

[2] We assume graphs are connected for simplicity although the proposed algorithm works without the assumption.

$\langle I_0, I_1, \ldots, I_\ell \rangle$ is defined to be ℓ. We now define the independent set reconfiguration problem under the k-Jump rule (k-ISReconf) and its optimization variant, as follows.

Definition 1. *Given a graph G and independent sets $I_s, I_t \in \mathcal{I}(G)$, the problem k-ISReconf asks to determine whether $I_s \overset{k}{\rightleftharpoons} I_t$ or not.*

Definition 2. *Given a graph G and independent sets $I_s, I_t \in \mathcal{I}(G)$, the problem Shortest-$k$-ISReconf asks to determine whether $I_s \overset{k}{\rightleftharpoons} I_t$ or not, and to compute the minimum length of any reconfiguration sequence between I_s and I_t if $I_s \overset{k}{\rightleftharpoons} I_t$.*

In the following, we use *tokens* to represent the vertices in an independent set $I \in \mathcal{I}(G)$; imagine that a token is placed on each vertex in I. If an independent set I' of G can be obtained from another independent set I such that $I' = (I \setminus \{u\}) \cup \{v\}$, we say that the token on u *moves* to v. To specify the token on a vertex $u \in I$, we often say the token u. For an independent set $I \in \mathcal{I}(G)$, we say a vertex $v \in V(G)$ is *blocked* (by I) when $N_G(v) \cap I \neq \emptyset$.

3   Equivalence of the k-Jump Rule ($k \geq 3$)

The goal of this section is to prove Theorem 1.

Theorem 1. *Let G be a connected graph, and $k \geq 3$ be an arbitrary integer. An instance (G, I_s, I_t) is reconfigurable under the k-Jump rule if and only if (G, I_s, I_t) is reconfigurable under the $D(G)$-Jump rule.*

When independent sets I_1 and I_2 of G satisfy $I_1 \overset{k}{\rightleftharpoons} I_2$, they also satisfy $I_1 \overset{k'}{\rightleftharpoons} I_2$ for any $k' > k$. The converse also holds as the following lemma shows.

Lemma 1. *Let $k' > k \geq 3$. If $I_1 \overset{k'}{\rightleftharpoons} I_2$ holds for independent sets I_1, I_2 of G, then $I_1 \overset{k}{\rightleftharpoons} I_2$ holds.*

Proof. Without loss of generality, we consider only the case of $I_1 \overset{k'}{\leftrightarrow} I_2$ (that is, only a single token moves in the transition from I_1 to I_2). Let t be the token which moves from vertex u_0 to u_ℓ in transition from I_1 to I_2, and $P = u_0, u_1, \ldots, u_\ell$ be a shortest path from u_0 to u_ℓ. The proof is by induction on ℓ. (Basis) $\ell \leq k$: the lemma obviously holds. (Inductive Step) Assuming that the lemma holds for any $\ell' < \ell$, consider the case of ℓ. When $u_{\ell-k+1}$ is not blocked and has no token, then t can move to u_ℓ via $u_{\ell-k+1}$ by induction assumption (Fig. 2(a)) since $\mathsf{dist}_G(u_0, u_{\ell-k+1}) \leq \ell'$ and $\mathsf{dist}_G(u_{\ell-k+1}, u_\ell) \leq k$ hold. Thus, $I_1 \overset{k}{\rightleftharpoons} I_2$ holds. When $u_{\ell-k+1}$ has a token or is blocked, then a vertex $u' \in \{u_{\ell-k+1}\} \cup N_G(u_{\ell-k+1})$ has a token, say t'. From $dist_G(u_0, u') < \ell$ and $dist_G(u', u_\ell) \leq k$, t' can move to u_ℓ and then t can move to u' by induction assumption (Fig. 2(b) and 2(c)). Thus $I_1 \overset{k}{\rightleftharpoons} I_2$ holds. $\square$

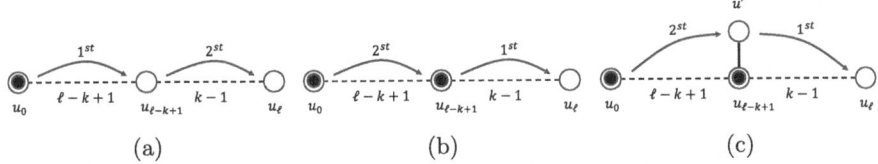

(a)   (b)   (c)

Fig. 2. Illustration for proof of Lemma 1. Tokens move along the arrows in the described order. (a) Case where $u_{\ell-k+1}$ is not blocked and has no token, (b) Case where $u_{\ell-k+1}$ has a token. (c) Case where $u_{\ell-k+1}$ is blocked.

4   Algorithm for 2-ISReconf on Split Graphs

In this section, we give a polynomial-time algorithm to solve 2-ISReconf for split graphs, and prove Theorem 2.

Theorem 2. *There exists a polynomial-time algorithm for ISReconf under the 2-Jump rule for split graphs.*

4.1   Split Graphs and Fundamental Properties

In this section, we present fundamental properties use in the algorithm. A *split graph* is a graph such that its vertex set can be partitioned into a clique and an independent set. Then, a split graph G can be seen as the graph consisting of a complete graph $G^A = (V^A, E^A)$ and a bipartite graph $G^B = (V^B, U^B, E^B)$ such that $V(G) = V^A \cup U^B$, $V^B \subseteq V^A$, $U^B \cap V^A = \emptyset$, and $E(G) = E^A \cup E^B$. (See Fig. 3). Since one can identify the vertex sets V^A and U^B for a given split graph G in polynomial time, the following argument assumes that the information on those sets are available. For simplicity, we assume that no isolated vertex exists in G^B. Each connected component of the bipartite Graph G^B is called a *cluster*. In the following, let the cluster set of the split graph G be $\mathcal{C} = \{C_0, C_1, \ldots, C_{m-1}\}$ with $C_i = (U_i, V_i, E_i)$ ($U_i \subseteq U^B$, $V_i \subseteq V^B$, and E_i is the set of edges induced from E^B). For convenience, when there exists a vertex set $V' \subseteq V^A$ not having any neighbor in U^B, we treat $(\emptyset, V', \emptyset)$ as a bipartite graph cluster and is included in $\mathcal{C}$.

In the following, we exclude the obvious case of $|I_s| = |I_t| > |U^B|$. In this case, the size of the maximum independent set of G is at most $|U^B| + 1$: tokens are placed on all vertices in U^B in both I_s and I_t and another token is placed on a node in V^A that is possibly different in I_s and I_t. Thus, $I_s \overset{2}{\rightleftharpoons} I_t$ always holds. In the case of $|I_s| = |I_t| \leq |U^B|$, we assume that I_s and I_t contain no vertex in V^A, which does not lose generality from the following reason. Note that the diameter of the split graph is at most 3, and the distance from the vertex in V^A to the vertex in U^B is always at most 2. When I_s contains a node v in V^A, we can obtain an independent set I'_s by moving the token on v to an arbitrary empty vertex in U^B. Similarly, we can obtain I'_t from I_t. It is clear that $I_s \overset{2}{\rightleftharpoons} I_t$

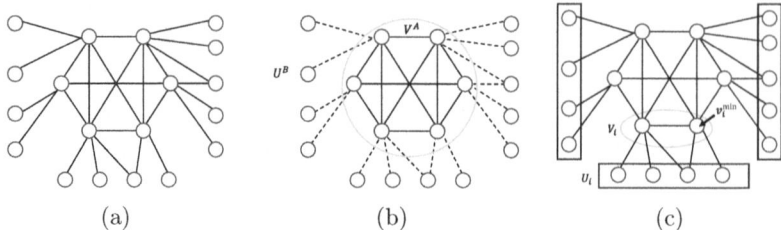

Fig. 3. (a) An example of a split graph. This graph consists of a complete subgraph composed of the solid lines in (b) and a bipartite graph composed of the solid lines in (c). The vertices in the complete subgraph in (b) constitute V^A, and the vertices in the bipartite graph contained in the squares in (c) constitute U^B. Each cluster C_i is represented by a rectangle containing U_i and its neighbor vertices.

holds if and only if $I'_s \overset{2}{\rightleftharpoons} I'_t$ holds. In the following, an independent set I such that $I \cap V^A = \emptyset$ is called a *typical* independent set.

4.2   Token Distribution

In this section, we state a lemma and its corollary. The following lemma shows that tokens can be freely moved within each cluster (independent of token placement outside the cluster).

Lemma 2. *Let I_1, I_2 ($I_1 \neq I_2$) be typical independent sets of G with the same size (i.e., $|I_1| = |I_2|$) such that their token placements are different only in some cluster $C_i = (U_i, V_i, E_i)$, that is, $|I_1 \cap U_i| = |I_2 \cap U_i|$ and $I_1 \backslash U_i = I_2 \backslash U_i$ are satisfied. Then, $I_1 \overset{2}{\rightleftharpoons} I_2$ holds.*

Given a typical independent set I, we define the *distribution* of I as vector $(|I \cap U_i|)_{0 \leq i \leq m-1}$. The following corollary is derived from Lemma 2.

Corollary 1. *If typical Independent sets I and I' have the same distribution, then $I \overset{2}{\rightleftharpoons} I'$.*

4.3   Cluster Types

In this section, we introduce the crucial notion of cluster types. For each i, let $v_i^{\min}$ be any vertex in V_i with the minimum degree in C_i and let $N_i = N_{C_i}(v_i^{\min})$. In the following, we assume without loss of generality that $|N_0| \leq |N_1| \leq \cdots \leq |N_{m} - 1|$ (recall that m is the number of clusters of G). Given an independent set I, we call $f_i(I) = |N_i \cap I|$ the *occupancy* of cluster C_i on I. By definition, the occupancy of a cluster C_i satisfying $U_i = \emptyset$ is 0. For a typical independent set I of G, let I^* be the typical independent set with the same distribution as I such that $f_i(I^*)$ of each cluster C_i is minimum among all typical independent sets with the same distribution as I. We now consider the following classification of clusters.

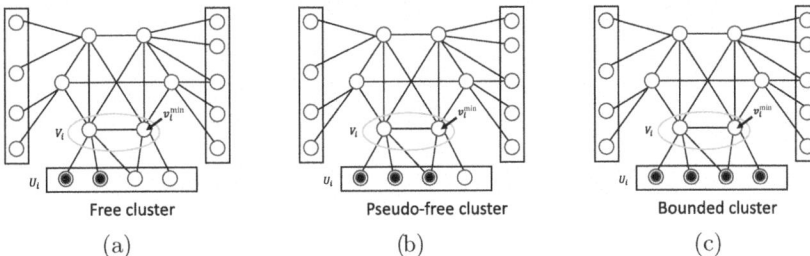

Fig. 4. An example of the cluster types. The cluster type of C_i is (a) a free cluster, (b) a pseudo-free cluster, and (c) a bounded cluster.

Definition 3. *Given a typical independent set I, a cluster C_i is called* Free *if $f_i(I^*) = 0$,* Pseudo-Free *if $f_i(I^*) = 1$, and* Bounded *otherwise. This is referred to as the* type *of the cluster C_i, and only depends on the distribution of C_i.*

The properties of each cluster type for a typical independent set I are intuitively described next.

Free Cluster. If there exists an i such that C_i is Free, then $N_G(v_i^{\min}) \cap I^* = \emptyset$ by definition (recall that I^* contains no vertex in V^A). Also, since the distance between any vertex in G and $v_i^{\min} \in V^A$ is at most 2, after transforming I to I^* by Corollary 1, any token in any cluster C_j can be moved via $v_i^{\min}$ to any vertex in any distinct cluster C_h. This move is possible even if $h = i$, which possibly makes C_i become Pseudo-free.

Pseudo-free Cluster. If C_i is Pseudo-free, after transforming I to I^*, the token in $N_i \cap I^*$ can be moved via $v_i^{\min}$ to any cluster vertex, which makes C_i become Free.

Bounded Cluster. If C_i is Bounded, tokens can move into C_i from other clusters (and vice versa) only if there exists a Free cluster.

Examples of cluster types are shown in Fig. 4. We say that cluster C_i is *full* in a typical independent set I if $I \cap U_i = U_i$. By definition, no free cluster is full and a Pseudo-free cluster C_i is full only if $|N_i| = 1$. Also, for any two independent sets I and I' with the same distribution, if the type of C_i is X for I, then its type is also X for I'. Similarly, if C_i is full for I, then C_i is full for I'. In the following, let $\mathcal{F}(I) \subseteq \mathcal{C}$ be the set of Free clusters for I. We show three lemmas.

Lemma 3. *If a cluster $C_h \in \mathcal{C}$ is Pseudo-free or Bounded for a typical independent set I, then the vertices of V_h are all blocked by I.*

Proof. We prove the lemma by contradiction. If a vertex $v \in V_h$ is unblocked, then $N_{C_h}(v) \cap I = \emptyset$, which implies $|I \cap U_h| \leq |U_h| - |N_{C_h}(v)|$. Since $|N_h| \leq |N_{C_h}(v)|$ by definition, $I^* \cap N_i = \emptyset$ and thus C_h is free, which is a contradiction. $\square$

Lemma 4. *Let I be a typical independent set satisfying one of the following conditions.*

(C1) *All clusters are Bounded for I.*
(C2) *For any Pseudo-free cluster C_i for I, all clusters in $\mathcal{C}\backslash\{C_i\}$ are full.*

Then, any typical independent set I' such that $I \overset{2}{\rightleftharpoons} I'$ has the same distributions as I.

Lemma 5. *If $\mathcal{F}(I_s) \cap \mathcal{F}(I_t) \neq \emptyset$, then $I_s \overset{2}{\rightleftharpoons} I_t$.*

4.4   Technical Ideas for the Algorithm

In this section, we explain some technical ideas for the algorithm (fully described in Sect. 4.5). We first consider the following three cases.

1. For I_s, all clusters are Bounded.
2. For I_s, there exists no Free cluster, and one or more clusters are Pseudo-free.
3. For I_s, a Free cluster exists.

For case 1, by Lemma 4, $I_s \overset{2}{\rightleftharpoons} I_t$ holds if and only if I_s and I_t have the same distribution. For case 2 satisfying condition C2 of Lemma 4, $I_s \overset{2}{\rightleftharpoons} I_t$ holds if and only if I_s and I_t have the same distribution. Thus, in the above cases, whether $I_s \overset{2}{\rightleftharpoons} I_t$ holds or not can be determined in polynomial time. For case 2 not satisfying condition C2, we can make a Pseudo-free cluster C_i Free by moving one token from C_i to a non-full cluster, which leads us to Case 3. So the remaining case we need to consider is case 3. By a similar argument for I_t, the only case we need to consider is the one where a Free cluster exists in I_t. So we consider only the case where I_s and I_t has a Free cluster respectively.

When I_s and I_t have a common Free cluster, Lemma 5 guarantees $I_s \overset{2}{\rightleftharpoons} I_t$. Otherwise, $I_s \overset{2}{\rightleftharpoons} I_t$ holds if there exist I'_s and I'_t such that $I_s \overset{2}{\rightleftharpoons} I'_s$, $I_t \overset{2}{\rightleftharpoons} I'_t$ and $\mathcal{F}(I'_s) \cap \mathcal{F}(I'_t) \neq \emptyset$. The following lemma holds.

Lemma 6. *Let C_i be any Free cluster for a typical independent set I. Let C_j $(j \neq i)$ be any cluster for I satisfying $|N_i| \geq k$ and $|U^B| \geq |I| + |N_i| + |N_j| - k$ for some $k \in \{0, 1, 2\}$, then there exists a typical independent set I' such that $I \overset{2}{\rightleftharpoons} I'$ and C_j is Free for I'. Furthermore, I' can be found in polynomial time.*

When a cluster C_j satisfies the condition of Lemma 6, any cluster $C_{j'}$ $(j' \leq j)$ also satisfies the condition because of $|N_{j'}| \leq |N_j|$. Similarly, when a Free cluster C_i for I satisfies the condition for some j, any Free cluster C'_i $(i' < i)$ also satisfies the condition for j. Thus, without loss of generality, the lemma can assume that i is the smallest such that C_i is Free for I and $j = 0$. By combining with Lemma 5, the following corollary is derived. In the corollary, $i(I)$ denotes the minimum i such that C_i is Free for I.

Corollary 2. *Let I_s and I_t be any typical independent sets having a Free cluster respectively and $i(I_s) \neq i(I_t)$ holds. If both of the following two conditions hold, then $I_s \overset{2}{\rightleftharpoons} I_t$.*

- For some $k_1 \in \{0,1,2\}$, $|N_{i(I_s)}| \geq k_1$ and $|U^B| \geq |I_s| + |N_{i(I_s)}| + |N_0| - k_1$,
- For some $k_2 \in \{0,1,2\}$, $|N_{i(I_t)}| \geq k_2$ and $|U^B| \geq |I_t| + |N_{i(I_t)}| + |N_0| - k_2$

This corollary gives us a sufficient condition for $I_s \overset{2}{\rightleftharpoons} I_t$, but in fact, the following lemma shows that it is also a necessary condition.

Lemma 7. *Let I_s and I_t be any typical independent sets having a Free cluster respectively and $i(I_s) \neq i(I_t)$ holds. Both of the following two conditions hold if $I_s \overset{2}{\rightleftharpoons} I_t$.*

- *For some $k_1 \in \{0,1,2\}$, $|N_{i(I_s)}| \geq k_1$ and $|U^B| \geq |I_s| + |N_{i(I_s)}| + |N_0| - k_1$,*
- *For some $k_2 \in \{0,1,2\}$, $|N_{i(I_t)}| \geq k_2$ and $|U^B| \geq |I_t| + |N_{i(I_t)}| + |N_0| - k_2$.*

4.5   Putting All Together: the Algorithm

In summary, we show a polynomial-time decision algorithm for 2-ISReconf on split graphs. We assume, without loss of generality, that given I_s and I_t are typical independent sets.

For a given split graph $G = (V^A \cup U^B, E^A \cup E^B)$, we first obtain clusters $C_0, \ldots, C_{m-1}$ by deleting all edges in E^A (or edges in the complete subgraph). Then, we obtain the following elements for each cluster C_i.

- $v_i^{\min}$: the vertex with the minimum degree in V_i.
- $|N_i|$: the degree of $v_i^{\min}$ in C_i.
- $|I_s \cap U_i|$ and $|I_t \cap U_i|$ for each i ($0 \leq i \leq m - 1$): the numbers of tokes in cluster C_i for I_s and I_t respectively.

We then classify the clusters for I_s and I_t into Free clusters, Pseudo-free clusters, and Bounded clusters: C_i is Free for $I(\in \{I_s, I_t\})$ if $|U_i| - |I \cap U_i| \geq |N_i|$, Pseudo-free if $|U_i| - |I \cap U_i| = |N_i| - 1$, or Bounded otherwise. Similarly, we determine whether C_i is full or not for I.

After the classification, we check whether $I_s \overset{2}{\rightleftharpoons} I_t$ or not. If all the clusters are Bounded, or only a single cluster is Pseudo-free and all other clusters are full, then we can determine, following Lemma 4, whether $I_s \overset{2}{\rightleftharpoons} I_t$ or not by checking whether they have the same distribution or not. If I_s and I_t have a common Free cluster, then we can determine, following Lemma 5, that $I_s \overset{2}{\rightleftharpoons} I_t$ holds. Finally, if no cluster is Free for at least one of I_s and I_t, then we can determine, following Lemma 6 and Lemma 7, whether $I_s \overset{2}{\rightleftharpoons} I_t$ or not by checking whether both the following condition are satisfied or not.

- $N_i = 1$ and $|U^B| \geq |I_s| + |N_i| + |N_0| - 1$, or $|N_i| > 1$ and $|U^B| \geq |I_s| + |N_i| + |N_0| - 2$.
- $N_{i'} = 1$ and $|U^B| \geq |I_t| + |N_{i'}| + |N_0| - 1$, or $|N_{i'}| > 1$ and $|U^B| \geq |I_t| + |N_{i'}| + |N_0| - 2$.

It is obvious that the procedure described above can be executed in polynomial time.

5   NP-Hardness of **Shortest-k-ISReconf**

In this section, we prove Theorem 3.

Theorem 3. *Let $k \geq 3$ be any integer. Then, there exists a graph class $\mathcal{G}_k$ such that $\mathcal{G}_k$ is a subclass of chordal graphs of diameter at most $2(k+1)$ and Shortest-ISReconf under the k-Jump rule is NP-hard for $\mathcal{G}_k$.*

To prove the theorem, we give a polynomial-time reduction from the E3-SAT problem. The E3-SAT problem is a special case of the SAT problem, where each clause contains exactly three literals. We reduce any instance Φ of E3-SAT to the instance $\Phi' = (G, I_s, I_t)$ of Shortest-k-ISReconf whose shortest reconfiguration sequence has a length at most $2(m + n)$ if and only if Φ is satisfiable, where m and n is the number of clauses and variables in Φ.

Consider any instance Φ of E3-SAT consisting of m clauses $c_0, c_1, \ldots, c_{m-1}$ and n variables $x_0, x_1, \ldots, x_{n-1}$. We construct the *clause gadget* C_i for each clause c_i in Φ, and construct the *variable gadget* L_j for each variable x_j in Φ.

Clause Gadget. We define the clause gadget C_i. The gadget C_i under the k-Jump rule is defined as follows (see Fig. 5):

- Create a path $P = (v_0, v_1, \ldots, v_{2k-1}, v_{2k})$, and define aliases s, k_1, and t as $s = v_0$, $k_1 = v_k$, and $t = v_{2k}$.
- Add two vertices k_0 and k_2, and add four edges $\{k_0, v_{k-1}\}$, $\{k_0, v_{k+1}\}$, $\{k_2, v_{k-1}\}$ and $\{k_2, v_{k+1}\}$.

For any vertex v in C_i, v^i represents the vertex v in the clause gadget C_i. Let $K = \bigcup_{i=0}^{m-1} \{k_0^i, k_1^i, k_2^i\}$. We further augment some edges crossing different clause gadgets.

- Connect any two vertices in K, i.e., K forms a clique.

Variable Gadget. The variable gadget L_i under the k-Jump rule is constructed as follows (see also Fig. 6):

- Create a path $P = (u_0, u_1, \ldots, u_{k-2}, u_{k-1})$. We give aliases t_0 and t_1 as $t_0 = u_0, t_1 = u_{k-1}$.
- Add two vertices s_0, s_1, and add two edges $\{s_0, t_0\}$ and $\{s_1, t_0\}$.

Similarly to the clause gadgets, for any vertex v in L_i, v^i represents the vertex v in L_i.

Whole Construction. We obtain the whole graph G by adding the edges connecting clause gadgets and variable gadgets, as follows:

- We perform the following process for each clause $c_i = (a \vee b \vee c)$. Let L_a (resp. L_b, L_c) be the variable gadgets corresponding to a (resp. b, c) and $\rho_i : \{a, b, c\} \to \{k_0^i, k_1^i, k_2^i\}$ be the function such that $\rho_i(a) = k_0^i$, $\rho_i(b) = k_1^i$, and $\rho_i(c) = k_2^i$. For all $\alpha \in \{a, b, c\}$. If α is a positive literal, we add two edges $e_0 = \{s_0^\alpha, \rho_i(\alpha)\}$ and $e_1 = \{t_0^\alpha, \rho_i(\alpha)\}$. Otherwise, we add two edges $e_0 = \{s_1^\alpha, \rho_i(\alpha)\}$ and $e_1 = \{t_0^\alpha, \rho_i(\alpha)\}$.

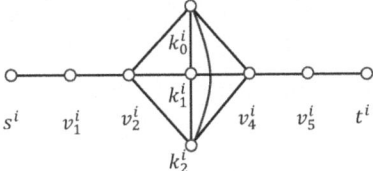

Fig. 5. Example of clause gadget C_i when $k = 3$. Note that edges of $\{k_0, k_1\}$, $\{k_1, k_2\}$ and $\{k_2, k_0\}$ are added in the last step of making K a clique.

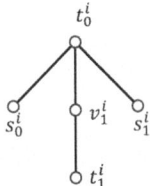

Fig. 6. Example of variable gadget L_i when $k = 3$.

We finish the construction of Φ' by defining the initial independent set I_s and the target independent set I_t as follows:

- $I_s = \bigcup_{i=0}^{m-1} v_0^i \cup \bigcup_{j=0}^{n-1} (s_0^j \cup s_1^j)$; and
- $I_t = \bigcup_{i=0}^{m-1} v_{2k}^i \cup \bigcup_{j=0}^{n-1} (t_0^j \cup t_1^j)$.

6   Concluding Remarks

In this paper, we proposed a new reconfiguration rule for the independent set reconfiguration problem, and investigated the relationship between the value of k and the computational complexity of k-ISReconf. We conclude this paper with some open problems related to our new rule.

- The complexity of 2-ISReconf for graph classes other than split graphs: A major class left as an open problem is chordal graphs, as shown in Table 1.
- The complexity of Shortest-2-ISReconf for split graphs: Is it solvable in poly-nomial time?
- The approximability of Shortest-k-ISReconf ($k \geq 3$) for even-hole-free graphs: Using the polynomial-time algorithm by Kamiński et al. [6], we can solve Shortest-k-ISReconf for connected even-hole-free graphs G when $k = D(G)$. Does it give any non-trivial approximation factor?

Acknowledgments. The second author (Naoki Kitamura) was supported in part by JSPS KAKENHI Grant Numbers JP23K16838 and JP22K21277. The third author (Taisuke Izumi) was supported in part by JSPS KAKENHI Grant Numbers

JP22H03569 and JP23H04385, and JST CRONOS. The fourth author (Takehiro Ito) was supported in part by JSPS KAKENHI Grant Numbers JP19K11814, JP24H00686 and JP24H00690.

References

1. Belmonte, R., Kim, E.J., Lampis, M., Mitsou, V., Otachi, Y., Sikora, F.: Token sliding on split graphs. Theory Comput. Syst. **65**(4), 662–686 (2021). https://doi.org/10.1007/S00224-020-09967-8
2. Bousquet, N., Mouawad, A.E., Nishimura, N., Siebertz, S.: A survey on the parameterized complexity of reconfiguration problems. Comput. Sci. Rev. **53** (2024). https://doi.org/10.1016/j.cosrev.2024.100663
3. Hatano, H., Kitamura, N., Izumi, T., Ito, T., Masuzawa, T.: Independent set reconfiguration under bounded-hop token. Arxiv (2024). https://doi.org/10.48550/arXiv.2407.11768
4. van den Heuvel, J.: The complexity of change. In: Surveys in Combinatorics 2013, vol. 409 London Mathematical Society Lecture Note Series, pp. 127–160. Cambridge University Press (2013). https://doi.org/10.1017/CBO9781139506748.005
5. Ito, T., et al.: On the complexity of reconfiguration problems. Theoret. Comput. Sci. **412**(12–14), 1054–1065 (2011). https://doi.org/10.1016/j.tcs.2010.12.005
6. Kamiński, M., Medvedev, P., Milanič, M.: Complexity of independent set reconfigurability problems. Theoret. Comput. Sci. **439**, 9–15 (2012). https://doi.org/10.1016/j.tcs.2012.03.004
7. Lokshtanov, D., Mouawad, A.E.: The complexity of independent set reconfiguration on bipartite graphs. ACM Trans. Algorithms, **15**(1), 7:1–7:19 (2019). https://doi.org/10.1145/3280825
8. Nishimura, N.: Introduction to reconfiguration. Algorithms, **11**(4), 52 (2018). https://doi.org/10.3390/a11040052
9. van der Zanden, T.C..: Parameterized complexity of graph constraint logic. In: Husfeldt, T., Kanj, I.A. (eds.), 10th International Symposium on Parameterized and Exact Computation, IPEC 2015, September 16-18, 2015, Patras, Greece, LIPICs, vol. 43 pp. 282–293. Schloss Dagstuhl - Leibniz-Zentrum für Informatik (2015). https://doi.org/10.4230/LIPICS.IPEC.2015.282

Approximation Algorithms
for Non-sequential Star Packing Problems

Mengyuan Hu[1], An Zhang[1(✉)], Yong Chen[1], Mingyang Gong[2],
and Guohui Lin[2(✉)]

[1] Department of Mathematics, Hangzhou Dianzi University, Hangzhou, China
{2341070105,anzhang,chenyong}@hdu.edu.cn
[2] Department of Computing Science, University of Alberta, Edmonton, Canada
{mgong4,guohui}@ualberta.ca

Abstract. For a positive integer $k \geq 1$, a k-star (k^+-star, k^--star, respectively) is a connected graph containing a degree-ℓ vertex and ℓ degree-1 vertices, where $\ell = k$ ($\ell \geq k$, $1 \leq \ell \leq k$, respectively). The k^+-star packing problem is to cover as many vertices of an input graph G as possible using vertex-disjoint k^+-stars in G; and given $k > t \geq 1$, the k^-/t-star packing problem is to cover as many vertices of G as possible using vertex-disjoint k^--stars but no t-stars in G. Both problems are NP-hard for any fixed $k \geq 2$. We present a $(1 + \frac{k^2}{2k+1})$- and a $\frac{3}{2}$-approximation algorithms for the k^+-star packing problem when $k \geq 3$ and $k = 2$, respectively, and a $(1+\frac{1}{t+1+1/k})$-approximation algorithm for the k^-/t-star packing problem when $k > t \geq 2$. They are all local search algorithms and they improve the best known approximation algorithms for the problems, respectively.

1   Introduction

For a positive integer $k \geq 1$, a k-star (k^+-star, k^--star, respectively) is a connected graph containing a degree-ℓ vertex, called the *center*, and ℓ degree-1 vertices, called *satellites*, where $\ell = k$ ($\ell \geq k$, $1 \leq \ell \leq k$, respectively).

We study the problem to cover the maximum number of vertices in an input graph using vertex-disjoint stars, referred to as *star packing*. Such a problem, and many variants, have a rich literature due to their theoretical importance and numerous networking applications such as wireless sensor network [23]. Specially, we study two *non-sequential* star packing problems from the approximation algorithm perspective. For one of them, the candidate stars each has at least k satellites for a given constant $k \geq 2$, called the k^+-*star packing* problem, and we propose a $(1 + \frac{k^2}{2k+1})$- and a $\frac{3}{2}$-approximation algorithms when $k \geq 3$ and $k = 2$, respectively; for the other, the candidate stars each has at most k but not exactly t satellites where $k > t \geq 2$, called the k^-/t-*star packing* problem, and we propose a $(1 + \frac{1}{t+1+1/k})$-approximation algorithm. All of them improve the previous best results, respectively.

S. Nakano and M. Xiao (Eds.): WALCOM 2025, LNCS 15411, pp. 229–243, 2025.
https://doi.org/10.1007/978-981-96-2845-2_15

The studied problems fall in the general $\mathcal{H}$-packing problem. Given a collection $\mathcal{H}$ of (non-isomorphic) graphs, an $\mathcal{H}$-*packing* of an input graph $G = (V, E)$ is a set of *vertex-disjoint* subgraphs of G in which each subgraph is isomorphic to a graph of $\mathcal{H}$. A vertex of G is *covered* by the $\mathcal{H}$-packing if it appears in a subgraph inside the packing. The goal of the $\mathcal{H}$-packing problem is to find an $\mathcal{H}$-packing that covers the maximum number of vertices. In particular, a *perfect* $\mathcal{H}$-packing or a so-called $\mathcal{H}$-*factor* refers to an $\mathcal{H}$-packing that covers all the vertices in the input graph. The existence of an $\mathcal{H}$-factor depends on $\mathcal{H}$ and the input graph; $\mathcal{H}$ and sometimes the class of the input graphs determine the complexity and approximability of the optimization $\mathcal{H}$-packing problem.

We first review some closely related $\mathcal{H}$-packing problems, and then the star packing problems for which $\mathcal{H}$ consists of only stars.

For a positive integer i, K_i (C_i, P_i, respectively) denotes the complete graph (the single cycle, the single path, respectively) on i vertices. One sees that when $\mathcal{H} = \{K_2\}$, the $\mathcal{H}$-packing problem is exactly the classical *maximum matching* problem, which is solvable in polynomial time [9]. When $\mathcal{H} = \{K_3\}$, the $\mathcal{H}$-packing problem is the classical *triangle packing* problem, which is polynomially solvable on subcubic graphs (i.e., graphs of maximum degree at most 3) [3] but becomes APX-hard on graphs of maximum degree 4 or larger [18]. When $\mathcal{H} = \{P_3\}$, the $\mathcal{H}$-packing problem or the *maximum P_3-packing* problem is NP-hard on subcubic graphs [26], and NP-hard on claw-free cubic planar graphs [29]. More generally, Kirkpatrick and Hell [20] showed that, when $\mathcal{H}$ contains a single graph H, the $\mathcal{H}$-packing problem is NP-hard as long as H is connected and contains at least three vertices; and this NP-hardness result holds even when restricted to planar graphs [2].

When $\mathcal{H} = \{K_2, H\}$ for some H non-isomorphic to K_2, Loebl and Poljak [24] proved that the $\mathcal{H}$-packing problem is NP-hard unless H is perfect matchable (i.e., H admits a perfect matching), or hypomatchable (i.e., the remainder graph $H - v$ of H after removing the vertex v admits a perfect matching, for any vertex v in H), or a propeller (i.e., H can be obtained from a hypomatchable graph by adding a new pair of vertices u and v, a new edge (u, v), and some new edges connecting u to at least one vertex in the hypomatchable graph). When $\mathcal{H}$ consists of K_2 and some other non-isomorphic graphs $H_1, H_2, \ldots, H_\ell$, then the $\mathcal{H}$-packing problem is polynomially solvable if every H_i is hypomatchable [13].

When $\mathcal{H}$ consists of only complete graphs, the $\mathcal{H}$-packing problem is NP-hard unless $K_2 \in \mathcal{H}$ [13]. When $\mathcal{H}$ contains only cycles, Hell et al. [15] showed that the $\mathcal{H}$-packing problem is NP-hard unless $\mathcal{H}$ contains all cycles, or all cycles but C_3, or all cycles but C_4. When $\mathcal{H}$ is composed of all paths of order k or above, the $\mathcal{H}$-packing problem or the k^+-*path packing* problem is polynomially solvable for $k \leq 3$ [4] while becomes NP-hard for $k \geq 4$ [21].

Closely related to the $\mathcal{H}$-packing problem, in the *weighted set packing* problem one is given a ground set and a collection of subsets of the ground set each associated with a non-negative weight, and the goal is to find a sub-collection of pairwise disjoint sets of the maximum total weight. The weighted set packing problem is a central optimization problem [8] that has received numerous studies from the approximation algorithm perspective, and the state-of-the-art includes a $(\frac{k}{2} + \epsilon)$-approximation algorithm [27], where k is the maximum cardinality of the given sets and $\epsilon > 0$, a 1.786-approximation algorithm when $k = 3$ [28], and a $(\frac{k+1}{3} + \epsilon)$-approximation algorithm when the sets are uni-weighted [5,7], all of which are based on local improvement.

Since the graph isomorphism problem is still not polynomially solvable [12], one cannot afford to reduce the $\mathcal{H}$-packing problem to the weighted set packing problem if the graphs in $\mathcal{H}$ are arbitrary, but otherwise such as the triangle packing and the P_3-packing problems can be reduced to the uni-weighted 3-set packing problem to obtain their first $(\frac{4}{3}+\epsilon)$-approximation algorithms. Nevertheless, typically directly designing local operations for the $\mathcal{H}$-packing problem leads to better performance analysis. To name a few, the triangle packing problem on graphs of maximum degree 4 admits a $\frac{6}{5}$-approximation algorithm [25], and the P_3-packing problem on cubic graphs admits a $\frac{4}{3}$-approximation algorithm [19]. For the k^+-path packing problem, Gong et al. [11] proposed a local improvement algorithm which achieves a ratio of $\rho(k) \leq 0.4394k + 0.6576$ for any fixed $k \geq 4$; when $k = 4$, Kobayashi et al. [21] presented a simple local improvement algorithm which turns out to be a 4-approximation, Gong et al. [11] proposed a 2-approximation algorithm using more local operations; and most recently Gong et al. [10] proposed to begin with a maximum matching and then apply multiple local improvement operations, resulting in a 1.874-approximation algorithm.

In this paper, we study the star packing problem, i.e., the $\mathcal{H}$-packing problem where $\mathcal{H}$ is a collection of various stars. In the literature, an i-star is also written as $K_{1,i}$, that is, the complete bipartite graph with a singleton on one side and i vertices on the other side. In particular, the k^+-star packing problem is the $\mathcal{H}$-packing problem where $\mathcal{H} = \{K_{1,k}, K_{1,k+1}, K_{1,k+2}, \ldots\}$, for a given $k \geq 1$, and the k^-/t-star packing problem is the one where $\mathcal{H} = \{K_{1,1}, \ldots, K_{1,t-1}, K_{1,t+1}, \ldots, K_{1,k}\}$, for given $k > t \geq 1$. Hell and Kirkpatrick [14] showed that the star packing problem is polynomially solvable if $\mathcal{H}$ is composed of *sequential* stars, including the cases where $\mathcal{H}$ contains all stars, or $\mathcal{H}$ contains all the stars with up to k satellites for any $k \geq 1$; otherwise, the same authors showed that the problem is strongly NP-hard [13]. Furthermore, the k^+-star packing problem for any $k \geq 2$ is NP-hard even on bipartite graphs [22], and the 2^+-star packing problem is NP-hard even on bipartite graphs with maximum degree 4 [22] or on cubic graphs [31]. On the positive side, the 2^+-star packing problem admits a 2-approximation algorithm [22]; and on cubic graphs it admits a better $\frac{7}{6}$-approximation algorithm [31]. The 2-approximation is recently improved to a $\frac{9}{5}$-approximation algorithm by Huang et al. [17], who also presented a local search $(1 + \frac{k}{2})$-approximation algorithm for the k^+-star packing problem for any $k \geq 3$. For the k^-/t-star packing problem with $k > t \geq 2$, Li and Lin [22] observed a simple $(1 + \frac{1}{t})$-approximation algorithm. There is not much result on inapproximability, except the APX-completeness of the 3-star packing on cubic graphs, which is equivalent to the maximum distance-3 independent set problem [6,30].

We continue to study the k^+-star packing problem with $k \geq 2$ and the k^-/t-star packing problem with $k > t \geq 2$ from the perspective of approximation algorithms. Our algorithms are all based on local search or local improvement. For the first problem, we design operations to increase the number of vertices covered by the current feasible k^+-star packing; for the second problem, the operations are designed to decrease the number of t-stars in the current optimal sequential k^--star packing, or to keep the number of t-stars but increase the number of other stars. These algorithms are presented in Sects. 2 and 3, respectively. Table 1 below summarizes the previous approximation results and the improved ones achieved in this paper. We conclude the paper in Sect. 4.

Table 1. Previous approximation results and the improved ones achieved for the two non-sequential star packing problems.

Problem	Approximation ratios
k^+-star packing with $k \geq 3$	$1 + \frac{k}{2}$ [17] $\longrightarrow 1 + \frac{k^2}{2k+1}$ (Theorem 1)
2^+-star packing	2 [22] $\longrightarrow \frac{9}{5}$ [17] $\longrightarrow \frac{3}{2}$ (Theorem 2)
k^-/t-star packing with $k > t \geq 2$	$1 + \frac{1}{t}$ [22] $\longrightarrow 1 + \frac{1}{t+1+1/k}$ (Theorem 3)
k^-/t-star packing with $k = \infty$ and $t \geq 2$	$1 + \frac{1}{t+2}$ (Extended from Theorem 3)

2   Approximating the k^+-Star Packing Problem

The previous best results for the k^+-star packing problem when $k \geq 3$ is the $(1 + \frac{k}{2})$-approximation algorithm, and when $k = 2$ is the $\frac{9}{5}$-approximation algorithm, both by Huang et al. [17] (see Table 1). While the former is a relatively simple local search algorithm, the latter $\frac{9}{5}$-approximation algorithm is involved and, it first computes a maximal collection of trees, then connects the other vertices to these trees, and lastly decomposes the resulting trees carefully into a solution.

We first present our algorithm for any $k \geq 2$, which is a local search algorithm denoted as LOCALSEARCH-k^+, in the next subsection, and then its performance analysis in Subsect. 2.2. When $k = 2$, the general algorithm LOCALSEARCH-k^+ works out to be a $\frac{9}{5}$-approximation; we then add one more operation to the general algorithm to become the final LOCALSEARCH-2^+, and show that it is a $\frac{3}{2}$-approximation.

We fix an input graph $G = (V, E)$ for presentation; a feasible solution $\mathcal{P}$ is a collection of vertex-disjoint k^+-stars in G, and a star in $\mathcal{P}$ is called an *internal* star. We use $V(\mathcal{P})$ to denote the set of the vertices in the internal stars, i.e., the covered vertices by $\mathcal{P}$. A vertex in $V \setminus V(\mathcal{P})$ is *uncovered*, and the subgraph R induced by all the uncovered vertices is the *remainder* graph with respect to $\mathcal{P}$. A star in R is called an *outside* star (with respect to $\mathcal{P}$—which is not repeated in the sequel).

2.1   The Algorithm for $k \geq 2$

Similar to many other local search algorithms such as in Huang et al. [17], we define several operations to be repeatedly executed inside our algorithm LOCALSEARCH-k^+. Note that the algorithm is iterative, and during each iteration a feasible solution is assumed at the beginning, referred to as the *current* solution or packing, an operation is applied and the solution is *updated*, and then the iteration ends. When none of the operations is applicable, the algorithm terminates and returns the current solution as the final solution.

Definition 1 (Operation COLLECT(v)). *Given a vertex v in the current remainder graph, the operation* COLLECT*(v) extracts an ℓ-star centered at v from the remainder graph, where $\ell \geq k$ is the degree of v in the remainder graph, and adds it to the current solution $\mathcal{P}$.*

The algorithm starts with the empty solution, and in the first phase (or the first a few iterations), COLLECT(v) is repeatedly applied on a vertex v of degree at least k in the remainder graph until impossible. Note that at this moment, the maximum degree of the remainder graph is at most $k - 1$ and no uncovered vertex is adjacent to the center of any internal star. This property is stated in the following Lemma 1. In fact, at the end of each iteration during the algorithm, COLLECT(v) is *re-applied* to the center v of every internal k^+-star to pick up the uncovered vertices adjacent to v, if any.

Lemma 1. *When operation* COLLECT *is not applicable, the maximum degree of the remainder graph with respect to the feasible solution $\mathcal{P}$ is at most $k - 1$ and no uncovered vertex is adjacent to the center of any internal star.*

Definition 2 (Operation PULL-BY-$(k+1)^+$). *Given a satellite v of an internal $(k+1)^+$-star, the operation removes v from the internal $(k+1)^+$-star and applies* COLLECT *to extract a k^+-star.*

Note that such an operation PULL-BY-$(k + 1)^+$ trades a satellite v of an internal $(k + 1)^+$-star for a k^+-star, which covers v since prior to this operation no COLLECT is applicable. There are two possible scenarios: In one scenario v is adjacent to at least k uncovered vertices and thus v is the center of the extracted star; in the other scenario v is adjacent to the center of an outside $(k - 1)$-star and thus v is a satellite of the extracted star. We remark that to apply PULL-BY-$(k+1)^+$, only one satellite of an internal $(k+1)^+$-star is examined, disregarding how large the internal star is. For example, even if there are two satellites of an internal $(k + 2)^+$-star adjacent to the center of an outside $(k - 2)$-star, no PULL-BY-$(k + 1)^+$ is applied.

Definition 3 (Operation PULL-BY-k). *Given an internal k-star, the operation removes the star from $\mathcal{P}$ and applies* COLLECT *to extract a $(k + 1)^+$-star, or if not possible then to extract two vertex-disjoint k-stars.*

Note that such an operation PULL-BY-k trades an internal k-star for a larger $(k + 1)^+$-star, or for two vertex-disjoint k-stars. We remark that right after the operation, some vertices in the remainder graph might have their degree greater than or equal to k, and thus COLLECT operations might be applicable to extract more k^+-stars. There are many possible scenarios to apply PULL-BY-k, for example, when a satellite v of the internal k-star is adjacent to k or more uncovered vertices, or when two satellites of the internal k-star are both adjacent to the center of an outside $(k - 1)$-star, or when two satellites of the internal k-star each is adjacent to the center of one of the two vertex-disjoint outside $(k - 1)$-stars, respectively; and so on.

After applying the PULL-BY-$(k + 1)^+$ and PULL-BY-k operations, we have the following properties stated in Lemma 2 for the achieved solution $\mathcal{P}$.

Lemma 2. *When* COLLECT, PULL-BY-$(k + 1)^+$ *and* PULL-BY-k *are not applicable,*

- *a satellite of any internal star is adjacent to at most $k-1$ uncovered vertices;*
- *no satellite of any internal $(k + 1)^+$-star is adjacent to the center of an outside $(k - 1)$-star;*
- *an internal k-star S has at most one satellite that is adjacent to the center of an outside $(k-1)$-star T, and if this happens, then no other satellite of S can be adjacent to $k - 1$ uncovered vertices excluded from T and no two other satellites of S can both be adjacent to the center of an outside $(k - 2)$-star vertex-disjoint from T.*

We next define two more operations that work on two internal stars to increase the covered vertices. They can be deemed as combinations of the above two PULL operations. For convenience, a $k_{v\text{-}u}$-*star* is an internal k-star of which the satellite v is adjacent to the center u of an outside $(k - 1)$-star, and a k_v^+-*star* is an internal k^+-star of which the satellite v is adjacent to exactly $k - 1$ uncovered vertices (see for an illustration in Fig. 1).

Definition 4 (Operation PULL-BY-$(k, (k+1)^+)$). *Given two internal k-star and $(k + 1)^+$-star, the operation removes the k-star from $\mathcal{P}$ and a satellite from the $(k + 1)^+$-star, and applies COLLECT operations to extract two vertex-disjoint k^+-stars.*

Definition 5 (Operation PULL-BY-(k, k)). *Given two internal k-stars, the operation removes both of them from $\mathcal{P}$, and applies COLLECT operations to extract either two vertex-disjoint k^+-star and $(k + 1)^+$-star, or three vertex-disjoint k-stars.*

Below, we sometimes use simply PULL to mean any one of the above four PULL-BY-$(k+1)^+$, PULL-BY-k, PULL-BY-$(k, (k+1)^+)$, and PULL-BY-(k, k) operations. Using COLLECT and PULL operations, our algorithm LOCALSEARCH-k^+ repeatedly applies one of them; when none of them is applicable, the algorithm outputs the current solution $\mathcal{P}$ to the k^+-star packing problem. A high-level description of the algorithm is depicted in Fig. 2.

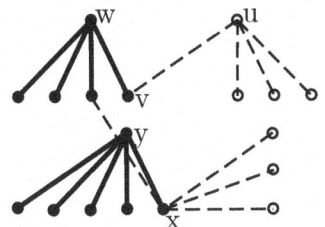

Fig. 1. An illustration to apply operation PULL-BY-$(k, (k+1)^+)$ on two internal k-star centered at w, which is a $k_{v\text{-}u}$-star, and $(k+1)^+$-star centered at y, which is a $(k+1)_x^+$-star. The filled vertices are covered, the empty vertices are uncovered, the edges in the internal stars are solid while the dashed edges are in the input graph G. In this case, $k = 4$, a vertex of the $k_{v\text{-}u}$-star other than v is adjacent to x. After removing the $k_{v\text{-}u}$-star from $\mathcal{P}$ and the satellite x from the $(k+1)_x^+$-star, two vertex-disjoint k-stars centered at u and x, respectively, are extracted.

LOCALSEARCH-k^+ for k^+-star packing:
Input: A connected graph $G = (V, E)$;
Output: A collection $\mathcal{P}$ of vertex-disjoint k^+-stars in G.

1. Initialize $\mathcal{P} = \emptyset$ and the remainder graph $R = G$;
2. While (a COLLECT or PULL operation is applicable) do
   (a) apply the operation to update $\mathcal{P}$;
   (b) re-apply COLLECT on the center of every internal star;
   (c) update the remainder graph $R = G[V \setminus V(\mathcal{P})]$;

Fig. 2. A high-level description of LOCALSEARCH-k^+ for k^+-star packing.

2.2 Performance Analysis

We fix an optimal k^+-star packing, denoted as $\mathcal{Q}^*$, for discussion, and denote by $Opt = |V(\mathcal{Q}^*)|$ the number of vertices covered by $\mathcal{Q}^*$. For ease of presentation, the stars inside $\mathcal{Q}^*$ are referred to as *optimal stars*. We start with some observations on $\mathcal{Q}^*$ (some proofs and more illustrations can be found in the full paper [16]).

Lemma 3. *At most $k - 1$ satellites of an optimal star are uncovered by $\mathcal{P}$.*

By Lemma 3, any optimal star contains at most k uncovered vertices by the computed solution $\mathcal{P}$, and if exactly k then the center of the optimal star is uncovered. An optimal star containing exactly $k - i + 1$ uncovered vertices is called a Type-i optimal star, for $i = 1, 2$; an optimal star containing at most $k - 2$ uncovered vertices is called a Type-3 optimal star.

Let $Apx = |V(\mathcal{P})|$ denote the number of vertices covered by the computed solution $\mathcal{P}$. Let Opt_i (Apx_i, respectively) denote the total number of vertices (covered vertices by $\mathcal{P}$, respectively) in all the Type-i optimal stars, for $i = 1, 2, 3$;

and let Apx_4 denote the total number of covered vertices that fall outside of any optimal star. It follows that

$$Opt = Opt_1 + Opt_2 + Opt_3, \quad Apx = Apx_1 + Apx_2 + Apx_3 + Apx_4. \quad (1)$$

When $k \geq 3$, for each $i = 1, 2, 3$, $k - i + 1 \geq 1$; $Opt_i - Apx_i$ is the total number of uncovered vertices by $\mathcal{P}$ in all the Type-i optimal stars, and thus there are at least $\frac{Opt_i - Apx_i}{k-i+1}$ such Type-i optimal stars. On the other hand, there are at most $\frac{Apx_i}{i}$ such Type-i optimal stars. Therefore, $\frac{Opt_i - Apx_i}{k-i+1} \leq \frac{Apx_i}{i}$, i.e.,

$$Opt_i \leq \frac{k+1}{i} Apx_i, i = 1, 2, 3, \quad (2)$$

where the inequality becomes an equality when $i = 1, 2$. When $k = 2$ and $i = 3$, $k - i + 1 = 0$ and thus $Opt_i = Apx_i$, i.e., Eq. (2) still holds.

A satellite of a Type-1 optimal star is called a *critical* vertex (or simply a c-vertex) if it is covered by $\mathcal{P}$. There are a total of Apx_1 c-vertices, stated as the first half in the next lemma.

Lemma 4. *There are a total of Apx_1 c-vertices, and each c-vertex is a satellite of a distinct internal k-star in $\mathcal{P}$.*

Lemma 4 states that a c-vertex maps injectively to an internal k-star. Continue from the proof of Lemma 4 where the c-vertex v is a satellite of a Type-1 optimal star centered at u, and exactly $k - 1$ satellites and the center u are uncovered by $\mathcal{P}$. One thus sees (from the perspective of $\mathcal{P}$) that the internal k-star to which v belongs (or maps) is a $k_{v\text{-}u}$-star. We call all the other vertices in the internal $k_{v\text{-}u}$-star the *adjugates* of v, or simply the a-vertices associated with the c-vertex v. This way, each c-vertex is associated with exactly k a-vertices, and thus by Lemma 4 there are exactly $k \times Apx_1$ a-vertices in total.

We next examine where these a-vertices sit in the optimal k^+-star packing solution $\mathcal{Q}^*$. We first do not distinguish whether $k \geq 3$ or $k = 2$, while treating a $(k - 2)$-star as a single vertex (i.e., the center vertex attached with no satellite) if $k = 2$. Then we pay special attention when $k = 2$, as some details become very different, for example, Lemma 5 versus Lemma 6; we also add another PULL operation to achieve a better performance guarantee.

2.2.1   Case 1 Where $k \geq 2$

Lemma 5. *Each Type-2 optimal star contains at most one a-vertex.*

Theorem 1. *Algorithm LOCALSEARCH-k^+ is an $O(|V|^3|E|)$-time $(1 + \frac{k^2}{2k+1})$-approximation for the k^+-star packing problem for any $k \geq 2$.*

Proof. We first analyze the time complexity of the algorithm. Note that a COLLECT operation needs to scan for the degrees of all the vertices in the remainder graph, which can be obtained in $O(|E|)$ time (noting that by maintaining certain data structure such as a heap, this can be done faster), followed by extracting a k^+-star and updating the remainder graph in $O(|V|)$ time. A PULL operation requires first trying one vertex out of a $(k+1)^+$-star, or trying two k-stars, returning a maximum number $2(k+1)$ of vertices back to the remainder graph, then extracting a maximum number 3 of k^+-stars, and thus is done in $O(|V|^2|E|)$ time. Since each iteration increases the number of covered vertices by at least one, there are $O(|V|)$ iterations. At the end of each operation, the algorithm re-applies COLLECT operations on the centers of internal k^+-stars in $O(|E|)$ time. These together imply the overall running time of the algorithm in $O(|V|^3|E|)$.

To prove the performance ratio, we recall that a Type-2 optimal star contains exactly $k-1$ uncovered vertices by $\mathcal{P}$, and thus at least two covered vertices. Lemma 5 states that at most one of the covered vertices in a Type-2 optimal star is an a-vertex. It follows from Lemma 4 that the total number of a-vertices is $k \times Apx_1 \le \frac{1}{2}Apx_2 + Apx_3 + Apx_4$. Combining this upper bound with Eqs. (1, 2), and setting $\alpha = 1 + \frac{k^2}{2k+1}$, for any $k \ge 2$ we have

$$
\begin{aligned}
Opt &= Opt_1 + Opt_2 + Opt_3 \le (k+1)Apx_1 + \tfrac{k+1}{2}Apx_2 + \tfrac{k+1}{3}Apx_3 \\
&= \alpha Apx_1 + (k+1-\alpha)Apx_1 + \tfrac{k+1}{2}Apx_2 + \tfrac{k+1}{3}Apx_3 \\
&\le \alpha Apx_1 + (k+1-\alpha) \cdot \tfrac{1}{k} \cdot (\tfrac{1}{2}Apx_2 + Apx_3 + Apx_4) + \tfrac{k+1}{2}Apx_2 + \tfrac{k+1}{3}Apx_3 \\
&= \alpha Apx_1 + (\tfrac{k+1-\alpha}{2k} + \tfrac{k+1}{2})Apx_2 + (\tfrac{k+1-\alpha}{k} + \tfrac{k+1}{3})Apx_3 + \tfrac{k+1-\alpha}{k}Apx_4 \\
&\le \max\{\alpha, \tfrac{k+1-\alpha}{2k} + \tfrac{k+1}{2}, \tfrac{k+1-\alpha}{k} + \tfrac{k+1}{3}\}(Apx_1 + Apx_2 + Apx_3 + Apx_4) \\
&\le (1 + \tfrac{k^2}{2k+1})Apx.
\end{aligned}
$$

This proves the theorem. $\qquad\square$

2.2.2   Case 2 Where $k=2$

The algorithm LOCALSEARCH-k^+ works for every $k \ge 2$, and it is referred to as the general algorithm. One sees that when $k = 2$, the general algorithm works out to be a $\frac{9}{5}$-approximation algorithm, which ties the $\frac{9}{5}$-approximation algorithm by Huang et al. [17] designed specifically for the 2^+-star packing problem. We note that the main design idea in the algorithm by Huang et al. is to compute a maximal number of balanced trees in the input graph and then to acquire 2^+ stars from the trees.

For convenience, we continue to use k inside the name of the algorithm LOCALSEARCH-k^+ and most notations as in the above, but keep in mind that in this subsection $k = 2$. We design an additional PULL operation that works on three internal k-stars to increase the covered vertices.

Definition 6 (Operation PULL-BY-(k, k, k)). *Given three internal k-stars, the operation removes all of them from $\mathcal{P}$ and applies COLLECT operations to extract three stars, of which two are k-stars and the other is a $(k+1)^+$-star.*

Lemma 6. *Each Type-2 optimal star contains no a-vertex.*

Recall that a Type-1 optimal star contains exactly two uncovered vertices by $\mathcal{P}$. If it contains only one c-vertex, i.e., it is a k-star, then it is called a Type-11 optimal star; the other Type-1 optimal stars each containing at least two c-vertices are called Type-12 optimal stars (which are $(k+1)^+$-stars). Also recall that a Type-3 optimal star contains no uncovered vertex by $\mathcal{P}$. If it contains at most two a-vertices, then it is called a Type-31 optimal star; the other Type-3 optimal stars each containing at least three a-vertices are called Type-32 optimal stars. Correspondingly, let Opt_{ij} (Apx_{ij}, respectively) denote the total number of vertices (covered vertices by $\mathcal{P}$, respectively) in all the Type-ij optimal stars, for $i = 1, 3$ and $j = 1, 2$. It follows that, besides Eqs. (1, 2),

$$Opt_i = Opt_{i1} + Opt_{i2}, \quad Apx_i = Apx_{i1} + Apx_{i2}, \text{ for } i = 1, 3; \tag{3}$$

$$Opt_{11} = 3Apx_{11}, \quad Opt_{12} \le 2Apx_{12}, \text{ and } Opt_{3j} = Apx_{3j}, \text{ for } j = 1, 2. \tag{4}$$

Lemma 7. *Each Type-32 optimal star contains no a-vertex associated with a c-vertex in any Type-11 optimal star.*

Theorem 2. *Algorithm* LOCALSEARCH-2^+ *is an* $O(|V|^4|E|)$*-time* $\frac{3}{2}$*-approximation for the* 2^+*-star packing problem.*

Proof. Note that we have an additional operation PULL-BY-(k, k, k) that works on three internal k-stars, and thus requires trying three k-stars, returning a maximum number $3(k+1)$ of vertices back to the remainder graph, then extracting a maximum number 3 of k^+-stars, and thus is done in $O(|V|^3|E|)$ time. That is, the overall time complexity increases by an order, which is $O(|V|^4|E|)$.

Since each Type-31 optimal star contains at most two a-vertices, at most $\frac{2}{3}$ of vertices of Type-31 optimal stars are a-vertices. Among these a-vertices, assume p_j of them are associated with the c-vertices in the Type-1j optimal stars, for $j = 1, 2$. That is,

$$p_1 + p_2 \le \frac{2}{3} Opt_{31}.$$

Among those Apx_4 covered vertices by $\mathcal{P}$ but falling outside of the optimal solution $\mathcal{Q}^*$, assume there are q_j a-vertices which are associated with the c-vertices in the Type-1j optimal stars, for $j = 1, 2$. That is,

$$q_1 + q_2 \le Apx_4.$$

Using Lemma 6 and 7, the total number of a-vertices associated with those c-vertices in the Type-1j optimal stars, for $j = 1, 2$, is

$$2Apx_{11} \le p_1 + q_1, \text{ and } 2Apx_{12} \le p_2 + q_2 + Opt_{32} \le \frac{2}{3} Opt_{31} - p_1 + q_2 + Opt_{32},$$

respectively. Combining the above with Eqs. (1–4), we obtain

$$
\begin{aligned}
Opt &= Opt_{11} + Opt_{12} + Opt_2 + Opt_3 \le 3Apx_{11} + 2Apx_{12} + \tfrac{3}{2}Apx_2 + Apx_3 \\
&= \tfrac{3}{2}Apx_{11} + \tfrac{1}{2}Apx_{12} + \tfrac{3}{2}(Apx_1 + Apx_2) + Apx_3 \\
&\le \tfrac{3}{4}(p_1 + q_1) + \tfrac{1}{4}(\tfrac{2}{3}Opt_{31} - p_1 + q_2 + Opt_{32}) + \tfrac{3}{2}(Apx_1 + Apx_2) + Apx_3 \\
&= \tfrac{1}{2}p_1 + \tfrac{3}{4}q_1 + \tfrac{1}{4}q_2 + \tfrac{1}{6}Opt_{31} + \tfrac{1}{4}Opt_{32} + \tfrac{3}{2}(Apx_1 + Apx_2) + Apx_3 \\
&\le \tfrac{1}{3}Opt_{31} + \tfrac{3}{4}Apx_4 + \tfrac{1}{6}Opt_{31} + \tfrac{1}{4}Opt_{32} + \tfrac{3}{2}(Apx_1 + Apx_2) + Apx_3 \\
&\le \tfrac{1}{2}Opt_3 + \tfrac{3}{4}Apx_4 + \tfrac{3}{2}(Apx_1 + Apx_2) + Apx_3 \\
&\le \tfrac{3}{2}(Apx_1 + Apx_2 + Apx_3 + Apx_4) = \tfrac{3}{2}Apx.
\end{aligned}
$$

This finishes the proof. □

3   Approximating the k^-/t-Star Packing Problem

Unlike the NP-hard k^+-star packing problem we consider in the last section, the star packing problem using vertex-disjoint k^--stars, i.e., stars with 1 up to k satellites, to cover the maximum number of vertices, is the *sequential k^--star packing* problem and solvable in $O(\sqrt{|V|}|E|)$ time for any $k \geq 1$ [1,14].

Let $\mathcal{Q}_0$ denote an optimal k^--star packing for an input graph $G = (V, E)$, and let $V_0 = V \setminus V(\mathcal{Q}_0)$ denote the set of uncovered vertices by $\mathcal{Q}_0$. One sees that each vertex $v \in V_0$ can be adjacent to only the centers of the internal k-stars in $\mathcal{Q}_0$, since otherwise v would be covered to achieve a better packing. These internal k-stars in $\mathcal{Q}_0$ of which the center is adjacent to a vertex in V_0 are called *critical*. For the same reason, each satellite of a critical k-star can be adjacent to only the centers of the internal k-stars in $\mathcal{Q}_0$, and recursively, those internal k-stars in $\mathcal{Q}_0$ of which the center is adjacent to a satellite of critical k-star become *critical* too. We denote by $\mathcal{K}(v) \subseteq \mathcal{Q}_0$ the collection of all critical k-stars associated with the vertex $v \in V_0$, and $\mathcal{K}(V_0) = \cup_{v \in V_0} \mathcal{K}(v)$.

Note that $v \in V_0$ and all the satellites covered by $\mathcal{K}(v)$ together form an independent set in the input graph G, and their neighbors are the centers of the critical k-stars in $\mathcal{K}(v)$.

Given $k > t \geq 1$, in the k^-/t-star packing problem one uses vertex-disjoint k^--stars except t-stars to cover the maximum number of vertices. The problem is NP-hard for any $k > t \geq 1$ [14], and admits a straightforward $(1 + \frac{1}{t})$-approximation algorithm, for $t \geq 2$, by first computing the $\mathcal{Q}_0$ and then removing one satellite from each t-star in $\mathcal{Q}_0$ [22]. Below we continue to use the idea in this simple approximation, but we seek for a better $\mathcal{Q}_0$ in which there are relatively fewer t-stars than the other stars. For instance, in the extreme case where $\mathcal{Q}_0$ contains no t-star, it is an optimal solution to the k^-/t-star packing problem.

So we consider $k > t \geq 2$, fix an optimal k^-/t-star packing $\mathcal{Q}^*$ for discussion and let $V_0^* = V \setminus V(\mathcal{Q}^*)$ denote the set of uncovered vertices by $\mathcal{Q}^*$. We observe the optimal packing in the next lemma.

Lemma 8. *There exists an optimal sequential k^--star packing $\mathcal{Q}_0$ for the input graph in which the uncovered vertex set $V_0 \subseteq V_0^*$.*

Lemma 8 can be proved by a construction (in the full paper [16]), which tells that, given an optimal k^-/t-star packing $\mathcal{Q}^*$ and an optimal sequential k^--star packing $\mathcal{Q}_0$, how to modify those critical k-stars in $\mathcal{K}(V_0)$ into one such that $V_0 \subseteq V_0^*$. Note that $\mathcal{Q}_0$ can be computed in polynomial time and below we assume without loss of generality that for the optimal k^-/t-star packing $\mathcal{Q}^*$, $V_0 \subseteq V_0^*$.

In our algorithm LOCALSEARCH-k^-/t for the k^-/t-star packing problem (see for a high-level description in Fig. 3), where $k > t \geq 2$, we propose local search operations to work with those stars in $\mathcal{Q}_0 \setminus \mathcal{K}(V_0)$ to reduce the number of t-stars, or to keep the number of t-stars while increasing the number of other stars. This way, only $O(|V|)$ iterations are done and at the end one satellite of each remaining t-star in $\mathcal{Q}_0$ is discarded to become a solution $\mathcal{P}$ for the k^-/t-star packing problem.

3.1   The Algorithm

The algorithm works with $\mathcal{Q}_0$, which maintains to be an optimal k^--star packing and is referred to as the *current k^--star* packing. Below we describe local search operations at the presence of at least a t-star inside $\mathcal{Q}_0$, as otherwise $\mathcal{Q}_0$ is an optimal k^-/t-star packing. These operations are collectively called REVISE, but they separately work on different combinations of stars in $\mathcal{Q}_0$ of which at least one is a t-star. An operation is *applicable* means it reduces the number of t-stars while keeping the same vertices covered, or it keeps the number of t-stars while increasing the number of other stars.

Definition 7 (Operation REVISE-t). *Given a t-star, the operation revises it into two stars, of which one is a $(t-2)$-star and the other is a 1-star.*

Note that an operation REVISE-t applies to a t-star if two satellites of the star are adjacent in the input graph G (and there is no other possibility since $\mathcal{Q}_0$ is an optimal k^--star packing), and thus the operation takes $O(1)$ time. Specially, when $t = 2$, no REVISE-t operation is applicable, since we do not want to form the two satellites into a 1-star while leaving out the center uncovered.

Definition 8 (Operation REVISE-(t, i)). *Given a t-star and an i-star, the operation revises them into two k^--stars of which none is a t-star, or into three k^--stars of which at most one is a t-star.*

We remark that there are multiple scenarios for REVISE-(t, i) to be applicable, for example, when a satellite of the t-star is adjacent to the center of the i-star and $i \neq t - 1, k$, or when two satellites of the t-star are both adjacent to the center of the i-star and $i = t - 1$, or when the center of the t-star is adjacent to a satellite of the i-star and $i \neq 1, t + 1$. Nevertheless, since at most $t + k + 2$ vertices are involved, the operation takes $O(1)$ time.

Definition 9 (Operation REVISE-(t, t, i)). *Given two t-stars and an i-star with $i \in \{1, t-1, t+1\}$, the operation revises them into two or three k^--stars of which none is a t-star.*

Again, we remark that there are multiple scenarios for REVISE-(t, t, i) to be applicable, for example, when the centers of the two t-stars are adjacent to different vertices of the 1-star, respectively, then the 1-star can be split so that two $(t + 1)$-stars are formed. Since at most $3t + 4$ vertices are involved, the operation takes $O(1)$ time.

Lemma 9. *When none of the REVISE operation is applicable, the optimal sequential k^--star packing $\mathcal{Q}_0$ satisfies the following:*

(1) *$V_0 = V \setminus V(\mathcal{Q}_0)$ and $\mathcal{K}(V_0)$ remain unchanged throughout the algorithm.*
(2) *Each satellite of every t-star is adjacent to only the centers of $(t-1)$-stars or k-stars.*
(3) *The center of every t-star is adjacent to only the satellites of $(t+1)$-stars or the centers of the other stars.*

Using all the above defined REVISE operations, our algorithm LOCALSEARCH-k^-/t applies one of them in each iteration; when none of them is applicable, the algorithm removes one satellite from each t-star in the current packing $\mathcal{Q}_0$ and returns it as the solution, denoted as $\mathcal{P}$, to the k^-/t-star packing problem. A high-level description of the algorithm is depicted in Fig. 3. We summarize the main conclusion in Theorem 3, while leave all the detailed performance analysis in the full paper [16].

LOCALSEARCH-k^-/t for k^-/t-star packing when $k > t \geq 2$:
Input: A connected graph $G = (V, E)$;
Output: A collection $\mathcal{P}$ of vertex-disjoint k^-/t-stars in G.

1. Compute an optimal k^--star packing $\mathcal{Q}_0$, set $V_0 = V \setminus V(\mathcal{Q}_0)$ and $\mathcal{K}(V_0)$;
2. While (a REVISE operation is applicable) do
   (a) apply the operation to update $\mathcal{Q}_0$;
3. Remove one satellite from each t-star in $\mathcal{Q}_0$ to become the final packing $\mathcal{P}$.

Fig. 3. A high-level description of LOCALSEARCH-k^-/t for k^-/t-star packing.

Theorem 3. *Algorithm* LOCALSEARCH-k^-/t *is an* $O(|V|^4|E|)$-*time* $(1 + \frac{1}{t+1+1/k})$-*approximation for the* k^-/t-*star packing problem for any* $k > t \geq 2$.

We remark that for the k^-/t-star packing problem where $k = \infty$ and $t \geq 2$, i.e., all but t-stars are candidates, the same argument above by removing the star size constraint proves that the algorithm LOCALSEARCH-k^-/t is a $(1 + \frac{1}{t+2})$-approximation for $t \geq 2$.

4   Conclusion

We studied two non-sequential star packing problems using large stars or small stars only, specially, the k^+-star packing for $k \geq 2$ and the k^-/t-star packing for $k > t \geq 2$, from the perspective of approximation algorithms. We designed a local search algorithm for each of them, which is $(1 + \frac{k^2}{2k+1})$-approximation and $(1 + \frac{1}{t+1+1/k})$-approximation, respectively; and furthermore by adding another local operation, the algorithm for k^+-star packing problem becomes a $\frac{3}{2}$-approximation for the 2^+-star packing problem. They improve the state-of-the-art $(1 + \frac{k}{2})$-, $(1 + \frac{1}{t})$-, and $\frac{9}{5}$-approximation, respectively. As pointed out by Li and Lin [22], it would be interesting to design approximation algorithms for the $\mathcal{H}$-packing problem where $\mathcal{H}$ contains two stars only. We also note that there seems no existing approximation algorithm for the $k^-/1$-star packing, i.e., $k > t = 1$, unless when $k = 2$ the problem becomes the P_3-packing problem.

Acknowledgments. The research is supported by the NSERC Canada, the NNSF of China (No. 12371316, 12471301), the PNSF of Zhejiang, China (No. LZ25A010001, LY21A010014), and the MST of China (No. G2022040004L, G2023016016L).

References

1. Bahenko, M., Gusakov, A.: New exact and approximation algorithms for the star packing problem in undirected graphs. In: Proceedings of STACS 2011, pp. 519–530 (2011)
2. Berman, F., Johnson, D., Leighton, T., Shor, P.W., Snyder, L.: Generalized planar matching. J. Algorithms **11**, 153–184 (1990)
3. Caprara, A., Rizzi, R.: Packing triangles in bounded degree graphs. Inf. Process. Lett. **84**, 175–180 (2002)
4. Chen, Y., et al.: Path cover with minimum nontrivial paths and its application in two-machine flow-shop scheduling with a conflict graph. J. Comb. Optim. **43**, 571–588 (2022)
5. Cygan, M., Grandoni, F., Mastrolilli, M.: How to sell hyperedges: the hypermatching assignment problem. In: Proceedings of SODA 2013, pp. 342–351 (2013)
6. Eto, H., Ito, T., Liu, Z., Miyano, E.: Approximation algorithm for the distance-3 independent set problem on cubic graphs. In: Proceedings of WALCOM 2017, pp. 228–240 (2017)
7. Fürer, M., Yu, H.: Approximating the k-set packing problem by local improvements. In: Proceedings of ISCO 2014, pp. 408–420 (2014)
8. Garey, M.R., Johnson, D.S.: Computers and Intractability: A Guide to the Theory of NP-Completeness. W. H. Freeman and Company, San Francisco (1979)
9. Goldberg, A.V., Karzanov, A.V.: Maximum skew-symmetric flows and matchings. Math. Program. **100**, 537–568 (2004)
10. Gong, M., Chen, Z.-Z., Lin, G., Wang, L.: An approximation algorithm for covering vertices by 4^+-paths. In: Proceedings of COCOA 2023. LNCS, vol. 14461, pp. 459–470 (2023)
11. Gong, M., Fan, J., Lin, G., Miyano, E.: Approximation algorithms for covering vertices by long paths. In: Proceedings of MFCS 2022, pp. 53:1–53:14 (2022)
12. Helfgott, H.A., Bajpai, J., Dona, D.: Graph isomorphisms in quasi-polynomial time (2017). https://arXiv.org/abs/1710.04574
13. Hell, P., Kirkpatrick, D.G.: Packing by cliques and by finite families of graphs. Discret. Math. **49**, 45–59 (1984)
14. Hell, P., Kirkpatrick, D.G.: Packing by complete bipartite graphs. SIAM J. Algebraic Discret. Methods **45**, 199–209 (1986)
15. Hell, P., Kirkpatrick, D.G., Kratochvíl, J., Kříž, I.: On restricted two-factors. SIAM J. Discret. Math. **1**, 472–484 (1988)
16. Hu, M., Zhang, A., Chen, Y., Gong, M., Lin, G.: Approximation algorithms for non-sequential star packing problems (2024). https://arxiv.org/abs/2411.11136
17. Huang, Z., Zhang, A., Gao, M., Sun, J., Chen, Y.: Approximation algorithms for the k^+-star packing problem. Oper. Res. Lett. (2024). Revision requested
18. Kann, V.: Maximum bounded 3-dimensional matching is MAX SNP-complete. Inf. Process. Lett. **37**, 27–35 (1991)
19. Kelmans, A., Mubayi, D.: How many disjoint 2-edge paths must a cubic graph have? J. Graph Theory **45**, 57–79 (2004)

20. Kirkpatrick, D.G., Hell, P.: On the complexity of general graph factor problems. SIAM J. Comput. **12**, 601–609 (1983)
21. Kobayashi, K., et al.: Path cover problems with length cost. Algorithmica **85**, 3348–3375 (2023)
22. Li, M., Lin, W.: On star family packing of graphs. RAIRO-Oper. Res. **55**, 2129–2140 (2021)
23. Lin, C., Cui, L., Coit, D.W., Lv, M.: Performance analysis for a wireless sensor network of star topology with random nodes deployment. Wireless Pers. Commun. **97**, 3993–4013 (2017)
24. Loebl, M., Poljak, S.: Efficient subgraph packing. J. Combin. Theory Ser. B **59**, 106–121 (1993)
25. Manic, G., Wakabayashi, Y.: Packing triangles in low degree graphs and indifference graphs. Discret. Math. **308**, 1455–1471 (2008)
26. Monnot, J., Toulouse, S.: The path partition problem and related problems in bipartite graphs. Oper. Res. Lett. **35**, 677–684 (2007)
27. Neuwohner, M.: The limits of local search for weighted k-set packing. Math. Program. (2023). https://doi.org/10.1007/s10107-023-02026-3
28. Thiery, T., Ward, J.: An improved approximation for maximum weighted k-set packing (2023). https://arXiv.org/abs/2301.07537
29. Xi, W., Lin, W.: On maximum P3-packing in claw-free subcubic graphs. J. Comb. Optim. **41**, 694–709 (2021)
30. Xi, W., Lin, W.: The maximum 3-star packing problem in claw-free cubic graphs. J. Combin. Optim. **47**, Article 73 (2024)
31. Xi, W., Lin, W., Lin, Y.: Packing 2- and 3-stars into cubic graphs. Appl. Math. Comput. **460**, 128287 (2024)

Reconfiguration Using Generalized Token Jumping

Jan Matyáš Křišťan[1]([⊠])(ID) and Jakub Svoboda[2]

[1] Faculty of Information Technology, Czech Technical University in Prague, Prague, Czech Republic
kristja6@fit.cvut.cz
[2] Institute of Science and Technology, Austria, Klosterneuburg, Austria

Abstract. In reconfiguration, we are given two solutions to a graph problem, such as VERTEX COVER or DOMINATING SET, with each solution represented by a placement of tokens on vertices of the graph. Our task is to reconfigure one into the other using small steps while ensuring the intermediate configurations of tokens are also valid solutions. The two commonly studied settings are TOKEN JUMPING and TOKEN SLIDING, which allows moving a single token to an arbitrary or an adjacent vertex, respectively.

We introduce new rules that generalize TOKEN JUMPING, parameterized by the number of tokens allowed to move at once and by the maximum distance of each move. Our main contribution is identifying minimal rules that allow reconfiguring any possible given solution into any other for INDEPENDENT SET, VERTEX COVER, and DOMINATING SET. For each minimal rule, we also provide an efficient algorithm that finds a corresponding reconfiguration sequence.

We further focus on the rule that allows each token to move to an adjacent vertex in a single step. This natural variant turns out to be the minimal rule that guarantees reconfigurability for VERTEX COVER. We determine the computational complexity of deciding whether a (shortest) reconfiguration sequence exists under this rule for the three studied problems. While reachability for VERTEX COVER is shown to be in P, finding a shortest sequence is shown to be NP-complete. For INDEPENDENT SET and DOMINATING SET, even reachability is shown to be PSPACE-complete.

1 Introduction

Reconfiguration problems arise whenever we want to transform one feasible solution into another in small steps while keeping all intermediate solutions also feasible. The problem is widely studied in the context of graph problems, such as INDEPENDENT SET [1,2,4,7,13,19,25,35], VERTEX COVER [20,27], DOMINATING SET [5,9,17,24,26], SHORTEST PATHS [14,21], and COLORING [3,6,8,11]. Moreover, outside of graphs, many results concern SATISFIABILITY [16,28]. See [29] for a general survey.

© The Author(s), under exclusive license to Springer Nature Singapore Pte Ltd. 2025
S. Nakano and M. Xiao (Eds.): WALCOM 2025, LNCS 15411, pp. 244–265, 2025.
https://doi.org/10.1007/978-981-96-2845-2_16

In graph settings, the problem is characterized by a condition, a reconfiguration rule, an input graph, and an initial and a target configuration. Configurations (including initial and target) are characterized by tokens placed on the vertices of the graph. The condition restricts the placement of the tokens, for instance, the tokens need to form an independent set or a dominating set. The reconfiguration rule allows only certain movements of the tokens to obtain a new configuration. The goal is to move the tokens to the target configuration following the prescribed reconfiguration rule and ensuring all intermediate configurations satisfy the condition.

There are two main questions about reconfiguration problems. Is the reconfiguration from the initial to the target configuration possible? And if so, what is the smallest number of steps that reconfigure the initial to the target configuration?

Two reconfiguration rules are most commonly studied, TOKEN JUMPING and TOKEN SLIDING. In TOKEN JUMPING, only one token is allowed to be removed from the graph and then placed on another vertex. In the case of TOKEN SLIDING, one token can be moved only to a neighboring vertex. We can understand TOKEN SLIDING as a restriction of TOKEN JUMPING to moves of distance 1.

We generalize these reconfiguration rules. First, we examine the case where all k tokens can move by distance at most d and we call this rule (k, d)-TOKEN JUMPING. Second, we allow all k tokens to move to distance 1, and out of those, k' tokens can move even farther to distance d. We denote this rule as $\{(k, 1), (k', d)\}$-TOKEN JUMPING. Note that TOKEN SLIDING is equivalent to $(1, 1)$-TOKEN JUMPING and on connected graphs the standard TOKEN JUMPING is equivalent to $(1, n)$-TOKEN JUMPING.

In this work, we present minimal rules that for a given condition (VERTEX COVER, DOMINATING SET, and INDEPENDENT SET) ensure that the reconfiguration problem is feasible for any pair of initial and target configurations in a connected graph. Conversely, we provide instances of initial and target configurations such that weaker rules do not allow reconfiguring one to the other.

We further consider the complexity of the reconfiguration problems under $(k, 1)$-TOKEN JUMPING where k is the size of the reconfigured solution, that is, each token can move to a neighbor. We show that the reachability question, i.e. whether a reconfiguration sequence exists, is easy for VERTEX COVER under $(k, 1)$-TOKEN JUMPING. On the other hand, deciding whether a reconfiguration sequence of length at most ℓ exists is NP-hard, even for fixed $\ell \geq 2$. Even just the reachability question for DOMINATING SET and INDEPENDENT SET is shown to be PSPACE-hard.

It is known that reachability is PSPACE-complete for INDEPENDENT SET under both TOKEN JUMPING and TOKEN SLIDING even on graphs of bounded bandwidth [34]. The same result applies also for VERTEX COVER under TOKEN JUMPING and TOKEN SLIDING if we do not allow multiple tokens on one vertex. It is also known that reachability is PSPACE-complete for DOMINATING SET under TOKEN SLIDING even on bipartite graphs, split graphs, and bounded bandwidth graphs [5]. The same problem is also PSPACE-complete under TOKEN JUMPING [5].

Reconfiguration by exchanging k consecutive vertices at once has been investigated for reconfiguration of shortest paths [14]. Parallel reconfiguration has also been previously studied in the context of distributed computation [10]. The rule of moving all tokens at once (or in other contexts, agents) appears in various other problems, such as in multi-agent path finding [31] and graph protection problems [23].

In the context of multi-agent pathfinding, the $(k, 1)$-TOKEN JUMPING rule is equivalent to the anonymous setting, meaning the agents are not distinguished. A reconfiguration sequence under $(k, 1)$-TOKEN JUMPING is thus equivalent to a solution to anonymous multi-agent pathfinding with an additional constraint on the positions of agents, such as requiring that the agents form a vertex cover or an independent set. A certain case of such a global constraint has been studied, the condition where labeled agents must form a connected subgraph has received a lot of attention recently [12,32].

Results. First, we explore which generalizations of TOKEN JUMPING allow us to guarantee that any two solutions can be reached from each other. We show that if all k tokens are not allowed to move at once, then we can not guarantee reachability for any of the three studied conditions (VERTEX COVER, DOMINATING SET, and INDEPENDENT SET). That is, each of these problems requires at least $(k, 1)$-TOKEN JUMPING.

Thus, we further focus on rules that allow all tokens to move at once and search for minimal possible extensions that guarantee reachability. Table 1 shows how much we have to extend the distance to which each token can move to ensure this guarantee. We also provide corresponding examples where a weaker rule is not sufficient, thus the established results are tight with regard to the allowed distance.

Table 1. This table displays whether it is guaranteed that any solution can be reached from any other under (k, d)-TOKEN JUMPING, where k is the size of the solution. That is, each token can move to distance at most d.

Condition/Distance	$d = 1$	$d = 2$	$d \geq 3$
VERTEX COVER	Yes (Theorem 2)		
DOMINATING SET	No (Proposition 2)	Yes (Theorem 3)	
INDEPENDENT SET	No (Proposition 3)		Yes (Theorem 4)

We further explore if it suffices that only a part of the tokens moves by distance at most d, while the rest moves only by distance at most 1. For VERTEX COVER, the $(k, 1)$-TOKEN JUMPING rule is already optimal, as any rule that allows moves of only $k - 1$ or fewer tokens has instances that can not be reconfigured, even if we allow any greater distance.

The case of DOMINATING SET and INDEPENDENT SET is more complicated, the overview of results for DOMINATING SET and INDEPENDENT SET is shown in Fig. 1. For DOMINATING SET, we establish that $\{(k, 1), (k - 3, d)\}$-TOKEN JUMPING is not sufficient for all d. On the other hand, we show that $\{(k, 1), (k - 2, 2)\}$-TOKEN JUMPING is sufficient to guarantee reachability. For INDEPENDENT SET, we show that $\{(k, 1), (1, 3)\}$-TOKEN JUMPING is sufficient and we show that $(k, 2)$-TOKEN JUMPING is not sufficient for every solution size $k \geq 3$, i.e. allowing a move to distance 3 is necessary. For each of these reachability guarantees, we also provide a polynomial algorithm that finds a reconfiguration sequence.

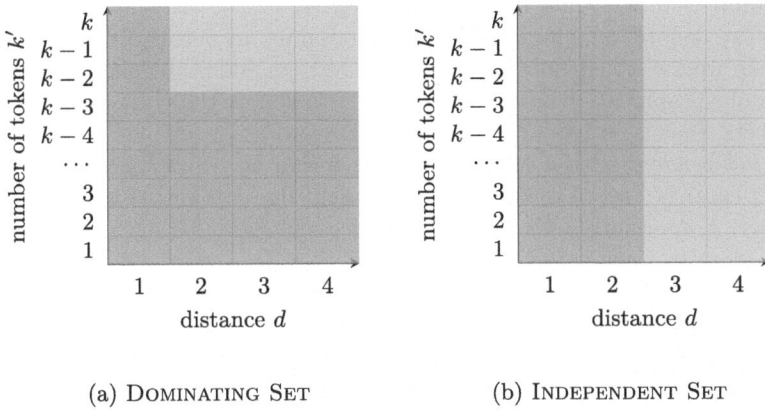

(a) DOMINATING SET   (b) INDEPENDENT SET

Fig. 1. Results for $\{(k, 1), (k', d)\}$-TOKEN JUMPING, where k is the size of the reconfigured solution. Green shows where the reachability is guaranteed. Red means that reachability can not be guaranteed and a counterexample is shown. (Color figure online)

When reachability is not guaranteed, it is natural to ask what is the complexity of the corresponding decision problem: are the two given solutions reachable under a given rule? And if so, what is the minimum length of a reconfiguration sequence? We establish the complexity of these two problems under $(k, 1)$-TOKEN JUMPING, see Table 2 for an overview.

Note that deciding if any sequence exists can be reduced to deciding whether a sequence of length at most 2^n exists, as any reconfiguration sequence without repeating configurations will have at most 2^n moves. Thus, if deciding whether a sequence exists is PSPACE-complete, deciding if a sequence of length at most ℓ exists is PSPACE-complete as well.

2   Preliminaries

Graphs and Graph Problems. All graphs are assumed to be simple. By $N(v)$ we denote the set of neighbors of v, and $N[v] = N(v) \cup \{v\}$ denotes the closed

Table 2. Results on complexities of reconfiguration problems under $(k, 1)$-TOKEN JUMPING.

Condition/Deciding if exists	Any sequence	Sequence with at most ℓ moves
VERTEX COVER	P (Theorem 2)	NP-complete (Theorem 5)
INDEPENDENT SET	PSPACE-complete (Proposition 4)	
DOMINATING SET	PSPACE-complete (Theorem 6)	

neighborhood of v. By $dist_G(v, u)$ we denote the distance between u and v in G and omit G if it is clear from the context. By $dist(C, u)$ where C is a set of vertices, we denote the minimum distance from u to any vertex of C. Throughout the paper, we use n to denote the number of vertices and m to denote the number of edges of a graph or a hypergraph. We use $\triangle$ to denote the symmetric difference of two sets, that is $A \triangle B = (A \backslash B) \cup (B \backslash A)$.

An *independent set* is a set of vertices such that no two vertices are adjacent. A *vertex cover* is a set of vertices such that each edge contains at least one vertex in the set. A *dominating set* is a set of vertices D, such that each vertex is in D or is incident to some vertex in D.

Let $H = (V, E)$ be a hypergraph, we define a path in H as a sequence of vertices $v_1, v_2, \ldots, v_\ell$ where v_i and v_{i+1} share an edge for all $1 \leq i \leq n - 1$ and the length of such path is $\ell - 1$. The distance of two vertices in H is the minimum length of a path between them.

Reconfiguration. Several previous works on TOKEN SLIDING have allowed multiple tokens to occupy a single vertex [5,9,24]. We provide our results for the variant where tokens can not share vertices, which in case of upper bounds leads to stronger results. At the same time, it is easy to verify that our lower bounds apply also for variants where multiple tokens can occupy the same vertex.

Let $\Pi(G)$ be the system of sets, which are feasible solutions of size k of a given problem on G. We say that sequence $D_1, \ldots, D_\ell$ is a *reconfiguration sequence* under the rule (k', d)-TOKEN JUMPING if $D_i \in \Pi(G)$ for all i and there is a *move* under (k', d)-TOKEN JUMPING from D_i to D_{i+1} for all $i < \ell$.

A move from D_i to D_{i+1} under (k', d)-TOKEN JUMPING is a bijection $f : D_i \to D_{i+1}$ such that $f(v) \neq v$ for at most k' distinct $v \in D_i$ and $dist_G(v, f(v)) \leq d$ for all v. A move from D_i to D_{i+1} under $\{(k, 1), (k', d)\}$-TOKEN JUMPING is a bijection $f : D_i \to D_{i+1}$ such that there is $D' \subseteq D_i$ with $|D'| \leq k'$ such that $dist_G(v, f(v)) \leq 1$ for all $v \in D_i \backslash D'$ and $dist_G(v, f(v)) \leq d$ for all $v \in D'$.

Note that the standard definition of TOKEN SLIDING coincides with $(1, 1)$-TOKEN JUMPING, and the standard definition of TOKEN JUMPING coincides with $(1, n)$-TOKEN JUMPING on connected graphs.

We study the complexity of the two following problems.

RECONFIGURATION OF Π UNDER $(k,1)$-TOKEN JUMPING
Input: Graph $G = (V, E)$ and two solutions $V_s, V_t \in \Pi(G)$ of size k.
Output: Whether there exists a reconfiguration sequence between V_s and V_t with condition Π under $(k,1)$-TOKEN JUMPING

SHORTEST RECONFIGURATION OF Π UNDER $(k,1)$-TOKEN JUMPING
Input: Graph $G = (V, E)$, two solutions $V_s, V_t \in \Pi(G)$ of size k and integer ℓ.
Output: Whether there exists a reconfiguration sequence between V_s and V_t of length at most ℓ with condition Π under $(k,1)$-TOKEN JUMPING.

Toolbox. Many of the results use the well-known Hall's Theorem, which we recall here. A matching is a set of edges such that no two share an endpoint. We say that a matching M saturates X if every vertex in X is in some edge of M.

Theorem 1 (Hall's Marriage Theorem [18]). *Let $G = (X \cup Y, E)$ be a bipartite graph. A matching that saturates X exists if and only if for every subset $S \subseteq X$, it holds $|N(S)| \geq |S|$.*

Recall that if a matching that saturates X exists, it can be found in polynomial time by a well-known algorithm [33]. If a matching saturating X does not exist, it follows that there exists $S \subseteq X$ such that $|S| > |N(S)|$, such S is called a *Hall violator*. A Hall violator (if exists) can be found in polynomial time by a well-known algorithm, well described for instance in [15].

Given a hypergraph H and its two configurations of tokens C_1, C_2, finding a move or determining that no move exists under $(k,1)$-TOKEN JUMPING can be done efficiently by reducing to bipartite matching in the following auxiliary graph. We define $B(C_1, C_2, H) = (V_1 \cup V_2, E)$, where the vertices $V_i = \{v_1^i, v_2^i, \ldots, v_k^i\}$ correspond to vertices $C_i = \{v_1, v_2, \ldots, v_k\}$ and $\{v_x^1, v_y^2\} \in E$ if and only if $dist_H(v_x, v_y) \leq 1$.

Observation 1. *Let C_1 and C_2 be two token configurations on a hypergraph H and d an integer. We can either find a move from C_1 to C_2 under $(k,1)$-TOKEN SLIDING or determine that no such move exists in $\mathcal{O}(n^{2.5} + nm)$ time.*

Proof. There is a move from C_1 to C_2 if and only if $B(C_1, C_2, H)$ has a perfect matching. Indeed, if the perfect matching exists, it describes the move and every move can be described by a perfect matching. We can construct $B' = B(C_1, C_2, H)$ in time $\mathcal{O}(n^2 + nm)$ by iterating all edges of H. Note that B' has $\mathcal{O}(n)$ vertices and $\mathcal{O}(n^2)$ edges its maximum matching can be found in $\mathcal{O}(n^{2.5})$ [33] time.

3 Guaranteeing Reconfigurability

In this section, we examine under which reconfiguration rules are solutions to problems VERTEX COVER, DOMINATING SET, and INDEPENDENT SET always reconfigurable, and for which rules it may be impossible. First, we show that $k-1$ tokens moving at once is not enough for any of these problems, even if we allow any greater distance. Then, we describe a procedure that finds a reconfiguration sequence for vertex covers on hypergraphs and we use it to show that $(k, 1)$-TOKEN JUMPING is sufficient for vertex cover. With slight modification, a similar proof is used also for DOMINATING SET, where $\{(k, 1), (k-2, 2)\}$-TOKEN JUMPING rule is shown to be sufficient. For that case, we also present a matching lower bound. Finally, we describe a way how to reconfigure INDEPENDENT SET with $\{(k, 1), (1, 3)\}$-TOKEN JUMPING rule, and again, we show that this is tight.

Proposition 1. *For every integer $k \geq 2$ and problems VERTEX COVER, DOMINATING SET, INDEPENDENT SET, there exists a connected graph G that has two unreachable solutions of size k under $(k-1, d)$-TOKEN JUMPING for every d.*

Proof. For VERTEX COVER and INDEPENDENT SET, consider a cycle of size $2k$. Such a cycle has exactly two minimum vertex covers of size k, both vertex disjoint. Thus, to reconfigure between them, each token has to move at once. Such a cycle also has exactly two maximum independent sets of size k, again both vertex disjoint.

For DOMINATING SET, consider a cycle of size $3k$. Such a cycle has exactly three minimum dominating sets, all vertex disjoint.

First, we make an observation regarding the case when no condition is placed on the configurations, except that no two tokens can share a vertex.

Observation 2. *Let V_s and V_t be two sets of vertices of size k of a hypergraph H of diameter d. We can compute a reconfiguration sequence between V_s and V_t under $(1, 1)$-TOKEN JUMPING that has at most $\mathcal{O}(n \cdot d)$ moves in $\mathcal{O}(n^2 m)$ time.*

For the proof, refer to Appendix A.1.

Observation 3. *Let V_s and V_t be two sets of vertices of size k of a hypergraph H of diameter d. We can compute a reconfiguration sequence between V_s and V_t under $(k, 1)$-TOKEN JUMPING that has at most $\mathcal{O}(n \cdot d)$ moves and has $V_s \cap V_t$ always occupied in $\mathcal{O}(n^2 m)$ time.*

For the proof, refer to Appendix A.2.

3.1   Vertex Cover and Dominating Set

We present an efficient procedure that constructs a reconfiguration sequence between vertex covers in hypergraphs under $(k, 1)$-TOKEN JUMPING. We consider the more general case of hypergraphs as that is useful in the later section on DOMINATING SET. The idea is to find an intermediate vertex cover V_m, which can be reached by some tokens from both V_s and V_t in one move. If this is possible, we can have some tokens occupy V_m in all intermediate configurations and let the remaining tokens reconfigure while ensuring that V_m is occupied, until we reach V_t.

Lemma 1. *Let V_s, V_t be two vertex covers of a hypergraph $H = (V, F)$. Then we can compute in $\mathcal{O}(n^{3.5} + n^2 m)$ time a vertex cover V_m such that both $B(V_s, V_m, H)$ and $B(V_t, V_m, H)$ have matchings saturating their copies of V_m.*

Proof. We say that a vertex cover V_i is *good* if both $B(V_s, V_i, H)$ and $B(V_t, V_i, H)$ have matchings that saturate the vertices corresponding to V_i. We present an algorithm which in each step either yields a good vertex cover V_i or decreases the size of V_i. Thus, after at most n steps, it yields a good vertex cover.

Let V_i be the vertex cover in i-th iteration, initially set $V_1 := V_s$. Now given V_i, check by Observation 1 if both $B(V_s, V_i, H)$ and $B(V_t, V_i, H)$ have matchings that saturate their copies of V_i and if they have, output V_i as good.

Otherwise, one of the bipartite graphs does not satisfy Hall's condition, without loss of generality, suppose it is $B' = B(V_s, V_i, H) = (V_s' \cup V_i', E_B)$. Note that for any $A' \subseteq V_i'$, A' corresponds to some $A \subseteq V_i$ and $N_{B'}(A')$ corresponds exactly to $N_H[A] \cap V_s$. Thus, by Theorem 1, there is $A \subseteq V_i$ such that $|A| > |N_H[A] \cap V_s|$. Now we set $V_{i+1} := (V_i \setminus A) \cup (N_H[A] \cap V_s)$, note that $|V_{i+1}| < |V_i|$.

Claim. V_{i+1} is a vertex cover of H.

Proof. We claim that for each $e \in F$, if e intersects A, then it intersects $N_H[A] \cap V_s$, from which the claim immediately follows. Suppose there is e which intersects A but does not intersect $N_H[A] \cap V_s$. Let $v \in e \cap A$ and let $v' \in e \cap V_s$, such v' must exist as V_s is a vertex cover. Then $dist(v, v') \leq 1$ and $v' \in N_H[A] \cap V_s \cap e$, a contradiction.

In each iteration, the algorithm either yields a good V_m or produces a vertex cover of smaller size, thus after at most n iterations, it yields a good V_m. Finding a maximum matching in a bipartite graph and finding a Hall violator can be done in time $\mathcal{O}(n^{2.5} + nm)$, thus the total runtime is $\mathcal{O}(n^{3.5} + n^2 m)$.

The following lemma shows that we can always find a V' such that $V' \supseteq V_m$ and there is a move from V_s to V'.

Lemma 2. *Let V_s and V_m be sets of vertices of hypergraph H such that there exists a matching in $B(V_s, V_m, H)$ saturating the copy of V_m. Then we can compute $V' \subseteq V(H)$ such that $V' \supseteq V_m$ along with a move from V_s to V' in $\mathcal{O}(n^2 m)$ time.*

Proof. First, let M be the matching saturating the copy of V_m in $B(V_s, V_m, H)$. Let B' be the result of taking $B(V_s, V_m, H)$ and assigning to each edge a cost equal to the distance of the corresponding vertices in G. Let M^* be the minimum cost matching saturating the copy of V_m in B', and let $V_s' \subseteq V_s$ be the vertices whose copy for V_s is not in any edge of M^*.

Let $V' = V_s' \cup V_m$, we need to show that $V_s' \cap V_m = \emptyset$. Suppose that there is $v \in V_s' \cap V_m$. The edge $e \in E(B')$ consisting of two copies of v has weight 0 and $e \notin M^*$. As M^* saturates the copy of V_m, there is $e' \in M^*$ of weight 1 containing the copy of v for V_m. Thus, $(M^* \setminus \{e'\}) \cup \{e\}$ is a matching saturating the copy of V_m with a smaller weight, a contradiction.

The move from V_s to V' has token on V'_s stay and each token of $V_s \backslash V'_s$ will move along M^* and thus reaching V_m. We can compute B' in $\mathcal{O}(n^2 m)$ by finding the distances from each vertex of H. A minimum cost matching of B' can be computed in time $\mathcal{O}(n^3)$ [33]. $\blacksquare$

Theorem 2 (Vertex cover reachability). *Let $H = (V, F)$ be a connected hypergraph of diameter d and V_s, V_t two vertex covers of H with size k. We can compute a reconfiguration sequence under $(k, 1)$-TOKEN JUMPING that has at most $\mathcal{O}(n \cdot d)$ moves in $\mathcal{O}(n^{3.5} + n^2 m)$ time.*

Proof. First, compute V_m by Lemma 1 in time $\mathcal{O}(n^{3.5} + n^2 m)$ and then compute V'_s and V'_t with $V_m \subseteq V'_s, V_m \subseteq V'_t$ by Lemma 2. We also have a move from V_s to V'_s and as each move is symmetric, also a move from V'_t to V_t. It remains to obtain a reconfiguration sequence between V'_s and V'_t. By Observation 3, we can compute a reconfiguration sequence from V'_s to V'_t while keeping $V'_s \cap V'_t \supseteq V'_m$ always occupied. Thus, every intermediate configuration contains V_m or is V_s or V_t and therefore is a vertex cover. $\blacksquare$

A similar idea is used in an algorithm that constructs a reconfiguration sequence between two dominating sets under $\{(k, 1), (k - 2, 2)\}$-TOKEN JUMPING.

Theorem 3 (Dominating set reachability). *Let $G = (V, E)$ be a connected graph of diameter at most d and D_s, D_t two dominating sets of G with size k. We can compute a reconfiguration sequence under $\{(k, 1), (k - 2, 2)\}$-TOKEN JUMPING that has at most $\mathcal{O}(n \cdot d)$ moves in $\mathcal{O}(n^{3.5} + n^2 m)$ time.*

Proof. Let $H = (V, F)$ be a hypergraph, where F is the set of all closed neighborhoods of G. Observe that D is a dominating set of G if and only if D is a vertex cover of H. Furthermore, note that for $u, v \in V$ $dist_G(v, u) \leq 2$ holds if and only if there is $f \in F$ such that $u, v \in f$.

Suppose that by Lemma 1 applied on H, D_s, D_t we computed D_m and by Lemma 2 we computed V'_s and V'_t. Let $B'_s = B(D_s, V'_s, H)$ and $B'_t = B(D_t, V'_t, H)$ and let M_s, M_t be their matchings corresponding to the move from D_s to V'_s and the move from D_t to V'_t, also obtained by Lemma 2. We say that the distance of edge $\{u, v\}$ in B'_s or B'_t is the distance of the vertices corresponding to u and v in G. If both M_s and M_t have at most $k - 2$ edges with distance 2, then they are valid moves under $\{(k, 1), (k - 2, 2)\}$-TOKEN JUMPING. In the opposite case, we show how to modify each matching separately, so that at most $k - 2$ edges with distance 2 are used.

Without loss of generality, suppose that M_s has at most 1 edge with distance at most 1. Let $W_{\leq 1}$ be the edges of B'_s with distance at most one. Observe that as D_s, V'_s are dominating sets, for each $v \in D_s$, there is $u \in V'_s$ with $dist_G(v, u) \leq 1$. Hence, each vertex of B'_s is incident to some edge in $W_{\leq 1}$.

We show how to obtain a cycle of size at least 4 consisting of alternating edges of M_s and $W_{\leq 1}$, which is then used to exchange the edges of the matching with edges of $W_{\leq 1}$. If every vertex of B'_s is incident to some edge in $W_{\leq 1} \backslash M_s$, we can easily find an alternating cycle by a standard argument. If this is not the case, note that by assumption $|M_s \cap W_{\leq 1}| \leq 1$ and hence we reduce to the previous case by considering B'_s with the vertices of $M_s \cap W_{\leq 1}$ deleted.

Hence way obtain a cycle C of size at least 4 and with alternating edges of M_s and $W_{\leq 1}$. We construct a new matching $M'_s := M_s \triangle E(C)$ exchanging $|C|/2 \geq 2$ edges for edges of distance 1. The obtained M'_s corresponds to a move from D_s to V'_s under $\{(k,1),(k-2,2)\}$-TOKEN JUMPING. The same argument can be applied for D_t and V'_t. By Observation 3, we can obtain a reconfiguration sequence from V'_s to V'_t under $(k,1)$-TOKEN JUMPING, which has $V'_s \cap V'_t \supseteq D_m$ always occupied. Hence, the resulting sequence is valid under $\{(k,1),(k-2,2)\}$-TOKEN JUMPING.

The following proposition shows that $\{(k,1),(k-2,2)\}$-TOKEN JUMPING is a minimal rule guaranteeing reachability.

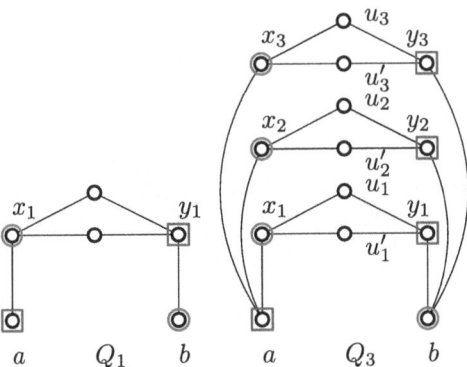

Fig. 2. Construction of a graph, such that moving all tokens at once by distance at least 1 and at least $k - 2$ of them by distance at least 2 is necessary to reconfigure between the blue and red dominating sets. (Color figure online)

Proposition 2 (Dominating set unreachability). *For every $k \geq 3$, there exists a connected graph G such that to reconfigure between two distinct minimum dominating sets of size k, we need to allow all tokens to move at once and at least $k - 2$ of them by distance at least 2.*

Proof. Consider the graphs Q_1 and Q_3 shown in Fig. 2. We can inductively construct Q_i for $i \geq 2$ by taking a copy of Q_{i-1} and Q_1 and identifying its vertices a and b.

We denote the size of a minimum dominating set of Q_i as $\gamma(Q_i)$. First, we show that $\gamma(Q_i) \leq i + 1$ by constructing a dominating set on $i + 1$ vertices: observe that $\{a, y_1, y_2, \ldots, y_i\}$ is a dominating set of size $i + 1$. On the other hand, from each set $S_j = \{x_j, u_j, u'_j, y_j\}$ we have to include at least one vertex.

At the same time, no vertex from S_j dominates all the other vertices in S_j or any vertices from any S_p for $p \neq j$, thus $\gamma(Q_i) \geq i + 1$.

Now we claim that if D is a minimum dominating set, then exactly one of a, b is in D. Suppose that $a, b \in D$, then we need one additional vertex for each u_i, thus $|D| \geq 2 + i$, a contradiction. Suppose that $a, b \notin D$, then D must contain at least 2 vertices from each set S_j. Then we have $|D| \geq 2i$ and as $k \geq 3$ and hence $i \geq 2$, we have $|D| > i + 1$, a contradiction.

Note that if $a \in D$ then $y_j \in D$ for every j. Thus, we have exactly two different minimum dominating sets, $D_a = \{a, y_1, y_2, \ldots, y_i\}$ and $D_b = \{b, x_1, x_2, \ldots, x_i\}$. Now observe that to move from D_a to D_b, at least $k - 2$ tokens must move by distance at least 2. Any matching between D_a and D_b such that matched tokens are at distance at most 2 will match $k - 2$ tokens on some x_j to some y_p. As $dist(x_j, y_p) \geq 2$ for all j, q, those tokens must move by distance at least 2.

3.2   Independent Sets

The following will allow us to reduce reconfiguring independent sets to moving tokens to vertices of some shared maximal independent set.

Lemma 3. *Let G be a connected graph, I^* a maximal independent set of G, and $I_s \subseteq I^*$, $I_t \subseteq I^*$ two independent sets of G of size k. Then we can compute a reconfiguration sequence of length $\mathcal{O}(n^2)$ between I_s and I_t under $(1, 3)$-Token Jumping in $\mathcal{O}(n^2 m)$ time.*

Proof. Let $G' = (I^*, E' = \{\{u, v\} \mid dist_G(u, v) \leq 3 \text{ and } u, v \in I^*\})$. Note that a reconfiguration sequence under $(1, 1)$-Token Sliding on G' corresponds to a sequence under $(1, 3)$-Token Sliding on G and any configuration (not necessarily an independent set) of tokens on G' where no two tokens share a vertex is an independent set on G.

First, observe that G' is connected, to see that, suppose G' has connected components $C_1, C_2, \ldots, C_\ell$. Let $P = (p_1, \ldots, p_q)$ be a shortest path in G from C_1 to C_2 with $p_1 \in C_1$ and $p_q \in C_2$. Note that $q \geq 5$, $dist_G(p_3, C_1) = 2$, and $dist_G(p_3, C_i) \geq 2$ for all i, therefore $I^* \cup \{p_3\}$ is an independent set, a contradiction with I^* being maximal. By Observation 2, we can reconfigure I_s to I_t under $(1, 1)$-Token Jumping on G' using $\mathcal{O}(n^2)$ moves.

Note that at no point do two tokens share a vertex. Therefore, we construct a sequence that is a valid reconfiguration sequence on G under $(1, 3)$-Token Sliding.

Theorem 4 (Independent set reachability). *Let G be a connected graph and I_s, I_t two independent sets of size k of G. Then we can compute a reconfiguration sequence between I_s and I_t of length $\mathcal{O}(n^2)$ under $\{(k, 1), (1, 3)\}$-Token Jumping in $\mathcal{O}(n^{3.5} + n^2 m)$ time.*

Proof. We describe an algorithm that reconfigures I_s to I_t. The idea is to first construct a maximal independent set I^* such that I_s and I_t can move into a subset of I^* in one move. The idea of construction of I^* is similar to that of Lemma 1.

We start by constructing an independent set I_m such that $B(I_s, I_i, G)$ and $B(I_t, I_i, G)$ have a matching that saturates the copies of I_s and I_t respectively, in that case, we say that I_m is *good*. Initially set $I_1 := I_s$. Now in each step, we show that either I_i is good or we can construct a larger independent set I_{i+1}.

Suppose that I_i is not good, then one of the bipartite graphs does not have a matching saturating the copies of I_s or I_t, suppose it is $B' = B(I_s, I_i, G) = (I'_s \cup I'_i, E_B)$. By Hall's condition, there is $A' \subseteq I'_s$ such that $|A'| > |N_{B'}(A')|$, which in G corresponds to some $A \subseteq I_s$ such that $|A| > |N_G[A] \cap I_i|$.

We set $I_{i+1} := (I_i \backslash N_G[A]) \cup A$, observe that $|I_{i+1}| > |I_i|$ as $(I_i \backslash N_G[A]) \cap A = \emptyset$. We claim that I_{i+1} is an independent set. To see that, suppose there is $\{u, v\} \in E$ such that $\{u, v\} \subseteq I_{i+1}$, note that $u, v \in A \subseteq I_s$ or $u, v \in I_i \backslash N_G[A] \subseteq I_i$ may not hold by assumption. In the remaining case $u \in I_i \backslash N_G[A], v \in A$ it holds $dist_G(u, v) \geq 2$.

Thus, after at most n iterations we obtain a good independent set I_m. Now we can greedily add vertices to I_m until we obtain a maximal independent set I^* with $I_m \subseteq I^*$. Given matchings M_s, M_t saturating I_s in $B(I_s, I_m, G)$ and I_t in $B(I_t, I_m, G)$ respectively, we can move the tokens on I_s along M_s to some $I'_s \subseteq I_m$ in one move under $(k, 1)$-TOKEN SLIDING. Similarly, we can move from I_t to $I'_t \subseteq I_m$ in one move. By Lemma 3, we can now construct a reconfiguration sequence of length $\mathcal{O}(n^2)$ between I'_s and I'_t under $(1, 3)$-TOKEN JUMPING in time $\mathcal{O}(nm)$. Thus, we obtain a reconfiguration sequence between I_s and I_t under $\{(k, 1), (1, 3)\}$-TOKEN JUMPING.

We check at most n times whether I_i is good, which by Observation 1 takes $\mathcal{O}(n^{2.5} + nm)$ time, thus the total runtime is $\mathcal{O}(n^{3.5} + n^2 m)$.

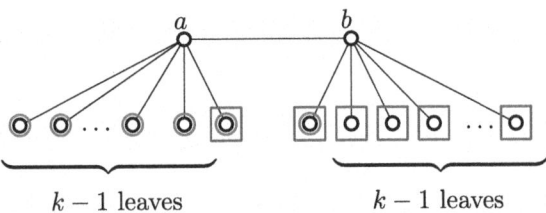

Fig. 3. Sliding tokens to distance 3 is necessary when reconfiguring between the red independent set and the blue independent set. (Color figure online)

We have shown in Proposition 1 that at least $(k, 1)$-TOKEN JUMPING is necessary to guarantee reachability. Now, we show that $(1, 3)$-TOKEN JUMPING is also necessary.

Proposition 3 (Independent set unreachability). *For every $k \geq 3$, there exists a connected graph G such that to reconfigure between two distinct independent sets of size k, we need to allow at least one token to move by distance at least 3.*

Proof. Consider the graph in Fig. 3 with $k \geq 3$. Note that the tokens of the red independent set incident to vertex a may not move, unless they jump to distance 3. This holds for all the possible positions of the token incident to b, hence a move to distance 3 is necessary.

4   Computational Complexities

In this section, we determine the complexities of deciding whether a reconfiguration sequence exists under $(k, 1)$-TOKEN JUMPING for the three studied problems, INDEPENDENT SET, VERTEX COVER, and DOMINATING SET. It follows from Theorem 2 that the reachability problem for VERTEX COVER is in P for $(k, 1)$-TOKEN JUMPING, as the answer is always YES. In contrast, the problem of finding the *shortest* reconfiguration sequence in this setting is shown to be NP-complete. We show that for INDEPENDENT SET and DOMINATING SET, the reachability question is PSPACE-complete.

Observation 4. *RECONFIGURATION OF Π UNDER $(k, 1)$-TOKEN JUMPING, where k is the size of the reconfigured solution, is in PSPACE for any problem Π in NP.*

Proof. There is a simple NPSPACE algorithm. For a given configuration V_s, we can nondeterministically enumerate all solutions of Π and traverse into those that are reachable in a single move, which can be checked in polynomial time by Observation 1. The computation stops after traversing 2^n configurations, as that is an upper bound for the length of the shortest reconfiguration sequence. By Savitch's Theorem [30], this NPSPACE algorithm can be converted into a PSPACE one.

4.1   Independent Sets

In this section, we show that INDEPENDENT SET under $(k, 1)$-TOKEN JUMPING, where k is the size of the reconfigured solution, is PSPACE-complete. This depends on the result of Belmonte et al., which shows that INDEPENDENT SET reconfiguration under TOKEN SLIDING is PSPACE-complete even on split graphs [2]. A graph is a *split* graph if its vertices can be partitioned into a clique and an independent set. We show that allowing multiple tokens to move at once provides no advantage on split graphs.

Lemma 4. *Let $G = (C \cup I, E)$ be a split graph with clique C and independent set I and I_s and I_t two independent sets of G of size k. There is a reconfiguration sequence between I_s and I_t under TOKEN SLIDING if and only if there is one under $(k, 1)$-TOKEN JUMPING.*

Proof. If there is a reconfiguration sequence under TOKEN SLIDING, the same sequence is valid under $(k, 1)$-TOKEN JUMPING as well.

Now suppose that there is a reconfiguration sequence between I_s and I_t under $(k, 1)$-TOKEN JUMPING. We show that such a sequence can be transformed so that each move consists of only one token slide and is thus valid under TOKEN SLIDING. Note that in each configuration, C can contain at most one token, and therefore we can move at most one token from I to C and move at most one token from C to I during one move. Therefore, each move consists of at most two token slides and each move with two token slides has one token move from C and one move to C.

Let $s = ((c_1, i_1), (i_2, c_2))$ be a move consisting of two token slides. Without loss of generality, let $c_1, c_2 \in C, i_1, i_2 \in I$. Note that if $i_1 = i_2$, we may reach the same configuration with a single token slide from c_1 to c_2. Hence, assume that $i_1 \neq i_2$. We claim that we can replace s with two sequential moves (c_1, i_1) and (i_2, c_2) and the reconfiguration sequence remains valid.

Let I_i be the configuration in the reconfiguration sequence that is reached after applying s, that is $I_i = (I_{i-1} \setminus \{c_1, i_2\}) \cup \{i_1, c_2\}$. First note that $I' = (I_{i-1} \setminus \{c_1\}) \cup \{i_1\}$ (reached by sliding token from c_1 to i_1) is independent as only vertices of I have tokens. Thus, in the reconfiguration sequence, we can place I' between I_{i-1} and I_i to ensure that reaching I_i from I_{i-1} is done using single token slides. After modifying each move with two token slides, we arrive at a reconfiguration sequence under TOKEN SLIDING.

Proposition 4. INDEPENDENT SET *under* $(k, 1)$-TOKEN JUMPING *is* PSPACE-*complete even on split graphs.*

Proof. Reconfiguration of INDEPENDENT SET under TOKEN SLIDING is PSPACE-complete [2]. By Lemma 4, reconfiguration of INDEPENDENT SET under $(k, 1)$-TOKEN JUMPING is PSPACE-complete as well.

4.2 Vertex Covers

In this section, we show that determining the length of the shortest reconfiguration sequence for VERTEX COVER under $(k, 1)$-TOKEN JUMPING is an NP-complete problem.

Theorem 5. *Deciding if a reconfiguration sequence of length at most ℓ between two vertex covers exists under* $(k, 1)$-TOKEN JUMPING *is* NP-*complete, even if $\ell \geq 2$ is fixed.*

For the proof see Appendix A.3

4.3 Dominating Sets

For DOMINATING SET, we show that the problem remains PSPACE-complete under $(k, 1)$-TOKEN JUMPING. For that end, we define reconfiguration rule with additional allowed moves, which is useful in the subsequent hardness proof. Its definition is in Appendix A.4.

Lemma 5. *There exists a reconfiguration sequence between vertex covers V_s and V_t under relaxed token jumping if and only if there exists a reconfiguration sequence under* TOKEN JUMPING.

For the proof, see Appendix A.4. This is used in the reduction from reconfiguration of VERTEX COVER under TOKEN JUMPING.

Theorem 6. *Deciding if one dominating set can be reconfigured into another under $(k, 1)$-*TOKEN JUMPING *, where k is the size of the given dominating sets, is* PSPACE-*complete.*

For the proof, see Appendix A.4

5   Conclusion

For DOMINATING SET and INDEPENDENT SET, we proved that $\{(k, 1), (k-2, 2)\}$-TOKEN JUMPING and $\{(k, 1), (1, 3)\}$-TOKEN JUMPING respectively ensure reachability, here k denotes the size of the given reconfigured solutions. If we decrease any of the parameters of these rules, we necessarily lose this guarantee. A natural next step is to determine the complexity of finding a shortest reconfiguration sequence under these rules.

It is also an interesting question which graph classes allow weaker rules to guarantee reachability. Furthermore, the investigated question is important for other graph problems as well, for instance, generalizations of DOMINATING SET and INDEPENDENT SET seem to be tractable with the tools we provide.

Acknowledgments. J. M. Křišťan acknowledges the support of the Czech Science Foundation Grant No. 24-12046S. This work was supported by the Grant Agency of the Czech Technical University in Prague, grant No. SGS23/205/OHK3/3T/18. J. Svoboda acknowledges the support of the ERC CoG 863818 (ForM-SMArt) grant.

A   Omitted proofs

A.1   Proof of Observation 2

Proof. We can reconfigure V_s to V_t under $(1, 1)$-TOKEN JUMPING on H as follows. In each iteration, we increase the size of $V_s \cap V_t$.

First, we pick $s \in V_s \backslash V_t, t \in V_t \backslash V_s$ and over several token slides, ensure that the token on s is moved to t. Let $P = p_1 = s, p_2, \ldots, p_\ell = t$ be the vertices of a shortest path between s and t. Suppose the token is currently placed on p_i, if p_{i+1} has no token, then we may directly move the token from p_i to p_{i+1}. Otherwise, suppose $p_{i+1}, p_{i+2}, \ldots, p_{i+j}$ are occupied by tokens and p_{i+j+1} is not occupied, such p_{i+j+1} must exist as t is not occupied by assumption. We may reach the same configuration corresponding to jumping with the token from p_i to p_{i+j+1} by performing a sequence of token slides from p_{i+j-q} to $p_{i+j+1-q}$, for all $0 \leq q \leq i$. Note that each such move is valid as for $q = 0$, we have that

p_{i+j+1} is unoccupied by assumption and for $q \geq 1$, we have that $p_{i+j+1-q}$ was left unoccupied by the preceding move.

A simple implementation of this process will in each iteration pick two arbitrary vertices in $V_s \backslash V_t$ and $V_t \backslash V_s$ and find a shortest path between them, which can be done in $\mathcal{O}(nm)$ on hypergraphs. This is done at most n times, thus the total runtime is $\mathcal{O}(n^2 + m)$. The total number of moves in each iteration is the length of the shortest path, hence we perform at most $\mathcal{O}(n \cdot d)$ moves, where d is the diameter of H.

A.2   Proof of Observation 3

Proof. Let H' be a hypergraph so that $V(H') = V(H) \backslash (V_s \cap V_t)$ and $E(H')$ is constructed by taking $E(H)$ and adding to H' the set of edges E' connecting by an edge every pair of vertices reachable by a path with internal vertices in $V_s \cap V_t$. By Observation 2, we can obtain a reconfiguration sequence under $(1,1)$-TOKEN JUMPING between $V_s \backslash V_t$ and $V_t \backslash V_s$ on H'. This can be transformed to a reconfiguration sequence under $(k,1)$-TOKEN JUMPING between V_s and V_t on H by replacing any move of a token over an edge of $E' = \{u,v\}$ by a move consisting of sliding tokens along the path connecting u with v through $V_s \cap V_t$. Note that this ensures that $V_s \cap V_t$ is always occupied.

A.3   Proof of Theorem 5

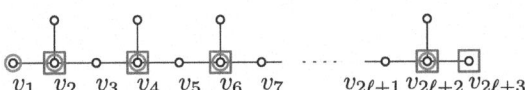

Fig. 4. Gadget T^ℓ. The vertices with red circles are in V_s^ℓ and the vertices with blue squares are in V_t^ℓ. (Color figure online)

Proof. First we show that the problem is in NP. Let G be the input graph and d its diameter. For $\ell > C \cdot n \cdot d$, the answer is always YES for some fixed C by Theorem 2. Also, any reconfiguration sequence of length at most $\ell = \mathcal{O}(n \cdot d)$ can be verified in polynomial time using Observation 1.

We follow with a reduction from VERTEX COVER. First let us define graph T^ℓ, see Fig. 4 for an illustration. To construct T^ℓ, first we create a path of length $2\ell + 3$ on vertices $v_1, v_2, \cdots, v_{2\ell+3}$. We connect a new vertex u_{2k} with v_{2k} for all integer k such that $2 \leq 2k \leq 2\ell + 2$. On T^ℓ we will later use as the starting configuration $V_s^\ell = \{v_1, v_2, v_4, v_6, \ldots, v_{2\ell+2}\}$ and as the ending configuration $V_t^\ell = \{v_2, v_4, v_6, \ldots, v_{2\ell+2}, v_{2\ell+3}\}$.

Claim. Reconfiguring V_s^ℓ to V_t^ℓ on T^ℓ under $(k,1)$-TOKEN JUMPING requires at least $\ell + 1$ turns.

Proof. The proof is by induction on ℓ. First note that for reconfiguring V_s^1 to V_t^1 on T^1 we need at least 2 moves.

Now suppose that $\ell \geq 2$. Observe that exactly two different configurations are reachable from V_s^ℓ in one move: $S_1 = \{u_2, v_2, v_4, \ldots, v_{2\ell+2}\}$ and $S_2 = \{v_2, v_3, v_4, \ldots, v_{2\ell+2}\}$. Only from S_2 can we reach any other configuration than V_s^ℓ, S_1, S_2, thus that is the only possible next configuration in a shortest reconfiguration sequence.

Note that $T \backslash \{v_1, v_2, u_2\}$ is isomorphic to $T^{\ell-1}$ and $S_2 \backslash \{v_2\}$ corresponds to $V_s^{\ell-1}$ on $T^{\ell-1}$. The token on v_2 can not move in subsequent moves, otherwise, the edge connecting to its leaf would be uncovered. Thus by the induction hypothesis, we need at least ℓ additional moves to reconfigure from S_2 to V_t^ℓ.

The idea of the subsequent reduction is that if we connect v_4 to a new vertex, which can supply an additional token to v_4, we can make the reconfiguration sequence shorter by one move. This additional token will be available to every copy of T^ℓ only if a given graph a has small enough vertex cover.

Now, we construct the instance (G', V_s, V_t) of SHORTEST RECONFIGURATION OF VERTEX COVER UNDER $(k, 1)$-TOKEN JUMPING given an instance of VERTEX COVER (G, k) and an integer $\ell \geq 2$. To construct G', we first add $n - k$ copies of the gadget T^ℓ, say $T_1, \ldots, T_k$. We will distinguish the vertices of T_i by superscript, meaning that path of T_i consists of $v_1^i, \ldots, v_{2\ell+3}^i$. For each T_i, we connect v_4^i with all vertices in G. This finishes the construction of G'. We set

$$V_s := \{v_k^i \mid i \in \{1, \ldots, n-k\}, k \in \{1, 2, 4, 6, \ldots, 2\ell+2\}\} \cup V(G)$$

$$V_t := (V_s \backslash \{v_1^1, v_1^2, \ldots, v_1^{n-k}\}) \cup \{v_{2\ell+3}^1, v_{2\ell+3}^2, \ldots, v_{2\ell+3}^{n-k}\}$$

Claim. If G has a vertex cover of size k then there is a reconfiguration sequence from V_s to V_t using at most ℓ moves.

Proof. Let V_c be a vertex cover of G of size k and $\overline{V_c} = V(G) \backslash V_c = \{c_1, \ldots, c_{n-k}\}$. We construct the reconfiguration sequence as follows. The first move is $(v_1^i, v_2^i), (v_2^i, v_3^i), (c_i, v_4^i), (v_4^i, v_5^i)$ for all $i \in \{1, \ldots, n - k\}$ simultaneously. The second move is $(v_3^i, v_4^i), (v_4^i, c_i), (v_5^i, v_6^i), (v_6^i, v_7^i)$ for all $i \in \{1, \ldots, n - k\}$ simultaneously. Now for $j \in \{3, \ldots, \ell\}$, the j-th move is $(v_{2j+1}^i, v_{2j+2}^i), (v_{2j+2}^i, v_{2j+3}^i)$ for all $i \in \{1, \ldots, n - k\}$. Note that each resulting state is a vertex cover, as V_c always remains on G and for all i, j we have a token on v_j^i for all $j \in \{2, 4, \ldots, 2\ell+2\}$. Thus, we reach V_t in exactly ℓ moves.

Claim. If G has no vertex cover on k vertices, then any reconfiguration sequence between V_s and V_t has at least $\ell + 1$ moves.

Proof. As at least $k + 1$ tokens must always be on G, there is a T_q such that it has the same number of tokens before and after the first move. As observed in the proof of the previous claim, the only viable move on T_q is $(v_1^q, v_2^q), (v_2^q, v_3^q)$. Then the token slides between v_4^q and G can not decrease the subsequent number of moves required to reconfigure the tokens on T_q and by the previous claim we require $\ell + 1$ moves in total.

Hence, G has a vertex cover of size k if and only if there is a reconfiguration sequence between V_s and V_t of length at most ℓ, which completes the proof.

A.4   Proofs of the Section 4.3

We first consider a reconfiguration rule with additional allowed moves, which is useful in the subsequent hardness proof.

Definition 1. *Reconfiguration under* relaxed token jumping *allows moves to remove a token, add a token, or jump with a token from one vertex to another. For each intermediate configuration V_i, it must hold $|V_i| \leq |V_s|$, where V_s is the starting configuration. In one turn, we may perform at most $|V_s| - |V_i|$ additions and at most $|V_s| - |V_i| + 1$ jumps and additions in total. Furthermore, it is allowed to jump or add a token to an already occupied vertex.*

Formally, the relaxed sequence is a sequence of *moves*, each consisting of a sequence of *steps*. Thus, we denote a relaxed token jumping sequence as

$$((o_{1,1}, \ldots, o_{1,\ell_1}), (o_{2,1}, \ldots, o_{2,\ell_2}), \ldots, (o_{p,1}, \ldots, o_{p,\ell_p}))$$

where p is the total number of moves, ℓ_i is the number of steps in the i-th move, and $o_{i,j}$ is one of the steps done in parallel in the i-th move. Each $o_{i,j} \in \{R(v), A(v), J(u, v)\}$, where $R(v)$ denotes a removal of a token on vertex v, $A(v)$ denotes an addition of a token to vertex v and $J(u, v)$ denotes a jump of a token from vertex u to vertex v.

Proof of Lemma 5

Proof. Observe that a reconfiguration sequence under TOKEN JUMPING is valid under relaxed token jumping. The rest of the proof shows how to transform a relaxed token jumping sequence into a TOKEN JUMPING sequence. We will show a set of operations that modify a valid relaxed sequence into another valid relaxed sequence, eventually reaching a form of only singleton jumps. Suppose we process the moves of a reconfiguration sequence S under relaxed token jumping between V_s into V_t one by one. Assume that S contains moves with more than one step. First, note that if a move contains a pair of an addition $A(v)$ and a removal $R(u)$, we may replace it with a jump $J(v, u)$. Hence, we may assume that all moves do not contain additions and removals at the same time.

Now, note that whenever a move m_i contains more than one step and one of them is a removal, say $R(v)$, we can remove $R(v)$ from m_i and place a new move $m' = (R(v))$ directly after m_i. Similarly, if a move m_i contains an addition, say $A(v)$, we may remove $A(v)$ from m_i and place a new move $m' = (A(v))$ directly before m_i. The resulting configurations remain supersets of what they were before and the following moves remain valid.

Thus, now we assume that removals and additions appear only in moves with one step. Now, any move with multiple steps consists of only jumps, note that all such jumps $J(u, u')$ can be replaced by a pair $R(u), A(u')$ and the previous reduction can be applied again. Hence, now all moves contain only one step.

Now, if any removals remain, let $m_i = (R(v))$ be such move with maximum i. Let m_j be the move with $j > i$ with minimum j which would not be allowed without m_i and hence contains an addition. First, note that we may place m_i so that it directly precedes m_j and the reconfiguration sequence remains valid by the choice of j and the fact that each configuration remains a superset of what it was before the change. As m_j contains an addition $A(u)$, we may replace $m_{j-1} = (R(v))$ with $m'_{j-1} = (J(v, u))$. Note that the reconfiguration sequence remains valid for all subsequence moves. After repeating this procedure, the resulting sequence consists of only single jumps, but we may have jumped with a token to an occupied vertex.

Let $m_i = (J(u, v))$ such that in V_i, v already has a token and such that i is minimum possible. There must be $m_j = (J(v, y))$ such that $i < j$, as V_t has no two tokens sharing a vertex. Assume m_j is such with minimum j. Let x be some vertex without a token in V_i. Then replace m_i with $m'_i = (J(u, x))$ and m_j with $m_j = (J(x, y))$. Note that each such replacement increases the index of the first move which places a token on an already occupied vertex, thus eventually all jumps are only to unoccupied vertices. Thus the resulting reconfiguration sequence is valid under TOKEN JUMPING.

Proof of Theorem 6

Proof. The problem is in PSPACE by Observation 4.

We show PSPACE-hardness by a reduction from reconfiguring INDEPENDENT SET under TOKEN JUMPING, which is known to be PSPACE-complete [22]. Note that it is equivalent to reconfiguring VERTEX COVER under TOKEN JUMPING (shortened to VC-TJ), as the complement of an independent set is always a vertex cover. Furthermore, the jump from u to v in an independent set is equivalent to a jump from v to u in the corresponding vertex cover. Thus, the two problems are equivalent and we continue our reduction from the reconfiguring VERTEX COVER under TOKEN JUMPING. The construction used in the reduction is similar to the standard reduction of VERTEX COVER to DOMINATING SET in bipartite graphs.

Let (G, V_s, V_t) be the VC-TJ instance on the input with $k = |V_s| = |V_t|$. We construct the instance of reconfiguration of DOMINATING SET under $(k, 1)$-TOKEN JUMPING (shortened to DS-kTJ) (G', D_s, D_t) as follows. Let $G' := (V(G) \cup E_1 \cup \cdots \cup E_{k+1} \cup \{x, y\}, E')$, where each E_i for $1 \leq i \leq k + 1$ is a distinct copy of $E(G)$. For each $e_i \in E_i$ corresponding to $\{u, v\} \in E(G)$, connect u and v with e_i in G'. Also, connect x with all vertices in $V(G) \cup \{y\}$. Now set $D_s := V_s \cup \{x\}$ and $D_t := V_t \cup \{x\}$.

We will show that there is a reconfiguration sequence for (G, V_s, V_t) under VC-TJ if and only if there is a reconfiguration sequence for (G', D_s, D_t) under DS-kTJ. Suppose there is a sequence of states $V_s = V_1, V_2, \cdots, V_t = V_k$ reconfiguring under VC-TJ. We show how to construct the sequence for (G', D_s, D_t). Each jump from u to v is translated to two parallel moves, one from u to x and another from x to v. Now each resulting configuration of tokens on G' is a dominating set, as each $v \in V(G) \cup \{y\}$ is dominated from x and each $e \in E(G)$ is incident to some $v \in V(G)$ with a token, as each corresponding configuration on G is a vertex cover. Thus, we can reconfigure D_s into D_t.

Now suppose there is a sequence S' of configurations $D_s = D_1, D_2, \cdots, D_t = D_k$ reconfiguring under DS-kTJ. Note we can assume that $x \in D_i$ for all i as a token must be on x and y to dominate y and $N_{G'}[y] \subseteq N_{G'}[x]$. Furthermore, note that each D_i must have a token on one vertex of each $e = \{u, v\} \in E(G)$, otherwise a copy of e_i in some E_i would be left undominated. Any move of a token to such $e_i \in E_i$ in S' can be rerouted to x as its neighborhood is a strict superset, u and v is already dominated from x, and one of u and v must have a token and hence e_i is already dominated. Thus, we will assume that moves of tokens from $V(G)$ are only to x. We show how to construct the sequence S for (G, V_s, V_t) under relaxed VC-TJ.

Each step in S is constructed from one step in S'. Whenever in S' a token moves from v to x and in parallel another token moves from x to u, we translate that into a move in S with a jump from v to u. We pick such a matching of parallel moves in each step of S' from x and to x arbitrarily. If, after considering the matching, a step in S' has any moves from x to $v_1, \cdots, v_p$ left, we translate that to additions of a token and placing it on $v_1, \cdots, v_p$ in S. Similarly, if after the matching, the current step has any moves to x from $v_1, \cdots, v_p$ left, we translate that to removal of tokens on $v_1, \cdots, v_p$ in S. Note that each state reached using S is still a vertex cover of G: the corresponding dominating set reached using S' must dominate the vertices of all E_i of G' from $V(G)$. Furthermore, to perform p jumps and additions in parallel, there must be at least p tokens on x, thus on $V(G)$ is at most $k - p + 1$. Also, there will always be at most k tokens on $V(G)$. Thus, S is a valid reconfiguration between V_s and V_t under relaxed token jumping and by Lemma 5, there exists a reconfiguration sequence under TOKEN JUMPING for (G, V_s, V_t). This concludes the proof.

References

1. Bartier, V., Bousquet, N., Dallard, C., Lomer, K., Mouawad, A.E.: On girth and the parameterized complexity of token sliding and token jumping. Algorithmica **83**(9), 2914–2951 (2021). https://doi.org/10.1007/s00453-021-00848-1

2. Belmonte, R., Kim, E.J., Lampis, M., Mitsou, V., Otachi, Y., Sikora, F.: Token sliding on split graphs. Theory Comput. Syst. **65**(4), 662–686 (2020). https://doi.org/10.1007/s00224-020-09967-8

3. Bonamy, M., Bousquet, N.: Recoloring bounded treewidth graphs. Electron. Notes Discrete Math. **44**, 257–262 (2013). https://doi.org/10.1016/j.endm.2013.10.040

4. Bonamy, M., Bousquet, N.: Token sliding on chordal graphs. In: Bodlaender, H.L., Woeginger, G.J. (eds.) WG 2017. LNCS, vol. 10520, pp. 127–139. Springer, Cham (2017). https://doi.org/10.1007/978-3-319-68705-6_10

5. Bonamy, M., Dorbec, P., Ouvrard, P.: Dominating sets reconfiguration under token sliding. Discret. Appl. Math. **301**, 6–18 (2021). https://doi.org/10.1016/j.dam.2021.05.014

6. Bonsma, P., Cereceda, L.: Finding paths between graph colourings: PSPACE-completeness and superpolynomial distances. Theoret. Comput. Sci. **410**(50), 5215–5226 (2009). https://doi.org/10.1016/j.tcs.2009.08.023

7. Bonsma, P., Kamiński, M., Wrochna, M.: Reconfiguring independent sets in claw-free graphs. In: Ravi, R., Gørtz, I.L. (eds.) SWAT 2014. LNCS, vol. 8503, pp. 86–97. Springer, Cham (2014). https://doi.org/10.1007/978-3-319-08404-6_8

8. Bonsma, P., Mouawad, A.E., Nishimura, N., Raman, V.: The complexity of bounded length graph recoloring and CSP reconfiguration. In: Cygan, M., Heggernes, P. (eds.) IPEC 2014. LNCS, vol. 8894, pp. 110–121. Springer, Cham (2014). https://doi.org/10.1007/978-3-319-13524-3_10

9. Bousquet, N., Joffard, A.: TS-reconfiguration of dominating sets in circle and circular-arc graphs. In: Bampis, E., Pagourtzis, A. (eds.) FCT 2021. LNCS, vol. 12867, pp. 114–134. Springer, Cham (2021). https://doi.org/10.1007/978-3-030-86593-1_8

10. Censor-Hillel, K., Rabie, M.: Distributed reconfiguration of maximal independent sets. J. Comput. Syst. Sci. **112**, 85–96 (2020). https://doi.org/10.1016/j.jcss.2020.03.003

11. Cereceda, L., van den Heuvel, J., Johnson, M.: Finding paths between 3-colorings. J. Graph Theory **67**(1), 69–82 (2010). https://doi.org/10.1002/jgt.20514

12. Charrier, T., Queffelec, A., Sankur, O., Schwarzentruber, F.: Complexity of planning for connected agents. Auton. Agents Multi-Agent Syst. **34**(2) (2020). https://doi.org/10.1007/s10458-020-09468-5

13. Demaine, E.D., et al.: Linear-time algorithm for sliding tokens on trees. Theoret. Comput. Sci. **600**, 132–142 (2015). https://doi.org/10.1016/j.tcs.2015.07.037

14. Gajjar, K., Jha, A.V., Kumar, M., Lahiri, A.: Reconfiguring shortest paths in graphs. In: Proceedings of the AAAI Conference on Artificial Intelligence, vol. 36. no. (9), pp. 9758–9766 (2022). https://doi.org/10.1609/aaai.v36i9.21211

15. Gan, J., Suksompong, W., Voudouris, A.A.: Envy-freeness in house allocation problems. Math. Soc. Sci. **101**, 104–106 (2019). https://doi.org/10.1016/j.mathsocsci.2019.07.005

16. Gopalan, P., Kolaitis, P.G., Maneva, E.N., Papadimitriou, C.H.: The connectivity of Boolean satisfiability: computational and structural dichotomies. In: Bugliesi, M., Preneel, B., Sassone, V., Wegener, I. (eds.) ICALP 2006. LNCS, vol. 4051, pp. 346–357. Springer, Heidelberg (2006). https://doi.org/10.1007/11786986_31

17. Haddadan, A., et al.: The complexity of dominating set reconfiguration. Theoret. Comput. Sci. **651**, 37–49 (2016). https://doi.org/10.1016/j.tcs.2016.08.016

18. Hall, P.: On representatives of subsets. J. London Math. Soc. **s1-10**(1), 26–30 (1935). https://doi.org/10.1112/jlms/s1-10.37.26

19. Hoang, D.A., Khorramian, A., Uehara, R.: Shortest reconfiguration sequence for sliding tokens on spiders. In: Heggernes, P. (ed.) CIAC 2019. LNCS, vol. 11485, pp. 262–273. Springer, Cham (2019). https://doi.org/10.1007/978-3-030-17402-6_22

20. Ito, T., Nooka, H., Zhou, X.: Reconfiguration of vertex covers in a graph. IEICE Trans. Inf. Syst. **E99.D**(3), 598–606 (2016). https://doi.org/10.1587/transinf.2015fcp0010

21. Kamiński, M., Medvedev, P., Milanič, M.: Shortest paths between shortest paths. Theoret. Comput. Sci. **412**(39), 5205–5210 (2011). https://doi.org/10.1016/j.tcs.2011.05.021

22. Kamiński, M., Medvedev, P., Milanič, M.: Complexity of independent set reconfigurability problems. Theoret. Comput. Sci. **439**, 9–15 (2012). https://doi.org/10.1016/j.tcs.2012.03.004

23. Klostermeyer, W., Mynhardt, C.: Protecting a graph with mobile guards. Appl. Anal. Discrete Math. **10**(1), 1–29 (2016). https://doi.org/10.2298/aadm151109021k

24. Křišťan, J.M., Svoboda, J.: Shortest dominating set reconfiguration under token sliding. In: Fernau, H., Jansen, K. (eds.) FCT 2023. LNCS, vol. 14292, pp. 333–347. Springer, Cham (2023). https://doi.org/10.1007/978-3-031-43587-4_24

25. Lokshtanov, D., Mouawad, A.E.: The complexity of independent set reconfiguration on bipartite graphs. ACM Trans. Algorithms **15**(1), 1–19 (2018). https://doi.org/10.1145/3280825

26. Lokshtanov, D., Mouawad, A.E., Panolan, F., Ramanujan, M., Saurabh, S.: Reconfiguration on sparse graphs. J. Comput. Syst. Sci. **95**, 122–131 (2018). https://doi.org/10.1016/j.jcss.2018.02.004

27. Mouawad, A., Nishimura, N., Raman, V., Siebertz, S.: Vertex cover reconfiguration and beyond. Algorithms **11**(2), 20 (2018). https://doi.org/10.3390/a11020020

28. Mouawad, A.E., Nishimura, N., Pathak, V., Raman, V.: Shortest reconfiguration paths in the solution space of Boolean formulas. SIAM J. Discret. Math. **31**(3), 2185–2200 (2017). https://doi.org/10.1137/16m1065288

29. Nishimura, N.: Introduction to reconfiguration. Algorithms **11**(4), 52 (2018). https://doi.org/10.3390/a11040052

30. Savitch, W.J.: Relationships between nondeterministic and deterministic tape complexities. J. Comput. Syst. Sci. **4**(2), 177–192 (1970). https://doi.org/10.1016/s0022-0000(70)80006-x

31. Stern, R.: Multi-agent path finding – an overview. In: Osipov, G.S., Panov, A.I., Yakovlev, K.S. (eds.) Artificial Intelligence. LNCS (LNAI), vol. 11866, pp. 96–115. Springer, Cham (2019). https://doi.org/10.1007/978-3-030-33274-7_6

32. Tateo, D., Banfi, J., Riva, A., Amigoni, F., Bonarini, A.: Multiagent connected path planning: PSPACE-completeness and how to deal with it. In: Proceedings of the AAAI Conference on Artificial Intelligence, vol. 32. no. 1 (2018). https://doi.org/10.1609/aaai.v32i1.11587

33. West, D.B.: Introduction to Graph Theory. Prentice Hall, Upper Saddle River (2001)

34. Wrochna, M.: Reconfiguration in bounded bandwidth and tree-depth. J. Comput. Syst. Sci. **93**, 1–10 (2018). https://doi.org/10.1016/j.jcss.2017.11.003

35. Yamada, T., Uehara, R.: Shortest reconfiguration of sliding tokens on subclasses of interval graphs. Theoret. Comput. Sci. **863**, 53–68 (2021). https://doi.org/10.1016/j.tcs.2021.02.019

Parameterized Algorithms
for the Spanning Forest Isomorphism
(or Containment) on Tree Problems

Jingyi Liu[1], Xian Chen[1], Yicheng Zheng[1], Jianxin Wang[1], and Feng Shi[1,2]([✉])

[1] School of Computer Science and Engineering, Central South University,
Changsha 410083, China
`fengshi@csu.edu.cn`
[2] Xiangjiang Laboratory, Changsha 410205, China

Abstract. The Minimum Tree Cut/Paste Distance problem is a well-known NP-hard problem in computational biology. However, its fixed-parameter tractability with respect to the distance remains an open problem. Within the paper, we study a simplified variant of the problem, called the Spanning Forest Isomorphism on Tree problem (alternatively called the Tree Assembly problem). Its input contains a rooted target tree T^* and a rooted forest F, and the goal is to decide whether F is a spanning forest of T^*. The problem has been shown to be NP-hard and fixed-parameter tractable with respect to the number k of trees in F. Within our investigation, we present several FPT algorithms for the problem and its general variant, where n is the size of the input instance.

- We present an FPT algorithm with runtime $O(4^k k^2 n^2 + n^3)$ for the Spanning Forest Isomorphism on Tree problem, improving the previous best FPT algorithm for the problem with runtime $2^{O(k \log k)} n^{O(1)}$.
- We study a generalized variant called the Spanning Forest Containment on Tree problem, which decides whether F contains k trees that can comprise a spanning forest of T^*. We present an FPT algorithm with runtime $O(25.6^k k n^4)$ using the color coding technique.

Keywords: Spanning Forest Isomorphism on Tree problem · Spanning Forest Containment on Tree problem · Fixed-parameter algorithm

1 Introduction

The Minimum Tree Cut/Paste Distance problem [15] is a well-known NP-hard problem in computational biology, whose goal is to transform a given tree T_1 into another given tree T_2 using the minimum number of edge-deletion and edge-addition operations. To our best knowledge, the *fixed-parameter tractability* [8] of the Minimum Tree Cut/Paste Distance problem remains an open problem. If all edge-deletion operations are specified, the problem is the Spanning Forest Isomorphism on Tree problem (abbr. SFITP). Given a rooted target tree T^* and

© The Author(s), under exclusive license to Springer Nature Singapore Pte Ltd. 2025
S. Nakano and M. Xiao (Eds.): WALCOM 2025, LNCS 15411, pp. 266–280, 2025.
https://doi.org/10.1007/978-981-96-2845-2_17

a rooted forest F, the SFITP asks whether F is a spanning forest of T^*. In other words, the SFITP asks whether F (resp., T^*) can be transformed into T^* (resp., F) using only $|F| - 1$ edge-addition operations (resp., edge-deletion operations). Unfortunately, this problem was shown to be NP-complete, even if each tree in H is a path or star [2,12].

The SFITP provides a theoretical foundation for the study of related problems in bioinformatics and databases [13–15]. Specifically, a particularly related problem is the Minimum Common String Partition problem [7]. A series of works have explored the fixed-parameter tractability of the problem with respect to combinations of various parameters [4,9,11], whereas the one showed that the problem is fixed-parameter tractable with respect to the sole parameter, partition size k, by presenting a fixed-parameter algorithm (alternatively called FPT algorithm or parameterized algorithm) with runtime $O(k^{21k^2}\mathrm{poly}(n))$ [5]. The possibility of further improving the algorithm's parameterized time complexity for this problem is a popular topic of interest. Notably, the problem remains NP-hard even if one of the two strings is already partitioned, which is called the Exact Block Cover problem [10]. The same work presented a fixed-parameter algorithm for this problem with runtime $2^k n^{O(1)}$, which is currently the best-known result. Further research due to You et al. [15] shows that the SFITP is a generalization of the Exact Block Cover problem. Due to the close relationship with the problems mentioned above, we believe that the study of the SFITP and its variants is critical for advancing algorithmic research in this area.

It is worthy to point out that You et al. [16] have studied a restricted version of the SFITP, in which the input forest F consists of k many stars (or paths). For the case that the input forest F consists of stars, they showed the problem has a kernel with size $O(k^3)$ and proposed a series of FPT algorithms for it, where the best one has runtime $O((2 + \epsilon)^k n^{O(1)})$ in terms of the exponential part. For the case that the input forest F consists of paths, they developed an FPT algorithm with runtime $O((3.47)^k n^{O(1)})$. Prior to our work, Shi et al. [14] have studied the SFITP, which they referred to as the Tree Assembly problem. They showed that the SFITP is fixed-parameter tractable with respect to the number k of trees in F by giving an FPT algorithm with an extremely high runtime $2^{O(k \log k)} n^{O(1)}$, where n is the number of nodes in T^* and F. To the end, we present a new FPT algorithm for the SFITP with an improved time complexity $O(4^k k^2 n^2 + n^3)$.

Theorem 1. *Given an instance $(T^*, F; k)$ of the SFITP, it can be solved in runtime $O(4^k k^2 n^2 + n^3)$, where $n = |V(F)| + |V(T^*)|$ equals the number of nodes in T^* and F, and $k = |F|$ equals the number of trees in F.* □

The algorithm proposed for the SFITP adopts the dynamic programming technique. It considers whether the sub-instance $(T^*(u), F(v, S); |F(v, S)|)$ of the SFITP is YES (v must map to u), where $T^*(u)$ is the *pendent subtree* of T^* with root u, and $F(v, S)$ is the forest comprising the trees in the forest $S \subseteq F$ and the pendent subtree $T(v)$, rooted at v, with $T \in F \setminus S$. The term $|F(v, S)|$ refers to the number of trees in $F(v, S)$. This sub-instance corresponds to the entry $f(u, v, S)$ in the dynamic programming table f. Therefore, the table f has a size $O(2^k n^2)$ $(k = |F|)$, and $\exists f(r^*, *, *) = 1$ (r^* is the root of T^*) if and only if $(T^*, F; k)$ is YES. As usual in dynamic programming over trees, the f-values are computed in a bottom-up manner, processing nodes from the leaves of T^* towards the root. A straightforward approach to get the value of a fixed entry $f(u, v, S)$, is to check all combinations of the entrys related to the children of u, resulting an expensive time complexity $2^{O(k \log k)}$. To solve the issue efficiently, we design a dynamic programming subroutine and improve the time complexity from $2^{O(k \log k)}$ to $O(2^k k^2)$, by using the fast subset convolution technique [3].

Next, we study a generalized variant of the SFITP, named the Spanning Forest Containment on Tree problem (abbr. SFCTP). In the SFITP, the input forest F has exactly $|F| = k$ many trees. But in the SFCTP, the number of trees in the input forest F is unbounded, and the goal is to determine whether F contains a forest F^* with at most k trees that is a spanning forest of T^*. As a result, we cannot directly extend the algorithm for the SFITP to the SFCTP, while preserving its fixed-parameter tractability with respect to k. Fortunately, by introducing the color coding technique [6] and cooperating it with dynamic programing technique, we successfully propose an FPT algorithm for the problem, whose main idea is given as follows.

Theorem 2. *Given an instance $(T^*, F; k)$ of the SFCTP, it can be solved in runtime $O(25.6^k k n^4)$, where $n = |V(F)| + |V(T^*)|$ equals the number of nodes in T^* and F, and k is at most the number of trees in F.*

Consider a YES-instance $(T^*, F; k)$ of the SFCTP, and a solution F^* to it. We first apply the color coding technique on the forest F to generate $O(6.4^k n)$ coloring functions (specifically, using k colors to color the trees in F), ensuring that at least one of these functions assigns different colors to the trees in F^*. Then we can reduce the SFCTP to the *Colorful* SFCTP. In this variant, the input consists of a target tree T^* with root r^*, a tree T rooted at r with the color $c(T)$ (T may be NULL), a forest F where each tree has a color (in S), and a color set S with $c(T) \notin S$. The parameter of the problem is k ($k = |S|+1$ if $T \neq$ NULL; otherwise, $k = |S|$). The goal of the problem is to decide whether F contains a forest F' such that $F' \cup \{T\}$ is a spanning forest of T^* (r maps to r^* if $T \neq$ NULL), where each tree in F' has a different color. Obviously, the SFCTP instance $(T^*, F; k)$ corresponds to a Colorful SFCTP instance $(T^*, \text{NULL}, F, R; |R|)$, where R is the initial color set with size k.

To solve the colorful SFCTP, we again utilize dynamic programming technique and decide whether the sub-instance $(T^*(u), T(v), F_S, S; k')$ of the colorful SFITP is YES (v must map to u), where $T^*(u)$ is the pendent subtree of T^* with root u, $T(v)$ is the pendent subtree of T with root v (or $T(v) =$ NULL), $F_S \subseteq F$ is a forest where each tree has a color in S, $S \subseteq R$ with $c(T) \notin S$, and $k' = |S|+1$ if $T(v) \neq$ NULL; otherwise $k' = |S|$. The sub-instance corresponds to the entry $f(u, v, S)$ in the dynamic programming table f, and $\exists f(r^*, *, *) = 1$ if and only if the input instance $(T^*, F; k)$ of the SFCTP is YES. Using the reasoning similar to the algorithm for the SFITP, we can bound the size of f and calculate a specified f-value in runtime $O(2^k k^2)$.

The paper is organized as follows. Section 2 gives the related notions and formulations of the considered problems. Sections 3 and 4 present FPT algorithms for the SFITP and SFCTP, respectively. Section 5 summarizes the paper.

2 Preliminary

Given a tree T, denote by $V(T)$ and $E(T)$ the node set and edge set of T, respectively, where $|V(T)|$ is called the *size* of T. For any subset $U \subseteq V(T)$, denote by $T[U]$ the *induced subforest* of T comprising all nodes of U and all edges with both endpoints in U. In particular, $T[U]$ is a *subtree* of T if $T[U]$ is connected. Additionally, the tree T is *rooted* if there is a unique ancestor-descendant relationship defined in it.

Given a rooted tree T, the common ancestor r of the nodes in $V(T)$, is the *root* of the tree. Thus for two nodes $u, v \in V(T)$, we say that u is an *ancestor* of v (or v is a *descendant* of u) if u is on the unique simple path from r to v. Any node is an ancestor and descendant of itself. Denote by $C_T(v)$ the set containing all descendants of v that are adjacent to v (i.e., $C_T(v)$ contains all *children* of v), and $d_v = |C_T(v)|$; and by $T(v)$ the *pendent subtree* with root v, which is induced by all descendants of v in T. For an edge $e \in E(T)$ with two endpoints u and v, it can be represented as a node pair $[u, v]$, where the first node is an ancestor of the second one by default. Given two rooted trees T_1 and T_2, they are *isomorphic*, denoted by $T_1 = T_2$, if there is a bijection Φ from $V(T_1)$ to $V(T_2)$ such that $[v, v'] \in E(T_1)$ if and only if $[\Phi(v), \Phi(v')] \in E(T_2)$, for any two nodes $v, v' \in V(T_1)$. Let $T_1 \neq T_2$ denotes its negation.

A forest is *rooted* if all its trees are rooted. Denote by $V(F)$ and $E(F)$ the node set and edge set of F, respectively; and by $|F|$ the number of trees in F. For two forests F_1 and F_2, denote by $F_1 \cup F_2$ the forest consisting of all the trees in F_1 and F_2. For a forest F and an edge set E with $V(E) \subseteq V(F)$, denote by $F + E$ the simple graph obtained by adding the edges of E into F. A forest F is called the *spanning forest* of T if there is an edge set E such that $T = F + E$.

Given a tree T, denote by $Q^R(T)$ the *reverse DFS ordering* of T. For two trees T_1 and T_2, denote by $Q^R(T_1) + Q^R(T_2)$ the sequence obtained by concatenating $Q^R(T_2)$ to the end of $Q^R(T_1)$. Similarly, for a forest F comprising trees $T_1, T_2, \ldots, T_{|F|}$, $Q^R(F)$ is defined as $Q^R(T_1) + Q^R(T_2) + \cdots + Q^R(T_{|F|})$.

All the trees and forests discussed in this paper are rooted. The definitions of the problems considered in the paper are given as follows.

Spanning Forest Isomorphism on Tree Problem (abbr. SFITP)
Input: A rooted target tree T^* and a rooted forest F;
Parameter: $k = |F|$;
Output: Return YES if F is a spanning forest of T^*; otherwise, return NO.

Spanning Forest Containment on Tree Problem (abbr. SFCTP)
Input: A rooted target tree T^* and a rooted forest F;
Parameter: $k \leq |F|$;
Output: Return YES if F contains a forest F^* with $|F^*| \leq k$ that is a spanning forest of T^*; otherwise, return NO.

The following lemma is critical to solve the SFITP.

Lemma 1. *An instance $(T^*, F; k)$ of the SFITP is* YES *if and only if there is a bijection Φ from $V(F)$ to $V(T^*)$ such that, if $[v, v']$ is an edge of F, then $[\Phi(v), \Phi(v')]$ is an edge of T^*.*

Proof. $\Rightarrow$. The fact that $(T^*, F; k)$ is YES indicates that there is an edge set E^+ such that $F + E^+ = T^*$. Thus the claimed bijection Φ can be constructed by the isomorphism between $F + E^+$ and T^*.

$\Leftarrow$. For the bijection Φ from $V(F)$ to $V(T^*)$, we remove all the edges $[\Phi(v), \Phi(v')]$ from T^* such that $[v, v'] \notin E(F)$ but $[\Phi(v), \Phi(v')] \in E(T^*)$. Observe that there are $|F| - 1$ such edges. For the resulting forest F', by the bijection Φ, we have that $[v, v']$ is an edge of F if and only if $[\Phi(v), \Phi(v')]$ is an edge of F', i.e., $F = F'$. Consequently, $(T^*, F; k)$ is YES. $\square$

By Lemma 1, for the SFITP, our aim can be transformed as looking for a bijection Φ (if any) from $V(F)$ to $V(T^*)$ such that if $[v, v']$ is an edge of F then $[\Phi(v), \Phi(v')]$ is an edge of T^*. In the remaining text, such a Φ is called a *feasible bijection* for $(T^*, F; k)$. With respect to Φ, we simply say that $\Phi(v)$ is the *image* of v (or v *maps to* $\Phi(v)$), and conversely, v is the *pre-image* of $\Phi(v)$.

It is easy to see that $(T^*, F; k)$ is a YES-instance of the SFCTP if and only if $(T^*, F^*; |F^*|)$ is a YES-instance of the SFITP for some subset $F^* \subseteq F$ with $|F^*| \leq k$. Thus, similar to Lemma 1, we have the following lemma.

Lemma 2. *An instance $(T^*, F; k)$ of the SFCTP is* YES *if and only if there is a forest $F^* \subseteq F$ with $|F^*| \leq k$ and a feasible bijection Φ from $V(F^*)$ to $V(T^*)$.*

3   An Algorithm for the SFITP

For the completeness of the paper, the section starts with some common sense for the SFITP. Given two trees with size n, we have the following lemma for their isomorphism (for more details, please refer to pp. 84–85 in [1]).

Lemma 3 ([1]). *It takes runtime $O(n)$ to decide whether or not two trees with size n are isomorphic.*

Given an instance $(T^*, F; k = |F|)$ of the SFITP, by Lemma 1, our aim can be transformed as looking for a feasible bijection Φ for this instance (if any). Now we introduce the algorithm ALG-SFITP given in the paper for the SFITP, which adopts the dynamic programming strategy. The algorithm maintains a boolean table f with entry $f(u, v, S)$, $u \in V(T^*)$, $v = $ NULL or $v \in V(T)$ for some $T \in F$, and $S \subseteq F$ if $v = $ NULL; otherwise $S \subseteq F \backslash \{T\}$. Thus the table f has a size upper bounded by $|V(T^*)| \cdot (|V(F)| + 1) \cdot 2^k$. Let $F(v, S)$ be the forest comprising the trees in S and the one $T(v)$ if $v \neq $ NULL ($T \notin S$). The entry $f(u, v, S)$ indicates whether there is a feasible bijection Φ for the instance $(T^*(u), F(v, S); |F(v, S)|)$ of the SFITP, where $\Phi(v) = u$ if $v \neq $ NULL. If exists, then $f(u, v, S) = 1$;

otherwise, $f(u, v, S) = 0$. Consequently, if $f(r^*, \text{NULL}, F) = 1$ then the instance $(T^*, F; k)$ is YES, where r^* is the root of T^*.

Now we consider the setting of $f(u, v, S)$. If u and v are leaves of T^* and T, respectively, then we have $f(u, v, \emptyset) = 1$. Additionally, if $f(u, r, S) = 1$ for some tree T in F with root r, then we have $f(u, \text{NULL}, S \cup \{T\}) = 1$. In the following, we assume that $v \in V(T)$ (i.e., $v \neq \text{NULL}$) and all the elements $f(u_i, v_j, S')$ have been given, where $u_i \in C_{T^*}(u) = \{u_1, u_2, \ldots, u_{d_u}\}$ and $v_j \in C_T(v) \cup \{\text{NULL}\} = \{v_1, v_2, \ldots, v_{d_v}, \text{NULL}\}$. The following lemma given in literature [14] is critical to the correctness of ALG-SFITP.

Lemma 4 ([14]). *If there is a feasible bijection Φ for $(T^*(u), F(v, S); |F(v, S)|)$ of the SFITP, where $\Phi(v) = u$, and there are two nodes $u_i \in C_{T^*}(u)$ and $v_j \in C_T(v)$ such that $T^*(u_i) = T(v_j)$, then there is a feasible bijection Φ' for $(T^*(u), F(v, S); |F(v, S)|)$ such that $\Phi'(v) = u$ and $\Phi'(v') \in V(T^*(u_i))$ for any node $v' \in V(T(v_j))$.*

Rule 1 ([14]). *For any child $v_j \in C_T(v)$, if there is a child $u_i \in C_{T^*}(u)$ such that $T(v_j) = T^*(u_i)$, then remove all nodes of $T(v_j)$ and $T^*(u_i)$ from T and $T^*(u)$, respectively.*

It is noteworthy that Rule 1 is unrelated with S. In the following discussion, we assume that Rule 1 is not applicable on the instance $(T^*(u), F(v, S); |F(v, S)|)$.

Lemma 5. *If the instance $(T^*(u), F(v, S); |F(v, S)|)$ of the SFITP has a feasible bijection Φ such that $\Phi(v) = u$, and Rule 1 is not applicable on it, then $d_v \leq d_u \leq |F(v, S)| - 1 = |S|$.*

Proof. It is easy to see that $d_v \leq d_u$; otherwise, $(T^*(u), F(v, S); |F(v, S)|)$ has no feasible bijection with respect to which v maps to u. In the following, we show $d_u \leq |S|$ by contradiction. Assume that $d_u > |S|$. Due to $|F(v, S) \setminus \{T(v)\}| = |S|$, there is at least one child u_i of u such that $T^*(u_i)$ contains no image of any of the trees in $F(v, S) \setminus \{T(v)\}$ with respect to Φ. Consequently, there must be a child v_j of v such that $T(v_j) = T^*(u_i)$, which leads to a contradiction to the fact that Rule 1 is not applicable on $(T^*(u), F(v, S); |F(v, S)|)$.   □

By Lemma 5, for the instance $(T^*(u), F(v, S); |F(v, S)|)$, if it has a feasible bijection with respect to which v maps to u, then there are $d_u - d_v$ many children of u, whose pre-images do not contain any nodes in $T(v)$. That is, for such a child u_i, there is a subset $S' \subseteq S$, such that the pre-images of the nodes in $V(T^*(u_i))$ are all in the trees of S', implying that $f(u_i, \text{NULL}, S') = 1$. Let $\mathcal{C}$ denote a multi-set consisting of set $C_T(v)$ and $d_u - d_v$ extra NULL elements.

Lemma 6. $f(u, v, S) = 1$ *if and only if there is a permutation $P_v = \{v'_1, v'_2, \ldots, v'_{d_u}\}$ of vertices in $\mathcal{C}$ and an ordered partition $P_S = \{S_1, S_2, \ldots, S_{d_u}\}$ of S, such that $f(u_i, v'_i, S_i) = 1$ for all $1 \leq i \leq d_u$.*

Proof. ⇒. We have that there exists a feasible bijection Φ for the instance $(T^*(u), F(v, S); |F(v, S)|)$ such that $\Phi(v) = u$, and construct P_v and P_S as follows. For each child u_i of u with $1 \le i \le d_u$, if $\Phi^{-1}(u_i) = v_j$ with $v_j \in C_T(v)$, then $v_i' = v_j$; otherwise, $v_i' = $ NULL. Meanwhile, by traversing the nodes in $V(T^*(u_i))$ and their pre-images with respect to Φ, we can find the trees in S_i. Because of Φ, we have $f(u_i, v_i', S_i) = 1$.

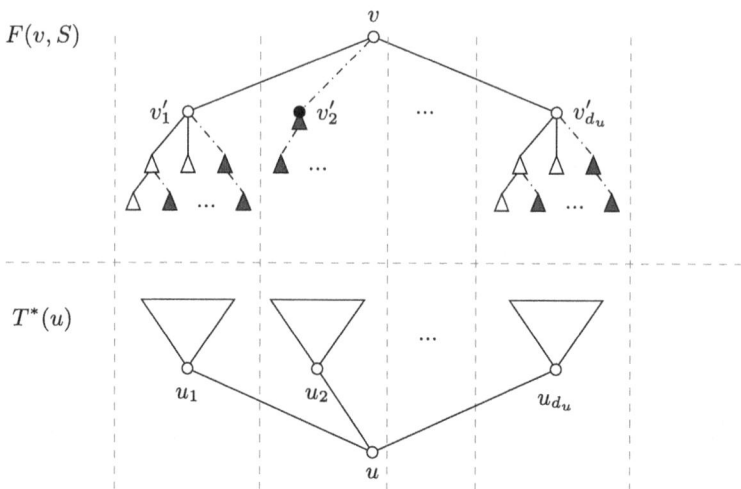

Fig. 1. An illustration of Lemma 6. The dash line in the middle indicates that there exists a feasible bijection from $V(F(v, S))$ to $V(T^*(u))$. Above the dash line, black triangles denote trees in S, while white triangles denote subtrees in $T(v)$. For $1 \le i \le d_u$, S_i comprises all black triangles below $T(v_i')$. Black nodes such as v_2' denote NULL elements. Actually, they means that the pre-images of some children of u are roots of some black triangles. The white nodes like v_1' are children of v in T.

⇐. For each $1 \le i \le d_u$, let Φ_i be a feasible bijection for $(T^*(u_i), F(v_i', S_i); |F(v_i', S_i)|)$ such that $\Phi_i(v_i') = u_i$. Combining all bijections Φ_i $(1 \le i \le d_u)$ altogether and mapping v to u, we can construct a feasible bijection Φ for $(T^*(u), F(v, S); |F(v, S)|)$ (specifically, $\Phi(v) = u$ and $\Phi(v') = \Phi_i(v')$ if v' is in $T(v_i)$ or some tree of S_i), implying that $f(u, v, S) = 1$ (Fig. 1). □

By exhaustive enumerating all possibilities to find the permutation P_v and the ordered partition P_S claimed in Lemma 6 is obviously expensive. In the following discussion, we design a subroutine to decide the existence of P_v and P_S, where the related notions are given below.

A triplet (u_i, v_i', S_i) is *feasible* if $f(u_i, v_i', S_i) = 1$, where $u_i \in C_{T^*}(u)$, $v_i' \in C$, and $S_i \subseteq F \setminus \{T\}$. Two feasible triplets (u_i, v_i', S_i) and (u_j, v_j', S_j) are *disjoint* if $u_i \ne u_j$, $v_i' \ne v_j'$ or at least one of v_i' and v_j' is NULL, and $S_i \cap S_j = \emptyset$. Thus if there are d_u pairwise disjoint feasible triplets such that $\bigcup_{i=1}^{d_u} S_i = S$, then we can derive that $f(u, v, S) = 1$ by Lemma 6.

Algorithm 1. ALG-SFITP(T^*, F; k)

Input: A target tree T^*, a forest F, and an integer $k = |F|$;
Output: Return YES if $(T^*, F; k)$ is YES; otherwise, return NO.
1: **if** $|V(F)| \neq |V(T^*)|$ **then**
2:     return NO;
3: let $Q^R(T^*)$ and $Q^R(F)$ be the reverse DFS orderings of T^* and F, respectively;
4: initialize the table f with initial value 0;
5: **for** each $u \in Q^R(T^*)$ **do**
6:     **for** each $v \in Q^R(F)$ **do**   ▷ $v \in V(T)$, $T \in F$
7:         exhaustively apply Rule 1 to $T^*(u)$ and $T(v)$;
8:         **if** $d_u > k - 1$ or $d_u < d_v$ **then** continue;
9:         **if** $d_u = 0$ **then**
10:             $f(u, v, \emptyset) \leftarrow 1$;
11:             **if** $v = r$ **then** $f(u, \text{NULL}, \{T\}) \leftarrow 1$;   ▷ r is the root of T
12:             continue;
13:         initialize the table g based on f;
14:         **if** $d_u \geq 2$ **then**
15:             **for** $i \leftarrow 2$ **to** d_u **do**
16:                 **for** each multi-subset $\mathcal{V} \in 2^{\mathcal{C}}$ with size i **do**
17:                     **for** each $v' \in \mathcal{V}$ **do**
18:                         $g(i, \mathcal{V}, *) \leftarrow \mathcal{FSC}(f(u_i, v', *), g(i - 1, \mathcal{V} \setminus \{v'\}, *))$;
19:         **for** each $S \in 2^{F \setminus \{T\}}$ **do**
20:             $f(u, v, S) \leftarrow g(d_u, \mathcal{C}, S)$;
21:             **if** $f(u, v, S) = 1$ and $v = r$ **then**
22:                 $f(u, \text{NULL}, S \cup \{T\}) \leftarrow 1$;
23: **if** $f(r^*, \text{NULL}, F) = 1$ **then**
24:     return YES;
25: return NO;

Now we try to find the d_u disjoint feasible triplets one by one. When considering the ith feasible triplet (u_i, v_i', S_i), where $i \geq 2$, we do not need to record the specific compositions of the preceding $i - 1$ feasible triplets. But we need to record the unions of v_j' and S_j separately for $1 \leq j \leq i - 1$. Thus the subroutine adopts the dynamic programming strategy and maintains a boolean table g. The entry $g(i, \mathcal{V}, S)$ in g indicates whether the forest comprising the pendent subtrees $T^*(u_j)$ for all $1 \leq j \leq i$, can be assembled by the trees in S and the pendent subtrees $T(v')$ for all $v' \in \mathcal{V}$. Specially, we have $1 \leq i \leq d_u$, $\mathcal{V}$ is a multi-subset of $\mathcal{C}$, and $S \subseteq F$. We initialize the table g by setting $g(1, \{v_1'\}, S) = 1$ if there is a feasible triplet (u_1, v_1', S); all other entries are set to 0. For the case $2 \leq i \leq d_u$, we have the following transition function.

$$g(i, \mathcal{V}, S) = \begin{cases} 1 & \text{if } \exists v_i' \in \mathcal{V} \wedge S_i \subseteq S \text{ s.t. } f(u_i, v_i', S_i) = 1 \wedge g(i - 1, \mathcal{V} \setminus \{v_i'\}, S \setminus S_i) = 1, \\ 0 & \text{otherwise.} \end{cases}$$

Observe that the above transition function can be represented in the form of subset convolution over an integer ring a

llows:

$$g(i, \mathcal{V}, S) = \begin{cases} 1 & \text{if } \exists v_i' \in \mathcal{V} \text{ s.t. } \sum_{S_i \subseteq S} f(u_i, v_i', S_i) \cdot g(i-1, \mathcal{V} \setminus \{v_i'\}, S \setminus S_i) > 0, \\ 0 & \text{otherwise.} \end{cases}$$

Referring to the work in [3], we can derive the following lemma regarding the Fast Subset Convolution algorithm.

Lemma 7 ([3]).   *The subset convolution of an n-element ground set over an arbitrary ring can be evaluated in $O(2^n n^2)$ ring operations.*

By the definition of $g(i, \mathcal{V}, S)$, we can get that $f(u, v, S) = g(d_u, \mathcal{C}, S)$. In the algorithm ALG-SFITP (given in Algorithm 1), the function $\mathcal{FSC}$ given in Step 18 represents the Fast Subset Convolution algorithm. The symbol '*' on a particular dimension of a table indicates that all elements along that dimension are extracted to form a vector for collective operations.

Proof (of Theorem 1). We first show the correctness of the algorithm ALG-SFITP. The correctness of Steps 1–6 and 9–12 are obvious. Lemma 4 and Lemma 5 show the correctness of Steps 7–8. If u and v are both leaves, then we have $d_u = d_v = 0$ and Steps 7–8 are not be executed. Lemma 6 indicates the correctness of Steps 13–22. Note that after Step 13, we have $g(1, \{v_1'\}, S) = 1$ if and only if there is a feasible triplet (u_1, v_1', S). The definition of $f(u, \text{NULL}, S \cup \{T\}) = 1$ demonstrates the correctness of Steps 23–25.

Next, we analyze the runtime of the algorithm. Step 3 and Step 4 have runtime $O(n)$ and $O(2^k n^2)$, respectively. The runtime bottleneck is the computation of tables f and g in Steps 5–22. Steps 5–6 require runtime $O(n^2)$. According to the algorithm referenced in Lemma 3, the runtime that Step 7 requires to exhaustively apply Rule 1 can be bounded by

$$\sum_{u \in V(T^*)} \sum_{v \in V(F)} \sum_{u_i \in C_{T^*}(u)} \sum_{v_j \in C_T(v)} O\left(|V(T^*(u_i))| + |V(T(v_j))|\right) = O(n^3).$$

Step 13 takes runtime $O(4^k k)$ to initialize g. In Steps 15–16, enumerating all subs

of $\mathcal{C} = \{v_1, v_2, \ldots, v_{d_u}\}$ with size $2 \leq i \leq d_u$ takes runtime $O(2^k)$. Step 17 requires runtime $O\left(2^k \cdot \sum_{u \in V(T^*)} \sum_{v \in V(F)} d_u\right) = O(2^k n^2)$ in total. By Lemma 7, the runtime of Step 18 is bounded by $O(2^k k^2)$. Additionally, it is evident that Steps 19–22 take runtime $O(2^k)$. As a result, the total runtime of ALG-SFITP is $O(4^k k^2 n^2 + n^3)$.

4 An Algorithm for the SFCTP

Consider an instance $(T^*, F; k)$ of the SFCTP. Based on the close relationship between the SFCTP and the SFITP, it is easy to see that $(T^*, F; k)$ is a YES-instance if and only if $(T^*, F^*; |F^*|)$ is a YES-instance of the SFITP for some subset $F^* \subseteq F$ with $|F^*| \leq k$. We say such an F^* a *solution* of $(T^*, F; k)$.

The number of trees in F cannot be bounded by a function $f(k)$ that depends only on k, i.e., directly applying the method from the previous section cannot give a parameterized algorithm for the SFCTP. Fortunately, we find that the color coding technique [6] can be introduced to narrow the search space. Specifically, we use k distinct colors to color the trees in F. If the considered instance $(T^*, F; k)$ of the SFCTP is YES with a solution F^*, then there is a coloring approach for F such that the trees in F^* have pairwise distinct colors, and the coloring approach can be found in a runtime that is only exponential in k.

We are ready to introduce the color coding technique. Let S be any set and W a subset of S. A function f on S is *injective* from W if for any two different elements $x, y \in W$, we have $f(x) \neq f(y)$. Let $R = \{1, 2, \ldots, k\}$ be a color set. A *coloring function* of S is a function from S to R. A *coloring family* $\mathcal{F}$ of S is a set of coloring functions such that for any subset $W \subseteq S$ with size k, there is a coloring function $c \in \mathcal{F}$ that is injective from W.

Theorem 3 ([6]). *Given a forest F and a color set $R = \{1, 2, \ldots, k\}$, the minimum size of a coloring family of F is $O(6.4^k |F|)$ and it can be constructed in runtime $O(6.4^k |F|)$.*

Theorem 3 shows that we can obtain a coloring function c that is injective from F^* (specifically, from the trees in F^*) in runtime $O(6.4^k |F|)$. From now on, w.l.o.g., we assume that the considered instance $(T^*, F; k)$ is YES and we have obtained such a coloring function c (specifically, $c(T)$ denotes the color with respect to the coloring function c for each tree T in F). To simplify the discussion, we have the following notion. A forest $F' \subseteq F$ is *colorful* if all the trees in F' have distinct colors. A colorful forest $F' \subseteq F$ is with the color set $S \subseteq R$ if $|F'| = |S|$ and each tree of F' has a unique color of S.

By Lemma 2, our aim can be transformed to find a colorful forest F^* such that there is a feasible bijection Φ from $V(F^*)$ to $V(T^*)$. To find such a colorful forest F^*, we construct a boolean table f with entry $f(u, v, S)$, $u \in V(T^*)$, $v = \text{NULL}$ or $v \in V(T)$ for some $T \in F$, and $S \subseteq R$ if $v = \text{NULL}$; otherwise $S \subseteq R \setminus \{c(T)\}$. To provide a clear and more concise definition of the element $f(u, v, S)$, we formulate a new problem given below.

Colorful Spanning Forest Containment on Tree Problem (abbr. CSFCTP)

Input: A target tree T^* with root r^*, a tree T rooted at r with a color $c(T)$, a forest F in which each tree has a color (in S), and a color set S with $c(T) \notin S$;

Parameter: $k = |S| + 1$ if $T \neq \text{NULL}$; otherwise, $k = |S|$;

Output: Return YES if F contains a colorful forest F' with color set S such that (1) $F' \cup \{T\}$ is a spanning forest of T^*, and (2) there is a feasible bijection Φ such that $\Phi(r) = r^*$ when $T \neq$ NULL; otherwise, return NO.

If $T =$ NULL, then $r =$ NULL and $c(T) =$ NULL $\notin S$. If $v \in V(T)$ for some tree $T \in F$, the element $f(u, v, S)$ with $c(T) \notin S$ indicates whether the instance $(T^*(u), T(v), F_S, S; |S| + 1)$ of the CSFCTP is YES ($F_S \subset F$ is the forest with each tree has a color in S): If yes then $f(u, v, S) = 1$; otherwise, $f(u, v, S) = 0$. Similarly, if $v =$ NULL, the element $f(u, \text{NULL}, S)$ indicates whether the instance $(T^*(u), \text{NULL}, F_S, S; |S|)$ of the CSFCTP is YES : If yes then $f(u, \text{NULL}, S) = 1$; otherwise, $f(u, \text{NULL}, S) = 0$. Note that $(T^*(u), \text{NULL}, F_S, S; |S|)$ of the CSFCTP is YES if and only if $(T^*(u), F_S; |S|)$ for the SFCTP is YES.

Now we consider the setting of $f(u, v, S)$. Specifically, if both u and v are leaves of T^* and T, respectively, then $f(u, v, \emptyset) = 1$. If $f(u, r, S) = 1$ for some tree $T \in F$, then $f(u, \text{NULL}, S \cup \{c(T)\}) = 1$. In the following, we suppose that $v \in V(T)$ for some $T \in F$ (i.e., $v \neq$ NULL).

Lemma 8 ([14]). *For the instance $(T^*(u), T(v), F_S, S; |S| + 1)$ of the CSFCTP, if it is a YES-instance with a solution F^* and there are two nodes $u_i \in C_{T^*}(u)$ and $v_j \in C_T(v)$ with $T^*(u_i) = T(v_j)$, then there is a feasible bijection Φ' from $V(\{T(v)\} \cup F^*)$ to $V(T^*(u))$ such that $\Phi'(v) = u$ and $\Phi'(v') \in V(T^*(u_i))$ for any node $v' \in V(T(v_j))$.*

Then we can apply Rule 1 to bound the size of d_u and d_v.

Lemma 9. *If the instance $(T^*(u), T(v), F_S, S; |S| + 1)$ of the CSFCTP is YES and Rule 1 cannot be applied on it, then $d_v \leq d_u \leq |S|$.*

The correctness of Lemma 9 follows from the proof of Lemma 5. In the following, we assume that Rule 1 is not applicable on the instance $(T^*(u), T(v), F_S, S; |S| + 1)$ of the CSFCTP. Let $\mathcal{C}$ be the multi-set comprising the nodes in $C_T(v)$ and $d_u - d_v$ extra NULL elements. Suppose that all the elements $f(u_i, v_i, S')$ are given, where $u_i \in C_{T^*}(u)$, $v_i \in \mathcal{C}$, and $S' \subseteq S$.

Lemma 10. $f(u, v, S) = 1$ *if and only if there is a permutation $P_v = \{v_1', v_2', \ldots, v_{d_u}'\}$ of elements in $\mathcal{C}$, and an ordered partition $P_S = \{S_1, S_2, \ldots, S_{d_u}\}$ of S such that $f(u_i, v_i', S_i) = 1$ for all $1 \leq i \leq d_u$.*

Proof. $\Rightarrow$. Since $f(u, v, S) = 1$, we have $(T^*(u), T(v), F_S, S; |S| + 1)$ of the CSFCTP is YES, i.e., there is a colorful forest $F' \subseteq F_S$ with color set S such that there is a feasible bijection Φ from $V(F') \cup V(T(v))$ to $V(T^*(u))$ with $\Phi(v) = u$. We can construct P_v and P_S as follows. For each child u_i of u with $1 \leq i \leq d_u$, if $\Phi^{-1}(u_i) = v_j$ with $v_j \in C_T(v)$, then $v_i' = v_j$; otherwise, $v_i' =$ NULL. To construct P_S, we first construct an ordered partition $\{F_1', \ldots, F_{d_u}'\}$ of F': For each $1 \leq i \leq d_u$, by traversing nodes in $V(T^*(u_i))$ and their pre-images with respect to Φ, we can find trees in F_i'. Let S_i be the color set of the trees in F_i'. Because $\{F_1', \ldots, F_{d_u}'\}$ is an ordered partition of F', the corresponding

$P_S = \{S_1, \ldots, S_{d_u}\}$ is an ordered partition of S. Then for each $1 \leq i \leq d_u$, there is a colorful forest F_i' with color set S_i such that there is a bijection Φ from $V(F_i') \cup V(T(v_i'))$ to $V(T^*(u_i))$ with $\Phi(v_i') = u_i$. Thus $f(u_i, v_i', S_i) = 1$. □

$\Leftarrow$. For each $1 \leq i \leq d_u$, there is a colorful forest F_i' with color set S_i such that there is a feasible bijection Φ_i from $V(F_i') \cup V(T(v_i'))$ to $V(T^*(u_i))$ with $\Phi_i(v_i') = u_i$. Let F' and S be the unions of F_i' and S_i for all $1 \leq i \leq d_u$, respectively. Then F' is a colorful forest with the color set S. Combining all the individual bijections Φ_i ($1 \leq i \leq d_u$) altogether and the mapping from v to u, we can construct a bijection Φ from $V(F') \cup V(T(v))$ to $V(T^*(u))$ (specifically, $\Phi(v) = u$ and $\Phi(v') = \Phi_i(v')$ if v' is in $V(T(v_i))$ or some tree of F_i'). Because $c(T) \notin S$, we have $f(u, v, S) = 1$.

Algorithm 2. ALG-SFCTP(T^*, F; k)

Input: A target tree T^*, a forest F, and an integer k;
Output: Return YES if $(T^*, F; k)$ is YES; otherwise, return NO.
  1: let $Q^R(T^*)$ and $Q^R(F)$ be the reverse DFS orderings of T^* and F, respectively;
  2: $R \leftarrow \{1, \ldots, k\}$;   ▷ R is the color set
  3: $\mathcal{F} \leftarrow$ the color family outputted by color coding;
  4: **for** each function $c \in \mathcal{F}$ **do**
  5:     initialize the table f with initial value 0;
  6:     **for** each $u \in Q^R(T^*)$ **do**
  7:         **for** each $v \in Q^R(F)$ **do**   ▷ $v \in V(T)$, $T \in F$
  8:             exhaustively apply Rule 1 to $T^*(u)$ and $T(v)$;
  9:             **if** $d_u > k - 1$ or $d_u < d_v$ **then** continue;
 10:             **if** $d_u = 0$ **then**
 11:                 $f(u, v, \emptyset) \leftarrow 1$;
 12:                 **if** $v = r$ **then** $f(u, \text{NULL}, \{c(T)\}) \leftarrow 1$;   ▷ r is the root of T
 13:                 continue;
 14:             initialize the table g based on f;
 15:             **if** $d_u \geq 2$ **then**
 16:                 **for** $i \leftarrow 2$ **to** d_u **do**
 17:                     **for** each multi-subset $\mathcal{V} \in 2^{\mathcal{C}}$ with size i **do**
 18:                         **for** each $v' \in \mathcal{V}$ **do**
 19:                             $g(i, \mathcal{V}, *) \leftarrow \mathcal{FSC}(f(u_i, v', *), g(i-1, \mathcal{V} \setminus \{v'\}, *))$;
 20:             **for** each $S \in 2^{R \setminus \{c(T)\}}$ **do**
 21:                 $f(u, v, S) \leftarrow g(d_u, \mathcal{C}, S)$;
 22:                 **if** $f(u, v, S) = 1$ and $v = r$ **then**
 23:                     $f(u, \text{NULL}, S \cup \{c(T)\}) \leftarrow 1$;
 24:     **if** $f(r^*, \text{NULL}, R) = 1$ **then**
 25:         return YES;
 26: return NO.

Then we accelerate the computation of $f(u, v, S)$. Note that if there are d_u pairwise disjoint feasible triplets $(u_1, v_1', S_1), \ldots, (u_{d_u}, v_{d_u}', S_{d_u})$ such that

$\bigcup_{i=1}^{d_u} S_i = S$, then we can derive that $f(u, v, S) = 1$ by Lemma 10. Now we try to find these d_u pairwise disjoint feasible triplets one by one. Observe that when we find the ith feasible triplet (u_i, v_i', S_i), we only need to record the unions of v_j' and S_j separately for $1 \le j \le i - 1$. Thus, we construct a boolean table g such that the entry $g(i, \mathcal{V}, S)$ indicates whether the forest comprising the pendent subtrees $T^*(u_j)$ for all $1 \le j \le i$, can be assembled by the pendent subtrees $T(v')$ for all $v' \in \mathcal{V}$ and $|S|$ many colorful trees such that each tree has a unique color in S. Specially, $1 \le i \le d_u$, $\mathcal{V}$ is a multi-subset of $\mathcal{C}$, and $S \subseteq R$.

We initialize the table g by setting $g(1, \{v_1'\}, S) = 1$ if there is a feasible triplet (u_1, v_1', S); all other entries are set to 0. For the case $2 \le i \le d_u$, we have

$$g(i, \mathcal{V}, S) = \begin{cases} 1 & \text{if } \exists v_i' \in \mathcal{V} \text{ s.t. } \sum_{S_i \subseteq S} f(u_i, v_i', S_i) \cdot g(i - 1, \mathcal{V} \setminus \{v_i'\}, S \setminus S_i) > 0, \\ 0 & \text{otherwise.} \end{cases}$$

By Lemma 7, we can update the table $g(i, \mathcal{V}, S)$ in runtime $O(2^k k^2)$. By the definition, $f(u, v, S) = g(d_u, \mathcal{C}, S)$. Once the table f is obtained, we can verify whether the instance $(T^*, F; k)$ of the SFCTP is YES by the following lemma.

Lemma 11. *The instance* $(T^*, F; k)$ *of the*

P *is* YES *if and only if* $f(r^*, \text{NULL}, R) = 1$.

Proof. $\Rightarrow$ holds obviously by the corresponding definitions.
$\Leftarrow$. Since $f(r^*, \text{NULL}, R) = 1$, the instance $(T^*, \text{NULL}, F, R; k)$ of the CSFCTP is YES, which is equivalent to the instance $(T^*, F; k)$ of the SFCTP. $\square$

Proof (of Theorem 2). We first show the correctness of ALG-SFCTP. Since the algorithm returns NO when no solution is found, it suffices to consider that the input instance is YES with a solution F^*. Theorem 3 ensures the correctness of Steps 3–4. Then each tree in F^* has a unique color. Steps 5–7 and 10–13 are clearly correct. The correctness of Steps 8–9 is supported by Lemma 8 and Lemma 9, respectively. If u and v are leaves, then Steps 8–9 are not executed. Lemma 10 justifies the correctness of Steps 14–23. Finally, by Lemma 11, we have $f(r^*, \text{NULL}, R) = 1$ and the algorithm outputs YES. Steps 24–26 are correct.

Now we analyze the runtime of ALG-SFCTP. Step 1 takes runtime $O(n)$. Steps 3–4 each takes runtime $O(6.4^k |F|)$. For each coloring function in $\mathcal{F}$, Step 5 requires runtime $O(2^k n^2)$. Steps 6–7, which enumerate all possible pairs of u and v, take runtime $O(n^2)$. Similar to the proof of Theorem 1, Step 8 takes runtime $O(n^3)$. Step 14 takes runtime $O(4^k k)$ to construct g. Steps 16–18 take runtime $O(2^k n^2)$ to find all possible $\mathcal{V}$ and v' in total. By Lemma 7, Step 19 requires runtime $O(2^k k^2)$. Additionally, Steps 20–23 take runtime $O(2^k)$. Thus, the tables f and g can be computed in runtime $O(4^k k^2 n^2 + n^3)$. Consequently, the runtime of ALG-SFCTP is $O(6.4^k n^3 |F| + 25.6^k k^2 n^2 |F|) = O(25.6^k k n^4)$. $\square$

5   Conclusion

Within the paper, we investigated the Spanning Forest Isomorphism on Tree problem and its generalized variant. By taking advantage of the dynamic programming technique, we proposed a parameterized algorithm for the Spanning Forest Isomorphism on Tree problem with runtime $O(4^k k^2 n^2 + n^3)$, which improves the previously best algorithm with runtime $2^{O(k \log k)} n^{O(1)}$, where n is the total number of nodes of the target tree and the forest. Additionally, we studied the Spanning Forest Containment on Tree problem and proposed an algorithm with runtime $O(25.6^k k n^4)$, under the help of the color coding technique. In particular, when the input target tree T^* is binary, the time complexity can be improved to $8^{k+12 \log^2 k} n^{O(1)}$ by introducing the divide-and-color technique. This result will be provided in the full version.

For the Spanning Forest Containment on Tree problem, if the input target tree is not binary, we have not found an efficient way to divide the instance into two sub-instances of nearly equal "size". Therefore, it is an interesting topic for future research. Moreover, investigating whether the algorithms proposed in the paper can be extended to the corresponding unrooted versions of the considered problems, while maintaining the same exponential parts of the time complexities is another intriguing area for further exploration.

Acknowledgments. This work is supported by the National Natural Science Foundation of China under Grants 62472449 and 62332020; the Open Project of Xiangjiang Laboratory (No. 22XJ03005).

References

1. Aho, A.V., Hopcroft, J.E., Ullman, J.D.: The Design and Analysis of Computer Algorithms. Addison-Wesley (1974)
2. Baumbach, J., Guo, J., Ibragimov, R.: Covering tree with stars. J. Comb. Optim. **29**(1), 141–152 (2015). https://doi.org/10.1007/S10878-013-9692-Y
3. Björklund, A., Husfeldt, T., Kaski, P., Koivisto, M.: Fourier meets möbius: fast subset convolution. In: Proceedings of the 39th Annual ACM Symposium on Theory of Computing, pp. 67–74 (2007). https://doi.org/10.1145/1250790.1250801
4. Bulteau, L., Fertin, G., Komusiewicz, C., Rusu, I.: A fixed-parameter algorithm for minimum common string partition with few duplications. In: Darling, A., Stoye, J. (eds.) WABI 2013. LNCS, vol. 8126, pp. 244–258. Springer, Heidelberg (2013). https://doi.org/10.1007/978-3-642-40453-5_19
5. Bulteau, L., Komusiewicz, C.: Minimum common string partition parameterized by partition size is fixed-parameter tractable. In: Proceedings of the 25th Annual ACM-SIAM Symposium on Discrete Algorithms, pp. 102–121 (2014). https://doi.org/10.1137/1.9781611973402.8
6. Chen, J., Lu, S., Sze, S.-H., Zhang, F.: Improved algorithms for path, matching, and packing problems. In: Bansal, N., Pruhs, K., Stein, C. (eds.) Proceedings of the 18th Annual ACM-SIAM Symposium on Discrete Algorithms, pp. 298–307 (2007)

7. Chen, X., et al.: Assignment of orthologous genes via genome rearrangement. IEEE/ACM Trans. Comput. Biol. Bioinform. **2**(4), 302–315 (2005). https://doi.org/10.1109/TCBB.2005.48

8. Cygan, M., et al.: Parameterized Algorithms, vol. 5. Springer (2015). https://doi.org/10.1007/978-3-319-21275-3

9. Damaschke, P.: Minimum common string partition parameterized. In: Crandall, K.A., Lagergren, J. (eds.) WABI 2008. LNCS, vol. 5251, pp. 87–98. Springer, Heidelberg (2008). https://doi.org/10.1007/978-3-540-87361-7_8

10. Jiang, H., Su, B., Xiao, M., Xu, Y., Zhong, F., Zhu, B.: On the exact block cover problem. In: Gu, Q., Hell, P., Yang, B. (eds.) AAIM 2014. LNCS, vol. 8546, pp. 13–22. Springer, Cham (2014). https://doi.org/10.1007/978-3-319-07956-1_2

11. Jiang, H., Zhu, B., Zhu, D., Zhu, H.: Minimum common string partition revisited. J. Comb. Optim. **23**, 519–527 (2012). https://doi.org/10.1007/S10878-010-9370-2

12. Kirkpatrick, B., Reshef, Y., Finucane, H.K., Jiang, H., Zhu, B., Karp, R.M.: Comparing pedigree graphs. J. Comput. Biol. **19**(9), 998–1014 (2012). https://doi.org/10.1089/CMB.2011.0254

13. Kul, G., Luong, D.T.A., Xie, T., Chandola, V., Kennedy, O., Upadhyaya, S.J.: Similarity metrics for SQL query clustering. IEEE Trans. Knowl. Data Eng. **30**(12), 2408–2420 (2018). https://doi.org/10.1109/TKDE.2018.2831214

14. Shi, F., You, J., Zhang, Z., Liu, J., Wang, J.: Fixed-parameter tractability for the tree assembly problem. Theoret. Comput. Sci. **886**, 3–12 (2021). https://doi.org/10.1016/J.TCS.2021.06.036

15. You, J., Shi, F., Wang, J., Feng, Q.: Fixed-parameter tractability for minimum tree cut/paste distance and minimum common integer partition. Theoret. Comput. Sci. **806**, 256–270 (2020). https://doi.org/10.1016/J.TCS.2019.04.003

16. You, J., Wang, J., Feng, Q., Shi, F.: Kernelization and parameterized algorithms for covering a tree by a set of stars or paths. Theoret. Comput. Sci. **607**, 257–270 (2015). https://doi.org/10.1016/J.TCS.2015.06.030

Min-Sum Disjoint Paths on Subclasses of Chordal Graphs

Bar Menashe[✉] and Meirav Zehavi

Ben-Gurion University of the Negev, Beersheba, Israel
barmena@post.bgu.ac.il, meiravze@bgu.ac.il

Abstract. We study the optimization version of the classic DISJOINT PATHS problem, known as MIN-SUM DISJOINT PATHS, as well as its restriction to shortest paths, known as DISJOINT SHORTEST PATHS. Both problems are notoriously hard in the sense that very few positive results are known in their context even when confined to grids, in contrast to the classic DISJOINT PATHS problem, despite significant research efforts in recent years. In light of this, we focus on restricted graph classes, being subclasses of chordal graphs: specifically, we consider the classes of split graphs, well-partitioned chordal graphs, and threshold graphs. For each of the two problems and each of these graph classes, we provide either a polynomial-time algorithm or a fixed-parameter algorithm (when a polynomial-time algorithm is unlikely to exist).

Keywords: Parameterized Complexity · Vertex-Disjoint Paths Problem · Chordal Graphs

1 Introduction

In the DISJOINT PATHS problem, the input is an undirected n-vertex graph G and a multiset of terminal pairs $T = \{(s_1, t_1), \ldots, (s_k, t_k)\}$, and the goal is to determine whether there exist k pairwise internally vertex-disjoint paths $P_1, \ldots, P_k$ in G such that P_i is a path with endpoints s_i and t_i, and where no terminal can appear as an internal vertex. The DISJOINT PATHS problem is a fundamental routing problem that finds applications in the areas of VLSI design and virtual circuit routing [1–4]. Notably, DISJOINT PATHS is a cornerstone of the groundbreaking Graph Minors project of Robertson and Seymour [5,6]. Robertson and Seymour established that any infinite sequence of graphs contains two graphs such that one is a minor of the other. The proof spans a series of 23 papers, and regarded as one of the most amazing feats of modern mathematics.

For over four decades, the DISJOINT PATHS problem has been extensively studied from the viewpoints of both parameterized complexity and approximation algorithms. Since the survey of these results is beyond the scope of this paper, we only mention a handful of developments most relevant for our exposition. First, on general graphs, Robertson and Seymour [5] gave an $\mathcal{O}(f(k) \cdot n^3)$-time algorithm. Subsequently, the polynomial dependency was reduced to n^2 by

© The Author(s), under exclusive license to Springer Nature Singapore Pte Ltd. 2025
S. Nakano and M. Xiao (Eds.): WALCOM 2025, LNCS 15411, pp. 281–295, 2025.
https://doi.org/10.1007/978-981-96-2845-2_18

Kawarabayashi et al. [7], and, very recently, it was further reduced to be almost linear by Krohnen et al. [8]. However, in all of these algorithms, the dependency on k is galactic, being a tower of 2's of height at least 5; until this date, this dependency remains the state-of-the-art. Thus, special attention was given to restricted graph classes, particularly planar graphs and subclasses of chordal graphs. Currently, on planar graphs, the state-of-the-art is that the problem is solvable in time $2^{\mathcal{O}(k^2)} \cdot n$ [9], and admits a polynomial kernel[1] with respect to k plus the treewidth of the graph (but not with respect to k alone unless the polynomial hierarchy collapses) [10].

For subclasses of chordal graphs, being the most relevant to our work, the following results are known from the perspectives of polynomial-time and parameterized algorithms. First, Kammer and Tholey [11] proved that DISJOINT PATHS is NP-hard on chordal graphs, but is solvable in time $k^{\mathcal{O}(k)} \cdot n$ on chordal graphs. Heggernes et al. [12] proved that DISJOINT PATHS is NP-hard also on the more restricted class of split graphs, but admits an $\mathcal{O}(k^2)$-vertex kernel on split graphs. In particular, it means that MIN-SUM DISJOINT PATHS is NP-hard on split graphs as well. Since well-partitioned chordal graphs are a generalization of the class of split graphs, MIN-SUM DISJOINT PATHS is also NP-hard on well-partitioned chordal graphs. Thus, for each case where we do not give a polynomial-time algorithm, it is because such algorithm is unlikely to exist. Afterward, Yang et al. [13] showed that the restricted version of DISJOINT PATHS where T excludes terminal repetitions admits a $4k$-vertex kernel. Then, Ahn et al. [14] introduced the class of *well-partitioned chordal graphs* (being a generalization of split graphs), and showed that DISJOINT PATHS on this graph class admits an $\mathcal{O}(k^3)$-vertex kernel. Lastly, Chaudhary et al. [15] proved that DISJOINT PATHS admits an $\mathcal{O}(k^2)$-vertex kernel on well-partitioned chordal graphs, a $4k$-vertex kernel on split graphs (lifting the aforementioned requirement of T to exclude terminal repetitions), and is solvable in polynomial time on threshold graphs and block graphs. We remark that the edge-version of DISJOINT PATHS has also been studied on subclasses of chordal graphs, and refer to the aforementioned papers and references therein for information.

We study the optimization version of the classic DISJOINT PATHS problem, known as MIN-SUM DISJOINT PATHS, as well as its restriction to shortest paths, known as DISJOINT SHORTEST PATHS. Both problems are notoriously hard in the sense that very few positive results are known in their context even when confined to grids, in contrast to the classic DISJOINT PATHS problem. Exactly as in the DISJOINT PATHS problem, the input of the MIN-SUM DISJOINT PATHS problem is an undirected n-vertex graph G and a multiset of terminal pairs $T = \{(s_1, t_1), \ldots, (s_k, t_k)\}$, and we need to determine whether there exist k pairwise internally vertex-disjoint paths $P_1, \ldots, P_k$ in G such that P_i is a path with endpoints s_i and t_i, and where no terminal can appear as an internal vertex. However, here, if the answer is positive, we further need to compute such a collection of paths $P_1, \ldots, P_k$ such that the sum of their lengths is minimized. Second, in the DISJOINT SHORTEST PATHS problem, the input is again an undirected

---

[1] Standard definitions in parameterized complexity are deferred to Sect. 2.

n-vertex graph G and a multiset of terminal pairs $T = \{(s_1, t_1), \ldots, (s_k, t_k)\}$, but now the goal is only to determine whether there exist k pairwise internally vertex-disjoint paths $P_1, \ldots, P_k$ in G such that P_i is a *shortest* path with endpoints s_i and t_i (i.e., there does not exist a path shorter than P_i in G with endpoints s_i and t_i), and where no terminal can appear as an internal vertex. Thus, DISJOINT SHORTEST PATHS is a restriction of MIN-SUM DISJOINT PATHS in the sense that if there exists a solution to the former, then that is a solution to the latter as well.

Lochet [16] proved that the DISJOINT SHORTEST PATHS problem (and, hence, also MIN-SUM DISJOINT PATHS) is W[1]-hard when parameterized by k. On the positive side, Lochet [16] proved that DISJOINT SHORTEST PATHS is in XP—specifically, it is solvable by an $n^{\mathcal{O}(k^{5^k})}$-time algorithm. This result subsumed several previous works, which asserted polynomial-time solvability (with different time complexities) for $k = 2$. The time bound of $n^{\mathcal{O}(k^{5^k})}$ was later sped-up to $n^{\mathcal{O}(k! \cdot k)}$ by Bentert et al. [17]. The state-of-the-art of MIN-SUM DISJOINT PATHS remains grimmer—it is still open whether it is in XP (when parameterized by k on general graphs); in fact, we do not even know whether it is NP-hard or not for $k = 3$. On the positive side, very recently, Mari et al. [18] showed that MIN-SUM DISJOINT PATHS is fixed-parameter tractable (FPT) on grids without holes, where the time complexity is $k^{2^{\mathcal{O}(k)}}$. Also, on the positive side, the problem is known to be solvable in polynomial time when $k = 2$ by a randomized algorithm [19]. For additional positive results on restricted cases (none of which concerns subclasses of chordal graphs), we refer to [18] and references therein.

Our Contributions. In light of this, we focus on subclasses of chordal graphs: split graphs, well-partitioned chordal graphs, and threshold graphs. For each of the two problems and each of these graph classes, we provide either a polynomial-time algorithm or a fixed-parameter algorithm (when a polynomial-time algorithm is unlikely to exist):

- **Split Graphs:** DISJOINT SHORTEST PATHS is solvable in time $\mathcal{O}(k \cdot (n+m))$, and MIN-SUM DISJOINT PATHS is solvable in time $\mathcal{O}(2^k \cdot (k^2 \cdot n + k \cdot m))$. (Section 3.)
- **Well-Partitioned Chordal Graphs:** DISJOINT SHORTEST PATHS is solvable in time $\mathcal{O}(k \cdot (n^{1.5} + m))$, and MIN-SUM DISJOINT PATHS is solvable in time $\mathcal{O}(2^k \cdot (k^3 \cdot n + k^2 \cdot m))$ when T is a set. (Section 4.)
- **Threshold Graphs:** MIN-SUM DISJOINT PATHS (and, hence, also DISJOINT SHORTEST PATHS) is solvable in time $\mathcal{O}(n + m + k)$. (Omitted due to space constraints.)

The full details of the omitted result (regarding threshold graphs) and all proofs will appear in the full version of this paper.

We remark that at the core of our proofs lie matching-based arguments. In particular, our algorithm for MIN-SUM DISJOINT PATHS on split graphs is based on a "guessing" step where we determine the number of internal vertices of each solution path, followed by a reduction to a computation of a maximum matching (a figure demonstrating this step will appear in the full version of this paper). Then, our algorithm for MIN-SUM DISJOINT PATHS on well-partitioned chordal graphs makes multiple uses of the one for split graphs, and its correctness is rooted in the following critical insight: given the "canonical" tree decomposition of the input well-partitioned chordal graph, the guesses required for each individual bag are independent from each other.

Lastly, we pose the following open questions. While it is easy to see that the algorithm by Kammer and Tholey [11] can be extended to solve MIN-SUM DISJOINT PATHS in time $n^{\mathcal{O}(k)}$ on chordal graphs, it remains open whether MIN-SUM DISJOINT PATHS is solvable in time $k^{\mathcal{O}(k)}$ on chordal graphs, as well as whether DISJOINT SHORTEST PATHS is NP-hard on chordal graphs.

2   Preliminaries

Due to space constraints, proofs marked by # are omitted and will appear in the full version of this paper. Additionally, standard definitions in parameterized complexity and graph theory will appear in the full version of this paper as well. Still, we briefly note that given a path in a graph, we use the term *length* to refer to its number of *vertices*. Now, let us present the graph classes relevant to our research. The most general graph class relevant to our paper is the class of *chordal graphs*, which is defined as follows.

Definition 1 (Chordal Graphs). Let G be a graph. If for any $\ell \geq 4$ the graph G does not contain an induced $C_\ell,^2$ then G is a *chordal graph*.

A more restricted graph class is the class of *split graphs*.

Definition 2 (Split Graphs). Let G be a graph. If $V(G)$ can be partitioned into two disjoint sets, K and I, such that $G[K]$ is a clique and $G[I]$ is an independent set, then G is a *split graph*.

It is worth mentioning that a split graph may have more than one *split partition* into a clique and an independent set. However, split graphs can be recognized in linear time, and a split partition can be found in linear time if one exists [20].

Another graph class, which generalizes the class of split graphs but still is a subclass of chordal graphs, is the class of *well-partitioned chordal graphs*.

Definition 3 (Well-Partitioned Chordal Graphs). A graph G is a *well-partitioned chordal graph* if it can be presented as a *partition-tree*, i.e., a tree $\mathcal{T} = (V_{\mathcal{T}}, E_{\mathcal{T}})$ such that:

1. $V_{\mathcal{T}}$ is a partition of $V(G)$. Equivalently, $V(G) = \bigcup_{b \in V_{\mathcal{T}}} b$ and for any $b_1, b_2 \in V_{\mathcal{T}}$ such that $b_1 \neq b_2$, it holds that $b_1 \cap b_2 = \emptyset$.
2. For any $b \in V_{\mathcal{T}}$, the induced subgraph $G[b]$ is a clique.

---

^{2} C_ℓ is a cycle with ℓ vertices.

3. For any edge $\{u,v\} \in E(G)$ such that $u \in b_1$ and $v \in b_2$ for different $b_1, b_2 \in V_\mathcal{T}$, it holds that $\{b_1, b_2\} \in E_\mathcal{T}$.
4. Let $\{b_1, b_2\} \in E_\mathcal{T}$, and let $E \subseteq E(G)$ be the set of all the edges of the form $\{u,v\}$ where $u \in b_1$ and $v \in b_2$. Then, the graph induced by E is a complete bipartite graph.

An example of a well-partitioned chordal graph and a corresponding partition-tree will appear in a figure in the full version of this paper. A partition-tree representation of a well-partitioned chordal graph can be found in polynomial time:

Lemma 1 (Ahn et al.) [21]. *Let G be a well-partitioned chordal graph. Then, a partition-tree of G can be found in polynomial time.*

Finally, the most restricted graph class relevant to our paper, which is a subclass of split graphs, is the class of *threshold graphs*. An example of a threshold graph will appear in a figure in the full version of this paper.

Definition 4 (Threshold Graphs). A split graph G, with a split partition (K, I) is called a *threshold graph* if there exists a linear ordering of the vertices in the independent set I, say, $v_1, \ldots, v_{|I|}$, such that $N_G(v_1) \subseteq \ldots \subseteq N_G(v_{|I|})$.

We present the following property of threshold graphs, whose proof is given for the sake of completeness.

Observation 1 (Folklore, #). *Let G be a threshold graph. Then, there exists an order $v_1, \ldots, v_n$ of the vertices of $V(G)$ such that for any $i < j$ it holds that:*

- *If $\{v_i, v_j\} \in E(G)$, then $N_G[v_i] \subseteq N_G[v_j]$.*
- *Otherwise: $N_G(v_i) \subseteq N_G(v_j)$.*

3   Split Graphs

3.1   Polynomial-Time Algorithm for Disjoint Shortest Paths

In this subsection, we present a polynomial-time algorithm for the DISJOINT SHORTEST PATHS problem on split graphs. At the core of our algorithm is a reduction to the problem of finding a maximum matching in bipartite graphs, which can be solved in polynomial time.

Let (G, T) be an instance of the DISJOINT SHORTEST PATHS problem where G is a split graph. Then, $V(G)$ is partitioned into two disjoint sets: a set K such that $G[K]$ is a clique and a set I such that $G[I]$ is an independent set. Now, let us show how we can find a solution to (G, T) in polynomial time.

First, let us note the following observation:

Observation 2 (#). *Let (G, T) be an instance of MIN-SUM DISJOINT PATHS. Let $(s, t) \in T$ be a pair of terminals such that $\{s, t\} \in E(G)$. Then any solution to MIN-SUM DISJOINT PATHS on (G, T) contains the edge $\{s, t\}$ as a path.*

Keeping in mind that any solution to DISJOINT SHORTEST PATHS is in particular a solution to MIN-SUM DISJOINT PATHS, we can get rid of such pair of terminals if it appears only once in the multiset T, or output "No" otherwise. More formally:

Reduction Rule 1. *Let (G, T) be an instance of the DISJOINT SHORTEST PATHS such that G is a split graph. Let $(s, t) \in T$ be a pair of terminals such that $\{s, t\} \in E(G)$. If the pair (s, t) appears more than once in T, then output "No". Otherwise, remove this pair from the set T. If either of s or t does not appear as a part of some pair in the set T anymore, remove it from G as well.*

After applying this rule exhaustively, one can note that for any pair of terminals (s, t) in the reduced set T, it holds that at least one of the terminals is in the set I, i.e., we have that either both are in I or that one is in I and the other is in K. In the latter case, it is clear that any shortest path between the terminals is of length 3, so the possible middle vertices of such a path are common neighbors of both terminals.

In the first case, if both terminals have a common neighbor, then any shortest path between them is of length 3, and the possible middle vertices of such a path are common neighbors of both terminals. Otherwise, we can observe that any shortest path between these terminals is of length 4, since the neighbors of each terminal are in K (so, any neighbor of one terminal has an edge connecting it with any neighbor of the other terminal).

However, by removing terminals in the application of Reduction Rule 1, we might have changed the distances between the remaining terminal pairs. Clearly, in this case, we should output "No". So, we define the following reduction rule.

Reduction Rule 2. *After applying Reduction Rule 1 exhaustively, if there exists a pair of terminals in the reduced instance whose distance is different from their distance in the original instance, output "No".*

This rule is applied exactly once, and can be computed in $\mathcal{O}(k \cdot (n + m))$ time by performing a breadth-first-search (BFS) twice for each pair of terminals, once on the original instance and once on the reduced instance.

Observation 3 (#). *Let (G, T) denote the original instance for DISJOINT SHORTEST PATHS, and let (G^*, T^*) denote the instance obtained by exhaustively applying Reduction Rule 1. If Reduction Rule 2 does not output "No", then (G^*, T^*) is an equivalent instance to DISJOINT SHORTEST PATHS such that any shortest path between a pair of terminals in the reduced instance contains either 3 or 4 vertices. Otherwise, (G, T) is a "No" instance.*

This allows us to reduce the problem to finding a matching in a bipartite graph. Roughly speaking, the "left" side of the constructed bipartite graph will contain only terminal vertices, which are connected to their possible adjacent internal vertices in the solution in the "right" side. Formally, we construct a bipartite graph B with a bipartition (L, R) in the following manner. For every terminal pair $(s, t) \in T$:

- **If a shortest path between s and t is of length 3:** We add to L a copy of the terminal s. To R, we add the common *non-terminal* neighbors of s and t,[3] and connect them to the new copy of the terminal s in L.
- **If a shortest path between s and t is of length 4:** We add to L a copy of the terminal s. To R, we add the *non-terminal* neighbors of s, and connect them to the new copy of the terminal s in L. Similarly, we add a copy of the terminal t to L, then add its *non-terminal* neighbors to R, and connect them to the new copy of t in L.

A figure demonstrating this construction will appear in the full version of this paper.

We will see that (G, T) admits a solution to the DISJOINT SHORTEST PATHS problem if and only if B admits a matching that matches all the vertices of L. Hence, we can immediately output "No" if we have $|L| > |R|$.

Reduction Rule 3. *If $|L| > |R|$, output "No".*

Lemma 2 (#). *If $|L| \leq |R|$, then the constructed bipartite graph B has $\mathcal{O}(n)$ vertices and $\mathcal{O}(m)$ edges.*

Then, we use as a black box an algorithm for finding a maximum matching in a bipartite graph.

Lemma 3 (#). *There exists an algorithm with running time of $\mathcal{O}(k \cdot (n + m))$ for computing a maximum matching in the bipartite graph B.*

If the algorithm described in Lemma 3 returns a matching that matches all terminals in L, we will claim (in Lemma 4) that there is a solution to (G, T). Otherwise, we will claim that there is no such solution.

Now, let us present and prove the correctness of our algorithm.

Lemma 4. *Let (G, T) be an instance for DISJOINT SHORTEST PATHS on split graphs. Our algorithm returns "yes" if and only if (G, T) admits a solution.*

Proof. Let $(G^\star, T^\star)$ denote the instance obtained by exhaustively applying Reduction Rule 1. From Observation 3, it suffices to show that the constructed bipartite graph B with the bipartition (L, R) for $(G^\star, T^\star)$ has a matching that matches all the vertices of L if and only if there exists a solution for $(G^\star, T^\star)$.[4]

First, let us prove the forward direction. Suppose that B has a matching M that matches all the vertices of L, and let $(s, t) \in T^\star$. From Observation 3, we know that a shortest path between s and t in $G^\star$ can only be either of length 3 or of length 4. Let us distinguish between these two cases:

- **Case 1 - length 3:** Then, only a copy of s appears in L, and is matched by M to some vertex $x \in R$. From the construction of B, x is a common neighbor of s and t, so $s - x - t$ is a shortest path between s and t in $G^\star$.

---

[3] Some of these neighbors may have been already added due to another terminal pair.

[4] Keep in mind that Reduction Rule 3 may be applied, but if it outputs "No", then there is no such matching in B. Hence, it indeed suffices to prove this statement to deduce the correctness of our algorithm.

 – **Case 2 - length 4:** Then, both copies of s and t appear in L, and are matched by M to some vertices x and y (respectively) in R. From the construction of B, x is a neighbor of s and y is a neighbor of t. Moreover, in this case $s, t \in I$, so we have that $x, y \in K$ since I is an independent set in G (and in particular in $G^\star$). So, $\{x, y\} \in E(G^\star)$, and $s - x - y - t$ is a shortest path between s and t in $G^\star$.

For any other $(s', t') \in T^\star$, the vertices matched to s' and t' in M are not those that are matched to s nor t, so applying this process for each pair of terminals constructs a solution for the DISJOINT SHORTEST PATHS on the instance $(G^\star, T^\star)$.

The proof of the reverse direction is omitted due to space constraints and will appear in the full version of this paper.   □

We conclude the subsection with its main theorem:

Theorem 4 (#). *There exists an algorithm solving* DISJOINT SHORTEST PATHS *on split graphs in* $\mathcal{O}(k \cdot (n + m))$ *time.*

3.2   FPT Algorithm for Min-Sum Disjoint Paths

In this subsection, we present an $\mathcal{O}(2^k \cdot (k^2 \cdot n + k \cdot m))$ time algorithm for the MIN-SUM DISJOINT PATHS problem on split graphs, which can be seen as a generalization of the previous subsection.

Similarly to the previous subsection, this algorithm uses as a black-box the polynomial-time algorithm for finding a maximum matching in bipartite graphs which is described in the proof of Lemma 3 (that will appear in the full version of this paper), but whose running-time bound is slightly different in this case, as described in the proof of Theorem 6 (that will appear in the full version of this paper as well).

Let (G, T) be an instance for the MIN-SUM DISJOINT PATHS problem where G is a split graph. Recall that $V(G)$ can be partitioned into two disjoint sets: a set K such that $G[K]$ is a clique and a set I such that $G[I]$ is an independent set. Now, let us show how we can find a solution to the MIN-SUM DISJOINT PATHS on (G, T) in $\mathcal{O}^*(2^k)$ time.

First, we apply to the instance a procedure similar to Reduction Rule 1 from the previous subsection. The only difference is that now if there is an edge between a pair of terminals, if this pair appears more than once in the multiset T, it is still possible that a solution for the problem exists. Formally, let us define the *Edge Procedure* as follows:

Procedure 1 (Edge Procedure). *Let (G, T) be an instance of the* MIN-SUM DISJOINT PATHS *such that G is a split graph. For every pair of terminals $(s, t) \in T$ such that $\{s, t\} \in E(G)$, remove one instance of this pair from the multiset T. That is, now the cardinality in the multiset T of each of these pairs is smaller by one. If either of s or t does not appear as a part of some pair in the set T anymore, remove it from G as well.*[5]

---

[5] This procedure, unlike reduction rules, is applied exactly once.

Now, for each of the remaining pairs in T, we should look for a path that contains at least three vertices. Let us define formally this new problem:

Definition 5 (Min-Sum Disjoint Paths*). Let G be a graph, and let T be a set of pairs of terminals. Let $k = |T|$ be a parameter. In the Min-Sum Disjoint Paths* problem, the objective is to find a set of k internally vertex disjoint paths connecting the pairs of terminals of T such that each path in the set contains at least three vertices, where the sum of the lengths of the paths is minimum (or output "No" if there is no possible solution).

In addition, let us introduce the following problem:

Definition 6 (Disjoint Paths*). Let G be a graph, and let T be a set of pairs of terminals. Let $k = |T|$ be a parameter. In the Disjoint Paths* problem, the objective is to find a set of k internally vertex disjoint paths connecting the pairs of terminals of T such that each path in the set contains at least three vertices (or output "No" if there is no possible solution).

Next, we define a subset of the possible solutions for Disjoint Paths*.

Definition 7. Let (G, T) be an instance for the Disjoint Paths* problem, and let S be a solution to this problem. Then, S is called a *minimal solution* if for every path $P \in S$, connecting some terminals s and t, there are no two non-consecutive vertices $u, v \in V(P)$ such that $\{u, v\} \in E(G)$ except for possibly s and t.

Observation 5. *Let (G, T) be an instance for the Min-Sum Disjoint Paths* problem, and let S be a solution to this problem on (G, T). Then, S is also a minimal solution to the Disjoint Paths* problem on (G, T).*

We claim that the following statement holds.

Lemma 5 (#). *Let (G, T) be an instance for the Disjoint Paths* problem where G is a split graph. Then, any minimal solution to this problem contains only paths with at most four vertices each.*

We conclude the subsection with its main theorem:

Theorem 6 (#). *There exists an algorithm solving Min-Sum Disjoint Paths on split graphs in $\mathcal{O}(2^k \cdot (k^2 \cdot n + k \cdot m)))$ time.*

4   Well-Partitioned Chordal Graphs

In this section, we present two algorithms, solving Disjoint Shortest Paths and Min-Sum Disjoint Paths on well-partitioned chordal graphs. Recall that such graphs can be decomposed using a partition-tree, and that this representation can be found in polynomial time. Our algorithms exploit this structure to generalize the idea presented in the previous section for split graphs and use a similar reduction to the problem of computing a maximum matching in bipartite graphs.

4.1   Polynomial-Time Algorithm for Disjoint Shortest Paths

In this subsection, we present a polynomial time algorithm solving the DISJOINT SHORTEST PATHS problem on well-partitioned chordal graphs.

First, let us recall Observation 2 from Subsect. 3.1, stating that if there exists a pair of terminals that are connected with an edge, then this edge appears as a path in any solution to DISJOINT SHORTEST PATHS.

Hence, we can get rid of such a pair of terminals if it appears only once in the multiset T, or output "No" otherwise. More formally:

Reduction Rule 4. *Let (G, T) be an instance of the* DISJOINT SHORTEST PATHS *such that G is a well-partitioned chordal graph. Let $(s, t) \in T$ be a terminal pair such that $\{s, t\} \in E(G)$. If (s, t) appears more than once in T, then output "No". Otherwise, remove this pair from the set T. If either of s or t does not appear as a part of some pair in the set T anymore, remove it from G as well.*

The reason why we need this reduction rule is as follows. Later, in the construction of the bipartite graph, we will ignore the terminals and consider only the possibilities for internal vertices. Thereby, we can accidentally create duplicate paths (in our solution) for shortest paths having only one edge. Since nonterminal vertices are not ignored, pairs whose distance is strictly greater than two are not vulnerable to this.

Similarly to Subsect. 3.1, by removing terminals from the graph this reduction rule might change the distances between the remaining pairs of terminals. But if this happens, then we can output "No".

Reduction Rule 5. *After applying Reduction Rule 4 exhaustively, if there exists a pair of terminals in the reduced instance whose distance is different from their distance in the original instance, output "No".*

This rule is applied exactly once, and can be computed in $\mathcal{O}(k \cdot (n + m))$ time by performing a breadth-first-search (BFS) twice for each pair of terminals, once on the original instance and once on the reduced instance.

Observation 7 (#). *Let (G, T) be the original instance for* DISJOINT SHORTEST PATHS, *and $(G^\star, T^\star)$ be the one obtained by exhaustively applying Reduction Rule 4. If Reduction Rule 5 does not output "No", then $(G^\star, T^\star)$ is an equivalent instance to* DISJOINT SHORTEST PATHS. *Otherwise, (G, T) is a "No" instance.*

Now, we describe in more detail the reduction to the problem of computing a maximum matching on bipartite graphs. First, we compute the lengths of shortest paths between the pairs of terminals (by performing a BFS for each pair).

Next, for the given instance (G, T), we construct a bipartite graph $\mathcal{B}_{(G,T)}$. Informally, it is defined as follows. First, denote the "left" side of $\mathcal{B}_{(G,T)}$ as L and the "right" side as R. Intuitively, the vertices of L represent pairs of numbers (i, j) that are connected to all the options for the j-th vertex in a shortest path connecting terminal pair $(s_i, t_i) \in T$. Since terminal vertices may appear as endpoints in several different paths and are always the endpoints of their corresponding paths in the solution, we will not consider them in the constructed bipartite graph. In other words, the vertices of L will represent the positions of only the internal vertices in the solution, and R will not contain terminals at all since an internal vertex in the solution cannot be a terminal.

Formally, we define the graph $\mathcal{B}_{(G,T)}$ as follows (see Fig. 1):

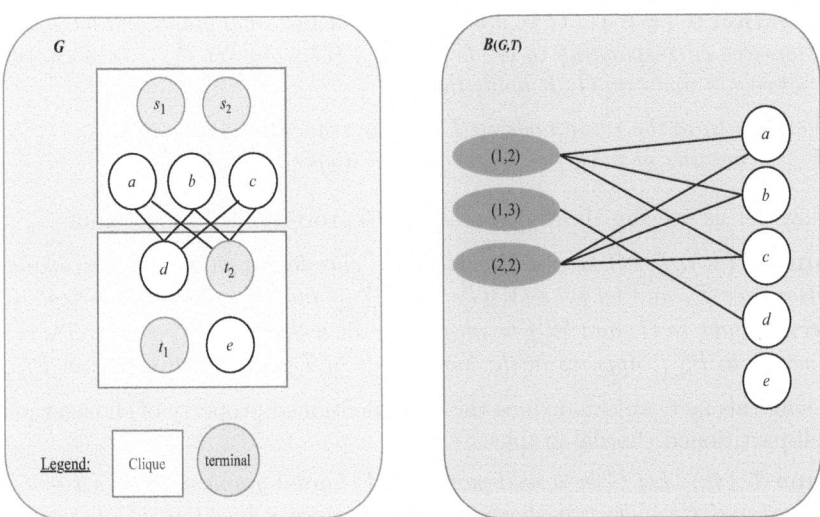

Fig. 1. Example of the construction of the bipartite graph.

Definition 8. Let (G, T) be an instance for the DISJOINT SHORTEST PATHS problem, such that G is a well-partitioned chordal graph. We define the bipartite graph $\mathcal{B}_{(G,T)}$ as follows.

- The "left" side of the bipartite graph: $L = \{(i, j) : 1 \leq i \leq k, 1 < j < d(s_i, t_i)\}$.
- The "right" side of the bipartite graph: $R = V(G) \setminus \bigcup T$.
- Now, let us define the set of edges of the bipartite graph. Each $(i, j) \in L$ is connected to all the vertices $v \in R$ such that there exists a shortest path P between s_i and t_i having v as its j-th vertex.

As we will see (in Lemma 7), the partition-tree structure of a well-partitioned chordal graph yields the following property: for every j, all the "options" for the j-th vertex in a shortest path between two terminals are connected to all the options for the $(j + 1)$-th vertex in such path. Intuitively, this property allows us to examine separately the options for each vertex in a shortest path between a pair of terminals. If we can match each "position" in an optimal solution to a unique option for this "position",[6] then we have a solution to DISJOINT SHORTEST PATHS, and vice versa.

To prove it formally, let us first observe the following property.

---
[6] Except for possibly the endpoints, since terminals can overlap.

Observation 8. *Let G be a well-partitioned chordal graph with a corresponding partition-tree $\mathcal{T}$. Let $s, t \in V(G)$, and let P be the path between the node containing s and the node containing t in $\mathcal{T}$. Denote $X = \bigcup_{b \in V(P)} b$, the union of the nodes along P. Let $P_{(s,t)}$ be a chordless path between s and t in G. Then, $V(P_{(s,t)}) \subseteq X$.*

Next, let us prove the following observation.

Observation 9 (#). *Let G be a well-partitioned chordal graph and let $\mathcal{T}$ be the partition-tree corresponding to G. Let $s, t \in V(G)$, and let $P_{(s,t)}$ be a chordless path between s and t in G. It holds that:*

1. *Vertices from the same node in $\mathcal{T}$ appear sequentially along $P_{(s,t)}$.*
2. *$P_{(s,t)}$ contains at most two vertices from a specific node of $\mathcal{T}$.*

Now, let us use the above observations to prove the following lemma.

Lemma 6 (#). *Let G be a well-partitioned chordal graph with a corresponding partition-tree $\mathcal{T}$, and let $s, t \in V(G)$. Let $P^1_{(s,t)}$ and $P^2_{(s,t)}$ be two shortest paths between s and t in G, and let j be an integer such that $1 \leq j \leq d(s,t)$. Then, the j-th vertex in $P^1_{(s,t)}$ appears in the same node of $\mathcal{T}$ as the j-th vertex in $P^2_{(s,t)}$.*

Using this fact, we can deduce the aforementioned property of shortest paths in well-partitioned chordal graphs:

Lemma 7 (#). *Let G be a well-partitioned chordal graph, and let $s, t \in V(G)$. Let $P^1_{(s,t)}$ and $P^2_{(s,t)}$ be two shortest paths between s and t in G. Let j be an integer such that $1 \leq j < d(s,t)$. Then, the j-th vertex in $P^1_{(s,t)}$ is connected to the $(j+1)$-th vertex in $P^2_{(s,t)}$.*

Now, let us prove the central lemma on which our algorithm relies.

Lemma 8 (#). *Let (G, T) be an instance for the DISJOINT SHORTEST PATHS problem such that G is a well-partitioned chordal graph and for any $(s, t) \in T$ it holds that $d(s, t) > 2$. Then, (G, T) admits a solution for the DISJOINT SHORTEST PATHS problem if and only if the bipartite graph $\mathcal{B}_{(G,T)}$ admits a maximum matching that matches all the vertices of L.*

Finally, let us present and analyze our algorithm. It is worth to mention that for the sake of maximum bipartite matching computation we use the algorithm given by Hopcroft et al. [22] rather than the one presented in Lemma 3, since it has a better running time bound in this case.

Theorem 10 (#). *There is an $\mathcal{O}(k \cdot (n^{1.5} + m))$ time algorithm solving the DISJOINT SHORTEST PATHS problem on well-partitioned chordal graphs.*

4.2   FPT Algorithm for Min-Sum Disjoint Paths

In this subsection, we present an $\mathcal{O}(2^k \cdot (k^3 \cdot n + k^2 \cdot m))$ time algorithm solving the MIN-SUM DISJOINT PATHS problem on well-partitioned chordal graphs. However, here we assume that the given multiset T of the pairs is a set. In other words, any pair of terminals can appear at most once in T. The reason we need this assumption is because now it is guaranteed that any path in the solution is chordless (see Observation 11), so the properties from Observations 8 and 9 from the previous subsection, hold for the paths in a solution to MIN-SUM DISJOINT PATHS as well. We note that the algorithm in this subsection relies on the one presented in Subsect. 3.2, solving the problem on split graphs. First, we define a subset of the possible solutions for the DISJOINT PATHS problem.

Definition 9. Let (G, T) be an instance for the DISJOINT PATHS problem, and let S be a solution to this problem. Then, S is a *minimal solution* if for every path $P \in S$ connecting some terminals s and t, there are no two non-consecutive vertices $u, v \in V(P)$ such that $\{u, v\} \in E(G)$.

Please note that this definition for minimal solution is slightly different from the one presented in Subsect. 3.2. Here, we have an additional restriction that the endpoints of paths of length at least 3 in the solution cannot be connected with an edge. Since we assume that our set of terminals does not contain duplicates of the same pair, one can observe that any solution to MIN-SUM DISJOINT PATHS (under this assumption) is a minimal solution according to the new definition.

Observation 11. *Let (G, T) be an instance for the MIN-SUM DISJOINT PATHS problem such that each pair in T appears only once, and let S be a solution to this problem on (G, T). Then, S is also a minimal solution to the DISJOINT PATHS problem on (G, T).*

We stress that Observation 11 is correct because T is a set. Indeed, if T was not a set, a pair of terminals with distance 2 (i.e., the terminals are connected with an edge) that appears more than once in T will be associated with at least one path of length at least 3 in the solution, but the edge between these terminals will be a chord in this path as opposed to the definition of a minimal solution. Intuitively, the only situation where a path in the solution has a chord is if this chord prevents this path from being identical to some other path in the solution. However, this might happen only if the paths share both endpoints, since otherwise the paths are necessarily different. With the restriction that T is a set, any chord in a path in a solution yields a contradiction to the minimality of the solution since this chord can be used to shorten this path.

Now, we can deduce that Observations 8 and 9 from the previous subsection hold for the paths in a solution to MIN-SUM DISJOINT PATHS as well. These observations lead to the key insight of our algorithm. Keeping in mind that the graph induced by the edges between two adjacent bags in a partition-tree is a complete bipartite graph, and using the fact that we know exactly what are the bags from the partition-tree used by each path in the solution,[7] we can solve the problem independently for each node in the partition-tree. Roughly speaking, according to Observation 9 we know that each path in the solution uses at most two vertices from each of the relevant bags in the partition-tree: one for entering this bag (or the "starting" terminal), and one for leaving it (or the "ending"

---

[7] According to Observation 8, these are the bags along the path in the partition-tree connecting the bags that contain the relevant terminals.

terminal), and these two vertices can be the same. For each node in the partition-tree, we solve the problem independently by constructing a split graph whose clique is the bag corresponding to the node and the vertices in the independent set represent the next and previous nodes for each pair of terminals, and then solving MIN-SUM DISJOINT PATHS on each of the constructed split graphs. An example of this construction will appear in a figure in the full version of this paper.

Now, let us present and analyze our algorithm formally.

Theorem 12 (#). *Let (G,T) be an instance for* MIN-SUM DISJOINT PATHS *where G is a well-partitioned chordal graph and any pair of terminals in T appears exactly once. Given the partition tree $\mathcal{T}$ of G, there is an $\mathcal{O}(2^k \cdot (k^3 \cdot n + k^2 \cdot m))$ time algorithm solving the* MIN-SUM DISJOINT PATHS *problem on (G,T).*

References

1. Frank, A.: Packing paths, circuits, and cuts-a survey. Paths Flows VLSI-Layout **49**, 100 (1990)
2. Nishizeki, T., Vygen, J., Zhou, X.: The edge-disjoint paths problem is NP-complete for series-parallel graphs. Discret. Appl. Math. **115**(1–3), 177–186 (2001)
3. Schrijver, A.: Combinatorial Optimization: Polyhedra and Efficiency, vol. 24. Springer (2003)
4. Srinivas, A., Modiano, E.: Finding minimum energy disjoint paths in wireless ad-hoc networks. Wireless Netw. **11**, 401–417 (2005)
5. Robertson, N., Seymour, P.D.: Graph minors XIII. The disjoint paths problem. J. Combin. Theory Ser. B **63**(1), 65–110 (1995)
6. Lokshtanov, D., Saurabh, S., Zehavi, M.: Efficient graph minors theory and parameterized algorithms for (planar) disjoint paths. In: Treewidth, Kernels, and Algorithms - Essays Dedicated to Hans L. Bodlaender on the Occasion of His 60th Birthday, pp. 112–128 (2020)
7. Kawarabayashi, K., Kobayashi, Y., Reed, B.: The disjoint paths problem in quadratic time. J. Combin. Theory Ser. B **102**(2), 424–435 (2012)
8. Korhonen, T., Pilipczuk, M., Stamoulis, G.: Minor containment and disjoint paths in almost-linear time. In: Accepted to Proceedings of the 65th IEEE Symposium on Foundations of Computer Science (FOCS), FOCS 2024 (2024)
9. Cho, K., Oh, E., Oh, S.: Parameterized algorithm for the disjoint path problem on planar graphs: exponential in k^2 and linear in n. In: Proceedings of the 2023 ACM-SIAM Symposium on Discrete Algorithms, SODA 2023, Florence, Italy, 22–25 January 2023, pp. 3734–3758 (2023)
10. Wlodarczyk, M., Zehavi, M.: Planar disjoint paths, treewidth, and kernels. In: 64th IEEE Annual Symposium on Foundations of Computer Science, FOCS 2023, Santa Cruz, CA, USA, 6–9 November 2023, pp. 649–662 (2023)
11. Kammer, F., Tholey, T.: The k-disjoint paths problem on chordal graphs. In: Graph-Theoretic Concepts in Computer Science, 35th International Workshop, WG 2009, Montpellier, France, 24–26 June 2009, Revised Papers, pp. 190–201 (2009)

12. Heggernes, P., Hof, P.V., Leeuwen, E.J.V., Saei, R.: Finding disjoint paths in split graphs. Theory Comput. Syst. **57**(1), 140–159 (2015)
13. Yang, Y., Shrestha, Y.R., Li, W., Guo, J.: On the kernelization of split graph problems. Theoret. Comput. Sci. **734**, 72–82 (2018)
14. Ahn, J., Jaffke, L., Kwon, O., Lima, P.T.: Well-partitioned chordal graphs. Discret. Math. **345**(10), 112985 (2022)
15. Chaudhary, J., Gahlawat, H., Wodarczyk, M., Zehavi, M.: Kernels for the disjoint paths problem on subclasses of chordal graphs (2023)
16. Lochet, W.: A polynomial time algorithm for the k-disjoint shortest paths problem. In: Proceedings of the 2021 ACM-SIAM Symposium on Discrete Algorithms, SODA 2021, Virtual Conference, 10–13 January 2021, pp. 169–178 (2021)
17. Bentert, M., Nichterlein, A., Renken, M., Zschoche, P.: Using a geometric lens to find-disjoint shortest paths. SIAM J. Discret. Math. **37**(3), 1674–1703 (2023)
18. Mari, M., Mukherjee, A., Pilipczuk, M., Sankowski, P.: Shortest disjoint paths on a grid. In: Proceedings of the 2024 ACM-SIAM Symposium on Discrete Algorithms, SODA 2024, Alexandria, VA, USA, 7–10 January 2024, pp. 346–365 (2024)
19. Bjorklund, A., Husfeldt, T.: Shortest two disjoint paths in polynomial time. SIAM J. Comput. **48**(6), 1698–1710 (2019)
20. Hammer, P.L., Simeone, B.: The splittance of a graph. Combinatorica **1**, 275–284 (1981)
21. Ahn, J., Jaffke, L., Kwon, O.J., Lima, P.T.: Well-partitioned chordal graphs. Discret. Math. **345**(10), 112985 (2022)
22. Hopcroft, J.E., Karp, R.M.: A $n^{5/2}$ algorithm for maximum matchings in bipartite graphs. SIAM J. Comput. **2**, 225–231 (1971)

Maximize the Rightmost Digit:Gray Codes for Restricted Growth Strings

Yuan Qiu[1], Joe Sawada[2], and Aaron Williams[1(⊠)] (iD)

[1] Williams College,Williamstown, MA 01267, USA
{yq1,aaron.williams}@williams.edu
[2] University of Guelph,Guelph, ON N1G 2W1, Canada
jsawada@uoguelph.edu

Abstract. The term *restricted growth string* typically refers to strings of non-negative integers $a_1 a_2 \cdots a_n$ (with $a_1 = 0$) in which the next symbol is at most one more than the maximum of the previous symbols: $0 \leq a_i \leq \max(a_1 \cdots a_{i-1}) + 1$ for $2 \leq i \leq n$. These strings are counted by the Bell numbers $\mathcal{B}_n$ (OEIS A000110) and encode set partitions. Kerr showed that the following algorithm generates a Gray code starting from 0^n: greedily maximize the rightmost possible digit that creates a new string. For example, the result is 000, 001, 011, 012, 010 for $n = 3$; the last transition causes the rightmost digit to decrease to 0 because that is the largest value for that digit that creates a new string. Kerr's algorithm is a special case of more general results for **e**-restricted and **st**-restricted strings by Mansour and Vajnovszki (and Nassar), although those authors did not describe their results greedily. We show that the same greedy max-right algorithm generates restricted growth strings parameterized by $(s, f, \mathbf{c})$: $0 \leq a_1 \leq s-1$ and $0 \leq a_i \leq f(a_1 a_2 \cdots a_{i-1}) + c_i$ where f is any function with $f \geq 0$ and $c_i \geq 1$ are constants for each digit. The resulting Gray codes change a single digit by -1 or -2 (cyclically). Special cases include the binary reflected Gray code ($s = 2$, $f = 0$, $\mathbf{c} = 1^n$) and the aforementioned results. We also consider restricted growth string counted by the k-Catalan numbers and provide loopless algorithms for generating these k-Catalan strings and Bell strings.

Keywords: restricted growth strings · Bell numbers · set partitions · Catalan numbers · k-Catalan numbers · Gray codes · greedy Gray codes

1 Introduction

This paper efficiently orders and generates restricted growth strings. We first describe two common types of restricted growth strings and their significance.

1.1 Bell and Catalan Strings

The term *restricted growth string* is often defined as a string of integers (called *digits*) of the form $a_1 a_2 \cdots a_n$ that satisfies the following conditions,

$$a_1 = 0 \text{ and } 0 \leq a_i \leq max(a_1 a_2 \cdots a_{i-1}) + 1 \text{ for } 2 \leq i \leq n. \tag{1}$$

S. Nakano and M. Xiao (Eds.): WALCOM 2025, LNCS 15411, pp. 296–311, 2025.
https://doi.org/10.1007/978-981-96-2845-2_19

In other words, the first digit is 0, and each subsequent digit is at least 0 and at most one more than the maximum of the previous digits. For example, 0102 is a restricted growth string of length $n = 4$, but 0103 is not. The number of restricted growth strings of length $n \geq 0$ is the nth Bell number $\mathcal{B}_n$ (OEIS A000110): 1, 1, 2, 5, 15, 52, 203, 877, 4140, 21147, 115975, ...

Since they are enumerated by the Bell numbers, we refer to this type of restricted growth string as *Bell strings*. Bell strings provide a convenient representation for the set partitions of $[n] = \{1, 2, \ldots, n\}$, which are also Bell objects. The standard bijection puts i into the a_i^{th} part, as shown below for $n \leq 3$ [36].

0	00	01	000	001	010	011	012
{1}	{1,2}	{1},{2}	{1,2,3}	{1,2},{3}	{1,3},{2}	{1},{2,3}	{1},{2},{3}

Note that a small change in a set partition can lead to a large change in its Bell string. For example, the set partition $\{1, 2\}, \{3\}, \{4\}, \ldots, \{n\}$ corresponds to the Bell string $00123 \cdots (n-2)$. If we move the 2 into its own subset to create the set partition of singletons, then the corresponding Bell string becomes $0123 \cdots (n-1)$ (i.e., all digits change except the leading 0). On the other hand, changing a single digit in a Bell string always corresponds to moving a single value in its set partition. For this reason, when designing efficient orders of set partitions it can be preferable to instead work with Bell strings.

Perhaps the most well-known ordering of set partitions was created by Knuth and presented by Kaye [15]. Later work by Ruskey and Savage [29] adapted the approach to Bell strings. A student project by Kerr [16] provided an alternate ordering of Bell strings (see [26]) that uses a greedy approach [40]. This paper generalizes Kerr's result from Bell strings to other restricted growth strings.

Another type of restricted growth string is obtained by modifying (1),

$$a_1 = 0 \text{ and } 0 \leq a_i \leq a_{i-1} + 1 \text{ for } 2 \leq i \leq n. \tag{2}$$

Here the bound on a_i uses the previous digit a_{i-1} rather than all previous digits. For example, 0102 is not valid since (2) is not satisfied for $i = 4$ as $2 > 0 + 1$. These *Catalan strings* of length $n \geq 0$ are counted by the nth Catalan number $\mathcal{C}_n$ (OEIS A000108): 1, 1, 2, 5, 14, 42, 132, 429, 1430, 4862, 16796, 58786, ...

We let $\mathbf{B}(n)$ and $\mathbf{C}(n)$ be the sets of Bell and Catalan strings of length n, respectively. Figure 3 has lists of $\mathbf{B}(n)$ and $\mathbf{C}(n)$ for $n \leq 5$. In particular, the reader can confirm that $\mathbf{C}(n) \subseteq \mathbf{B}(n)$ and in particular $\mathbf{B}(4) \setminus \mathbf{C}(4) = \{0102\}$.

Catalan strings provide an alternate representation for the large number of other Catalan objects counted by $\mathcal{C}_n$ [37]. We will also provide a simple generalization to k-Catalan strings $\mathbf{C}_k(n)$ in Sect. 2. There are several other types of strings counted by Catalan and k-Catalan numbers (e.g., see [41,42]).

1.2   Generalized Restricted Growth Strings

Although the term restricted growth string often refers specifically to Bell strings, it is also used much more broadly in the literature. Here we consider a generalization that allows for flexibility in the first digit, the function applied to the previous digits, and the constant added to each digit. Formally, an $(s, f, \mathbf{c})$-*restricted growth string* is a string of integers of the form $a_1 a_2 \cdots a_n$ satisfying

$$0 \leq a_1 \leq s - 1 \text{ and } 0 \leq a_i \leq f(a_1 a_2 \cdots a_{i-1}) + c_i \text{ for } 2 \leq i \leq n \qquad (3)$$

with $s \geq 1$, $f \geq 0$, and $c_i \geq 1$. In other words, the *starting digit* a_1 has s possible values $0 \leq a_1 \leq s$. Then each subsequent digit a_i is a non-negative integer limited by the sum of a *function* f that maps the previous digits $a_1 a_2 \cdots a_{i-1}$ to a non-negative integer and a positive integer *constant* c_i that depends only on the index i. (For notational convenience, we write $c = w$ and $f = w$ when $c_i = w$ and $f(a_1 a_2 \cdots a_{i-1}) = w$ for all $2 \leq i \leq n$, respectively.)

Our generalization captures a wide variety of previously studied strings as seen in Table 1. In particular, st-restricted strings are considered by Mansour and Vajnovszki [20] and Sabri and Vajnovszki [30]. These strings start with $a_1 = 0$ and then bound each successive digit by a *prefix statistic* (e.g., number of ascents): $0 \leq a_i \leq \mathsf{st}(a_1 a_2 \cdots a_{i-1}) + 1$. By carefully tailoring the statistic they can also model our $(s, f, \mathbf{c})$-restricted growth strings. Both [19] and [20] use the greedy max-right strategy discussed in Sect. 3, although they do not observe this interpretation. For example, $\mathsf{succ}_{1,m}$ and $\mathsf{succ}_{2,m}$ in [19] mirror our g_0 and g_1 expansions (see Sect. 4).

Table 1. Types of $(s, f, \mathbf{c})$-restricted growth strings. Note that names differ across the literature (e.g., [1] uses (e) max-increment, (i) increment-i, (k) K-increment).

Type	Start s	Function $f(a_1 a_2 \cdots a_{i-1})$	Constant c_i	References
(a) binary strings	2	0	1	
(b) k-ary strings	k	0	$k - 1$	
(c) mixed-radix strings	b_1	0	$b_i - 1$	
(d) Bell strings	1	$\max(a_1, a_2, \ldots, a_{i-1})$	1	
(e) RGS of order d	1	$\max(a_1, a_2, \ldots, a_{i-1})$	d	[19]
(f) e-restricted growth functions	1	$\max(a_1, a_2, \ldots, a_{i-1})$	e_i	[19]
(g) restricted growth tails	k	$\max(a_1, a_2, \ldots, a_{i-1}, k)$	1	[29]
(h) Catalan strings	1	a_{i-1}	1	
(i) k-Catalan strings	1	a_{i-1}	$k - 1$	
(j) ascent sequences	1	$|\{j \mid 2 \leq j < i, a_{j-1} < a_j\}|$	1	[3,20]
(k) subexcedent sequences	1	i	0	[11,20]
(l) st-restricted strings	1	$\mathsf{st}(a_1 a_2 \cdots a_{i-1})$	1	[20,30]

1.3   Goals and Results

We are interested in creating Gray codes for restricted growth strings. That is, we want to list these sets so that consecutive strings differ in a small constant amount. Furthermore, we want to generate these lists efficiently. In this context, *constant amortized time (CAT)* and *loopless* algorithms generate successive strings in amortized and worst-case $O(1)$-time, respectively.

An initial roadblock is that Bell strings do not have a ± 1 Gray code when $n \equiv 4, 6, 7, 9 \pmod{12}$ [10, 28]. In other words, it is not possible to order the strings in an arbitrary $\mathbf{B}(n)$ so that consecutive strings differ in only one digit and only by ± 1. However, Ehrlich [10] constructed a Gray code for $\mathbf{B}(n)$ in which a single digit changes by ± 1 when considered cyclically[1] and provided a loopless implementation. On the other hand, Ruskey [28] created a CAT algorithm that allows ± 1 and ± 2 non-cyclicallyLi and Sawada provided a Gray code for $\mathbf{B}(n)$ as part of their *reflectable languages* framework [18], and their special values $x = 0$ and $y = 1$ arise naturally in our results.

Our goal is to present an approach to generating restricted growth string Gray codes with the following benefits:

(a) The approach is very easy to describe.
(b) The approach generalizes previous results.
(c) The approach works for all $(s, f, \mathbf{c})$-restricted growth strings.
(d) The approach leads to loopless generation algorithms.

We reach our goals using an approach that can be summarized in one sentence: **start a list with 0^n then repeatedly extend it to a new string by greedily changing the rightmost digit to the maximum possible value.** We refer to this approach as the *max-right algorithm*, and we note that "possible" depends on which type of string is being generated. As we will see, successive strings in the resulting *max-right orders* differ in a single digit by -1 or -2 (cyclically). In particular, the change from 0 to the maximum possible value is -1 taken cyclically, and 1 to the maximum value is -2 taken cyclically. Moreover, we provide loopless implementations and applications for $\mathbf{B}(n)$ and $\mathbf{C}_k(n)$. We also obtain the binary reflected Gray code for n-bit binary strings using $s = 2$, $f = 0$, and $c = 1$.

1.4   Outline

Sect. 2 discusses k-Catalan strings and proves that they are an example of $(s, f, \mathbf{c})$-restricted growth strings. Section 3 discusses Gray codes and combinatorial generation. Section 4 provides our Gray codes for $(s, f, \mathbf{c})$-restricted growth strings. Section 5 provides new loopless algorithms for mixed-radix, k-Catalan, and Bell strings. An online version of the paper includes appendices with additional figures and Python code.

2 k-Catalan Strings

In this section we provide a natural generalization of Catalan strings. Our generalization replaces the $+1$ in (2) with $+(k-1)$ (for any fixed $k \geq 2$) as follows:

$$a_1 = 0 \text{ and } 0 \leq a_i \leq a_{i-1} + (k-1) \text{ for } 2 \leq i \leq n. \tag{4}$$

---

[1] Ehrlich uses a flexible notion of cyclic ± 1: $a_i = 0 \leftrightarrow a_i = m$ and $a_i = 0 \leftrightarrow a_i = m+1$ are allowed when $m = max(a_1 a_2 \cdots a_{i-1})$. We would consider the former case as ± 2.

We refer to the resulting strings as k-*Catalan strings* and let $\mathbf{C}_k(n)$ be the set of length n. For example, when $n = 3$ and $k = 3$ we have the following set.

$$\mathbf{C}_3(3) = \{000, 001, 002, 010, 011, 012, 013, 020, 021, 022, 023, 024\}. \tag{5}$$

We prove that these sets are counted by the k-Catalan numbers $\mathcal{C}_{k,n}$. Other objects counted by $\mathcal{C}_{k,n}$ are found in OEIS sequences A000108, A001764, A002293, A002294, A002295 for $2 \leq k \leq 6$. For example, the $|\mathbf{C}_3(3)| = 12$ strings in (5) are in bijection with the ternary trees with 3 internal nodes (OEIS A001764).

Standard Catalan strings $\mathbf{C}(n)$ are obtained from (4) with $k = 2$, and Catalan numbers are also called 2-Catalan numbers (i.e., $\mathcal{C}_n = \mathcal{C}_{2,n}$). We prove that $\mathbf{C}_k(n)$ are an example of $(s, f, \mathbf{c})$-restricted growth strings (and st-restricted strings).

Theorem 1 ([1])[2]. $|\mathbf{C}_k(n)| = \mathcal{C}_{k,n}$ for all $n \geq 0$ and $k \geq 2$.

Proof. We prove that the members of $\mathbf{C}_k(n)$ are in bijective correspondence with the k-ary trees with n internal nodes, which are known to be counted by $\mathcal{C}_{k,n}$. The proof is by induction on n for a fixed k and is illustrated by Fig. 1. There is a single k-ary tree with one internal node and $\mathbf{C}_k(1) = \{0\}$, so the result is true for $n = 1$. Suppose the result holds for $n = t$. Now we extend the bijection to strings and trees with $n = t + 1$. By (4) each string in $\mathbf{C}_k(t)$ that ends with digit d is the prefix of $d + (k - 1)$ distinct strings in $\mathbf{C}_k(t + 1)$. Next consider a k-ary tree with t internal nodes and label them by a preorder traversal. Consider the location of the node x that is last in preorder; it is a leaf with label t. To grow this tree without changing the preorder traversal we can add a new leaf as a child of x or as a last child of any node on the path from the root to x that doesn't already have a kth child. Thus, if x had been added in the dth rightmost location, then the new node can added in $d + (k - 1)$ locations. So we can extend the bijection with the new node's position as a 0-based value. ∎

Theorem 2. *The set of k-Catalan strings $\mathbf{C}_k(n)$ are an example of $(s, f, \mathbf{c})$-restricted growth strings (as well as st-restricted strings).*

Proof. We claim this is true from $s = 1$, $f(a_1 a_2 \cdots a_{i-1}) = a_{i-1}$, and $c_i = k - 1$ for all $2 \leq i \leq n$. This follows from (3) as these choices force $0 \leq a_1 = s - 1 = 0$ (i.e., $a_1 = 0$) and the following bound for $2 \leq i \leq n$ that matches (4),

$$0 \leq a_i \leq f(a_1 a_2 \cdots a_{i-1}) + c_i = a_{i-1} + (k - 1). \tag{6}$$

Similarly, $\mathbf{C}_k(n)$ are st-restricted strings [20] using statistic $a_{i-1} + (k - 1)$. ∎

---

[2] This result was proven independently by the authors, however, a later literature review found that it was previously observed by Arndt [1] (Ch. 15.5). Arndt refers to k-Catalan strings as *i-increment RGS* and gives a bijection with k-ary Dyck words.

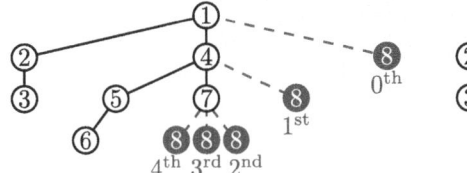 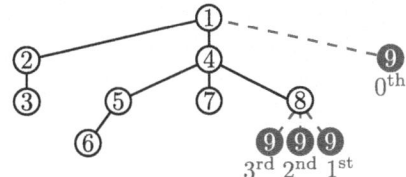

(a) The 3-ary tree with $n = 7$ nodes whose 3-ary Catalan string is $023135\underline{2}$. The location of node ⑦ is encoded as the last digit $\underline{2}$, so there are $\underline{2} + k = 5$ locations where a new leaf can be added and be last in preorder. Correspondingly, the digit d following $\underline{2}$ in a 3-ary Catalan string is one of the 5 values satisfying $0 \leq d \leq 4 = \underline{2} + (k - 1)$.

(b) The 3-ary tree with $n = 8$ nodes whose 3-ary Catalan string is $023135\underline{21}$. It is (a) with a leaf in the 1st position. So there are $\underline{1} + k = 4$ locations where a new leaf could be added and be last in preorder. Correspondingly, the digit d following $\underline{1}$ in a 3-ary Catalan string is one of the 4 values satisfying $0 \leq d \leq 3 = \underline{1} + (k - 1)$.

Fig. 1. The bijection between k-ary trees and k-Catalan words from Theorem 1.

3 Gray Codes and Combinatorial Generation

The term *Gray code* refers to an exhaustive list of some combinatorial object (parameterized by size) in which successive objects differ in some (small) way. They are named after the famous order of n-bit binary strings with Hamming distance one (i.e., a single bit's value is complemented or *flipped*) in Frank Gray's 1954 patent titled *Pulse Code Communication* [12]. The order is referred to as the *binary reflected Gray code (brgc)* and it appears below for $n = 3$, with overlines denoting the bit that changes to create the next string.

$$\mathbf{brgc}(3) = 00\overline{0}, 0\overline{0}1, 01\overline{1}, \overline{0}10, 11\overline{0}, 1\overline{1}1, 10\overline{1}, 100 \tag{7}$$

$$\mathbf{plain}(3) = 1\overleftrightarrow{23}, \overleftarrow{13}2, 3\overleftrightarrow{12}, \overrightarrow{32}1, 2\overrightarrow{31}, 213 \tag{8}$$

Plain changes predates the binary reflected Gray code by hundred years and is illustrated for $n = 3$ in (8). In this order, consecutive permutations of $[n] = \{1, 2, \ldots, n\}$ differ by a *swap* (i.e., adjacent entries are transposed) with the arrows in (8) showing a larger value "jumping over" a smaller value. The order was performed by bell-ringers in the 1600 s [38], and is known as the *Steinhaus-Johnson-Trotter algorithm* due to multiple rediscoveries in the mid-20th century.

Traditionally, Gray codes have been discovered and described recursively. For example, note that $\mathbf{brgc}(3)$ is obtained from two copies of $\mathbf{brgc}(2) = 00, 01, 11, 10$ by prefixing 0 to the strings in the first copy, and 1 to the strings in the second copy which is first *reflected* to $10, 11, 01, 00$. Similarly, $\mathbf{plain}(3)$ is obtained from $\mathbf{plain}(2) = 12, 21$ by sweeping 3 from right-to-left through 12 then left-to-right through 21. The first approach uses *global recursion* since $\mathbf{brgc}(n)$ is created from full copies of $\mathbf{brgc}(n-1)$, while the second uses *local recursion* since $\mathbf{plain}(n)$ expands individual objects in $\mathbf{plain}(n-1)$.

Countless Gray codes have been constructed over the years. Academic surveys have been written by Savage [31] and more recently Mütze [26], with Ruskey

[28] and Knuth [17] devoting extensive textbook coverage to the subject. In fact, one of the issues facing this research area is the sheer breadth of results and the recursive 'tricks' that have been used to obtain them. For an interactive introduction to the area, we recommend the *combinatorial object server* combos.org.

3.1 Greedy Gray Code Algorithm

This decade has seen the introduction of the *greedy Gray code algorithm* [40]. The algorithm eschews recursive schemes to focus on a simple idea: build an order one object at a time by prioritizing the possible changes. For example, $\mathbf{brgc}(n)$ can be constructed starting from 0^n (where exponentiation denotes concatenation) by greedily flipping the rightmost possible bit. Similarly, plain changes starts at $12\cdots n$ and then greedily swaps the largest possible valueTo clarify these descriptions, consider the partial orders below.

$$\mathbf{brgc}(3) = 00\overline{0}, 00\overline{1}, 01\overline{1}, 010, \ldots? \tag{9}$$

$$\mathbf{plain}(3) = 1\overleftarrow{2}3, 1\overleftarrow{3}2, 312, \ldots? \tag{10}$$

Which binary string should follow 010 in (9)? Flipping the rightmost bit gives $01\overline{0} = 011$ which is already in (9). Similarly, flipping the middle bit would repeat $0\overline{1}0 = 000$. But flipping the leftmost bit gives a new string $\overline{0}10 = 110$, so it is next in the order. In (10) we cannot swap 3 to the right as it would recreate $3\overrightarrow{1}2 = 132$, nor can it swap left as it is in the leftmost position. Thus, our highest priority change is to swap the next largest value 2 to the left to create $3\overleftarrow{1}2 = 321$.

These two greedy descriptions are not efficient in the sense of *combinatorial generation*, which is focused on efficiently generating exhaustive lists of combinatorial objects. More specifically, both algorithms require an exponential amount of space to determine if a specific change creates a new string or not. However, it is often possible to find an alternate description of a greedily defined order, such as the recursive descriptions of $\mathbf{brgc}(n)$ and $\mathbf{plain}(n)$ discussed earlier.

The greedy Gray code algorithm has also led to new results. In particular, the greedy description of plain change order was the impetus for the *permutation language* series [6,7,13,14,22], as well as new Gray codes for signed permutations [27] and Catalan objects [9]. Similarly, our new results generalize the binary reflected Gray code and other greedy generalizations of the 'original' Gray code include [24] and [23]. Greedy Gray code results also exist for de Bruijn sequences [21] and universal cycles [8,34], colored permutations [5], ballot sequences [39], and spanning trees [2,4]. Greedy Gray codes can often be translated into efficient history-free algorithms (c.f., [32,33]) but they typically do not produce sublist Gray codes (e.g., see [30,35]). The simplicity of the greedy approach belies the complexity of the general underlying problem [25].

3.2 Four Greedy Definitions of the Binary Reflected Gray Code

Here we provide four different greedy algorithms for generating the binary reflected Gray code starting from 0^n. The first approach was previously discussed, and it should be clear that the other three approaches produce identical results.

1. Greedily complement the rightmost bit.
2. Greedily increment or decrement the rightmost bit.
3. Greedily increment the rightmost bit cyclically modulo 2.
4. Greedily set the rightmost bit to the maximum possible value.

Fig. 2a illustrates the four interpretations for **brgc**(4). In the figure, we use the symbols ‾ for complement, ± 1 for increment / decrement, $\oplus$ for cyclic increment, and max for maximum possible value. While the four algorithms give the same order for binary strings, we will see that the last three produce different orders for other sets of strings; we henceforth ignore the first algorithm as complements cannot be applied to non-binary digits. Eventually, we will see that max approach has a particular advantage for restricted growth strings, as observed in [18, 19].

3.3 Three Greedy Gray Codes for Mixed-Radix Strings

Let $b_1, b_2, \cdots, b_n$ be a list of positive integers called *bases*. A *mixed-radix string* with these bases is any $a_1 a_2 \cdots a_n$ with $1 \leq a_i \leq b_i - 1$ for all i. In other words, each b_i provides the number of values that the ith digit can hold. Figure 2 illustrates how three of the greedy approaches mentioned in Sect. 3.2 generate Gray codes for the strings with bases $1, 2, 3, 4$. In each case, the reader's attention should be drawn to the different patterns created in the rightmost digit.

– When using increments and decrements (Fig. 2b) the rightmost digit ping-pongs back-and-forth: $0, 1, 2, 3, \ 3, 2, 1, 0, \ 0, 1, 2, 3, \ 3, 2, 1, 0, \ \ldots$ reflectively.
– When using cyclic increments (Fig. 2c) the rightmost digit's starting value climbs on each block $0, 1, 2, 3, \ 3, 0, 1, 2, \ 2, 3, 0, 1, \ 1, 2, 3, 0 \ 0, 1, 2, 3, \ 3, 0, 1, 2$.
– When using maximization (Fig. 2d) the rightmost digit alternately starts with 0 and ends with 1 or vice versa $0, 3, 2, 1, \ 1, 3, 2, 0, \ 0, 3, 2, 1, \ 1, 3, 2, 0, \ \ldots$.

The third pattern is quite useful in the context of restricted growth strings. This is because lower values are less likely to exceed their digit's upper bound, so having 0 and 1 as the first and last values (or vice versa) allows the greedy algorithm to uncover safer forms of recursion.

4 Main Result

Bell strings do not have ± 1 Gray codes (see Sect. 1.3), so greedily incrementing or decrementing the rightmost digit will not generate them. Similarly, greedily performing a cyclic increment of the rightmost possible digit does not work for $n \geq 7$ regardless of the start string. So of the greedy strategies discussed in Sect. 3.2, only maximizing the rightmost possible digit has the potential to generate all $(s, f, \mathbf{c})$-restricted growth strings. Now we prove that this is the case.

brgc(4)	$b_4b_3b_2b_1$	$^-$	$\pm$	$\oplus$	max
	0000	$\bar{1}$	$+_1$	$\oplus_1$	max_1
	0001	$\bar{2}$	$+_2$	$\oplus_2$	max_2
	0011	$\bar{1}$	$+_1$	$\oplus_1$	max_1
	0010	$\bar{3}$	$-_3$	$\oplus_3$	max_3
	0110	$\bar{1}$	$+_1$	$\oplus_1$	max_1
	0111	$\bar{2}$	$-_2$	$\oplus_2$	max_2
	0101	$\bar{1}$	$-_1$	$\oplus_1$	max_1
	0100	$\bar{4}$	$-_4$	$\oplus_4$	max_4
	1100	$\bar{1}$	$+_1$	$\oplus_1$	max_1
	1101	$\bar{2}$	$+_2$	$\oplus_2$	max_2
	1111	$\bar{1}$	$-_1$	$\oplus_1$	max_1
	1110	$\bar{3}$	$-_3$	$\oplus_3$	max_3
	1010	$\bar{1}$	$+_1$	$\oplus_1$	max_1
	1011	$\bar{2}$	$-_2$	$\oplus_2$	max_2
	1001	$\bar{1}$	$-_1$	$\oplus_1$	max_1
	1000				

(a) Greedily generating the binary reflected Gray code **brgc**(4) via complements ($^-$), increments/decrements ($\pm$), cyclic increments ($\oplus$), or digit maximizing (max). The complement operation specifies the bit index to change and is specific to binary strings.

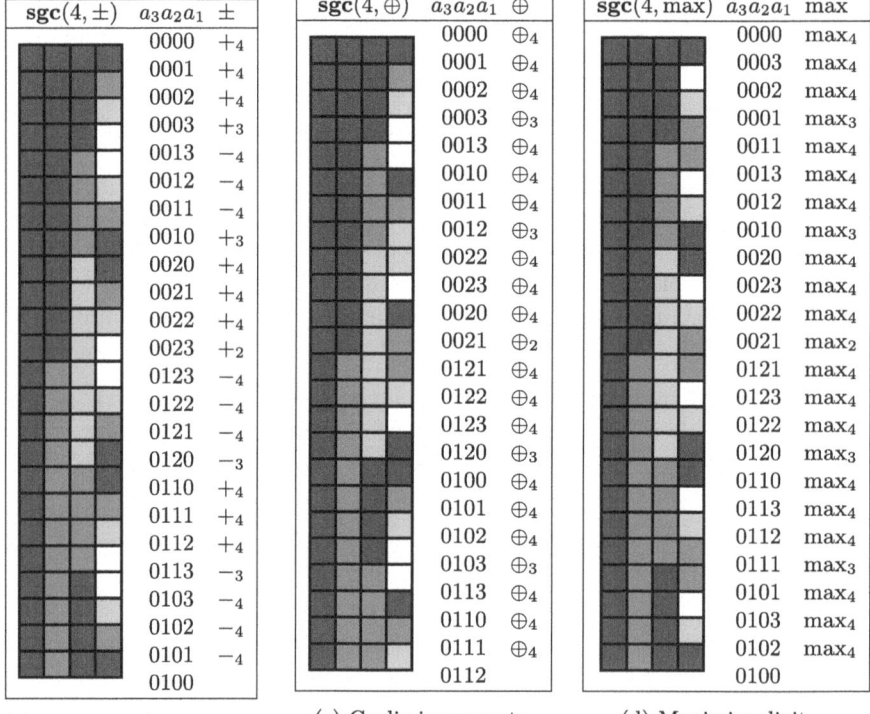

sgc(4, $\pm$)	$a_3a_2a_1$	$\pm$
	0000	$+_4$
	0001	$+_4$
	0002	$+_4$
	0003	$+_3$
	0013	$-_4$
	0012	$-_4$
	0011	$-_4$
	0010	$+_3$
	0020	$+_4$
	0021	$+_4$
	0022	$+_4$
	0023	$+_2$
	0123	$-_4$
	0122	$-_4$
	0121	$-_4$
	0120	$-_3$
	0110	$+_4$
	0111	$+_4$
	0112	$+_4$
	0113	$-_3$
	0103	$-_4$
	0102	$-_4$
	0101	$-_4$
	0100	

(b) Increment/decrement.

sgc(4, $\oplus$)	$a_3a_2a_1$	$\oplus$
	0000	$\oplus_4$
	0001	$\oplus_4$
	0002	$\oplus_4$
	0003	$\oplus_3$
	0013	$\oplus_4$
	0010	$\oplus_4$
	0011	$\oplus_4$
	0012	$\oplus_3$
	0022	$\oplus_4$
	0023	$\oplus_4$
	0020	$\oplus_4$
	0021	$\oplus_2$
	0121	$\oplus_4$
	0122	$\oplus_4$
	0123	$\oplus_4$
	0120	$\oplus_3$
	0100	$\oplus_4$
	0101	$\oplus_4$
	0102	$\oplus_4$
	0103	$\oplus_3$
	0113	$\oplus_4$
	0110	$\oplus_4$
	0111	$\oplus_4$
	0112	

(c) Cyclic increments.

sgc(4, max)	$a_3a_2a_1$	max
	0000	max_4
	0003	max_4
	0002	max_4
	0001	max_3
	0011	max_4
	0013	max_4
	0012	max_4
	0010	max_3
	0020	max_4
	0023	max_4
	0022	max_4
	0021	max_2
	0121	max_4
	0123	max_4
	0122	max_4
	0120	max_3
	0110	max_4
	0113	max_4
	0112	max_4
	0111	max_3
	0101	max_4
	0103	max_4
	0102	max_4
	0100	

(d) Maximize digit.

Fig. 2. (a) Four greedy interpretations of **brgc**(4). Each one greedily applies an operation (or operations) to the rightmost possible digit. Three of these greedy approaches also generate mixed-radix strings as seen for bases $1, 2, 3, 4$ in (b)–(d).

Theorem 3. *The greedy max-right algorithm starting from 0^n generates all $(s, f, \mathbf{c})$-restricted growth strings of length n, and successive strings differ by -1 or -2 in one digit where the subtractions are taken cyclically relative to (3).*

Proof. Recall from (3) that $a_1 a_2 \cdots a_n$ is an $(s, f, \mathbf{c})$-restricted growth string if

$$0 \leq a_1 \leq s - 1 \text{ and } 0 \leq a_i \leq f(a_1 a_2 \cdots a_{i-1}) + c_i \text{ for } 2 \leq i \leq n$$

with $s \geq 1$, $f \geq 0$, and $c \geq 1$. We prove the theorem by induction on $n \geq 1$.

For the base case of $n = 1$, notice that the conditions reduce to $0 \leq a_1 \leq s-1$. Therefore, the greedy max-right algorithm produces the list $0, s-1, s-2, \ldots, 1$.

Assume that the result holds for all valid choices and $n = k$. Now consider a specific choice of s, f, and $\mathbf{c}$ with $n = k + 1$. Let f' and $\mathbf{c}'$ be the restrictions of f and $\mathbf{c}$ to $n = k$, respectively. By induction, the greedy max-right algorithm creates a Gray code for the $(s, f', \mathbf{c}')$ strings of length k. Let this Gray code be $x_1, x_2, \ldots, x_p$ where p is the number of such strings. Now consider the greedy max-right algorithm for the $(s, f, \mathbf{c})$ strings of length $k + 1$. We claim that the algorithm will generate the strings in the following order,

$$g_0(x_1), \ g_1(x_2), \ g_0(x_1), \ g_1(x_2), \ \ldots, \ g_0(x_p) \text{ if } p \text{ is odd} \tag{11}$$
$$g_0(x_1), \ g_1(x_2), \ g_0(x_1), \ g_1(x_2), \ \ldots, \ g_1(x_p) \text{ if } p \text{ is even}$$

where the g_0 and g_1 functions expand each x_i string of length k into a sublist of strings of length $k + 1$ in a manner described below. Towards these definitions, let $x_i = a_1 a_2 \cdots a_k$. Therefore, $m = f(x_i) + c_{k+1}$ is the maximum value such that $x_i \cdot m$ is a $(s, f, \mathbf{c})$ string. We also know that $m \geq 1$ due to the conditions that $f \geq 0$ and $c \geq 1$. The two expansions of x_i are now defined as follows.

$$g_0(x_i) = x_i \cdot 0, \ x_i \cdot m, \ x_i \cdot (m-1), \ \ldots, \ x_i \cdot 2, \ x_i \cdot 1 \tag{12}$$
$$g_1(x_i) = x_i \cdot 1, \ x_i \cdot m, \ x_i \cdot (m-1), \ \ldots, \ x_i \cdot 2, \ x_i \cdot 0$$

In both cases, the expansion sets the last digit to the maximum value m and then repeatedly decrements it. The difference between the two expansions is that the last digit starts at 0 and ends at 1 in the g_0 expansion, and vice versa in the g_1 expansion. To complete the proof we need to argue the following points:

- The greedy max-right algorithm does indeed generate the list in (11).
- The list in (11) includes all $(s, f, \mathbf{c})$ strings of length $k + 1$.
- Successive strings in (11) differ in a single digit by -1 or -2 (cyclically).

To prove the first point, note that the greedy max-right algorithm prefers to change the rightmost digit to the maximum possible value that results in a new string. Therefore, if $x_i \cdot 0$ is the first string to be created with prefix x_i, then the algorithm will proceed by generating the list $g_0(x_i)$. Similarly, if $x_i \cdot 1$ is the first string to be created with prefix x_i, then the algorithm will proceed by generating the list $g_1(x_i)$. In both cases, all of the strings with prefix x_i are generated in succession. Therefore, when the expansion of x_i is completed, the algorithm will then set the rightmost possible digit in x_i to the maximum possible value. By induction, this means that the prefix x_i will be replaced by x_{i+1} by the next change. Finally, note that the sublist $g_0(x_i)$ ends with $x_i \cdot 1$, so the aforementioned change will result in $x_{i+1} \cdot 1$, which is the first string of $g_1(x_{i+1})$. Similarly, the sublist $g_1(x_i)$ ends with $x_i \cdot 0$, so the aforementioned change will result in $x_{i+1} \cdot 0$, which is the first string of $g_0(x_{i+1})$. Hence, the expanded sublists alternate as per (11). The second point follows from the fact that a digit's valid values are between 0 and m inclusively. The third point follows from (12) and induction. $\qquad\qquad\qquad\qquad\qquad\qquad\qquad\qquad\qquad\qquad\qquad\qquad$ $\square$

5   Loopless Algorithms

In this section we provide loopless algorithms for generating multi-radix strings, k-Catalan strings, and Bell strings according to our max-right Gray codes. This improves upon the excellent CAT implementations that follow from [20]. Here we generate the strings in reverse (i.e., right-to-left) to simplify the indexing. When the next string is ready we **yield** it and continue running. For each string, except the first, we also **yield** the index of the digit that was changed to create it.

5.1   Loopless Mixed-Radix Algorithms

The **MixedRadix** function in Algorithm 1 provides the well-known loopless algorithm for generating a mixed-radix Gray code using increment/decrement (i.e., ± 1) changes (see Knuth's description in [17]). The modified function **MixedRadixMax** in Algorithm 1 instead implements our mixed-radix Gray code using max changes. In this implementation, s_i keeps track of the starting value of the corresponding i-th digit: 0 or 1 (as per Sect. 4).

5.2   Loopless k-Catalan Strings

Our loopless implementation of our max-right k-Catalan Gray code is based on **MixedRadixMax**. One major differences is that the bases for each digit are not provided as inputs. Instead, they are computed as we generate the Gray code: the base of any position is the previous position's value plus $k - 1$.

Theorem 4. CatalanStrings(n) *in Algorithm 2 looplessly generates the max-right Gray code for k-Catalan strings of length n.*

5.3   Loopless Bell Strings

In our loopless implementation of the max-right Gray code for Bell strings, the concept of bases is not directly used, since computing the base of any digit (which is the maximum of previous digits plus 1) is a worst-case $\Theta(n)$ operation. Instead, we store the first positions of digits equal to successively larger values ≥ 2 (i.e., $2, 3, 4, \ldots$) in a stack **S**. If the stack is non-empty, then its size allows us

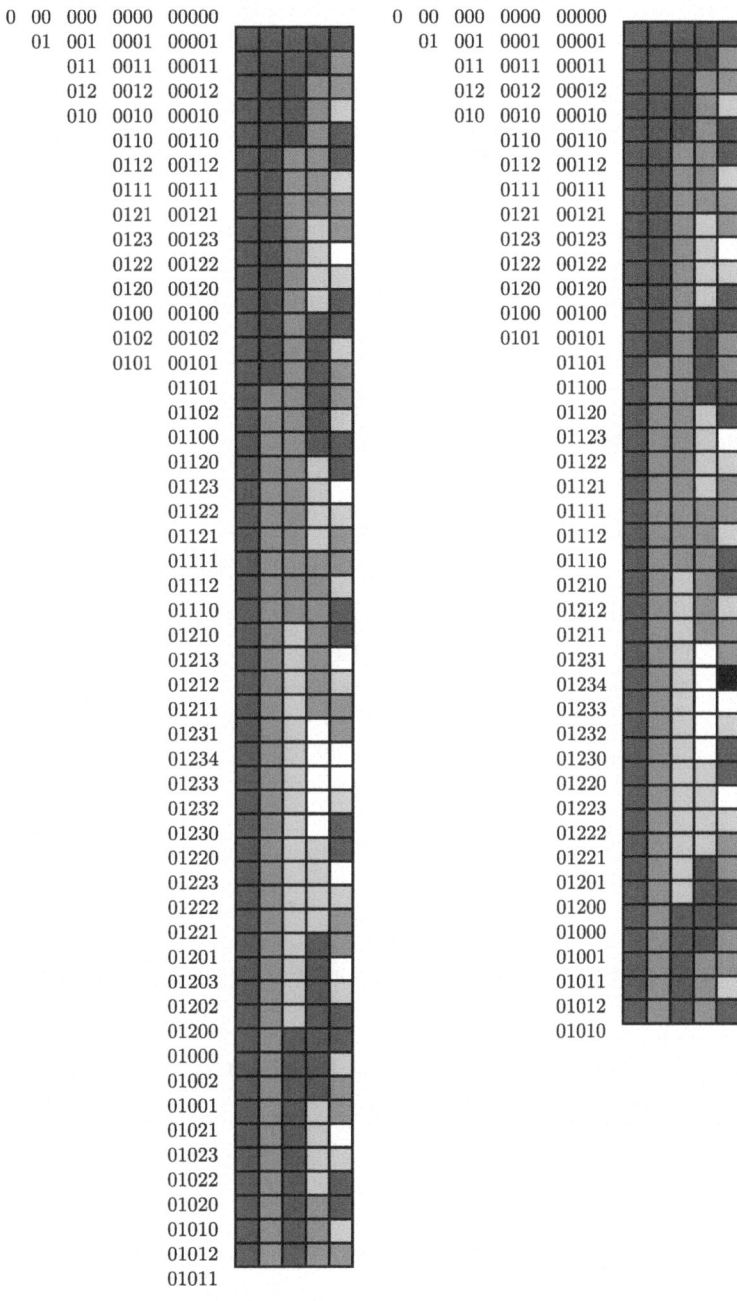

0	00	000	0000	00000
	01	001	0001	00001
		011	0011	00011
		012	0012	00012
		010	0010	00010
			0110	00110
			0112	00112
			0111	00111
			0121	00121
			0123	00123
			0122	00122
			0120	00120
			0100	00100
			0102	00102
			0101	00101
				01101
				01102
				01100
				01120
				01123
				01122
				01121
				01111
				01112
				01110
				01210
				01213
				01212
				01211
				01231
				01234
				01233
				01232
				01230
				01220
				01223
				01222
				01221
				01201
				01203
				01202
				01200
				01000
				01002
				01001
				01021
				01023
				01022
				01020
				01010
				01012
				01011

(a) Bell strings $\mathbf{B}(n)$ for $n \leq 5$.

0	00	000	0000	00000
	01	001	0001	00001
		011	0011	00011
		012	0012	00012
		010	0010	00010
			0110	00110
			0112	00112
			0111	00111
			0121	00121
			0123	00123
			0122	00122
			0120	00120
			0100	00100
			0101	00101
				01101
				01100
				01120
				01123
				01122
				01121
				01111
				01112
				01110
				01210
				01212
				01211
				01231
				01234
				01233
				01232
				01230
				01220
				01223
				01222
				01221
				01201
				01200
				01000
				01001
				01011
				01012
				01010

(b) Catalan strings $\mathbf{C}(n)$ for $n \leq 5$.

Fig. 3. Gray codes obtained from our max-right algorithm: start from 0^n then greedily maximize the rightmost possible digit. (Color figure online)

Algorithm 1. Loopless algorithms for generating mixed-radix Gray codes. The functions modify our target **a** and yield it every time it is modified. Focus pointers are stored in **f**. In **MixedRadix**, a_i is modified by +1 or -1 depending on the direction given by **d**. In **MixedRadixMax**, as discussed in Section 3.3, any position has 0 and 1 as the first and last value (or vice versa) in a loop. a_i is raised to maximum ($\mathbf{b_j} - \mathbf{1}$) when $a_i = s_i$ (except in some special cases), and is decreased otherwise (normally it decreases by 1, but decreases by 2 if it is 2 and the start value is 1, in this case it needs to become 0). The overall algorithms are loopless as each iteration runs in worst-case $\mathcal{O}(1)$-time.

MixedRadix(n)	MixedRadixMax(n)
1: $a_1 \, a_2 \, \cdots \, a_n \quad \leftarrow 0 \, 0 \, \cdots \, 0$	1: $a_1 \, a_2 \, \cdots \, a_n \quad \leftarrow 0 \, 0 \, \cdots \, 0$
2: $f_1 \, f_2 \, \cdots \, f_{n+1} \leftarrow 1 \, 2 \, \cdots \, n{+}1$	2: $f_1 \, f_2 \, \cdots \, f_{n+1} \leftarrow 1 \, 2 \, \cdots \, n{+}1$
3: $d_1 \, d_2 \, \cdots \, d_n \quad \leftarrow 1 \, 1 \, \cdots \, 1$	3: $s_1 \, s_2 \, \cdots \, s_n \quad \leftarrow 0 \, 0 \, \cdots \, 0$
4: **yield a**	4: **yield a**
5: **while** $f_1 \leq n$ **do**	5: **while** $f_1 \leq n$ **do**
6: $\quad j \leftarrow f_1$	6: $\quad j \leftarrow f_1$
7: $\quad f_1 \leftarrow 1$	7: $\quad f_1 \leftarrow 1$
8: $\quad a_j \leftarrow a_j + d_j$	8: $\quad$ **if** $a_j = s_j$ **then**
9: $\quad$ **yield** $j, \mathbf{a}$	9: $\quad\quad$ **if** $b_j = 2$ **and** $s_j = 1$ **then** $a_j \leftarrow 0$
10: $\quad$ **if** $a_j \in b_j - 1, 0$ **then**	10: $\quad\quad$ **else** $a_j \leftarrow b_j - 1$
11: $\quad\quad d_j \leftarrow -d_j$	11: $\quad$ **else if** $a_j = 2$ **and** $s_j = 1$ **then** $a_j \leftarrow 0$
12: $\quad\quad f_j \leftarrow f_{j+1}$	12: $\quad$ **else** $a_j \mathrel{-}= 1$
13: $\quad\quad f_{j+1} \leftarrow j + 1$	13: $\quad$ **yield** $j, \mathbf{a}$
	14: $\quad$ **if** $a_j = 1 - s_j$ **then**
	15: $\quad\quad s_j \leftarrow a_j$
	16: $\quad\quad f_j \leftarrow f_{j+1}$
	17: $\quad\quad f_{j+1} \leftarrow j + 1$

---

to determine a digit's maximum value. If the stack is empty, then the maximum is typically 2, since the earlier digits are comprised of 0s and 1s by (12). One exception is that these digits are all 0s precisely when the digit is being changed for the first time. To track this special case, we store whether or not a digit has ever been changed in a Boolean list **v**. Collectively, this additional information allows us to determine the maximum value for a digit in worst-case $\mathcal{O}(1)$-time.

Theorem 5. BellStrings(n) *in Algorithm 2 looplessly generates the max-right Gray code for Bell strings of length n.*

6 Final Remarks

We provided Gray codes for restricted growth strings parameterized by $(s, f, \mathbf{c})$. The orders change one digit by -1 or -2 (cyclically) and are generated from 0^n by a simple greedy rule. Our greedy max-right algorithms are not efficient, but the orders can be efficiently generated by other means. We showed this with loopless algorithms for mixed-radix strings, k-Catalan strings, and Bell strings.

Algorithm 2. Loopless algorithms for generating k-Catalan Gray codes and Bell Gray codes. The functions modify our target **a** and yield it every time it is modified. Focus pointers are stored in **f**. **CatalanStrings** largely replicates **MixedRadixMax**, except that the "bases" are calculated on the fly. In **BellStrings**, if the current digit is not visited, the maximum is set to 0 because all earlier digits are 0. If it is visited and the stack storing positions of large numbers is empty, the maximum is set to 1. If the stack is not empty, the maximum is set to the corresponding digit at the position determined by the top of stack. After calculating the maximum, it will be pushed onto the stack. When a digit is decreased, if it corresponds to the top of stack, the stack is popped.

CatalanStrings(n, k)	**BellStrings(n)**
1: $a_1 \, a_2 \, \cdots \, a_n \quad \leftarrow 0 \, 0 \, \cdots \, 0$	1: $a_1 \, a_2 \, \cdots \, a_n \quad \leftarrow 0 \, 0 \, \cdots \, 0$
2: $f_1 \, f_2 \, \cdots \, f_{n+1} \leftarrow 1 \, 2 \, \cdots \, n{+}1$	2: $f_1 \, f_2 \, \cdots \, f_{n+1} \leftarrow 1 \, 2 \, \cdots \, n{+}1$
3: $s_1 \, s_2 \, \cdots \, s_n \quad \leftarrow 0 \, 0 \, \cdots \, 0$	3: $s_1 \, s_2 \, \cdots \, s_n \quad \leftarrow 0 \, 0 \, \cdots \, 0$
4: **yield a**	4: $S \leftarrow empty$
5: **while** $f_1 < n$ **do**	5: $v_1 \, \cdots \, v_n \leftarrow$ **True** $\cdots$ **True**
6: $\quad j \leftarrow f_1$	6: **yield a**
7: $\quad f_1 \leftarrow 1$	7: **while** $f_1 < n$ **do**
8: $\quad$ **if** $a_j = s_j$ **then**	8: $\quad j \leftarrow f_1$
9: $\quad\quad$ **if** $a_j, a_{j+1}{=}1, 0$ **and** $k{=}2$ **then** $a_j{\leftarrow}0$	9: $\quad f_1 \leftarrow 1$
10: $\quad\quad$ **else** $a_j \leftarrow a_{j+1} + k - 1$	10: $\quad$ **if** $a_j = s_j$ **then**
	11: $\quad\quad$ **if** v_j **then** $m \leftarrow 0; \, v_j \leftarrow$ **False**
11: $\quad$ **else if** $a_j = 2$ **and** $s_j = 1$ **then** $a_j \leftarrow 0$	12: $\quad\quad$ **else if** S is empty **then** $m \leftarrow 1$
12: $\quad$ **else** $a_j \, {-}{=} \, 1$	13: $\quad\quad$ **else** $pos \leftarrow \textbf{top}(S); \, m \leftarrow a_{pos}$
13: $\quad$ **yield** $j, \textbf{a}$	14: $\quad$ **else if** $a_j = 2$ **and** $s_j = 1$ **then**
14: $\quad$ **if** $a_j = 1 - s_j$ **then**	15: $\quad\quad a_j \leftarrow 0$
15: $\quad\quad s_j \leftarrow a_j$	16: $\quad\quad$ **if** $\textbf{top}(S) = j$ **then** $\textbf{pop}(S)$
16: $\quad\quad f_j \leftarrow f_{j+1}$	17: $\quad$ **else**
17: $\quad\quad f_{j+1} \leftarrow j + 1$	18: $\quad\quad a_j \, {-}{=} \, 1$
	19: $\quad\quad$ **if** $\textbf{top}(S) = j$ **then** $\textbf{pop}(S)$
	20: $\quad$ **yield** $j, \textbf{a}$
	21: $\quad$ **if** $a_j = 1 - s_j$ **then**
	22: $\quad\quad s_j \leftarrow a_j$
	23: $\quad\quad f_j \leftarrow f_{j+1}$
	24: $\quad\quad f_{j+1} \leftarrow j + 1$

References

1. Arndt, J.: Matters Computational: Ideas, Algorithms, Source Code. Springer, Heidelberg (2010)
2. Behrooznia, N., Mütze, T.: Listing spanning trees of outerplanar graphs by pivot exchanges. arXiv preprint arXiv:2409.15793 (2024)
3. Bousquet-Mélou, M., Claesson, A., Dukes, M., Kitaev, S.: (2+ 2)-free posets, ascent sequences and pattern avoiding permutations. J. Comb. Theory Ser. A **117**(7), 884–909 (2010)
4. Cameron, B., Grubb, A., Sawada, J.: Pivot Gray codes for the spanning trees of a graph ft. the fan. Graphs Comb. **40**(4), 78 (2024)

5. Cameron, B., Sawada, J., Therese, W., Williams, A.: Hamiltonicity of k-sided pancake networks with fixed-spin: efficient generation, ranking, and optimality. Algorithmica **85**(3), 717–744 (2023)
6. Cardinal, J., Hoang, H.P., Merino, A., Mička, O., Mütze, T.: Combinatorial generation via permutation languages. V. Acyclic orientations. SIAM J. Disc. Math. **37**(3), 1509–1547 (2023)
7. Cardinal, J., Merino, A., Mütze, T.: Efficient generation of elimination trees and graph associahedra. In: Proceedings of the 2022 Annual ACM-SIAM Symposium on Discrete Algorithms (SODA), pp. 2128–2140. SIAM (2022)
8. DiMuro, J.: Classifying rotationally-closed languages having greedy universal cycles. Electron. J. Comb. P1–35 (2019)
9. Downing, E., Einstein, S., Hartung, E., Williams, A.: Catalan squares and staircases: relayering and repositioning Gray codes. In: Proceedings of the 35th Canadian Conference on Computational Geometry, CCCG (2023)
10. Ehrlich, G.: Loopless algorithms for generating permutations, combinations, and other combinatorial configurations. J. ACM (JACM) **20**(3), 500–513 (1973)
11. Foata, D., Han, G.N.: New permutation coding and equidistribution of set-valued statistics. Theor. Comput. Sci. **410**(38–40), 3743–3750 (2009)
12. Gray, F.: Pulse code communication. United States Patent Number 2632058 (1953)
13. Hartung, E., Hoang, H., Mütze, T., Williams, A.: Combinatorial generation via permutation languages. I. Fundamentals. Trans. Am. Math. Soc. **375**(4), 2255–2291 (2022)
14. Hoang, H.P., Mütze, T.: Combinatorial generation via permutation languages. II. Lattice congruences. Israel J. Math. **244**(1), 359–417 (2021)
15. Kaye, R.: A Gray code for set partitions. Inf. Process. Lett. **5**, 171–173 (1976)
16. Kerr, K.: Successor rule for a restricted growth string Gray code (2015)
17. Knuth, D.E.: The art of computer programming, Volume 4A: Combinatorial algorithms, Part 1. Pearson Education India (2011)
18. Li, Y., Sawada, J.: Gray codes for reflectable languages. Inf. Process. Lett. **109**(5), 296–300 (2009)
19. Mansour, T., Nassar, G., Vajnovszki, V.: Loop-free Gray code algorithm for the e-restricted growth functions. Inf. Process. Lett. **111**(11), 541–544 (2011)
20. Mansour, T., Vajnovszki, V.: Efficient generation of restricted growth words. Inf. Process. Lett. **113**(17), 613–616 (2013)
21. Martin, M.: A problem in arrangements. Bull. Am. Math. Soc. **40**(12), 859–864 (1934)
22. Merino, A., Mütze, T.: Combinatorial generation via permutation languages. III. Rectangulations. Disc. Comput. Geom. 1–72 (2022)
23. Merino, A., Mütze, T.: Traversing combinatorial 0/1-polytopes via optimization. SIAM J. Comput. **53**(5), 1257–1292 (2024)
24. . Merino, A., Mutze, T., Williams, A.: All your bases are belong to us: Listing all bases of a matroid by greedy exchanges. In: 11th International Conference on Fun with Algorithms (FUN 2022), vol. 226, p. 22. Schloss Dagstuhl–Leibniz-Zentrum für Informatik (2022)
25. Merino, A., Namrata, Williams, A.: On the hardness of Gray code problems for combinatorial objects. In: International Conference and Workshops on Algorithms and Computation, pp. 103–117. Springer, Heidelberg (2024). https://doi.org/10.1007/978-981-97-0566-5_9
26. Mütze, T.: Combinatorial Gray codes-an updated survey. Electron. J. Comb. **30**(3), DS26 (2022)

27. Qiu, Y., Williams, A.: Generating signed permutations by twisting two-sided ribbons. In: Latin American Symposium on Theoretical Informatics, pp. 114–129. Springer, Heidelberg (2024). https://doi.org/10.1007/978-3-031-55598-5_8

28. Ruskey, F.: Combinatorial generation. Preliminary working draft. University of Victoria, Victoria, BC, Canada **11**, 20 (2003)

29. Ruskey, F., Savage, C.D.: Gray codes for set partitions and restricted growth tails. Aust. J. Comb. **10**, 85–96 (1994)

30. Sabri, A., Vajnovszki, V.: Two reflected Gray code-based orders on some restricted growth sequences. Comput. J. **58**(5), 1099–1111 (2015)

31. Savage, C.: A survey of combinatorial Gray codes. SIAM Rev. **39**(4), 605–629 (1997)

32. Sawada, J., Williams, A.: Greedy flipping of pancakes and burnt pancakes. Disc. Appl. Math. **210**, 61–74 (2016)

33. Sawada, J., Williams, A.: Successor rules for flipping pancakes and burnt pancakes. Theor. Comput. Sci. **609**, 60–75 (2016)

34. Sawada, J., Williams, A., Wong, D.: Generalizing the classic greedy and necklace constructions of de Bruijn sequences and universal cycles. Electron. J. Comb. P1–24 (2016)

35. Sawada, J., Williams, A., Wong, D.: Flip-swap languages in binary reflected Gray code order. Theor. Comput. Sci. **933**, 138–148 (2022)

36. Stanley, R.P.: Enumerative combinatorics volume 1 second edition. Cambridge studies in advanced mathematics (2011)

37. Stanley, R.P.: Catalan Numbers. Cambridge University Press, Cambridge (2015)

38. Stedman, F.: Campanalogia: or the Art of Ringing Improved, With plain and easie rules to guide the Practitioner in the Ringing all kinds of Changes, To Which is added, great variety of New Peals. London (1677)

39. Vajnovszki, V., Wong, D.: Greedy Gray codes for Dyck words and ballot sequences. In: International Computing and Combinatorics Conference, pp. 29–40. Springer, Heidelberg (2023). https://doi.org/10.1007/978-3-031-49193-1_3

40. Williams, A.: The greedy gray code algorithm. In: Dehne, F., Solis-Oba, R., Sack, J.-R. (eds.) WADS 2013. LNCS, vol. 8037, pp. 525–536. Springer, Heidelberg (2013). https://doi.org/10.1007/978-3-642-40104-6_46

41. Williams, A.: Pattern avoidance for k-Catalan sequences. In: Proceedings of the 21st International Conference on Permutation Patterns, pp. 147–149 (2023)

42. Zaks, S.: Lexicographic generation of ordered trees. Theor. Comput. Sci. **10**(1), 63–82 (1980)

NP-Completeness and Physical Zero-Knowledge Proofs for Zeiger

Suthee Ruangwises[(✉)] [ID]

Department of Computer Engineering, Chulalongkorn University, Bangkok, Thailand
suthee@cp.eng.chula.ac.th

Abstract. Zeiger is a pencil puzzle consisting of a rectangular grid, with each cell having an arrow pointing in horizontal or vertical direction. Some cells also contain a positive integer. The objective of this puzzle is to fill a positive integer into every unnumbered cell such that the integer in each cell is equal to the number of different integers in all cells along the direction an arrow in that cell points to. In this paper, we prove that deciding solvability of a given Zeiger puzzle is NP-complete via a reduction from the not-all-equal positive 3SAT (NAE3SAT+) problem. We also construct a card-based physical zero-knowledge proof protocol for Zeiger, which enables a prover to physically show a verifier the existence of the puzzle's solution without revealing it.

Keywords: NP-completeness · zero-knowledge proof · card-based cryptography · Zeiger · puzzle

1 Introduction

Zeiger (also known as *Number Pointers* or *Arrows*) is a pencil puzzle regularly published in a German magazine Denksel [15]. This puzzle also appears in many other places online, including mobile apps [4]. The puzzle consists of a $k \times \ell$ rectangular grid, with each cell having an arrow pointing in horizontal or vertical direction. Some cells also contain a positive integer. The player has to fill a positive integer into every unnumbered cell such that the integer in each cell is equal to the number of different integers in all cells along the direction an arrow in that cell points to (see Fig. 1).

Solving Zeiger is difficult, as the puzzle's constraints involve relationships between multiple cells. Yuki Fujimoto, the developer of a Zeiger mobile app as well as many other pencil puzzle apps, stated that Zeiger is the most difficult pencil puzzle [4]. Unsurprisingly, the puzzle turns out to be NP-complete, as we will later prove in this paper.

Suppose Agnes, an expert in Zeiger, created a Zeiger puzzle and challenged her friend Brian to solve it. After several tries, he could not solve the puzzle and doubted whether it has a solution. Agnes needs to convince him that her puzzle has a solution *without* revealing the solution itself (which would make the challenge pointless). In this situation, she needs a *zero-knowledge proof (ZKP)*.

© The Author(s), under exclusive license to Springer Nature Singapore Pte Ltd. 2025
S. Nakano and M. Xiao (Eds.): WALCOM 2025, LNCS 15411, pp. 312–325, 2025.
https://doi.org/10.1007/978-981-96-2845-2_20

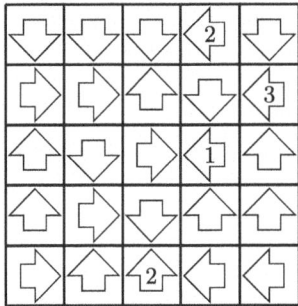 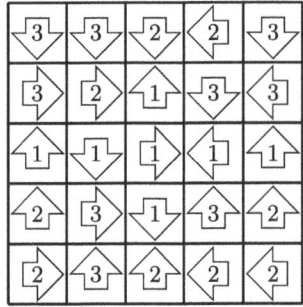

Fig. 1. An example of a 5×5 Zeiger puzzle (left) and its solution (right)

1.1 Zero-Knowledge Proof

A ZKP is an interactive proof between a prover P and a verifier V. Both P and V are given a computational problem x, but only P knows its solution w. A ZKP with perfect completeness and soundness must satisfy the following three properties.

1. **Perfect Completeness:** If P knows w, then V always accepts.
2. **Perfect Soundness:** If P does not know w, then V always rejects.
3. **Zero-knowledge:** V learns nothing about w. Formally, there exists a probabilistic polynomial time algorithm S (called a *simulator*) that does not know w but has access to V, and the outputs of S follow the same probability distribution as the ones of the real protocol.

The concept of a ZKP was introduced in 1989 by Goldwasser et al. [8]. Although a ZKP exists for every NP problem [7], it is more reasonable to construct ZKPs for NP-complete problems, as V cannot compute the solution by themselves in such problems. As one would expect, many popular pencil puzzles have been proved to be NP-complete, including Bridges [2], Five Cells [14], Goishi Hiroi [3], Heyawake [11], Kakuro [33], Makaro [13], Nondango [20], Nonogram [32], Numberlink [1], Nurikabe [10], Ripple Effect [31], Shikaku [31], Slitherlink [33], Sudoku [33], and Sumplete [9].

1.2 Physical Zero-Knowledge Proof Protocols

Instead of simply constructing computational ZKPs for these puzzles via reductions, many researchers have developed physical ZKPs using a deck of playing cards. These card-based protocols have the benefit that they do not require computers and also allow external observers to verify that the prover truthfully executes them (which is a challenging task for digital protocols). In addition, they are more intuitive and easier to verify the correctness and security, even for non-experts, and thus can be used for didactic purpose.

There is a line of work dedicated to constructing card-based physical ZKP protocols for pencil puzzles, such as Five Cells [26], Goishi Hiroi [21], Heyawake [18], Kakuro [17], Makaro [5], Masyu [16], Meadows [26], Nonogram [19], Numberlink [23], Nurikabe [18], Ripple Effect [24], Shikaku [22], Slitherlink [16], Sudoku [28], Sumplete [9], and Toichika [21].

1.3   Our Contribution

In this paper, we prove that deciding solvability of a given Zeiger puzzle is NP-complete via a reduction from the not-all-equal positive 3SAT (NAE3SAT+) problem. We also construct a physical ZKP protocol for Zeiger using a deck of playing cards.

2   NP-Completeness Proof of Zeiger

In this section, we will prove that deciding whether a given Zeiger puzzle has a solution is NP-complete.

As the problem clearly belongs to NP, the nontrivial part is to prove the NP-hardness. We will do so by constructing a reduction from the not-all-equal positive 3SAT (NAE3SAT+) problem. The NAE3SAT+ is the problem of deciding whether there exists a Boolean assignment such that every clause has at least one true literal and at least one false literal, in a setting where each clause contains exactly three positive literals (see Fig. 2). This problem is known to be NP-complete [29].

$$C_1 = x_1 \lor x_2 \lor x_3 \qquad\qquad x_1 = \text{TRUE}$$
$$C_2 = x_2 \lor x_3 \lor x_5 \qquad\qquad x_2 = \text{FALSE}$$
$$C_3 = x_1 \lor x_4 \lor x_5 \qquad\qquad x_3 = \text{TRUE}$$
$$C_4 = x_2 \lor x_4 \lor x_5 \qquad\qquad x_4 = \text{TRUE}$$
$$x_5 = \text{FALSE}$$

Fig. 2. An NAE3SAT+ instance (left) and one of its possible solutions (right)

Suppose we are given an NAE3SAT+ instance S with m clauses and n variables. Let $C_1, C_2, ..., C_m$ be the clauses of S and $x_1, x_2, ..., x_n$ be the variables of S. We assume that every variable appears in at least one clause (otherwise we can just remove unused variables).

We will transform S into a Zeiger puzzle grid G with size $(m + 3) \times (n + 5)$. The intuition is that each of the first m rows of G corresponds to each clause of S, each of the first n columns of G corresponds to each variable of S, and numbers 2 and 3 correspond to TRUE and FALSE, respectively.

Let $a(p, q)$ denotes the cell in the p-th row and q-th column of G. We define $a(p, q)$ as follows.

– If $q \leq n$,
  • if $p \leq m$,
    *if x_q appears in C_p, then $a(p,q)$ contains a down arrow with no number;
    * otherwise, $a(p,q)$ contains a right arrow with a number 4;

  • if $p = m + 1$, then $a(p,q)$ contains a right arrow with a number 4;
  • if $p = m + 2$, then $a(p,q)$ contains an up arrow with a number 2;
  • if $p = m + 3$, then $a(p,q)$ contains an up arrow with no number.
– If $q = n + 1$, then $a(p,q)$ contains a right arrow with a number 4.
– If $q = n + 2$,
  • If $p \leq m$, then $a(p,q)$ contains a left arrow with a number 3;
  • otherwise, $a(p,q)$ contains a right arrow with a number 3.
– If $q = n + 3$, then $a(p,q)$ contains a right arrow with a number 2.
– If $q = n + 4$, then $a(p,q)$ contains a right arrow with a number 1.
– If $q = n + 5$, then $a(p,q)$ contains a left arrow with a number 4 (see Fig. 3).

Fig. 3. A 7×10 Zeiger puzzle grid transformed from the NAE3SAT+ instance in Fig. 2

Clearly, this transformation can be done in polynomial time.

Consider each of the first n columns of G. We disregard constraints on right arrows in this column for now, and consider only the constraints on up and down arrows in this column. Next, we will prove that there are only two ways to fill numbers into this column to satisfy them: filling either all 2 s or all 3 s.

Lemma 1. *For each $q \leq n$, there are exactly two ways to fill numbers into all unnumbered cells in the q-th column of G to satisfy constraints of all up and down arrows in that column, which are filling all 2 s and filling all 3 s.*

Proof. Let $a(p, q)$ be the bottommost down arrow cell in the q-th column of G (there must be at least one down arrow since every variable appears in at least one clause). Observe that each cell below $a(p, q)$ contains either a 2 or 4 (and there is at least one 2 and at least one 4), except the cell $a(m + 3, q)$ which currently contains no number, so $a(p, q)$ must contain either a 2 or 3.

Case 1: $a(p, q)$ contains a 2.

Since $a(m + 2, q)$ contains a 2, all cells above it must contain either a 2 or 4 (and there is at least one 2 and at least one 4), which implies $a(m + 3, q)$ also contains a 2. Consider the second bottommost down arrow in this column. All cells below it contains either a 2 or 4 (and there is at least one 2 and at least one 4), so it must contain a 2. We then consider in the same way the third bottommost down arrow, the fourth bottommost down arrow, and so on. Finally, we can conclude that every down arrow in this column must contain a 2.

Case 2: $a(p, q)$ contains a 3.

Since $a(m + 2, q)$ contains a 2, all cells above it must contain either a 3 or 4 (and there is at least one 3 and at least one 4), which implies $a(m+3, q)$ contains a 3. Consider the second bottommost down arrow in this column. All cells below it contains either a 2, 3, or 4 (and there is at least one number of each type), so it must contain a 3. We then consider in the same way the third bottommost down arrow, the fourth bottommost down arrow, and so on. Finally, we can conclude that every down arrow in this column must contain a 3.

On the other hand, filling all 2 s or all 3 s clearly satisfies the constraints of all up and down arrows in the q-th column, so the converse is also true.   □

Now consider each p-th row of G $(p \leq m)$. Since C_p contains exactly three literals, there are exactly three unnumbered down arrow cells in the p-th row. We will prove that they must be filled with at least one 2 and at least one 3.

Lemma 2. *In any solution of G, the three unnumbered down arrow cells in each p-th row $(p \leq m)$ must be filled with at least one 2 and at least one 3.*

Proof. Consider the three unnumbered down arrow cells in the p-th row. From Lemma 1, each of them must be filled with either a 2 or 3. Also, as $a(p, n + 2)$ contains a left arrow with a number 3, exactly three different numbers must appear among the cells to the left of $a(p, n + 2)$. Besides these three cells, all other such cells contain all 4 s (and there is at least one 4 in $a(p, n+1)$). Therefore, these three cells must contain at least one 2 and at least one 3.   □

Finally, we will prove the following theorem, which implies NP-hardness of the puzzle.

Theorem 1. *S has a solution if and only if G has a solution.*

Proof. Suppose S has a solution T. From T, we construct a solution H of G as follows: for each $q \leq n$, we fill every unnumbered cell in the q-th column of G with a 2 if x_q is assigned to TRUE in T, and with a 3 otherwise (see Fig. 4).

From Lemma 1, the filled numbers satisfy constraints on all up and down arrows in the first n columns of G. Also, since each clause of S has at least one true literal and at least one false literal, the three unnumbered cells in each p-th row of G ($p \leq m$) are filled with at least one 2 and at least one 3, which satisfies the constraint on the arrow in $a(p, n+2)$. It is easy to verify that constraints on other arrows in G are also satisfied. Hence, we can conclude that H is a valid solution of G.

On the other hand, suppose G has a solution H'. From H', we construct a solution T' of S as follows: for each $q \leq n$, we assign x_q to TRUE if $a(m+3, q)$ contains a 2 in H', and to FALSE otherwise.

From Lemma 1, all unnumbered cells in each q-th column of G ($q \leq n$) must be filled with the same number as the one in $a(m+3, q)$. From Lemma 2, the three unnumbered cells in each p-th row of G ($p \leq m$) must be filled with at least one 2 and at least one 3, which means the three literals in C_p of S must contain at least one true literal and at least one false literal. Since this holds for every $p \leq m$, we can conclude that T' is a valid solution of S.   □

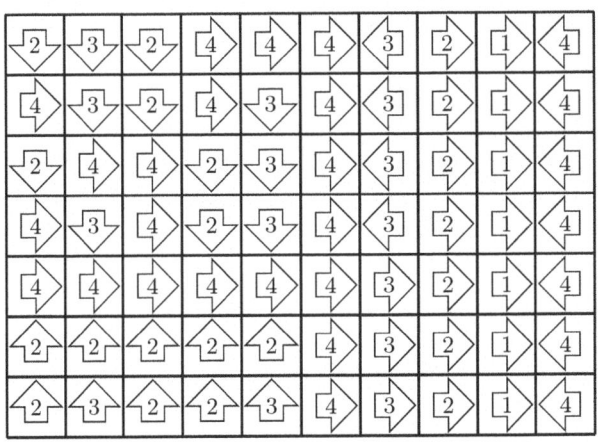

Fig. 4. A solution of the Zeiger puzzle in Fig. 3 transformed from the NAE3SAT+ solution in Fig. 2

From Theorem 1, we can conclude that deciding solvability of a given Zeiger puzzle is NP-complete.

3   Physical ZKP Protocol for Zeiger

Because of the NP-completeness of Zeiger, it is worth proposing a ZKP for it. In this section, we will construct a card-based physical ZKP protocol for the puzzle.

3.1   Preliminaries

Cards Each card used in our protocol has a ♣ or a ♡ written on the front side. All cards have indistinguishable back sides denoted by ?.

For $0 \leq x < q$, define $E_q^{\clubsuit}(x)$ to be a sequence of q cards, all of them being ♡s except the $(x+1)$-th leftmost card being a ♣. For example, $E_4^{\clubsuit}(1)$ is ♡ ♣ ♡ ♡. Analogously, define $E_q^{\heartsuit}(x)$ to be a sequence of q cards, all of them being ♣s except the $(x+1)$-th leftmost card being a ♡. For example, $E_4^{\heartsuit}(1)$ is ♣ ♡ ♣ ♣.

We may stack $E_q^{\clubsuit}(x)$ or $E_q^{\heartsuit}(x)$ into a single stack, with the leftmost card being the topmost card in the stack. Note that we refer to the "sequence form" and "stack form" of them interchangeably throughout the protocol.

Also, for $0 \leq x < q$, define $E_q(x)$ to be a sequence of q two-card stacks, all of them being $E_2^{\clubsuit}(0)$s except the $(x+1)$-th leftmost stack being an $E_2^{\clubsuit}(1)$. For example, $E_4(1)$ is ♣♡ ♡♣ ♣♡ ♣♡ (when extracting cards in each stack into a sequence). Note that the top cards of all stacks of $E_q(x)$ form a sequence $E_q^{\heartsuit}(x)$, and the bottom cards of all stacks of $E_q(x)$ form a sequence $E_q^{\clubsuit}(x)$.

Pile-Shifting Shuffle. A *pile-shifting shuffle* [30, §2.3] rearranges all columns of a matrix of cards (or a matrix of stacks) by a uniformly random cyclic shift unknown to V. It can be implemented by putting all cards in each column into an envelope, and repeatedly picking some envelopes from the bottom and putting them on the top of the pile of envelopes.

Pile-Scramble Shuffle. A *pile-scramble shuffle* [12, §3] rearranges all columns of a matrix of cards (or a matrix of stacks) by a uniformly random permutation unknown to V. It can be implemented by putting all cards in each column into an envelope, and scrambling all envelopes together on a table.

Copy Protocol. Given a sequence $E_q(x)$, a *copy protocol* creates an additional copy of the sequence without revealing x to V. This protocol was developed by Shinagawa et al. [30, §3.1] (using different encoding).

Subprotocol 1. Copy Protocol

Input: a face-down sequence $A = E_q(x)$ and $4q$ extra cards ($2q$ ♣ s and $2q$ ♡ s)
Output: two face-down sequences $E_q(x)$ and $E_q(x)$
Procedures: P performs the following steps.

1. Reverse the $q - 1$ rightmost stacks of A, i.e. move the $(i + 1)$-th leftmost stack to become the i-th rightmost stack for each $i = 1, 2, ..., q - 1$. The sequence A now becomes $E_q(-x \mod q)$.
2. Publicly construct two $E_q(0)$s from the $4q$ extra cards.
3. Construct a $3 \times q$ matrix of two-card stacks M by placing A in Row 1 and the two $E_q(0)$s in Rows 2 and 3. Turn all cards face-down.
4. Apply the pile-shifting shuffle to M.
5. Turn over all cards in Row 1 of M. Shift the columns of M cyclically such that the only $E_2^\clubsuit(1)$ in Row 1 moves to Column 1.
6. Return the two sequences in Rows 2 and 3 of M.

In Step 3, the sequence in Row 1 of M is $E_q(-x \mod q)$, and the sequences in Rows 2 and 3 of M are both $E_q(0)$s. Observe that the operations in Steps 4 and 5 only shift the columns of M cyclically, so the numbers encoded by sequences in all rows always shift together by the same constant modulo q. In Step 6, the sequence in Row 1 becomes an $E_q(0)$, thus the sequences in Rows 2 and 3 must both be $E_q(x)$s.

Note that this protocol also checks that the sequence A is in the correct format, i.e. consists of $q - 1$ $E_2^\clubsuit(0)$s and one $E_2^\clubsuit(1)$. In Step 5, if the sequence in Row 1 does not consist of $q - 1$ $E_2^\clubsuit(0)$s and one $E_2^\clubsuit(1)$, V immediately rejects the verification.

Set Size Protocol. Given p integers $x_1, x_2, ..., x_p \in \{0, 1, ..., q - 1\}$ unknown to V, a *set size protocol* computes the number of different values among $x_1, x_2, ..., x_p$, i.e. the size $|\{x_1, x_2, ..., x_p\}|$, without revealing any x_i to V.

This protocol was developed by Ruangwises et al. [27, §5.1] (using different encoding), and uses a similar idea to the set union protocol of Doi et al. [6, §5.1.2].

Subprotocol 2. Set Size Protocol

Input: p face-down sequences $E_q(x_1), E_q(x_2), ..., E_q(x_p)$

Output: q face-down sequences $E_2^\clubsuit(y_1), E_2^\clubsuit(y_2), ..., E_2^\clubsuit(y_q)$ such that $y_1, y_2, ..., y_q \in \{0, 1\}$ and $y_1 + y_2 + ... + y_q = |\{x_1, x_2, ..., x_p\}|$

Procedures: P performs the following steps.

1. Construct a $p \times q$ matrix of two-card stacks M by placing $E_q(x_i)$ in Row i of M for each $i = 1, 2, ..., p$. Let $M(i, j)$ denotes the stack at Row i and Column j of M.
2. Perform the following steps for $i = 2, 3, ..., p$.
   a) Apply the pile-scramble shuffle to M.
   b) Turn over all cards in Row i of M.
   c) For each $j = 1, 2, ..., q$, if $M(i, j)$ is an $E_2^\clubsuit(1)$, swap $M(i, j)$ and $M(1, j)$.
   d) Turn all cards face-down.
3. Apply the pile-scramble shuffle to M.
4. Return the q stacks in Row 1 of M as $E_2^\clubsuit(y_1), E_2^\clubsuit(y_2), ..., E_2^\clubsuit(y_q)$.

Observe that in Step 2(c), we only replace $M(1, j)$ with an $E_2^\clubsuit(1)$. Therefore, once a stack in Row 1 becomes an $E_2^\clubsuit(1)$, it will remain an $E_2^\clubsuit(1)$ throughout the protocol.

For any $r \in \{1, 2, ..., q\}$, consider the stack $M(1, r)$ at the beginning (we consider this exact stack no matter which column it later moves to). If it is an $E_2^\clubsuit(1)$ at the beginning, i.e. $x_1 = r - 1$, it will remain an $E_2^\clubsuit(1)$ at the end. On the other hand, if it is an $E_2^\clubsuit(0)$ at the beginning, it will become an $E_2^\clubsuit(1)$ at the end if and only if it has been replaced by an $E_2^\clubsuit(1)$ from some Row $i \geq 2$ during the i-th iteration, which occurs when $x_i = r - 1$.

Therefore, the stack $M(1, r)$ at the beginning is an $E_2^\clubsuit(1)$ at the end if and only if there is $i \in \{1, 2, ..., p\}$ such that $x_i = r - 1$. This implies the number of $E_2^\clubsuit(1)$s in Row 1 at the end is equal to the number of different values among $x_1, x_2, ..., x_p$.

Summation Protocol. Given q integers $y_1, y_2, ..., y_q \in \{0, 1\}$ unknown to V, a *summation protocol* computes the sum $y_1 + y_2 + ... + y_q$ without revealing any y_i to V. This protocol was developed by Ruangwises and Itoh [25, §3.1].

Subprotocol 3. Summation Protocol

Input: q face-down sequences $E_2^{\clubsuit}(y_1), E_2^{\clubsuit}(y_2), ..., E_2^{\clubsuit}(y_q)$ and two extra cards (a $\clubsuit$ and a $\heartsuit$)

Output: a face-down sequence $E_{q+1}^{\clubsuit}(y_1 + y_2 + ... + y_q)$

Procedures: P performs the following steps.

1. Let A be the sequence $E_2^{\clubsuit}(y_1)$.
2. Perform the following steps for $i = 2, 3, ..., q$.
   (a) Append a $\heartsuit$ to the right of A, making it an $E_{i+1}^{\clubsuit}(y_1 + y_2 + ... + y_{i-1})$.
   (b) Swap the two cards of $E_2^{\clubsuit}(y_i)$, making it $E_2^{\heartsuit}(y_i)$. Then, append $i - 1$ $\clubsuit$s between the two cards, making it $E_{i+1}^{\heartsuit}(-y_i \mod (i+1))$. Call this sequence B.
   (c) Construct a $2 \times (i + 1)$ matrix of cards M by placing A in Row 1 and B in Row 2. Turn all cards face-down.
   (d) Apply the pile-shifting shuffle to M.
   (e) Turn over all cards in Row 2 of M. Shift the columns of M cyclically such that the only $\heartsuit$ in Row 2 moves to Column 1.
   (f) Let A be the sequence in Row 1 of M, to be used in the next iteration. Note that A is $E_{i+1}^{\clubsuit}(y_1 + y_2 + ... + y_i)$.
3. Return the sequence A from the final iteration.

The purpose of the i-th iteration of Step 2 is to add $y_1 + y_2 + ... + y_{i-1}$ and y_i. In Step 2(c), the sequence in Row 1 of M is $E_{i+1}^{\clubsuit}(y_1 + y_2 + ... + y_{i-1})$, and the sequence in Row 2 of M is $E_{i+1}^{\heartsuit}(-y_i \mod (i + 1))$. Observe that the operations in Steps 2(d) and 2(e) only shift the columns of M cyclically, so the numbers encoded by sequences in both rows always shift together by the same constant modulo q. In Step 2(f), the sequence in Row 2 becomes $E_{i+1}^{\heartsuit}(0)$, thus the sequence in Row 1 must be $E_{i+1}^{\clubsuit}(y_1 + y_2 + ... + y_i)$.

Moreover, at the end of the i-th iteration, we get $i + 1$ extra cards (i $\clubsuit$s and one $\heartsuit$) from the unused sequence in Row 2 of M to use in Steps 2(a) and 2(b) in the next iteration without requiring additional cards.

After the q-th iteration, the sequence A will become $E_{q+1}^{\clubsuit}(y_1 + y_2 + ... + y_q)$.

Comparing Protocol. Given two integers $x_1, x_2 \in \{0, 1, ..., q - 1\}$ unknown to V, a *comparing protocol* lets V verify that $x_1 = x_2$ without revealing their value to V. This protocol was developed by Bultel et al. [5, §3.3].

Subprotocol 4. Comparing Protocol

Input: two face-down sequences $E_q^{\clubsuit}(x_1)$ and $E_q^{\clubsuit}(x_2)$
Output: V rejects the verification if $x_1 \neq x_2$, or does nothing if $x_1 = x_2$.
Procedures: P performs the following steps.

1. Construct a $2 \times q$ matrix of cards M by placing $E_q^{\clubsuit}(x_1)$ and $E_q^{\clubsuit}(x_2)$ in Row 1 and Row 2 of M, respectively.
2. Apply the pile-scramble shuffle to M.
3. Turn over all cards in M. If the $\boxed{\clubsuit}$ in Row 1 and the $\boxed{\clubsuit}$ in Row 2 are not in the same column, V rejects.

Observe that we only rearrange the columns of M, so the $\boxed{\clubsuit}$s in both rows are in the same column at the end if and only if they were in the same column at the beginning, i.e. $x_1 = x_2$.

3.2 Our Main Protocol

Suppose the puzzle grid has size $k \times \ell$. Let $b = \max\{k, \ell\}$. Observe that an integer in each cell in the puzzle grid can be at most $b - 1$.

On each cell with an integer x, P publicly places a face-up $E_b(x)$. On each unnumbered cell, P secretly places a face-down $E_b(x)$, where x is an integer in that cell in P's solution. Then, P turns all cards face-down.

For each cell c, P performs the following steps to check the constraint on c.

1. Let $c_1, c_2, ..., c_t$ be all cells along the direction the arrow in c points to. Let $E_b(d)$ be the sequence on c, and $E_b(d_1), E_b(d_2), ..., E_b(d_t)$ be the sequences on $c_1, c_2, ..., c_t$, respectively.
2. Apply the copy protocol to make a copy of each of $E_b(d), E_b(d_1), E_b(d_2), ..., E_b(d_t)$.
3. Apply the set size protocol to the copies of $E_b(d_1), E_b(d_2), ..., E_b(d_t)$ to get outputs $E_2^{\clubsuit}(z_1), E_2^{\clubsuit}(z_2), ..., E_2^{\clubsuit}(z_b)$, where $z_1 + z_2 + ... + z_b = |\{d_1, d_2, ..., d_t\}|$.
4. Apply the summation protocol to $E_2^{\clubsuit}(z_1), E_2^{\clubsuit}(z_2), ..., E_2^{\clubsuit}(z_b)$ to get output $E_{b+1}^{\clubsuit}(z)$, where $z = z_1 + z_2 + ... + z_b$. Note that z is the number of different integers in $c_1, c_2, ..., c_t$.
5. Append a $\boxed{\heartsuit}$ to the right of the copy of $E_b^{\clubsuit}(d)$ from Step 2 to make it an $E_{b+1}^{\clubsuit}(d)$. Then, apply the comparing protocol to $E_{b+1}^{\clubsuit}(d)$ and $E_{b+1}^{\clubsuit}(z)$ to check that $d = z$.

P performs these steps for every cell in the grid. If all verifications pass, then V accepts.

Our protocol uses $\Theta(bk\ell)$ cards and $\Theta(bk\ell)$ shuffles.

3.3 Proof of Correctness and Security

We will prove the perfect completeness, perfect soundness, and zero-knowledge properties of our main protocol.

Lemma 3 (Perfect Completeness). *If P knows a solution of the Zeiger puzzle, then V always accepts.*

Proof. Suppose P knows a solution of the puzzle. P can place sequences on all cells according to the solution.

Consider the verification of each cell c with a sequence $E_b(d)$. Because of the constraint of the puzzle, d is equal to the number of different integers in cells $c_1, c_2, ..., c_t$, i.e. $d = z$. Therefore, the verification for c will pass.

Since this is true for every cell, we can conclude that V always accepts.   □

Lemma 4 (Perfect Soundness). *If P does not know a solution of the Zeiger puzzle, then V always rejects.*

Proof. Suppose P does not know a solution of the puzzle. First, if a sequence on some cell does not have the correct format, i.e. does not consist of $b - 1$ $E_2^\clubsuit(0)$s and one $E_2^\clubsuit(1)$, it will be detected by the copy protocol, and V will immediately reject.

Suppose all sequences have the correct format. Since P does not know a solution, the sequence on at least one cell, say c, must violate the constraint of the puzzle. During the verification for the cell c, we will have $d \neq z$, and V will reject during the comparing protocol.

Hence, we can conclude that V always rejects.   □

Lemma 5 (Zero-Knowledge). *During the verification, V obtains no information about P's solution.*

Proof. We will prove that any interaction between P and V can be simulated by a simulator S that does not know P's solution. It is sufficient to show that all distributions of cards that are turned face-up can be simulated by S.

- In Step 5 of the copy protocol, because of the pile-shifting shuffle, the $E_2^\clubsuit(1)$ has probability $1/q$ to be at each of the q columns. Therefore, this step can be simulated by S.
- In Step 2(b) in each i-th iteration of the set size protocol, because of the pile-scramble shuffle, the $E_2^\clubsuit(1)$ has probability $1/q$ to be at each of the q columns. Therefore, this step can be simulated by S.
- In Step 2(e) in each i-th iteration of the summation protocol, because of the pile-shifting shuffle, the $\boxed{\heartsuit}$ has probability $1/(i+1)$ to be at each of the $i + 1$ columns. Therefore, this step can be simulated by S.
- In Step 3 of the set size protocol (in the case that V does not reject), because of the pile-scramble shuffle, the $\boxed{\clubsuit}$s in both rows are in the same column and have probability $1/q$ to be at each of the q columns. Therefore, this step can be simulated by S.

Hence, we can conclude that V obtains no information about P's solution.   □

4   Future Work

We proved the NP-completeness of a pencil puzzle Zeiger and constructed a card-based physical ZKP protocol for it. Future work includes proving the NP-completeness of other well-known pencil puzzles, as well as constructing card-based ZKP protocols for them.

References

1. Adcock, A., et al.: Zig-Zag numberlink is NP-complete. J. Inf. Process. **23**(3), 239–245 (2015)
2. Andersson, D.: Hashiwokakero is NP-complete. Inf. Process. Lett. **109**(9), 1145–1146 (2009)
3. Andersson, D.: HIROIMONO is NP-complete. In: Crescenzi, P., Prencipe, G., Pucci, G. (eds.) FUN 2007. LNCS, vol. 4475, pp. 30–39. Springer, Heidelberg (2007). https://doi.org/10.1007/978-3-540-72914-3_5
4. App Store: Number Pointers. https://apps.apple.com/us/app/number-pointers/id1065321061
5. Bultel, X., Dreier, J., Dumas, J.-G., Lafourcade, P., Miyahara, D., Mizuki, T., Nagao, A., Sasaki, T., Shinagawa, K., Sone, H.: Physical zero-knowledge proof for makaro. In: Izumi, T., Kuznetsov, P. (eds.) SSS 2018. LNCS, vol. 11201, pp. 111–125. Springer, Cham (2018). https://doi.org/10.1007/978-3-030-03232-6_8
6. Doi, A., et al.: Card-based protocols for private set intersection and union. N. Gener. Comput. **42**(3), 359–380 (2024)
7. Goldreich, O., Micali, S., Wigderson, A.: Proofs that yield nothing but their validity and a methodology of cryptographic protocol design. J. ACM **38**(3), 691–729 (1991)
8. Goldwasser, S., Micali, S., Rackoff, C.: The knowledge complexity of interactive proof systems. SIAM J. Comput. **18**(1), 186–208 (1989)
9. Hatsugai, K., Ruangwises, S., Asano, K., Abe, Y.: NP-completeness and physical zero-knowledge proofs for sumplete, a puzzle generated by ChatGPT. N. Gener. Comput. **42**(3), 429–448 (2024)
10. Holzer, M., Klein, A., Kutrib, M.: On the NP-completeness of the nurikabe pencil puzzle and variants thereof. In: Proceedings of the 3rd International Conference on Fun with Algorithms (FUN), pp. 77–89 (2004)
11. Holzer, M., Ruepp, O.: The troubles of interior design-a complexity analysis of the game heyawake. In: Proceedings of the 4th International Conference on Fun with Algorithms (FUN), pp. 198–212 (2007)
12. Ishikawa, R., Chida, E., Mizuki, T.: Efficient card-based protocols for generating a hidden random permutation without fixed points. In: Calude, C.S., Dinneen, M.J. (eds.) UCNC 2015. LNCS, vol. 9252, pp. 215–226. Springer, Cham (2015). https://doi.org/10.1007/978-3-319-21819-9_16
13. Iwamoto, C., Haruishi, M.,, Ibusuki, T.: Herugolf and Makaro are NP-complete. In: Proceedings of the 9th International Conference on Fun with Algorithms (FUN), pp. 24:1–24:11 (2018)
14. Iwamoto, C., Ide, T.: Five cells and tilepaint are NP-complete. IEICE Trans. Inf. Syst. **105.D**(3), 508–516 (2022)
15. Janko, A., Janko, O.: Zeiger (Logikrätsel). https://www.janko.at/Raetsel/Zeiger/index.htm

16. Lafourcade, P., Miyahara, D., Mizuki, T., Robert, L., Sasaki, T., Sone, H.: How to construct physical zero-knowledge proofs for puzzles with a "single loop" condition. Theor. Comput. Sci. **888**, 41–55 (2021)
17. Miyahara, D., Sasaki, T., Mizuki, T., Sone, H.: Card-based physical zero-knowledge proof for kakuro. IEICE Trans. Fundamentals **E102.A**(9), 1072–1078 (2019)
18. Robert, L., Miyahara, D., Lafourcade, P., Mizuki, T.: Card-based ZKP for connectivity: applications to nurikabe, hitori, and heyawake. N. Gener. Comput. **40**(1), 149–171 (2022)
19. Ruangwises, S.: An improved physical ZKP for nonogram and nonogram color. J. Comb. Optim. **45**(5), 122 (2023)
20. Ruangwises, S.: Nondango is NP-complete. In: Proceedings of the 40th European Workshop on Computational Geometry (EuroCG), pp. 1:1–1:10 (2024)
21. Ruangwises, S.: Verifying the first nonzero term: physical ZKPs for ABC end view, goishi hiroi, and toichika. J. Comb. Optim. **47**(4), 69 (2024)
22. Ruangwises, S., Itoh, T.: How to physically verify a rectangle in a grid: a physical ZKP for Shikaku. In: Proceedings of the 11th International Conference on Fun with Algorithms (FUN), pp. 24:1–24:12 (2022)
23. Ruangwises, S., Itoh, T.: Physical zero-knowledge proof for numberlink puzzle and k vertex-disjoint paths problem. N. Gener. Comput. **39**(1), 3–17 (2021)
24. Ruangwises, S., Itoh, T.: Physical zero-knowledge proof for ripple effect. Theor. Comput. Sci. **895**, 115–123 (2021)
25. Ruangwises, S., Itoh, T.: Securely computing the n-variable equality function with $2n$ cards. Theor. Comput. Sci. **887**, 99–110 (2021)
26. Ruangwises, S., Iwamoto, M.: Printing protocol: physical ZKPs for decomposition puzzles. N. Gener. Comput. **42**(3), 331–343 (2024)
27. Ruangwises, S., Ono, T., Abe, Y., Hatsugai, K., Iwamoto, M.: Card-based overwriting protocol for equality function and applications. In: Proceedings of the 21st International Conference on Unconventional Computation and Natural Computation (UCNC), pp. 18–27 (2024)
28. Sasaki, T., Miyahara, D., Mizuki, T., Sone, H.: Efficient card-based zero-knowledge proof for Sudoku. Theor. Comput. Sci. **839**, 135–142 (2020)
29. Schaefer, T.J.: The complexity of satisfiability problems. In: Proceedings of the 10th Annual ACM Symposium on Theory of Computing (STOC), pp. 216–226 (1978)
30. Shinagawa, K., et al.: Card-based protocols using regular polygon cards. IEICE Trans. Fund. **E100.A**(9), 1900–1909 (2017)
31. Takenaga, Y., Aoyagi, S., Iwata, S., Kasai, T.: Shikaku and ripple effect are NP-complete. Congr. Numer. **216**, 119–127 (2013)
32. Ueda, N., Nagao, T.: NP-completeness Results for NONOGRAM via Parsimonious Reductions. Technical Report TR96-0008, Department of Computer Science, Tokyo Institute of Technology (1996)
33. Yato, T., Seta, T.: Complexity and completeness of finding another solution and its application to puzzles. IEICE Trans. Fund. **86.A**(5), 1052–1060 (2003)

Abelian and Stochastic Sandpile Models on Complete Bipartite Graphs

Thomas Selig$^{(\boxtimes)}$ and Haoyue Zhu

Xi'an Jiaotong-Liverpool University, Suzhou, China
Thomas.Selig@xjtlu.edu.cn, Haoyue.Zhu18@student.xjtlu.edu.cn

Abstract. In the sandpile model, vertices of a graph are allocated grains of sand. At each unit of time, a grain is added to a randomly chosen vertex. If that causes its number of grains to exceed its degree, that vertex is called unstable, and *topples*. In the Abelian sandpile model (ASM), topplings are deterministic, whereas in the stochastic sandpile model (SSM) they are random. We study the ASM and SSM on complete bipartite graphs. For the SSM, we provide a stochastic version of Dhar's burning algorithm to check if a given (stable) configuration is recurrent or not, with linear complexity. We also exhibit a bijection between sorted recurrent configurations and pairs of *compatible* Ferrers diagrams. We then provide a similar bijection for the ASM, and also interpret its recurrent configurations in terms of labelled Motzkin paths.

Keywords: Sandpile model · Complete bipartite graphs · Recurrent configurations · Ferrers diagrams · Motzkin paths

1 Introduction

The Abelian sandpile model (ASM) is a dynamic process on a graph, where vertices are assigned a number of grains of sand. At each unit of time, a grain is added to a randomly chosen vertex. If this causes a vertex's number of grains to exceed its degree, the vertex is called *unstable*, and *topples*, sending one grain to each of its neighbours. A special vertex, the *sink*, absorbs grains, and so the process eventually stabilises. The model was originally introduced by Bak, Tang and Wiesenfeld [4,5] as an example of a model exhibiting a phenomenon known as *self-organised criticality*, before being formalised and generalised by Dhar [12].

Of central interest in the ASM are the *recurrent configurations* – those which appear infinitely often in its long-time running. A fruitful direction of ASM research has focused on combinatorial studies of these for graph families with high levels of symmetry, such as complete graphs [9], complete bipartite [15] and multi-partite [8] graphs, complete split graphs [11,14], wheel and fan graphs [18], Ferrers graphs [17,20], permutation graphs [16], and so on.

In the ASM, the only randomness lies in the choice of vertex where grains are added at each time step. After this, the toppling and stabilisation processes are entirely deterministic. The main focus of this paper is instead a stochastic variant of the ASM, called *stochastic sandpile model* (SSM), as introduced in [6],

© The Author(s), under exclusive license to Springer Nature Singapore Pte Ltd. 2025
S. Nakano and M. Xiao (Eds.): WALCOM 2025, LNCS 15411, pp. 326–345, 2025.
https://doi.org/10.1007/978-981-96-2845-2_21

in which topplings are also random. This was studied on complete graphs in [19], and on complete bipartite graphs in the special case where one of the components has exactly two vertices in the recent [1]. In this paper we consider the general complete bipartite case. We first set some notation and formally define the model.

As usual, $\mathbb{N}$ denotes the set of strictly positive integers. We let $\mathbb{Z}_+ := \mathbb{N} \cup \{0\}$ denote the set of non-negative integers. For $n \in \mathbb{N}$, we define $[n] := \{1, \ldots, n\}$. For a vector $a = (a_1, \ldots, a_n) \in \mathbb{R}^n$, we write $\mathrm{inc}\,(a) = (\tilde{a}_1, \ldots, \tilde{a}_n)$ for the non-decreasing rearrangement of a. In this paper, we consider the complete bipartite graph $K^0_{m,n}$ for $m \geq 0, n \geq 1$. This is the graph with vertex set $\{v^t_0, v^t_1, \ldots, v^t_m\} \sqcup \{v^b_1, \ldots, v^b_n\}$ and edge set $\{(v^t_i, v^b_j); i \in [m] \cup \{0\}, j \in [n]\}$. We refer to vertices v^t_i, resp. v^b_j, as *top*, resp. *bottom*, vertices in $K^0_{m,n}$. We will use the notation v^*_i to refer to any arbitrary vertex of $K^0_{m,n}$. The vertex v^t_0, called the *sink*, will play a special role in both the ASM and the SSM. Figure 1 shows the graph $K^0_{2,2}$, where the sink is represented as a black square. Finally, we fix a probability $p \in (0,1)$.

A (sandpile) *configuration* on $K^0_{m,n}$ is a vector $c = (c^t_1, \ldots, c^t_m; c^b_1, \ldots, c^b_n) \in \mathbb{Z}^{m+n}_+$. For simplicity, we write $c = (c^t; c^b)$. We think of c^*_i as the number of grains at vertex v^*_i. We denote by $\mathrm{Config}_{m,n}$ the set of all configurations on $K^0_{m,n}$. A top vertex v^t_i (for $i \in [m]$), resp. bottom vertex v^b_j (for $j \in [n]$), is *stable* if $c^t_i < n$, resp. $c^b_j < m + 1$ (i.e. the number of grains at the vertex is less than its degree). A configuration is stable if all of its vertices are stable. The set of stable configurations is denoted $\mathrm{Stable}_{m,n}$.

Unstable vertices *topple*. The toppling rules are different for the ASM and SSM, and we begin with the former. In the ASM, topplings are deterministic: an unstable vertex will send one grain to each of its neighbours (this includes the sink v^t_0 for bottom vertices). This may cause other vertices who receive grains to become unstable, and there topple in turn. Grains that are sent to the sink exit the system (we think of the sink as "absorbing excess grains").

From this last property, we can see that, starting from an unstable configuration c and successively toppling unstable vertices, we eventually reach a stable configuration c'. Moreover, Dhar [12] showed that the configuration c' reached does not depend on the order in which vertices are toppled. Note that in the ASM this process is entirely deterministic. We write $c' = \mathrm{DetStab}\,(c)$ and call it the *deterministic stabilisation* of c. Figure 1 shows the deterministic stabilisation process for the configuration $c = (2, 1; 0, 2)$. Here, blue vertices are unstable, green edges represent grains being sent from an unstable vertex to a neighbour.

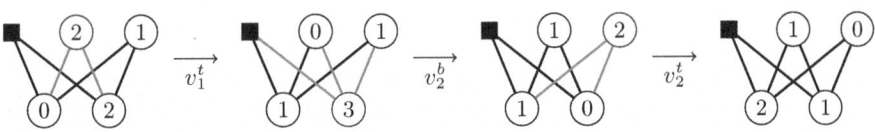

Fig. 1. Illustrating the deterministic stabilisation for $c = (2, 1; 0, 2) \in \mathrm{Config}_{2,2}$. Vertices under the arrows represent the vertex being toppled in that phase. (Colour figure online)

We now introduce the SSM. In this model, if a top vertex v_i^t is unstable, then for each bottom neighbour v_j^b we draw a Bernoulli random variable B_j with parameter p (the B_j's are independent of each other and of all prior topplings). If $B_j = 1$, then v_j^b receives one grain from v_i^t when it topples, otherwise vertex v_i^t keeps that grain. The process is the same for toppling a bottom vertex v_j^b.

As in the ASM, one can show (see [6, Theorem 2.2]) that, starting from an unstable configuration c and successively toppling unstable vertices, we will eventually reach a (random) stable configuration c' which does not depend on the order in which vertices are toppled. We write $c' = \text{StoStab}\,(c)$ and call it the *stochastic stabilisation* of c. Note that here $\text{StoStab}\,(c)$ is a random variable with values in $\text{Stable}_{m,n}$. Figure 2 shows one possible example of the stochastic stabilisation process for the configuration $c = (2, 1; 0, 2)$. Here, blue vertices are unstable, green edges represent grains being sent from an unstable vertex to a neighbour, while red edges represent no movement of grain.

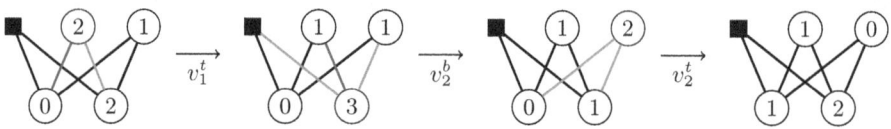

Fig. 2. Illustrating a possible stochastic stabilisation for $c = (2, 1; 0, 2) \in \text{Config}_{2,2}$. Vertices under the arrows represent the vertex being toppled in that phase. (Colour figure online)

For both ASM and SSM, we define a Markov chain on the set $\text{Stable}_{m,n}$. At each step, we add a grain to a non-sink vertex of $K_{m,n}^0$, chosen uniformly at random, and stabilise the resulting configuration according to the chosen model's rules. A configuration c is called *deterministically recurrent* (DR), resp. *stochastically recurrent* (SR) if it appears infinitely often in the long-time running of this Markov chain for the ASM, resp. SSM. We denote by $\text{DetRec}_{m,n}$, resp. $\text{StoRec}_{m,n}$, the set of DR, resp. SR, configurations on $K_{m,n}^0$.

A popular statistic for a recurrent configuration c is its *level*, defined by:

$$\text{level}\,(c) := \sum_{i \in [m]} c_i^t + \sum_{j \in [n]} c_j^b - m \cdot n, \tag{1}$$

which satisfies $0 \leq \text{level}\,(c) \leq m(n-1)$ (see e.g. [19, Equation (8)]).

The symmetries of $K_{m,n}^0$ make it natural to study both $\text{DetRec}_{m,n}$ and $\text{StoRec}_{m,n}$ up to re-ordering in each part. We therefore say that a configuration $c = (c^t; c^b) \in \text{Config}_{m,n}$ is *sorted* if c^t and c^b are both non-decreasing. We denote by $\text{DetSortedRec}_{m,n}$, resp. $\text{StoSortedRec}_{m,n}$, the set of sorted DR, resp. sorted SR, configurations for the ASM, resp. SSM.

For the reader's convenience, the longer proofs are omitted from the main body of the text. Proofs of results marked with (*) can instead be found in Appendix A.

2 Stochastically Recurrent Configurations on Complete Bipartite Graphs

In this section, we give a characterisation of SR configurations on $K^0_{m,n}$. We use this to describe a *stochastic burning algorithm*, which checks in linear time if a given configuration is SR or not.

2.1 Characterisation of StoRec$_{m,n}$

In [19, Theorem 2.6], the first author characterised SR configurations on general graphs. We begin by restating this in the complete bipartite graph case.

Theorem 1. *Let $c = (c^t; c^b) \in \text{Stable}_{m,n}$ be a stable configuration on $K^0_{m,n}$. Then $c \in \text{StoRec}_{m,n}$ if and only if for all subsets $A \subseteq [m], B \subseteq [n]$, we have*

$$\sum_{i \in A} c^t_i + \sum_{j \in B} c^b_j \geq |A| \cdot |B|. \tag{2}$$

If A, B do not satisfy Inequality (2), we call (A, B) a forbidden subconfiguration.

In fact, because of the structure of complete bipartite graph, it is unnecessary to check Inequality (2) for all subsets A, B. Instead, intuitively, we only need to check it for a linear proportion of such subsets. This is made precise in the following, which is the main result of this section.

Theorem 2 (*). *Let $c = (c^t; c^b) \in \text{Stable}_{m,n}$ be a stable configuration on $K^0_{m,n}$. For $j \in [n]$, define $k_j := |\{i \in [m]; c^t_i < j\}|$. Then $c \in \text{StoRec}_{m,n}$ if and only if*

$$\forall j \in [n], \tilde{c}^b_1 + \cdots + \tilde{c}^b_j \geq k_1 + \cdots + k_j, \tag{3}$$

where $\text{inc}(c^b) := (\tilde{c}^b_1, \ldots, \tilde{c}^b_n)$ *is the non-decreasing re-arrangement of c^b. Moreover, if c is SR, we have* $\text{level}(c) = c^b_1 + \cdots + c^b_n - (k_1 + \cdots + k_n)$.

Example 1. Consider the configuration $c = (c^t; c^b) = (3, 1, 3, 2, 3; 2, 0, 4, 3) \in \text{Stable}_{5,4}$. We first sort the bottom part c^b and get $c^b = \text{inc}(c^b) = (0, 2, 3, 4)$. Next, we calculate the vector k as in the statement of Theorem 2, yielding $k = (0, 1, 2, 5)$. We then check Condition (3) for $j = 1, 2, 3, 4$.

- For $j = 1$, we have $\tilde{c}^b_1 = 0 \geq 0 = k_1$.
- For $j = 2$, we have $\tilde{c}^b_1 + \tilde{c}^b_2 = 0 + 2 = 2 \geq 1 = 0 + 1 = k_1 + k_2$.
- For $j = 3$, we have $\tilde{c}^b_1 + \tilde{c}^b_2 + \tilde{c}^b_3 = 5 \geq 3 = k_1 + k_2 + k_3$.
- For $j = 4$, we have $\tilde{c}^b_1 + \tilde{c}^b_2 + \tilde{c}^b_3 + \tilde{c}^b_4 = 9 \geq 8 = k_1 + k_2 + k_3 + k_4$.

Therefore the configuration c is SR by Theorem 2, with $\text{level}(c) = 9 - 8 = 1$.

Algorithm 1. Pre-processing: calculating the vector $k = (k_1, \ldots, k_n)$

Require: $n \in \mathbb{N}$; $c^t = (c_1^t, \ldots, c_m^t) \in \{0, \ldots, n-1\}^m$

  Initialise: $k' = (k_0', \ldots, k_{n-1}') = (0, \ldots, 0)$; $k = (k_1, \ldots, k_n) = (0, \ldots, 0)$; sum $= 0$

  for i from 1 to m **do**

    $k_{c_i^t}' \leftarrow k_{c_i^t}' + 1$ $\triangleright$ Calculate $k_j' := |\{i \in [m]; c_i^t = j\}|$

  end for

  for j from 1 to n **do**

    sum $\leftarrow$ sum $+ k_{j-1}'$; $k_j \leftarrow$ sum

  end for

  return $k = (k_1, \ldots, k_n)$

2.2 A Stochastic Burning Algorithm for Complete Bipartite Graphs

We now exhibit an algorithm to check whether a given stable configuration $c \in \text{Stable}_{m,n}$ is SR or not, in two steps. Algorithm 1 first calculates the vector $k = (k_1, \ldots, k_n)$ as defined in Theorem 2. Algorithm 2 then gives the desired check. The terminology *burning algorithm* refers to Dhar's process for the ASM that checks whether a given configuration is DR or not (see [13, Section 6.2]).

Proposition 1. *Algorithm 1 outputs the vector $k = (k_1, \ldots, k_n)$ defined by $k_j = |\{i \in [m]; c_i^t < j\}|$ for any $j \in [n]$. Moreover, it runs in $O(m + n)$ time.*

Proof. The first loop calculates the vector $k' = (k_0', \ldots, k_{n-1}')$ defined by $k_j' = |\{i \in [m]; c_i^t = j\}|$ for $0 \leq j \leq n - 1$, in $O(m)$ time. The second loop then calculates the vector k defined by $k_j = k_0' + \cdots + k_{j-1}'$ for $j \in [n]$, from which the formula immediately follows, in $O(n)$ time.

Algorithm 2. Stochastic burning algorithm for complete bipartite graphs

---

Require: $c = (c^t; c^b) \in \text{Stable}_{m,n}$

  $c^b \leftarrow \text{inc}\left(c^b\right)$ $\triangleright$ Sort bottom part c^b of configuration c

  Pre-process: calculate vector $k = (k_1, \ldots, k_n)$ by Algorithm 1

  Initialise: sumK $= 0$; sumC $= 0$

  for j from 1 to n **do**

    sumK $\leftarrow$ sumK $+ k_j$; sumC $\leftarrow$ sumC $+ c_j^b$

    if sumC $<$ sumK **then**

      return False

    end if

  end for

  return True

Theorem 3. *Algorithm 2 returns True if and only if the input (stable) configuration c is stochastically recurrent. Moreover, it runs in $O(m + n)$ time.*

Proof. The recurrence check follows from Theorem 2. The pre-processing of k in Algorithm 1 has complexity $O(m + n)$. Now note that for all $j \in [n]$, we have

$0 \leq c_j^b \leq m$ (since c is stable). As such, sorting c^b can be done in $O(m + n)$ time by using the CountSort algorithm (also known as *sort by values*, see e.g. [7, proof of Proposition 13]). In brief, this algorithm first computes an auxiliary array $a = (a_0, \ldots, a_m)$ where $a_j = \left| \{ i \in [n]; c_i^b = j \} \right|$ for all $0 \leq j \leq m$. This is analogous to the calculation of the vector k' at the start of Algorithm 1, and thus takes $O(n)$ time. We then recover inc (c^b) from a in linear time by looping over a and appending a_j times the value j to inc (c^b) at each step, taking $O(m + n)$ time since $\sum_{j=0}^{m} a_j = n$. The rest of Algorithm 2 then runs in $O(n)$ time.

3   Recurrent Configurations as Pairs of Ferrers Diagrams

In this section, we present a combinatorial interpretation of StoSortedRec$_{m,n}$ in terms of *Ferrers diagrams*. A Ferrers diagram is a left-aligned collection of cells such that the number of cells in each row is non-decreasing from bottom to top (some rows may be empty). We denote by Ferrers$_{m,n}$, resp. Ferrers$_{\leq m,n}$, the set of Ferrers diagrams with m columns, resp. at most m columns, and n rows. The *area* Area(F) of a Ferrers diagram F is its number of cells. Given a non-decreasing sequence $s = (s_1, \ldots, s_n) \in \mathbb{Z}_+^n$, we denote $F(s) \in$ Ferrers$_{s_n,n}$ the Ferrers diagram with s_i cells in row i (rows are ordered from bottom to top).

For example, $F(0, 1, 4) :=$ ⌐⊓⊓⊓ is an element of Ferrers$_{4,3}$ with area 5.
    We consider the following two families of operations on Ferrers diagrams:

1. Shift which shifts a cell of the diagram in a given row to some row below;
2. Add which adds a cell to the right of a given row.

A Shift or Add operation is called *legal* if it still results in a Ferrers diagram (possibly with a different number of columns). Figure 3 shows these operations.

Definition 1. *An ordered pair* (F, F') *of Ferrers diagrams is called* compatible *if F' can be obtained from F through a sequence of legal Shift and Add operations.*

    We return to our study of StoRec$_{m,n}$. Note that the vectors $k = (k_1, \ldots, k_n)$ and inc $(c^b) = (\tilde{c}_1^b, \ldots, \tilde{c}_n^b)$ from Inequality (3) are non-decreasing. Moreover, we have $k_n = m$ and, since c is stable, $\tilde{c}_n^b \leq m$. This yields the following.

Theorem 4 (*). *For a* sorted *configuration* $c = (c^t; c^b) \in$ Config$_{m,n}$, *we define* $\Psi(c) := \left(F(k), F(c^b) \right) \in$ Ferrers$_{m,n} \times$ Ferrers$_{\leq m,n}$, *where the vector* $k = (k_1, \ldots, k_n)$ *is as in Theorem 2. Then* Ψ *is a bijection from* StoSortedRec$_{m,n}$ *to the set of compatible pairs* $(F, F') \in$ Ferrers$_{m,n} \times$ Ferrers$_{\leq m,n}$. *Moreover, we have* level $(c) =$ Area$\left(F(c^b) \right) -$ Area $(F(k))$, *which is also equal to the number of* Add *operations in a legal sequence* $F(k) \rightsquigarrow F(c^b)$.

Example 2. Consider the sorted configuration $c = (0, 2, 2; 2, 2, 2)$. We have $k = (1, 1, 3)$, so c is recurrent by Theorem 2. Figure 3 illustrates a possible legal sequence of Shift and Add operations to go from the k-diagram (left) to the c^b-diagram (right). Note that this sequence is not unique: we could instead first add a cell in the middle row, then shift a cell from the top to the bottom row.

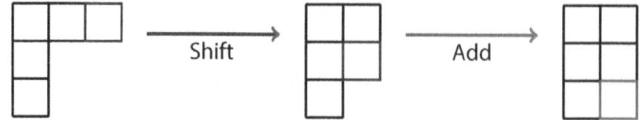

Fig. 3. Illustrating a legal sequence from $F(1,1,3)$ to $F(2,2,2)$, showing that the configuration $c = (0,2,2;2,2,2)$ is stochastically recurrent. (Colour figure online)

Remark 1. The top configuration inc (c^t) can be recovered from the Ferrers diagram $F(k)$ by taking the heights of East steps along the South-East border of the diagram, from the bottom-left corner to the top-right. For example, for the diagram $F(1,1,3)$ (Fig. 3, left), the South-East border can be written as $ENNEEN$, with E denoting an East step, and N a North one. Then we have $c^t = (0,2,2)$, corresponding to the heights of the E steps.

Finally, we propose a representation of the compatibility notion through a directed acyclic graph (DAG). The vertices of the graph are the Ferrers diagrams $F \in \text{Ferrers}_{\leq m,n}$ satisfying $\text{Area}(F) \geq m$. For every pair (F, F'), we put an edge from F to F' if $F' = \text{Shift}(F)$ or $F' = \text{Add}(F)$. As in Fig. 3, Shift edges are coloured blue, and Add edges are red. We denote $\text{DAG}^{\text{SR}}_{m,n}$ the DAG thus obtained. Note that $\text{DAG}^{\text{SR}}_{m,n}$ is *bipolar*: it has a unique source $F(0,\ldots,0,m)$ and a unique sink $F(m,\ldots,m)$. With this representation, Ψ is a bijection from $\text{StoSortedRec}_{m,n}$ to pairs of vertices (F, F') of $\text{DAG}^{\text{SR}}_{m,n}$ such that F has m columns and there is a (directed) path from F to F' in $\text{DAG}^{\text{SR}}_{m,n}$. The level of the configuration equals the number of red edges in such a path. Figure 4 illustrates this construction.

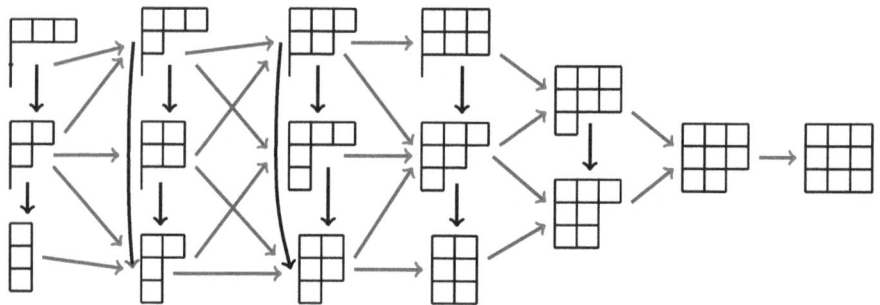

Fig. 4. The graph $\text{DAG}^{\text{SR}}_{3,3}$. (Colour figure online)

4   Deterministically Recurrent Configurations on Complete Bipartite Graphs

In this section, we return to the ASM, and its DR configurations. While the ASM on complete bipartite graphs has already been extensively studied in [2,3,15], our main goal here is to give a new description of DR configurations in terms of Ferrers diagrams, as we did for the SSM in Sect. 3.

4.1   Characterisation of DetRec$_{m,n}$

We begin with a characterisation of DR configurations in the spirit of Theorem 2.

Theorem 5. *Let* $c = (c^t; c^b) \in \text{Stable}_{m,n}$ *be a stable configuration on* $K^0_{m,n}$. *As previously, for* $j \in [n]$, *define* $k_j := |\{i \in [m]; c^t_i < j\}|$. *Then* $c \in \text{DetRec}_{m,n}$ *if and only if*

$$\forall j \in [n], \, \tilde{c}^b_j \geq k_j, \tag{4}$$

where $\tilde{c}^b_j$ *is the* j-*th item from the non-decreasing re-arrangement of* c^b.

Proof (sketch). It is sufficient to show the result for sorted configurations. Suppose that $c \in \text{Stable}_{m,n}$ is sorted, and that there exists $j \in [n]$ such that $c^b_j < k_j$. Then we have $c^b_{j'} < k_j$ for all $j' \in [j]$. Moreover, by definition, we have $c^t_i < j$ for all $i \in [k_j]$. It follows that the configuration c restricted to the induced subgraph on $([k_j], [j])$ is stable. This is exactly the definition of a forbidden subconfiguration for the ASM (see e.g. [13, Section 6]), and therefore c is not recurrent. The converse can be shown in an analogous manner to Theorem 2: if there is a forbidden subconfiguration (in the above sense) (A, B), we show that we can choose $A = [k_j]$ and $B = [j]$ for some $j \in [n]$, and the stability condition for c then implies that $c^b_j < k_j$, as desired. $\qed$

Theorem 5 immediately implies the following.

Corollary 1. *There exists an algorithm that checks in* $O(m + n)$ *time whether a configuration* $c \in \text{Config}_{m,n}$ *is recurrent or not.*

4.2   Ferrers Diagram Interpretation

As in Sect. 3, Theorem 5 allows for a characterisation of DR configurations in terms of pairs of Ferrers diagrams. Note that this time we must have $\tilde{c}^b_n = m$ since $\tilde{c}^b_n \geq k_n$ by Eq. (4), and $k_n = m$. Moreover, Eq. (4) means exactly that each row of $F' := F\left(\text{inc}\left(c^b\right)\right)$ has a number of cells that is greater than or equal to the number of cells in the same row of $F := F(k)$.

Definition 2. *An ordered pair* (F, F') *of Ferrers diagrams is called* strongly compatible *if the diagram* F' *can be obtained from* F *through a sequence of legal* Add *operations.*

We then get the following characterisation of DetSortedRec$_{m,n}$ in terms of Ferrers diagrams. The proof is analogous to that of Theorem 4, combined with the straightforward observation that Eq. (4) is equivalent to the pair $\left(F(k), F\left(c^b\right)\right)$ being strongly compatible.

Theorem 6. *For a* sorted *configuration* $c = (c^t; c^b) \in \text{Config}_{m,n}$, *we define*
$\Psi(c) := \left(F(k), F\left(c^b\right)\right) \in \text{Ferrers}_{m,n} \times \text{Ferrers}_{m,n}$, *where* $k = (k_1, \ldots, k_n)$ *is as in Theorem 5. Then* Ψ *is a bijection from* $\text{DetSortedRec}_{m,n}$ *to the set of strongly compatible pairs* $(F, F') \in \text{Ferrers}_{m,n} \times \text{Ferrers}_{m,n}$. *Moreover, we have* $\text{level}(c) = \text{Area}\left(F\left(c^b\right)\right) - \text{Area}(F(k))$, *which is also equal to the number of* Add *operations in a legal sequence* $F(k) \rightsquigarrow F(c^b)$.

Example 3. Consider the sorted configuration $c = (0, 2, 2; 2, 2, 3)$. We have $k = (1, 1, 3)$, $\tilde{c}^b = (2, 2, 3)$, and for all $j \in \{1, 2, 3\}$, $\tilde{c}^b_j \geq k_j$, so c is DR by Theorem 5. Figure 5 shows a possible legal sequence of Add operations to go from the k-diagram (left) to the c^b-diagram (right) for $\text{DetRec}_{m,n}$.

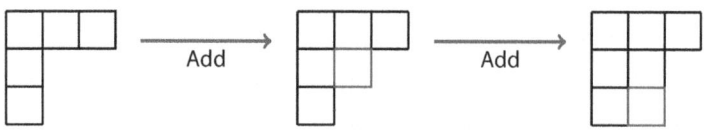

Fig. 5. Illustrating a legal sequence from $F(1, 1, 3)$ to $F(2, 2, 3)$, showing that the configuration $c = (0, 2, 2; 2, 2, 3)$ is deterministically recurrent. (Colour figure online)

As in the SSM case, we can give a DAG interpretation of Theorem 6. Here the vertices are all Ferrers diagrams in $\text{Ferrers}_{m,n}$, and we put an edge from F to F' if $F' = \text{Add}(F)$. For consistency with the SSM case we colour the edges red. We denote $\text{DAG}^{\text{DR}}_{m,n}$ the DAG thus obtained. $\text{DAG}^{\text{DR}}_{m,n}$ is also *bipolar* in the ASM case with the same source $F(0, \ldots, 0, m)$ and sink $F(m, \ldots, m)$. With this representation, Φ is a bijection from $\text{DetSortedRec}_{m,n}$ to pairs of vertices (F, F') of $\text{DAG}^{\text{DR}}_{m,n}$ such that there is a (directed) path from F to F' in $\text{DAG}^{\text{DR}}_{m,n}$. The level of the configuration is equal to the number of (red) edges in such a path. Figure 6 illustrates this construction.

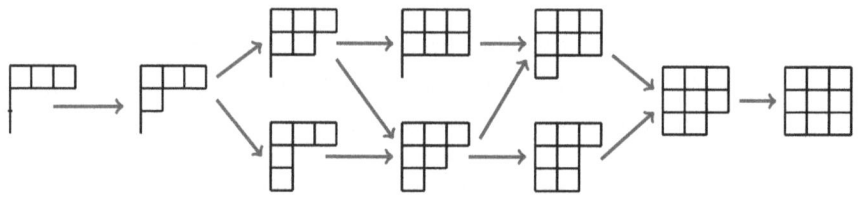

Fig. 6. The graph $\text{DAG}^{\text{DR}}_{3,3}$ for the ASM. (Colour figure online)

4.3   Connection to Parallelogram Polyominoes

In this section, we use our characterisation with Ferrers diagrams to recover results from Dukes and Le Borgne [15] that link DR configurations on complete bipartite graphs to *parallelogram polyominoes*. A parallelogram polyomino is a collection of cells in a bounding rectangle $[0, m] \times [0, n]$ which lies between two lattice paths $\mathcal{U}$ (the *upper* path) and $\mathcal{L}$ (the *lower* path) with the following properties:

- the paths $\mathcal{U}$ and $\mathcal{L}$ go from $(0, 0)$ to (m, n), with steps $N = (0, 1)$ and $E = (1, 0)$;
- the paths $\mathcal{U}$ and $\mathcal{L}$ do not intersect except at $(0, 0)$ and (m, n).

We write $P = P(\mathcal{U}, \mathcal{L})$ for the parallelogram polyomino with upper path $\mathcal{U}$ and lower path $\mathcal{L}$. We denote $\mathrm{ParaPoly}_{m,n}$ the set of parallelogram polyominoes with bounding box $[0, m] \times [0, n]$. The *area* of a parallelogram polyomino P, denoted $\mathrm{Area}(P)$, is its number of cells. Figure 7 shows an example of a parallelogram polyomino $P \in \mathrm{ParaPoly}_{5,3}$ with area 10.

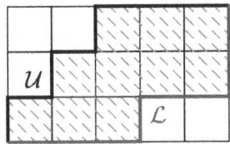

Fig. 7. An example of a parallelogram polyomino $P \in \mathrm{ParaPoly}_{5,3}$ with upper path $\mathcal{U}$ (in blue), lower path $\mathcal{L}$ (in purple), and $\mathrm{Area}(P) = 10$. (Colour figure online)

Dukes and Le Borgne in [15] described a bijection from $\mathrm{DetSortedRec}_{m,n}$ to $\mathrm{ParaPoly}_{m+1,n}$ as follows. Given a *sorted* configuration $c = (c^t; c^b) \in \mathrm{Stable}_{m,n}$, they defined two lattice paths.

- The upper path $\mathcal{U} = \mathcal{U}(c^t)$ is the path from $(0, 0)$ to $(m+1, n)$ whose E steps occur at heights $(1 + c_1^t, \ldots, 1 + c_m^t, n)$ above the x-axis. One may think of the final E step as corresponding to the sink vertex v_0^t, with the convention $c_0^t = n - 1$.
- The lower path $\mathcal{L} = \mathcal{L}(c^b)$ is the path from $(0, 0)$ to $(m+1, n)$ whose N steps occur at heights $(1 + c_1^b, \ldots, 1 + c_n^b)$ to the right of the y-axis.

Theorem 7 ([15], [Theorem 3.7]). *For a sorted configuration* $c = (c^t; c^b) \in \mathrm{Stable}_{m,n}$, *let* $\mathcal{U} = \mathcal{U}(c^t)$ *and* $\mathcal{L} = \mathcal{L}(c^b)$ *be defined as above. Then the map* $\Phi : c \mapsto P := P(\mathcal{U}, \mathcal{L})$ *is a bijection from* $\mathrm{DetSortedRec}_{m,n}$ *to* $\mathrm{ParaPoly}_{m+1,n}$. *Moreover, if c is DR, we have* $\mathrm{level}(c) = \mathrm{Area}(P) - m - n$.

Example 4. Consider the polyomino from Fig. 7. It has upper path $\mathcal{U} = NENENEEE$. The heights of its E steps are $(1, 2, 3, 3, 3)$, yielding the top configuration $c^t = (0, 1, 2, 2)$. Similarly, the lower path $\mathcal{L} = EEENEENN$ has N steps at heights $(3, 5, 5)$, yielding the bottom configuration $c^b = (2, 4, 4)$. Finally, we get the configuration $c = (0, 1, 2, 2; 2, 4, 4)$, which one can check is indeed DR by Theorem 5 (here we have $k = (1, 2, 4)$).

In fact, we can recover this bijection by using our characterisation as pairs of strongly compatible Ferrers diagrams. Indeed, for a pair $(F_1, F_2) \in$ Ferrers$_{m,n} \times$ Ferrers$_{m,n}$ of strongly compatible Ferrers diagrams, we construct the parallelogram polyomino as follows. Let $F_1' \in$ Ferrers$_{m+1,n+1}$ be the Ferrers diagram obtained from F_1 by adding one empty row at the bottom, and an extra cell to (the right of) the top row. Let $F_2' \in$ Ferrers$_{m+1,n+1}$ be the Ferrers diagram obtained from F_2 by adding one extra cell to each row, and one "full" row at the top (with $m+1$ cells). Note that the pair (F_1', F_2') is strongly compatible, and that both F_1' and F_2' have $m+1$ cells in their top row. In particular, the difference $F_2' \setminus F_1'$ consisting of cells that are in F_2' but not in F_1' is contained in the box $[0, m+1] \times [0, n]$. We denote Diff$(F_1, F_2) := F_2' \setminus F_1'$ the collection of cells thus obtained. We will see that Diff(F_1, F_2) turns out to always be a parallelogram polyomino.

Example 5. Consider the pair (F_1, F_2) of strongly compatible Ferrers diagrams given by $F_1 = F(1,1,3)$ and $F_2 = F(2,2,3)$ (Fig. 8, left). This yields $F_1' = F(0,1,1,4)$ and $F_2' = F(3,3,4,4)$ (Fig. 8, centre). Then taking the difference Diff(F_1, F_2) gives the collection of shaded cells in the right of Fig. 8. Note that Diff(F_1, F_2) is a parallelogram polyomino, as claimed above.

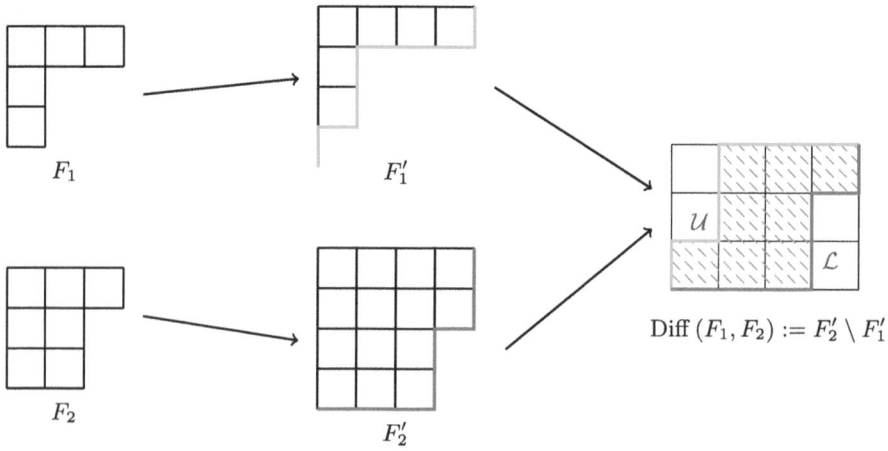

Fig. 8. Illustrating the construction from a pair (F_1, F_2) of strongly compatible Ferrers diagrams to the polyomino Diff(F_1, F_2). (Colour figure online)

Theorem 8. *We have $\Phi = $ Diff $\circ \, \Psi$, where Φ is the Dukes-Le Borgne bijection from sorted DR configurations to parallelogram polyominoes (Theorem 7) and Ψ is our bijection from sorted DR configurations to pairs of strongly compatible Ferrers diagrams (Theorem 6).*

Proof. This follows from the various constructions and from Remark 1 which details how the top configuration c^t can be read from the Ferrers diagram $F(k)$.

4.4 Connection to Motzkin Paths

Finally, in this section we exhibit a bijection between sorted DR configurations and a family of labelled *Motzkin paths*. A Motzkin path is a lattice path from $(0,0)$ to $(k,0)$ for some $k \geq 0$ taking steps $U = (1,1)$ (*up* step), $H = (1,0)$ (*horizontal* step), and $D = (1,-1)$ (*down* step), which never goes below the x-axis. The *area* of a Motzkin path w, denoted by Area(w), is the area between its curve and the x-axis (see Example 6). We will consider Motzkin paths whose H-steps are labelled either N or E, and we denote these steps H^N, H^E. We first briefly recall a classical bijection between parallelogram polyominoes and these labelled Motzkin paths.

Let $P \in \text{ParaPoly}_{m+1,n}$ be a parallelogram polyomino, with upper and lower paths $\mathcal{U} = (N, u_1, \ldots, u_{m+n-1}, E)$ and $\mathcal{L} = (E, \ell_1, \ldots, \ell_{m+n-1}, N)$. We define a lattice path $w = w(P) = w_1 \ldots w_{m+n-1}$ by setting, for each $i \in [m+n-1]$:

$$w_i := \begin{cases} U & \text{if } u_i = N \text{ and } \ell_i = E, \\ H^N & \text{if } u_i = N \text{ and } \ell_i = N, \\ H^E & \text{if } u_i = E \text{ and } \ell_i = E, \\ D & \text{if } u_i = E \text{ and } \ell_i = N. \end{cases} \tag{5}$$

Example 6. Consider the parallelogram polyomino P on the left of Fig. 9, with $\mathcal{U}(P) = (N)NNEEENNEE(E)$ and $\mathcal{L}(P) = (E)EENENENEN(N)$, putting the first and last steps in parentheses as they will be ignored in the construction of w. Equation (5) yields the path $w = UUDH^EDUH^NH^ED$ (Fig. 9, right). In Fig. 9 we have coloured the regions of P defined by pairs (u_i, ℓ_i) and the region below the corresponding step w_i in the same colour. For example, the red region is given by $u_7 = \ell_7 = N$, so the corresponding step is $w_7 = H^N$, and the region below this step in w has also been coloured red.

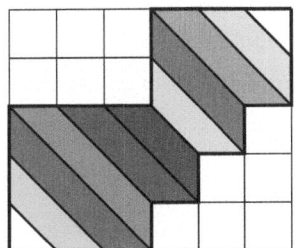
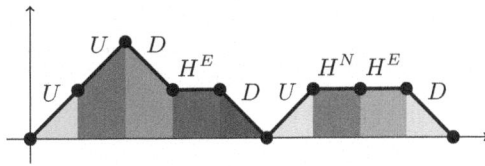

Fig. 9. Illustrating the construction from polyominoes to labelled Motzkin paths. (Colour figure online)

Note that the path w thus obtained is indeed a Motzkin path with labelled H-steps. Moreover, given the dimension of the parallelogram polyomino $((m+1) \times n)$, w has $m + n - 1$ steps in total, of which exactly m are either U or H^E steps (equivalently $n - 1$ steps D or H^N, equivalently m steps D or H^E, equivalently $n - 1$ steps U or H^N). We denote $\text{LabMotz}_{m,n-1}$ the set of such labelled Motzkin paths.

Theorem 9 ([10], Sect. 3). *The map* Ξ : $\mathrm{ParaPoly}_{m+1,n} \to \mathrm{LabMotz}_{m,n-1}$, $P \mapsto w(P)$, *where w is defined by Eq. (5), is a bijection.*

By combining this map Ξ and the bijection Φ from Theorem 7, we can get a bijection from $\mathrm{LabMotz}_{m,n-1}$ to $\mathrm{DetSortedRec}_{m,n}$. Let us conclude this paper by giving algorithms to directly compute this bijection and its converse.

Algorithm 3. From $\mathrm{LabMotz}_{m,n-1}$ to $\mathrm{DetSortedRec}_{m,n}$

Require: $w = w_1 \ldots w_{m+n-1} \in \mathrm{LabMotz}_{m,n-1}$
  Initialise: $c^t = c^b = ()$ (empty lists); $\mathrm{tVal} = \mathrm{bVal} = 0$; $i = j = 1$
  for k from 1 to $m + n - 1$ **do**
    if $w_k = U$ **then**
      $\mathrm{tVal} \leftarrow \mathrm{tVal} + 1$; $\mathrm{bVal} \leftarrow \mathrm{bVal} + 1$
    else if $w_k = H^E$ **then**
      c^t.append(tVal); $\mathrm{bVal} \leftarrow \mathrm{bVal} + 1$
    else if $w_k = H^N$ **then**
      c^b.append(bVal); $\mathrm{tVal} \leftarrow \mathrm{tVal} + 1$
    else   ▷ $w_k = D$
      c^t.append(tVal); c^b.append(bVal)
    end if
  end for
  c^b.append(m)   ▷ to get $c_n^b = m$
  return $c = (c^t; c^b) \in \mathrm{DetSortedRec}_{m,n}$

Theorem 10. *The map $w \mapsto c(w)$ described by Algorithm 3 is the bijection* $\Phi^{-1} \circ \Xi^{-1} : \mathrm{LabMotz}_{m,n-1} \to \mathrm{DetSortedRec}_{m,n}$.

Proof (sketch). By construction, we see that Algorithm 3 outputs a configuration $c = c(w) = (c_1^t, \ldots, c_m^t; c_1^b, \ldots c_n^b)$, which satisfies the following properties.

- For $i \in [m]$, let j be such that w_j is the i-th step of the form D or H^E in w. Then we set $c_i^t := \sharp \{ j' < j; w_{j'} \in \{ U, H^N \} \}$.
- For $j \in [n-1]$, let i be such that w_i is the j-th step of the form D or H^N in w. Then we set $c_j^b := \sharp \{ i' < i; w_{i'} \in \{ U, H^E \} \}$. Finally, we also set $c_n^b := m$.

From Theorem 7, Eq. (5) and Theorem 9, we then get that, for $w \in \mathrm{LabMotz}_{m,n-1}$, we have $c(w) = \Phi^{-1} \circ \Xi^{-1}(w)$, as desired.

For the converse bijection, given $c = (c^t; c^b) \in \mathrm{DetSortedRec}_{m,n}$, we put the sorted top and bottom configurations c^t and c^b into *sorted stacks*, with the head of each stack being its smallest element. In the stack c^t, we place an additional element $c_{m+1}^t = n - 1$ at the tail. The operation head() returns the element at the head of the stack, while pop() removes it. Finally, **1** denotes the all-1 vector.

Theorem 11 (*). *The map $c \mapsto w(c)$ described by Algorithm 4 is the bijection* $\Xi \circ \Phi : \mathrm{DetSortedRec}_{m,n} \to \mathrm{LabMotz}_{m,n-1}$. *Moreover, for $c \in \mathrm{DetSortedRec}_{m,n}$, we have* level $(c) = \mathrm{Area}(w(c))$.

Algorithm 4. The bijection from DetSortedRec$_{m,n}$ to LabMotz$_{m,n-1}$

Require: sorted stacks $c^t = (c_1^t \leq \cdots \leq c_m^t \leq c_{m+1}^t = n - 1)$, $c^b = (c_1^b \leq \cdots \leq c_n^b)$
1: **Initialise:** empty list $w = ()$
2: **while** $c^t \neq \emptyset$ & $c^b \neq \emptyset$ **do**
3:     **while** c^t.head() > 0 & c^b.head() > 0 **do**
4:         w.append(U)
5:         $c^t \leftarrow c^t - 1$; $c^b \leftarrow c^b - 1$ ▷ Decrease all values in c^t and c^b by one
6:     **end while**
7:     **if** $0 = c^t$.head() $< c^b$.head() **then**
8:         c^t.pop(); w.append(H^E)
9:         $c^b \leftarrow c^b - 1$
10:     **else if** $0 = c^b$.head() $< c^t$.head() **then**
11:         c^b.pop(); w.append(H^N)
12:         $c^t \leftarrow c^t - 1$
13:     **else**   ▷ c^b.head() $= c^t$.head() $= 0$
14:         c^t.pop(); c^b.pop(); w.append(D)
15:     **end if**
16: **end while**
17: **Delete** final D-step from $w = (w_1, \ldots, w_{m+n-1}, w_{m+n} = D)$
18: **return** $w = (w_1, \ldots, w_{m+n-1}) \in$ LabMotz$_{m,n-1}$

Acknowledgments. The authors have no competing interests to declare that are relevant to the content of this article. The research leading to these results is partially supported by the National Natural Science Foundation of China, grant number 12101505, by the Research Development Fund of Xi'an Jiaotong-Liverpool University, grant number RDF-22-01-089, and by the Postgraduate Research Scholarship of Xi'an Jiaotong-Liverpool University, grant number PGRS2012026.

A   Proof of Main Results

A.1   Proof of Theorem 2

It is sufficient to show the result when c is sorted, in which case, for $j \geq 0$, we have $c_i^t = j$ if and only if $k_j < i \leq k_{j+1}$ (with the convention $k_0 = 0$). Fix $j \in [n]$. We have:

$$\sum_{i=1}^{k_j} c_i^t = 0(k_1 - k_0) + 1 \cdot (k_2 - k_1) + \cdots + (j-1) \cdot (k_j - k_{j-1}) = j \cdot k_j - (k_1 + \cdots + k_j). \quad (6)$$

In particular, if we take $A = [k_j]$ and $B = [j]$, we have

$$\sum_{i \in A} c_i^t + \sum_{j' \in B} c_{j'}^b = \sum_{i=1}^{k_j} c_i^t + \sum_{j'=1}^{j} c_{j'}^b$$

$$= j \cdot k_j - (k_1 + \cdots + k_j) + (c_1^b + \cdots + c_j^b)$$

$$= |A| \cdot |B| + (c_1^b + \cdots + c_j^b) - (k_1 + \cdots + k_j).$$

It follows that Inequality (3) is equivalent to Inequality (2) when we take $A = [k_j], B = [j]$. It therefore suffices to show that if c is not recurrent (i.e. there exists some forbidden subconfiguration (A, B) for c), then there exists $j \in [n]$ such that $([k_j], [j])$ is a forbidden subconfiguration for c.

For this, take $B \subseteq [n]$ to be a minimal subset such that there exists $A \subseteq [m]$ with (A, B) forbidden, that is, $\sum_{i \in A} c_i^t + \sum_{j \in B} c_j^b < |A| \cdot |B|$. Let $q := \max B$. We claim that for any $q' < q$, we have $c_{q'}^b < |A|$. Otherwise, if $c_{q'}^b \geq |A|$, let $B' := [q'] \cap B \subsetneq B$. Since c is assumed to be sorted, by using $c_j^b \geq c_{q'}^b \geq |A|$ for $j \in (q', q]$, we get $\sum_{j \in B'} c_j^b = \sum_{j \in B} c_j^b - \sum_{j \in (q', q] \cap B} c_j^b \leq \sum_{j \in B} c_j^b - |A| \cdot |B \setminus B'|$. Then we have:

$$\sum_{i \in A} c_i^t + \sum_{j \in B'} c_j^b \leq \sum_{i \in A} c_i^t + \sum_{j \in B} c_j^b - |A| \cdot |B \setminus B'|$$
$$< |A| \cdot |B| - |A| \cdot |B \setminus B'|$$
$$= |A| \cdot |B'|,$$

which implies that (A, B') is a forbidden subconfiguration. But since $B' \subsetneq B$, this contradicts the minimality of B. This proves our claim that for any $q' < q$, we have $c_{q'}^b < |A|$, and in particular this holds for $q' \in [q] \setminus B$.

We therefore get:

$$\sum_{i \in A} c_i^t + \sum_{j \in [q]} c_j^b = \sum_{i \in A} c_i^t + \sum_{j \in B} c_j^b + \sum_{j \in [q] \setminus B} c_j^b$$
$$< \sum_{i \in A} c_i^t + \sum_{j \in B} c_j^b + |A| \cdot |[q] \setminus B|$$
$$< |A| \cdot |B| + |A| \cdot |[q] \setminus B|$$
$$= |A| \cdot |[q]|,$$

which implies that $(A, [q])$ is also forbidden, i.e., we can choose B of the form $B = [q]$. By a similar argument (considering A as a minimal subset such that $(A, [q])$ is forbidden), we can choose $A = [p]$. From there, it is straightforward to check that $([k_q], [q])$ is also forbidden, by distinguishing the cases $p > k_q$ and $p < k_q$.

Finally, we have shown that if there exists a forbidden subconfiguration (A, B) for c, then there exists j such that $([k_j], j)$ is forbidden for c. This completes the proof of the characterisation of SR configurations.

The level formula then follows from the definition of the level statistic in Eq. (1), and Eq. (6) with $j = n$, noting that $k_n = m$, which yields:

$$\text{level}(c) := \sum_{i \in [m]} c_i^t + \sum_{j \in [n]} c_j^b - m \cdot n$$
$$= n \cdot m - (k_1 + \cdots + k_n) + \sum_{j \in [n]} c_j^b - m \cdot n$$
$$= c_1^b + \cdots + c_n^b - (k_1 + \cdots + k_n),$$

as desired. $\square$

A.2   Proof of Theorem 4

By preceding remarks, if $c \in \text{StoRec}_{m,n}$, then $F(k) \in \text{Ferrers}_{m,n}$ and $F(c^b) \in$ $\text{Ferrers}_{\leq m,n}$. To show that Ψ is injective, we note that the vector k uniquely defines the non-decreasing re-arrangement inc(c^t) of the top part of c. Indeed, let $k'_j := |\{i \in [m]; c^t_i = j\}|$ for $0 \leq j \leq n-1$. Then we have $k'_j = k_{j+1} - k_j$ (with the convention $k_0 = 0$), and the vector inc(c^t) simply consists of the values j repeated k'_j times for $0 \leq j \leq n-1$. Moreover, the level formula follows from the level formula in Theorem 2, which gives level$(c) = c^b_1 + \cdots + c^b_n - (k_1 + \cdots + k_n) = \tilde{c}^b_1 + \cdots + \tilde{c}^b_n - (k_1 + \cdots + k_n) = \text{Area}(F(c^b)) - \text{Area}(F(k))$. Since Shift operations leave the area of a Ferrers diagram unchanged, and Add operations increase it by one, this is also the number of Add operations in a legal sequence $F(k) \leadsto F(c^b)$.

It remains to show that Ψ is surjective, i.e. that c is SR if and only if the pair $\left(F(k), F(c^b)\right)$ is compatible. Suppose first that $\left(F(k), F(c^b)\right)$ is compatible. This means that $F(c^b)$ can be reached from $F(k)$ through a sequence of legal Shift and Add operations. As such, by Theorem 2, it is sufficient to show the following. If $v = (v_1, \ldots, v_n)$ is such that $v_1 + \cdots + v_j \geq k_1 + \cdots + k_j$ for all $j \in [n]$, and $v' = (v'_1, \ldots, v'_n)$ is such that $F(v')$ is obtained from $F(v)$ through a single legal Shift or Add operation, then we have $v'_1 + \cdots + v'_j \geq k_1 + \cdots + k_j$ for all $j \in [n]$. In other words, Condition (3) is preserved through applying a single legal Shift or Add operation to the Ferrers diagram of the left-hand side vector. Adding a cell to row p in $F(v)$ simply increases v_p by one, leaving other values unchanged, so the inequalities are clearly preserved. Shifting a cell from row p to row $p' < p$ yields $v'_p = v_p - 1$, $v'_{p'} = v_{p'} + 1$, and other values unchanged. Thus the sum $v_1 + \cdots + v_j$ is increased by one if $j \in [p', p)$ and unchanged otherwise, so that Condition (3) is preserved.

We now show the converse, namely that if c is SR, then the pair of Ferrers diagrams $\left(F(k), F(c^b)\right)$ is compatible. Let c be a *sorted* SR configuration, and k the corresponding vector. First, note that if $c^b_j \geq k_j$ for all $j \in [n]$, then $F(c^b)$ can be obtained from $F(k)$ by adding the required cells in each row from top to bottom, which is a legal sequence, and the result follows. We therefore assume this is not the case, and define $p := \min\{j \in [n]; c^b_j < k_j\}$. Since c is SR, we have $c^b_1 + \ldots c^b_p \geq k_1 + \cdots + k_p$ by Theorem 2. Therefore there must exist $q \in [p]$ such that $c^b_q > k_q$, and we set p' to be the maximal such q. Now note that on the one hand, we have $k_p > c^b_p \geq c^b_{p-1} \geq k_{p-1}$, i.e. $k_p > k_{p-1}$, while on the other hand, we have $k_{p'} < c^b_{p'} \leq c^b_{p'+1} \leq k_{p'+1}$, i.e. $k_{p'} < k_{p'+1}$. These two conditions combined mean that shifting a cell from row p to row p' in $F(k)$ is a legal operation. But this shift yields a Ferrers diagram whose vector v satisfies $c^b_j \geq v_j$ for all $j < p$, and $v_p - c^b_p < k_p - c^b_p$. Iterating this construction will eventually allow us to reach a Ferrers diagram whose vector v satisfies $v_p - c^b_p = 0$. We then iterate on the index p to eventually reach a Ferrers diagram whose vector v satisfies $c^b_j \geq v^b_j$ for all $j \in [m]$, which brings us back to the previous case, from where a legal series of Add additions can be performed to reach $F(c^b)$, as desired. This completes the proof. $\square$

A.3   Proof of Theorem 11

First, note that Algorithm 4 always terminates for any sorted input stacks c^t, c^b, and outputs a finite word $w \in \{U, D, H^N, H^E\}^*$. We may therefore define the map $A : \text{DetSortedRec}_{m,n} \to \{U, D, H^N, H^E\}^*, c \mapsto w(c)$, given by the algorithm when the input corresponds to a sorted DR configuration on $K_{m,n}^0$. We wish to show that, for any $m \geq 0$ and $n \geq 1$, we have:

$$\forall c \in \text{DetSortedRec}_{m,n}, \ A(c) = \Xi \circ \Phi(c), \tag{7}$$

where Ξ and Φ are the bijections from Theorems 9 and 7 respectively.

We proceed by induction on m, n. The base case is $m = 0, n = 1$, for which $K_{m,n}^0$ is the graph consisting of a single edge from the sink v_0^t to the bottom vertex v_1^b. In this case there is a unique DR configuration $c = (\emptyset; 0)$ (which is the only stable configuration). For the input to Algorithm 4, we set the convention $c_1^t = n - 1 = 0$, leaving the stacks $c^t = c^b = (0)$. When running the algorithm, the inner while loop (lines 3 to 6) is empty, and we are in the case of the else branch (line 13). This yields $w = D$, and empties both stacks, so that we exit the outer while loop (lines 2 to 16). The D step is therefore immediately deleted, so that the algorithm outputs the empty word $w = \emptyset$. Now we also have $\Phi(c) = P \in \text{ParaPoly}_{1,1}$, where P is the parallelogram polyomino consisting of a single cell, and clearly $\Xi(P) = \emptyset$ as desired.

For the induction step, fix $m \geq 0$ and $n \geq 1$ such that $m + n > 1$. Suppose that Eq. (7) holds for all $m' \leq m$ and $n' \leq n$ such that $m' + n' < m + n$. Let $c \in \text{DetSortedRec}_{m,n}$, with the corresponding stacks c^t and c^b, and define $k := \min\{c_1^t, c_1^b\}$. We let $P := \Phi(c) \in \text{ParaPoly}_{m+1,n}$ be the parallelogram polyomino corresponding to c via the Dukes-Le Borgne bijection (Theorem 7), and write $\mathcal{U} = \mathcal{U}(P) = (N, u_1, \ldots, u_{m+n-1}, E)$ and $\mathcal{L} = \mathcal{L}(P) = (E, \ell_1, \ldots, \ell_{m+n-1}, N)$ for its upper and lower paths. Finally, we let $w = A(c)$ be the corresponding word output by Algorithm 4. We wish to show that $w = \Xi(P)$. By definition of k, the word w starts with exactly k steps U. Moreover, by construction of Φ, we have $u_1 = \cdots = u_k = N$ and $\ell_1 = \cdots = \ell_k = E$. We distinguish the three cases defining w_{k+1} given by the algorithm.

Case 1: $k = c_1^t < c_1^b$.

By definition, we have $w_{k+1} = H^E$. Moreover, the bijection Φ gives $u_{k+1} = \ell_{k+1} = E$. Next, we define $c' := (c_2^t, \ldots, c_m^t; c^b - 1) \in \text{Config}_{m-1,n}$. We also define $\mathcal{U}'$ and $\mathcal{L}'$ to be the paths $\mathcal{U}$ and $\mathcal{L}$ with their $(k+1)$-th step (which is E for both paths) deleted, i.e. $\mathcal{U}' := (N, u_1, \ldots, u_k, u_{k+2}, \ldots, u_{m+n-1}, E)$ and $\mathcal{L}' := (E, \ell_1, \ldots, \ell_k, \ell_{k+2}, \ldots, \ell_{m+n-1}, N)$. It is straighforward to check that these are the upper and lower paths of a new parallelogram polyomino $P' \in \text{ParaPoly}_{m,n}$, which satisfies $P' = \Phi(c')$. In particular, $c' \in \text{DetSortedRec}_{m-1,n}$, and the induction hypothesis gives $w' := A(c') = \Xi(P') \in \text{LabMotz}_{m-1,n-1}$.

Now by definition of c', the transformation $c \rightsquigarrow c'$ corresponds exactly to the first modification of the input stacks c^t, c^b in lines 8 and 9 of Algorithm 4, which is applied after the k iterations of the inner while loop (lines 4 and 5). In other words, if $w' = A(c') = w'_1 \ldots w'_{m+n-2}$ is output by the algorithm with input c', then we have $w = A(c) = w'_1 \ldots w'_k H^E w'_{k+1} \ldots w'_{m+n-2}$. The definition of Ξ and construction $P \rightsquigarrow P'$, combined with the induction hypothesis, then immediately imply that $w = \Xi(P)$, as desired.

Case 2: $k = c^b_1 < c^t_1$.
This case is entirely analogous to Case 1, exchanging the roles of c^t and c^b and changing the steps E and H^E to N and H^N respectively.

Case 3: $c^b_1 = c^t_1 = k$.
By definition, we have $w_{k+1} = D$, and the bijection Φ gives $u_{k+1} = E$ and $\ell_{k+1} = N$. First note that in this case we must have $k \geq 1$ since c is DR. Indeed, otherwise we would have $\mathcal{U}(P) = (N, E, \ldots)$ and $\mathcal{L}(P) = (E, N, \ldots)$, contradicting the definition of a parallelogram polyomino. We then define $c' := (c^t_2 - 1, \ldots, c^t_m - 1; c^b_2 - 1, \ldots, c^b_n - 1) \in \text{Config}_{m-1,n-1}$. Similarly to Case 1, we consider the corresponding parallelogram polyomino $P' = \Phi(c')$, which has upper and lower paths $\mathcal{U}' := (N, u_1, \ldots, u_k, u_{k+2}, \ldots, u_{m+n-1}, E)$ and $\mathcal{L}' := (E, \ell_1, \ldots, \ell_k, \ell_{k+2}, \ldots, \ell_{m+n-1}, N)$.

We apply the induction hypothesis to c', which implies that $w' := A(c') = \Xi(P') \in \text{LabMotz}_{m-1,n-2}$. Now note that, in this case, the transformation $c \rightsquigarrow c'$ corresponds exactly to initially applying the steps of the inner loop (lines 4 and 5) of Algorithm 4 to the input stacks c^t, c^b once, together with the first modification of the input stacks in line 14 (after all k iterations of the inner loop). In other words, if $w' = A(c') = w'_1 \ldots w'_{m+n-3}$ is output by the algorithm with input c', then we have $w = A(c) = U w'_1 \ldots w'_{k-1} D w'_k \ldots w'_{m+n-3}$. The result then follows as in Case 1 by applying the induction hypothesis to w' and using the definition of the bijection Ξ, together with the construction $P \rightsquigarrow P'$.

For the level formula, note that in the bijection Ξ from $\text{ParaPoly}_{m+1,n}$ to $\text{LabMotz}_{m,n-1}$, each step in the Motzkin path corresponds to a "diagonal band" in the parallelogram polyomino (see Fig. 9 for an illustration of these). It is straightforward to see that the area of each diagonal band is exactly one more than the area under the corresponding step. The result then follows from the level formula in Theorem 7. □

References

1. Alofi, A., Dukes, M.: A note on the lacking polynomial of the complete bipartite graph. Disc. Math. **348**(2), 8 (2025). ISSN-0012-365X. https://doi.org/10.1016/j.disc.2024.114323
2. Aval, J.C., D'Adderio, M., Dukes, M., Hicks, A., Le Borgne, Y.: Statistics on parallelogram polyominoes and a q, t-analogue of the Narayana numbers. J. Comb. Theory, Ser. A **123**(1), 271–286 (2014). https://doi.org/10.1016/j.jcta.2013.09.001

3. Aval, J.C., d'Adderio, M., Dukes, M., Le Borgne, Y.: Two operators on sandpile configurations, the sandpile model on the complete bipartite graph, and a cyclic lemma. Adv. Appl. Math. **73**, 59–98 (2016). https://doi.org/10.1016/j.aam.2015.09.018

4. Bak, P., Tang, C., Wiesenfeld, K.: Self-organized criticality: an explanation of the 1/f noise. Phys. Rev. Lett. **59**, 381–384 (1987). https://doi.org/10.1103/PhysRevLett.59.381

5. Bak, P., Tang, C., Wiesenfeld, K.: Self-organized criticality. Phys. Rev. A **38**(1), 364–374 (1988). https://doi.org/10.1103/PhysRevA.38.364

6. Chan, Y., Marckert, J.F., Selig, T.: A natural stochastic extension of the sandpile model on a graph. J. Comb. Theory Ser. A **120**(7), 1913–1928 (2013). https://doi.org/10.1016/j.jcta.2013.07.004

7. Cori, R., Le Borgne, Y.: On computation of Baker and Norine's rank on complete graphs. Electron. J. Comb. **23**(1), P1-31, 47pp. (2016). https://doi.org/10.37236/4350

8. Cori, R., Poulalhon, D.: Enumeration of (p, q)-parking functions. Disc. Math. **256**(3), 609–623 (2002). https://doi.org/10.1016/S0012-365X(02)00338-2

9. Cori, R., Rossin, D.: On the sandpile group of dual graphs. Eur. J. Comb. **21**(4), 447–459 (2000). https://doi.org/10.1006/eujc.1999.0366

10. Delest, M.P., Viennot, G.: Algebraic languages and polyominoes enumeration. Theor. Comput. Sci. **34**, 169–206 (1984). https://doi.org/10.1016/0304-3975(84)90116-6

11. Derycke, H., Dukes, M., Le Borgne, Y.: The sandpile model on the complete split graph: q, t-schröder polynomials, sawtooth polyominoes, and a cyclic lemma. ArXiv preprint (2024). https://doi.org/10.48550/arXiv.2402.15372

12. Dhar, D.: Self-organized critical state of sandpile automaton models. Phys. Rev. Lett. **64**(14), 1613–1616 (1990). https://doi.org/10.1103/PhysRevLett.64.1613

13. Dhar, D.: Theoretical studies of self-organized criticality. Phys. A Stat. Mech. Appl. **369**(1), 29–70 (2006). https://doi.org/10.1016/j.physa.2006.04.004

14. Dukes, M.: The sandpile model on the complete split graph, Motzkin words, and tiered parking functions. J. Comb. Theory Ser. A **180**, 15 (2021). https://doi.org/10.1016/j.jcta.2021.105418. id/No 105418

15. Dukes, M., Le Borgne, Y.: Parallelogram polyominoes, the sandpile model on a complete bipartite graph, and a q, t-Narayana polynomial. J. Comb. Theory Ser. A **120**(4), 816–842 (2013). https://doi.org/10.1016/j.jcta.2013.01.004

16. Dukes, M., Selig, T., Smith, J.P., Steingrímsson, E.: Permutation graphs and the abelian sandpile model, tiered trees and non-ambiguous binary trees. Electron. J. Comb. **26**(3), P3–29, 25pp. (2019). https://doi.org/10.37236/8225

17. Dukes, M., Selig, T., Smith, J.P., Steingrímsson, E.: The abelian sandpile model on Ferrers graphs - a classification of recurrent configurations. Eur. J. Comb. **81**, 221–241 (2019). https://doi.org/10.1016/j.ejc.2019.05.008

18. Selig, T.: Combinatorial aspects of sandpile models on wheel and fan graphs. Eur. J. Comb. **110**, 23 (2023). https://doi.org/10.1016/j.ejc.2022.103663. id/No 103663
19. Selig, T.: The stochastic sandpile model on complete graphs. Electron. J. Comb. **31**(3), P3-26, 29pp. (2024). https://doi.org/10.37236/12780
20. Selig, T., Smith, J.P., Steingrímsson, E.: EW-tableaux, Le-tableaux, tree-like tableaux and the abelian sandpile model. Electron. J. Comb. **25**(3), P3-14, 32pp. (2018). https://doi.org/10.37236/7480

Changing Induced Subgraph Isomorphisms Under Extended Reconfiguration Rules

Tatsuhiro Suga$^{(\boxtimes)}$ 🆔, Akira Suzuki 🆔, Yuma Tamura 🆔, and Xiao Zhou

Graduate School of Information Sciences, Tohoku University, Sendai, Japan
suga.tatsuhiro.p5@dc.tohoku.ac.jp, {akira,tamura,zhou}@tohoku.ac.jp

Abstract. In a reconfiguration problem, we are given two feasible solutions of a combinatorial problem and our goal is to determine whether it is possible to reconfigure one into the other, with the steps dictated by specific reconfiguration rules. Traditionally, most studies on reconfiguration problems have focused on rules that allow changing a single element at a time. In contrast, this paper considers scenarios in which $k \geq 2$ elements can be changed simultaneously. We investigate the general reconfiguration problem of isomorphisms. For the INDUCED SUBGRAPH ISOMORPHISM RECONFIGURATION problem, we show that the problem remains PSPACE-complete even under stringent constraints on the pattern graph when k is constant. We then give two meta-theorems applicable when k is slightly less than the number of vertices in the pattern graph. In addition, we investigate the complexity of the INDEPENDENT SET RECONFIGURATION problem, which is a special case of the INDUCED SUBGRAPH ISOMORPHISM RECONFIGURATION problem.

Keywords: combinatorial reconfiguration · extended reconfiguration rule · independent set reconfiguration · induced subgraph isomorphism · parameterized complexity · PSPACE-completeness

1 Introduction

In the field of combinatorial reconfigurations, we focus on the relationships between feasible solutions of combinatorial search problems. This field often addresses reconfiguration problems, which involve determining whether a current feasible solution can be changed to a target feasible solution in a step-by-step manner under a prescribed rule, called a *reconfiguration rule*. For example, in the INDEPENDENT SET RECONFIGURATION problem (ISR for short) under the token jumping rule, which is one of the most studied combinatorial problems [2,4,7,8,12,19,21,22,26,27], we are given a graph G and its two independent sets S_s and S_t. The goal is to find a sequence $\sigma = \langle S_s = S_0, S_1, ..., S_\ell = S_t \rangle$ of independent sets of G such that for each $i \in \{1, ..., \ell\}$, the size of the symmetric difference between S_i and S_{i-1} is 2, that is, $|S_i \setminus S_{i-1}| = |S_{i-1} \setminus S_i| = 1$.

Partially supported by WISE Program for AI Electronics, subsidized project by the Ministry of Education, Culture, Sports, Science and Technology.

Although the study of reconfiguration problems dates back to classical puzzles like the Tower of Hanoi and the 15-puzzle, research in this area has rapidly advanced since Ito et al. proposed a combinatorial reconfiguration framework in 2011 [22]. Various results on reconfiguration problems are summarized in several survey papers [20,30].

While reconfiguration problems have attracted attention in theoretical computer science, due to their fundamental properties, they are also well-studied for modeling continuously operating services that cannot afford downtime, such as infrastructure and monitoring systems. For instance, in an electrical power distribution network, it is not practical to halt the power supply to modify the switch configuration. Instead, it is essential to change the current configuration to the desired one step-by-step. To apply the theoretical insights developed in the field of combinatorial reconfigurations to real-world scenarios, various solvers for different reconfiguration problems have been developed. In particular, solvers for ISR have been created based on AI planning methods [10], answer set programming [33], zero-suppressed binary decision diagram (ZDD) [23], and bounded model checking [31]. Notably, all the above studies were published in 2023 and 2024. Thus, combinatorial reconfigurations are progressing from theoretical research to practical applications.

However, the existing reconfiguration rules face certain issues when reconfiguration problems are applied to real-world situations. Suppose that we are required to change the current configuration to perform maintenance on an electrical power distribution network, but we are aware that there is no way to reach the desired configuration under the existing reconfiguration rule. It makes no sense to abandon maintenance for that reason. We must somehow find a way to change to the desired configuration.

To resolve this, this paper deals with extended reconfiguration rules. Most of the reconfiguration problems in previous work iteratively change *one* element in each step. We extend this rule and allow the simultaneous change of *two or more* elements in each step. In reconfiguration problems involving changes to a vertex subset in a graph, we view such a vertex subset as a set of tokens. The token sliding rule (TS) and the token jumping rule (TJ) have been addressed in the literature. For each step, only one token can slide along an edge under TS, and it can jump to any vertex under TJ. We consider two reconfiguration rules k-TS and k-TJ, which are generalizations of TS and TJ, respectively. Each of them allows at most k tokens to be moved simultaneously. These reconfiguration rules may provide a way to reach the desired solution, even if the original rule could not (see also Fig. 1).

1.1   Our Contribution

This paper analyzes the computational complexity of a comprehensive reconfiguration problem, namely the INDUCED SUBGRAPH ISOMORPHISM RECONFIGURATION problem (ISIsoR for short), under the extended rules. Let $\mathcal{G}$ and $\mathcal{H}$ be graph classes. In ISIsoR, we are given two graphs $G \in \mathcal{G}$ and $H \in \mathcal{H}$, called a *host graph* and a *pattern graph*, respectively, two vertex subsets S_s and

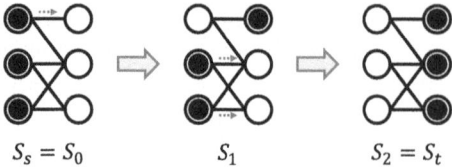

$$S_s = S_0 \qquad\qquad S_1 \qquad\qquad S_2 = S_t$$

Fig. 1. A sequence $\langle S_s = S_0, S_1, S_2 = S_t \rangle$ of independent sets within the same graph cannot be achieved under either TS or TJ, but it can be achieved under both 2-TS and 2-TJ. Tokens corresponding to the independent sets are marked in black.

S_t of G, each of which induces the subgraph isomorphic to H, and a recon-figuration rule R. The objective is to determine whether there is a sequence $\sigma = \langle S_s = S_0, S_1, ..., S_\ell = S_t \rangle$ of vertex subsets of G that satisfies the following two conditions: S_i induces the subgraph isomorphic to H for each $i \in \{0, ..., \ell\}$; and S_{i-1} and S_i follow the rule R for each $i \in \{1, ..., \ell\}$. Here, ISR can be viewed as a special case of ISIsoR where H is a null graph (i.e., a graph with no edges).

We first show that, for any additive graph class $\mathcal{H}$, there exists a fixed positive integer $k_{\mathcal{H}}$ such that, for any integer $k \geq k_{\mathcal{H}}$, ISIsoR under R $\in$ $\{k$-TS$, k$-TJ$\}$ is PSPACE-complete for $\mathcal{G}$ and $\mathcal{H}$. Furthermore, the problem remains PSPACE-complete even when $\mathcal{G}$ is the class of perfect graphs. In con-trast, if $k = |V(H)|$ for a pattern graph $H \in \mathcal{H}$, ISIsoR under k-TJ is trivially solvable in linear time. We thus focus on the case where k is slightly smaller than $|V(H)|$. To be precise, we consider a parameter $\mu = |V(H)| - k$, assum-ing $k < |V(H)|$. Such a parameter, so-called a (*below*) *guaranteed value*, was introduced by Mahajan and Raman [28] and has been studied in the literature. For ISIsoR under k-TJ, we give negative and positive meta-theorems when parameterized by $\mu = |V(H)| - k$ for a pattern graph $H \in \mathcal{H}$. The negative result is that, for the graph class $\mathcal{G}$ of general graphs and any additive graph class $\mathcal{H}$, if the INDUCED SUBGRAPH ISOMORPHISM problem (ISIso for short) is NP-complete for $\mathcal{G}$ and $\mathcal{H}$, then ISIsoR under k-TJ is also NP-complete for $\mathcal{G}$ and $\mathcal{H}$ when μ is any fixed positive integer. The positive result is that, if ISIso is solvable in polynomial time for a hereditary graph class $\mathcal{G}$ and a finite assorted graph class $\mathcal{H}$, then ISIsoR under k-TJ is in XP for $\mathcal{G}$ and $\mathcal{H}$ when parameterized by μ. As we will see later, a finite assorted graph class $\mathcal{H}$ is also additive. Combining the two meta-theorems, ISIsoR under k-TJ takes over the (in)tractability of ISIso for any finite assorted graph class $\mathcal{H}$.

The results for ISIsoR provide an interesting complexity change of ISR under k-TJ. Let I be an initial independent set of a given graph G. ISR under k-TJ is PSPACE-complete when $k = \Theta(1)$; NP-complete when $k = |I| - \Theta(1)$; and solvable in polynomial time when $k = |I|$. Although ISR under k-TJ on perfect graphs remains PSPACE-complete, we show the problem on perfect graphs is in XP when parameterized by $\mu = |I| - k$. The XP algorithm is the best possible under a reasonable assumption Gap-ETH. We also study the complexity of ISR under k-TS. We show that ISR under k-TS is essentially equivalent to ISR under TS if a given graph is even-hole-free. This result indicates that, for any integer k

with $1 \leq k \leq |I|$, ISR under k-TS is PSPACE-complete even for split graphs and solvable in polynomial time for interval graphs. In contrast to ISR under k-TJ, it is unlikely that an XP algorithm exists when parameterized by $\mu = |I| - k$ for ISR under k-TS on split graphs, a subclass of perfect graphs.

Due to the space limitation, proofs marked ($*$) are omitted.

1.2   Related Work

Only a few studies have addressed extended reconfiguration rules. Several works have explored the SHORTEST PATH RECONFIGURATION problem under the rule that allows at most k vertices of a shortest path to change simultaneously. This problem is known to be PSPACE-complete when k is constant [5,14,15,25], while there is an FPT algorithm when parameterized by $\mu = n/2 - k \geq 0$, where n is the number of vertices of an input graph [14]. Additionally, the research of the ISR problem has been conducted on how many vertices need to be changed simultaneously to ensure that any independent sets are reachable [3].

ISR can be viewed as a special case of ISIsoR where H is a null graph. It is known that ISR under $\mathsf{R} \in \{\mathsf{TS}, \mathsf{TJ}\}$ is PSPACE-complete in general [19,22,26], while the problem is solvable in polynomial time in several restricted cases [4,6–8,12,21,26]. The problems under $\mathsf{R} \in \{\mathsf{TS}, \mathsf{TJ}\}$ in the cases where pattern graphs are paths or cycles have also been explored, and PSPACE-completeness results established [18]. In the case where pattern graphs are (i, j)-biclique, the problem under TJ is also PSPACE-complete [18]. Another line of research has shown that the problem under $\mathsf{R} \in \{\mathsf{TS}, \mathsf{TJ}\}$ remains PSPACE-complete when a pattern graph comes from complete graphs (so-called CLIQUE RECONFIGURATION) [24]. It is important to note that these results for ISIsoR assume that exactly one vertex can be changed at a time.

2   Preliminaries

For a positive integer i, we write $[i] = \{1, 2, ..., i\}$. Let $G = (V, E)$ be a finite, simple, and undirected graph with the vertex set V and the edge set E. We denote by $V(G)$ and $E(G)$ the vertex set and the edge set of G, respectively. For a vertex v of G, we denote by $N_G(v)$ the *open neighborhood* of v, that is, $N_G(v) = \{u \in V \mid uv \in E\}$ and by $N_G[v]$ the *closed neighborhood*, that is, $N_G[v] = N_G(v) \cup \{v\}$. For a vertex set $X \subseteq V(G)$, we define $N_G(X) = \{v \notin X \mid uv \in E(G), u \in X\}$ and $N_G[X] = N_G(X) \cup X$. For two sets X and Y, we denote by $X \triangle Y$ the symmetric difference between X and Y, that is, $(X \setminus Y) \cup (Y \setminus X)$. A subgraph of G is a graph G' such that $V(G') \subseteq V(G)$ and $E(G') \subseteq E(G)$. For a subset $S \subseteq V(G)$, we denote by $G[S]$ the subgraph induced by S. The *complement* of G is the graph $\overline{G} = (V, \overline{E})$ such that $uv \in \overline{E}$ if and only if $uv \notin E$ for any pair of vertices $u, v \in V$. A sequence $\langle v_0, v_1, \dots, v_\ell \rangle_G$ of vertices of a graph G, where $v_{i-1} v_i \in E$ for each $i \in [\ell]$, is called a *path* if all the vertices are distinct. It is called a *cycle* if $v_0, v_1, \dots, v_{\ell-1}$ are distinct and $v_0 = v_\ell$. The value ℓ is referred to as the *length* of the path or cycle. A graph

G is said to be *connected* if there exists a path between u and v for any pair of vertices $u, v \in V$. The *diameter* of a connected graph G is the largest length of a shortest path between any two vertices of G. A *component* of G is a maximal connected subgraph of G. Let $\mathscr{C}_G$ denote the set of components of G. For two graphs G_1 and G_2, we denote by $G_1 + G_2$ the *disjoint union* of G_1 and G_2, that is, $G_1 + G_2 = (V(G_1) \cup V(G_2), E(G_1) \cup E(G_2))$. For an integer t and a graph G, we denote by tG the disjoint union of t copies of G. For a graph G and a vertex $u \in V(G)$, *duplicating* u is to add a new vertex v and add a new edge vw for each $w \in N_G(u)$. Similarly, for a graph G and a vertex subset $S \subseteq V(G)$, *duplicating* S is to duplicate each vertex in S. For two vertex sets $A, B \subseteq V(G)$, *joining* A and B is to add a new edge between each pair of vertices $u \in A$ and $v \in B$. For two graphs G and F, *replacing* a vertex $v \in V(G)$ with a graph F is to remove v from G, add the vertices and edges in F to G, and add new edges so that $N_G(v) = N_G(w)$ for each vertex $w \in V(F)$. For a positive integer i, we denote by K_i the complete graph with i vertices.

A graph class is a (non-empty) family of graphs satisfying a certain property. A graph class $\mathscr{G}$ is called *hereditary* if any induced subgraph G' of G is in $\mathscr{G}$ for any graph $G \in \mathscr{G}$. A graph class $\mathscr{G}$ is called *additive* if $G_1 + G_2$ is in $\mathscr{G}$ for any two graphs $G_1, G_2 \in \mathscr{G}$. We say that a graph class $\mathscr{G}$ is *finite assorted* if there exists a family $\mathscr{F}$ of connected graphs with constant size such that, for any positive integer $\ell \leq |\mathscr{F}|$, any ℓ positive integers $t_1, t_2, \cdots, t_\ell$, and any ℓ graphs $F_1, F_2, ..., F_\ell \in \mathscr{F}$, the graph $t_1 F_1 + t_2 F_2 + \cdots + t_\ell F_\ell$ belongs to $\mathscr{G}$. In other words, any graph in a finite assorted graph class $\mathscr{G}$ is obtained by the union of graphs in $\mathscr{F}$. For example, the class of graphs of maximum degree 1 is finite assorted, because there exists $\mathscr{F} = \{K_1, K_2\}$.

A *hole* is an induced cycle with a length at least four. An *odd hole* (resp. *even hole*) is a hole with an odd (resp. even) length. A graph G is *odd-hole-free* (resp. *even-hole-free*) if G has no odd hole (resp. even hole) as an induced subgraph. A graph G is called a *perfect graph* if both G and $\overline{G}$ are odd-hole-free [11]. A graph G is called a *bipartite graph* if $V(G)$ can be partitioned into two independent sets. It is well known that every bipartite graph is a perfect graph.

2.1   Our Problems

A graph H is *isomorphic* to a graph G if there exists a bijection ρ from $V(H)$ to $V(G)$ such that, for each $u, v \in V(H)$, $uv \in E(H)$ if and only if $\rho(u)\rho(v) \in E(G)$. A graph H is *induced subgraph isomorphic* to G if H is isomorphic to some induced subgraph of G. A vertex set $V' \subseteq V(G)$ is called an *H-induced subgraph isomorphic set* if H is isomorphic to $G[V']$. Let $\mathscr{G}$ and $\mathscr{H}$ be graph classes. The INDUCED SUBGRAPH ISOMORPHISM problem (ISIso for short) asks whether, given a *host graph* $G \in \mathscr{G}$ and a *pattern graph* $H \in \mathscr{H}$, there is an H-induced subgraph isomorphic set of G. INDEPENDENT SET and CLIQUE, which are well-known NP-complete problems [16], are special cases of ISIso. We define a reconfiguration variant of ISIso, that is, INDUCED SUBGRAPH ISOMORPHISM RECONFIGURATION (ISIsoR for short). In the problem, we are given a host graph $G \in \mathscr{G}$, a pattern graph $H \in \mathscr{H}$, two H-induced subgraph isomorphic sets $S_s \subseteq V(G)$ and $S_t \subseteq V(G)$, and a reconfiguration rule R. Then, the problem asks whether there exists a reconfiguration sequence $\sigma = \langle S_s = S_0, S_1, ...S_\ell = S_t \rangle$ of H-induced subgraph isomorphic sets and any two consecutive sets in σ follow the reconfiguration rule R.

For two vertex subsets S and S' of G, the change from S to S' can be viewed as the move of tokens on the vertices in $S \triangle S'$. In this paper, we consider the following extended reconfiguration rules for ISIsoR.

- k-**Token Jumping** (k-TJ): at most k tokens on vertices in G can be moved simultaneously.
- k-**Token Sliding** (k-TS): at most k tokens on vertices in G can be moved simultaneously and each token can be moved to an adjacent vertex in G.

For two H-induced subgraph isomorphic sets S and S', if we can change from S to S' by applying k-TJ (resp. k-TS) at a time, then we call that S and S' are *adjacent* under k-TJ (resp. k-TS). If there exists a reconfiguration sequence between S and S' under k-TJ (resp. k-TS), then we call that S and S' are *reconfigurable* under k-TJ (resp. k-TS). Note that, when $k = 1$, k-TJ and k-TS are equivalent to TJ and TS, which are fundamental and broadly studied reconfiguration rules in the field of combinatorial reconfigurations.

ISIsoR under k-TJ is trivial if k is the number of tokens, as we can move all tokens on an H-induced subgraph isomorphic set.

3   Induced Subgraph Isomorphism Reconfiguration

3.1   PSPACE-Completeness

In this subsection, we show that for a sufficiently large positive constant k, ISIsoR under $R \in \{k\text{-TS}, k\text{-TJ}\}$ is PSPACE-complete even for any additive graph class $\mathcal{H}$. The PSPACE-hardness of ISIsoR under $R \in \{k\text{-TS}, k\text{-TJ}\}$ is provided by a polynomial-time reduction from the W-WORD REACHABILITY problem, which is known to be PSPACE-complete [32]. Our reduction is inspired by the work of Kamiński et al., which proved the PSPACE-hardness of ISR for perfect graphs [26]. Given a pair $W = (\Sigma, A)$, where Σ is a set of symbols and $A \subseteq \Sigma^2$ is a binary relation between symbols, a string over Σ is a W-*word* if every two consecutive symbols are in A. The W-WORD REACHABILITY problem asks whether two given W-words w_s, w_t of equal length can be transformed into one another by changing one symbol at a time so that all intermediate strings are also W-words.

Theorem 1. *Let $\mathcal{G}$ be the graph class of general graphs, and let $\mathcal{H}$ be any additive graph class. Then there exists a fixed positive integer $k_{\mathcal{H}}$ depending on $\mathcal{H}$ such that, for any integer $k \geq k_{\mathcal{H}}$, ISIsoR under $R \in \{k\text{-TS}, k\text{-TJ}\}$ is PSPACE-complete for $\mathcal{G}$ and $\mathcal{H}$.*

Proof (Sketch). We only give the construction of our instance.

Consider an instance (w_s, w_t) of the W-WORD REACHABILITY problem where $W = (\Sigma, A)$. We will construct an instance (G, H, S_s, S_t, R) of ISIsoR under R $\in \{k\text{-TS}, k\text{-TJ}\}$ (see also Fig. 2). Let $\Sigma = \{\sigma_1, \sigma_2, ..., \sigma_{|\Sigma|}\}$ and $n = |w_s| = |w_t|$. Let $F \in \mathscr{H}$ be a graph with the smallest size in $\mathscr{H}$. Note that F is independent of the instance of ISIsoR and hence fixed. We set k to an arbitrary positive integer at least $k_{\mathscr{H}} = 2|V(F)|$. Let m be a positive integer such that $2^m|V(F)| \leq k < 2^{m+1}|V(F)|$ and $t = 2^m$. We construct a graph G' as follows. The vertex set $V(G')$ is the union of n vertex sets $L_1, L_2, ..., L_n$ such that each vertex set is with size $|\Sigma|$ and is a clique of G'. We call each of the vertex sets $L_1, L_2, ..., L_n$ a *layer*. For a positive integer $i \in [n-1]$ and two positive integers $j, j' \in [|\Sigma|]$, two vertices $v_j \in L_i$ and $v_{j'} \in L_{i+1}$ are joined by an edge if and only if $(\sigma_j, \sigma_{j'}) \notin A$. Then, G is a graph obtained from G' by replacing $v_j \in L_i$ with a graph tF_i^j, where F_i^j is isomorphic to F for each pair of $i \in [n]$ and $j \in [|\Sigma|]$. This completes the construction of G. Let $H = ntF$. Note that $H \in \mathscr{H}$ because $F \in \mathscr{H}$ holds and $\mathscr{H}$ is additive. For a W-word w and the graph G, we associate w with an H-induced subgraph isomorphic set S_w in G as follows. Denote by $w[i]$ the i-th symbol of w. For each $i \in [n]$, we add tokens on a vertex set $V(tF_i^j)$ if $w[i] = \sigma_j$ for $j \in [|\Sigma|]$. Lastly, let $S_s = S_{w_s}$ and $S_t = S_{w_t}$.

Lastly, we can show that (w_s, w_t) is a yes-instance of W-WORD REACHABILITY if and only if (G, H, S_s, S_t, R) is a yes-instance of ISIsoR. The details are omitted. □

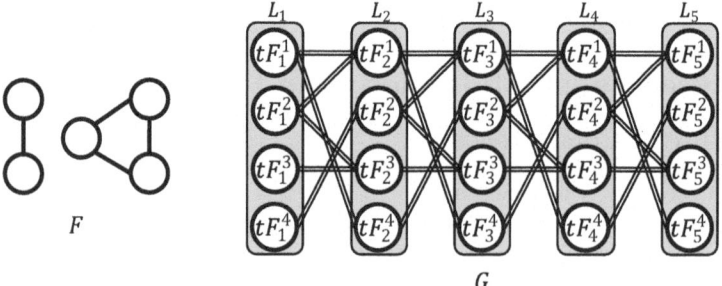

Fig. 2. An example of constructing G from $W = (\Sigma, A)$ and (w_s, w_t) such that $\Sigma = \{a, b, c, d\}$, $A = \{(a, b), (a, c), (b, b), (b, d), (c, a), (c, b), (c, d), (d, a), (d, c), (d, d)\}$, and $|w_s| = |w_t| = 5$. Each node in the layers corresponds to a subgraph isomorphic to tF, and each double line represents all possible edges between the two subgraphs.

From our construction of G in Theorem 1, we have the following corollary.

Corollary 1 (∗). *Let $\mathscr{G}$ be the graph class of perfect graphs, and let $\mathscr{H}$ be any additive graph class that contains some perfect graph. Then there exists a positive integer $k_{\mathscr{H}}$ depending on $\mathscr{H}$ such that, for any integer $k \geq k_{\mathscr{H}}$, ISIsoR is PSPACE-complete under R $\in \{k\text{-TS}, k\text{-TJ}\}$ for $\mathscr{G}$ and $\mathscr{H}$.*

It should be noted that many graph classes are additive and usually contain some perfect graph, such as K_1. Corollary 1 indicates that, for a sufficiently large constant k, ISIsoR is PSPACE-complete under R $\in \{k\text{-TS}, k\text{-TJ}\}$ for perfect graphs $\mathscr{G}$ and any "natural" graph class $\mathscr{H}$.

3.2 NP-Completeness

Recall that ISIsoR under k-TJ is trivially solvable in linear time if $k \geq |V(H)|$, where H is a pattern graph. Assuming $k < |V(H)|$, Sects. 3.2 and 3.3 are devoted to ISIsoR parameterized by $\mu = |V(H)| - k \geq 1$. In this subsection, we show the hardness result. Before providing it, we here define the notion of a *reconfiguration graph*, which will also be used to show the tractability of ISIsoR for special cases.

Let $(G, H, S_s, S_t, k\text{-TJ})$ be an instance of ISIsoR. We define the *reconfiguration graph* $\mathcal{C} = (\mathcal{V}, \mathcal{E})$ for ISIsoR as follows. Each vertex in $\mathcal{V}$ corresponds to an H-induced subgraph isomorphic set S of G with size exactly $|V(H)|$. We call each vertex of $\mathcal{C}$ a *node* in order to distinguish it from a vertex of G. We denote by w_S a node corresponding to an H-induced subgraph isomorphic set S. Then two H-induced subgraph isomorphic sets S and S' of G are adjacent under k-TJ if and only if the two corresponding nodes w_S and $w_{S'}$ are joined by an edge in the reconfiguration graph.

Theorem 2. *Let $\mathscr{G}$ be the graph class of general graphs, let $\mathscr{H}$ be any additive graph class, and let μ be any fixed positive integer. If* ISIso *is* NP-complete *for $\mathscr{G}$ and $\mathscr{H}$, then* ISIsoR *is* NP-complete *under k-TJ for $\mathscr{G}$ and $\mathscr{H}$ when $k = |V(H)| - \mu \geq 1$ for a pattern graph $H \in \mathscr{H}$.*

Proof (Sketch). We give the proof sketch of membership in NP and the construction of our instance for showing NP-hardness.

To show that ISIsoR under k-TJ is in NP when $k = |V(H)| - \mu$, we prove the diameter of each component of the reconfiguration graph C is $O(n^\mu)$, which implies that the length of a shortest reconfiguration sequence between any pair of two reconfigurable H-induced subgraph isomorphic sets is $O(n^\mu)$.

We next show that ISIsoR for $\mathscr{G}$ and $\mathscr{H}$ is NP-hard if ISIso for $\mathscr{G}$ and $\mathscr{H}$ is NP-hard by a polynomial-time reduction from ISIso to ISIsoR. Let (G', H') be an instance of ISIso with $2|V(H')| \geq \mu$. Even with this constraint on the instance, ISIso remains NP-hard because if $2|V(H')| < \mu$, then ISIso is solvable in polynomial time by enumerating all sets of constant size $|V(H')| < \mu/2$. We will construct an instance $(G, H, S_s, S_t, k\text{-TJ})$ of ISIsoR under k-TJ from (G', H').

We construct G from G' as follows (see also Fig. 3). Let A, B, X, and Y be four graphs such that each of them is isomorphic to $2H'$, and let H^* be a graph isomorphic to H'. Let G_0 be a graph such that $V(G_0) = \{g', a, b, h^*, x, y\}$ and $E(G_0) = \{g'a, g'b, ab, ah^*, bh^*, ax, by, xy\}$. Let G be a graph obtained from G_0 by replacing g', a, b, h^*, x and y with G', A, B, H^*, X, and Y, respectively. This completes the construction of G. Finally, set $H = 4H'$, $S_s = V(A) \cup V(Y)$, and $S_t = V(B) \cup V(X)$.

Lastly, we can prove that (G', H') is a yes-instance of ISIso if and only if $(G, H, S_s, S_t, k\text{-TJ})$ is a yes-instance of ISIsoR. The details are omitted. $\square$

Similar to Corollary 1, we can show that the above hardness holds even for perfect graphs.

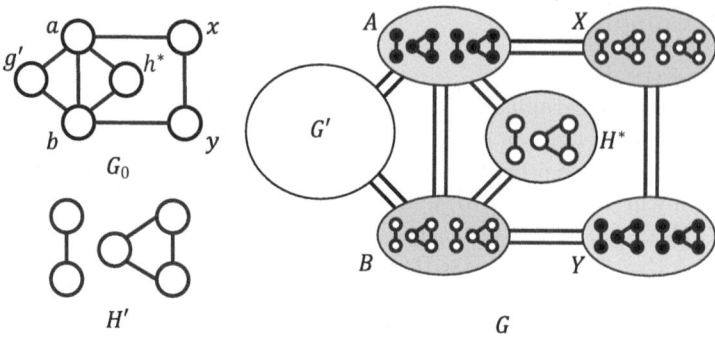

Fig. 3. Illustration of our construction of G. The double lines represent all possible edges between the two subgraphs. The black tokens are placed on the initial H-induced subgraph isomorphic set S_s.

Corollary 2 (∗)**.** *Let $\mathscr{G}$ and $\mathscr{H}$ be the graph class of perfect graphs and the additive graph subclass of perfect graphs, and let μ be any fixed positive integer. If* ISIso *is* NP-complete *for $\mathscr{G}$ and $\mathscr{H}$, then* ISIsoR *is* NP-complete *under k-TJ for $\mathscr{G}$ and $\mathscr{H}$ when $k = |V(H)| - \mu \geq 1$ for a pattern graph $H \in \mathscr{H}$.*

3.3   XP Algorithms

Theorem 3. *Let $\mathscr{G}$ be any hereditary graph class, and let $\mathscr{H}$ be any finite assorted graph class. If* ISIso *can be solved in polynomial time for $\mathscr{G}$ and $\mathscr{H}$, then* ISIsoR *under k-TJ is in* XP *for $\mathscr{G}$ and $\mathscr{H}$ when parameterized by $\mu = |V(H)| - k \geq 1$ for a pattern graph $H \in \mathscr{H}$.*

We give an XP algorithm that solves ISIsoR. To this end, we will construct a *clique-compressed reconfiguration graph* $\mathcal{C}' = (\mathcal{V}', \mathcal{E}')$, which is essentially equivalent to a reconfiguration graph $\mathcal{C} = (\mathcal{V}, \mathcal{E})$ but has fewer nodes than $\mathcal{C}$. We call each vertex of $\mathcal{C}'$ a *clique-node* in order to distinguish it from a node of $\mathcal{C}$ and a vertex of G. We denote by $(G, H, S_s, S_t, k\text{-TJ})$ an instance of ISIsoR under k-TJ.

We first define the clique-compressed reconfiguration graph $\mathcal{C}' = (\mathcal{V}', \mathcal{E}')$ as follows. Clique-nodes in $\mathcal{V}'$ are in one-to-one correspondence with vertex sets of $V(G)$ of size exactly μ. We denote by w'_T the clique-node that is assigned a vertex set $T \subseteq V(G)$. Then two clique-nodes w'_T and $w'_{T'}$ in $\mathcal{C}'$ are joined by an edge if and only if there exists an H-induced subgraph isomorphic set S such that $T \cup T' \subseteq S$.

Proposition 1 (∗)**.** *Let w_{S_s} and w_{S_t} be the two nodes in $\mathcal{C}$ corresponding to S_s and S_t, respectively. There is a path in $\mathcal{C}$ connecting w_{S_s} and w_{S_t} if and only if there is a path in $\mathcal{C}'$ connecting two clique-nodes w'_I and w'_J in $\mathcal{C}'$ for any $I \subseteq S_s$ and $J \subseteq S_t$ with size exactly μ.*

We now explain how to construct the clique-compressed reconfiguration graph $\mathcal{C}' = (\mathcal{V}', \mathcal{E}')$ without the reconfiguration graph $\mathcal{C}$.

For each vertex subset S of G with size exactly μ, we create a clique-node in $\mathcal{C}'$ to which S is assigned. We can construct the clique-node set $\mathcal{V}'$ in time $O(n^\mu)$ because $|\mathcal{V}'| = \binom{n}{\mu} = O(n^\mu)$, where $n = |V(G)|$.

We then construct the edge set $\mathcal{E}'$ of $\mathcal{C}'$. Consider two clique-nodes w'_A and w'_B, where A and B are vertex sets of G with $|A| = |B| = \mu$. Recall that the two clique-nodes w'_A and w'_B of $\mathcal{C}'$ are joined by an edge if and only if there exists an H-induced subgraph isomorphic set $S \subseteq V(G)$ such that $A \cup B \subseteq S$. In what follows, we show how to check whether such a vertex subset S exists. Let $C = A \cup B$. Since $\mathcal{H}$ is a finite assorted graph class, $H \in \mathcal{H}$ can be denoted as $H = t_1 F_1 + t_2 F_2 + \cdots + t_\ell F_\ell$ for some ℓ positive integers $t_1, t_2, \cdots, t_\ell$ and some graphs $F_1, F_2, ..., F_\ell$ with constant sizes. For each vertex $v \in C$, we assign a graph F_i for some $i \in [\ell]$ and we pick a set W_v of $|F_i|$ vertices in G such that $v \in W_v$ as a candidate for an F_i-induced subgraph isomorphic set in G. For $v_1, v_2, ..., v_{|C|} \in C$, let $W = W_{v_1} \cup W_{v_2} \cup \cdots \cup W_{v_{|C|}}$. Note that $A \cup B \subseteq W$. If $G[W]$ is not isomorphic to any components of H, then we abort the current choice and consider the next choice of F_i and W. Suppose that $G[W]$ is isomorphic to a graph consisting of some components F of H. Then we denote by G' the graph obtained from G by removing all vertices in $N_G[W]$. In addition, we denote by H' the graph obtained from H by removing the vertices of F. Since $\mathcal{G}$ is a hereditary graph class and $\mathcal{H}$ is a finite assorted graph class, we have $G' \in \mathcal{G}$ and $H' \in \mathcal{H}$. We solve the ISIso for the two graphs G' and H' in time $g(n)$, where $g(n)$ is a polynomial in n. We denote by (G', H') the instance of ISIso. If (G', H') is a yes-instance, then G' has an H'-induced subgraph isomorphic set $S_{H'}$. Since G' is a graph obtained from G by removing all vertices in $N_G[W]$, the graph G has an H-induced subgraph isomorphic set $W \cup S_{H'}$. Since $A \cup B \subseteq W$, in this case we have found an H-induced subgraph isomorphic set $W \cup S_{H'}$ of G such that $A \cup B \subseteq W \cup S_{H'}$. Therefore, the two clique-nodes w'_A and w'_B are joined by an edge. If (G', H') is a no-instance, we proceed to the next choice of F_i and W. We consider all possible choices of F_i and W. If there is no vertex set S such that $A \cup B \subseteq S$, we conclude that w'_A and w'_B are not joined by an edge. In this way, we construct $\mathcal{E}'$ by applying the above procedure for any two clique-nodes.

We lastly determine whether S_s and S_t are reconfigurable under k-TJ by checking whether there exists a path between two clique-node w'_A and w'_B for arbitrarily chosen vertex sets $A \subseteq S_s$ and $B \subseteq S_t$ of G with size exactly μ. The correctness follows from Proposition 1.

The total running time of this algorithm is $O(n^\mu + \ell^{2\mu} n^{2\mu(f_{\max}+1)} g(n) + n^{2\mu})$, where $f_{\max}$ denotes the largest number of vertices among the graphs $F_1, F_2, ..., F_\ell$, and $g(n)$ is a polynomial in n. The detailed explanation is omitted here. This completes the proof of Theorem 3.

4   Independent Set Reconfiguration

As applications of the results in Sect. 3, we discuss ISR under k-TJ. Given a graph G and a positive integer k, the INDEPENDENT SET problem (IS for short) asks whether G has an independent set of size at least k. IS is a well-known NP-complete problem. Recall that ISR is a reconfiguration variant of IS and is equivalent to ISISoR when $\mathcal{H}$ is the class of all null graphs. Combined with the PSPACE-completeness under TJ, Corollary 1 asserts the following corollary.

Corollary 3. *For any fixed integer $k \geq 1$, ISR under k-TJ is PSPACE-complete for perfect graphs.*

Let μ be a fixed positive integer, and let I be an initial independent set of given graph G. When $k = |I| - \mu$, since IS is NP-complete for general graphs and the class $\mathcal{H}$ of null graphs is finite assorted (more generally, additive), the NP-completeness of ISR under k-TJ follows from Theorem 2.

Corollary 4. *Let μ be any fixed positive integer. ISR under k-TJ is NP-complete for general graphs when $k = |I| - \mu \geq 1$, where I is an initial independent set of a given graph G.*

In contrast, it is known that IS is solvable in polynomial time for perfect graphs [17]. Furthermore, the class of perfect graphs is hereditary. Therefore, Theorem 3 gives the following tractability.

Corollary 5. *ISR under k-TJ is in XP for perfect graphs when parameterized by $\mu = |I| - k \geq 1$, where I is an initial independent set of a given graph G.*

We here show that, for perfect graphs, the XP algorithm by Corollary 5 is the best possible under a reasonable assumption. This result is based on Gap-ETH [13,29]. For a positive integer q, in the q-SAT problem, we are given a q-CNF formula ϕ in which each clause consists of at most q literals. Our goal is to decide whether ϕ is satisfiable. Max q-SAT is a maximization variant of q-SAT. For a positive integer q, the MAX q-SAT problem takes as input a q-CNF formula ϕ and our goal is to compute the maximum number SAT(ϕ) of clauses in ϕ that can be simultaneously satisfied. The Gap-ETH can now be expressed in terms of SAT as follows.

Conjecture 1 (Gap-Exponential Time Hypothesis (Gap-ETH) [13,29]). For some constant $\delta, \epsilon > 0$, no algorithm can, given a 3-SAT formula ϕ on n variables and $m = O(n)$ clauses, distinguish between the following cases correctly with probability at least $2/3$ in time $O(2^{\delta n})$: (i) SAT$(\phi) = m$ and (ii) SAT$(\phi) < (1 - \epsilon)m$.

Note that Gap-ETH with $\epsilon = 1/m$ means ETH (Exponential-Time Hypothesis), which is the widely accepted assumption in the field of parameterized complexity. Under Gap-ETH, we show the following Theorem 4.

Theorem 4. *Assuming Gap-ETH,* ISR *under* k-TJ *does not admit any* FPT *algorithm for bipartite graphs when parameterized by* $\mu = |I| - k \geq 1$, *where* I *is an initial independent set of a given graph* G.

Proof. We perform an FPT-reduction from MAXIMUM BALANCED BICLIQUE. Our reduction is inspired by the FPT-reduction of Agrawal et al. in [1], which proved that ISR under TJ for bipartite graphs does not admit an FPT algorithm parameterized by the number of tokens under Gap-ETH. In MAXIMUM BALANCED BICLIQUE, we are given a bipartite graph $G = (A \cup B, E)$ and an integer b. The goal is to decide whether G contains a complete bipartite subgraph (biclique) with b vertices on each side. MAXIMUM BALANCED BICLIQUE does not admit an FPT algorithm parameterized by b unless Gap-ETH fails [9].

We will first construct an instance (G', I, J, μ) of ISR under k-TJ parameterized by $\mu = |I| - k = b$ from an instance (G, b) of MAXIMUM BALANCED BICLIQUE (see also Fig. 4). Let $G^* = (A \cup B, \{uv \mid u \in A, v \in B, uv \notin E(G)\})$, and let $\tilde{G}$ be the graph obtained from G^* by duplicating B a total of $c = |A| - b$ times. We denote by $B_1, ..., B_c$ the vertex subsets obtained by duplicating B and denote $B_0 = B$. Let $B' = B_0 \cup B_1 \cup \cdots \cup B_c$. We perform the following operations on $\tilde{G}$: add a new biclique with two independent sets S and T, where $|S| = |T| = (c+2)b$; join S and B'; and join T and A. We denote by G' the graph obtained as above. Let $I = S$ and $J = T$, then we have $|I| = |J| = (c+2)b$.

To complete our FPT-reduction, we now show that (G, b) is a yes-instance if and only if (G', I, J, μ) is a yes-instance.

Suppose that G has a biclique with b vertices on each side. Then, we construct a reconfiguration sequence between I and J as follows. First, move $\mu = b$ tokens from S to A. Note that this move is successful because $k = |I| - \mu = (c+2)b - b = (c+1)b > \mu$ and there is no edge between any pair of a vertex in S and a vertex in A. Second, move $(c+1)b = k$ remaining tokens on S to B' so that, for each $i \in [c] \cup \{0\}$, a set of vertices in $A \cup B_i$ on which the tokens are placed is an independent set. This can be done because a biclique with b vertices on each side of G now corresponds to the independent set of $\tilde{G}[A \cup B_i]$ for every $i \in [c] \cup \{0\}$. Third, move all μ tokens on A to T. Lastly, move all $(c+1)b$ tokens on B' to T. Therefore, I and J are reconfigurable.

Conversely, suppose that there is a reconfiguration sequence $\sigma = \langle I = I_0, I_1, ..., I_\ell = J \rangle$ between I and J under k-TJ with $k = |I| - \mu$. Let $i > 0$ be the minimum integer such that $I_i \cap S = \emptyset$. Considering I_{i-1}, we have $I_{i-1} \subseteq S \cup A$ because $I_{i-1} \cap (B' \cup T) = \emptyset$ due to $I_{i-1} \cap S \neq \emptyset$. Recall that $|I_{i-1}| = |I| = (c+2)b$ and $k = |I| - b = (c+1)b$. Since $I_i \cap S = \emptyset$, this implies that $|I_i \cap A| \geq b$ and hence $I_i \cap T = \emptyset$ from the construction of G'. Furthermore, since $|A| = c + b$, we have $|I_i \cap B'| \geq |I_i \setminus A| = (c+2)b - (c+b) = (c+1)(b-1) + 1$. With the pigeonhole principle, this indicates that at least one vertex set among $B_0, B_1, ..., B_c$ contains b vertices in I_i. Therefore, $\tilde{G}[A \cup B_i]$ contains an independent set I_i such that $|I_i \cap A| \geq b$ and $|I_i \cap B_j| \geq b$ for some $j \in [c] \cup \{0\}$, that is, G has a biclique with b vertices on each side.   □

We next deal with ISR under k-TS. We show the equivalence between TS and k-TS for even-hole-free graphs.

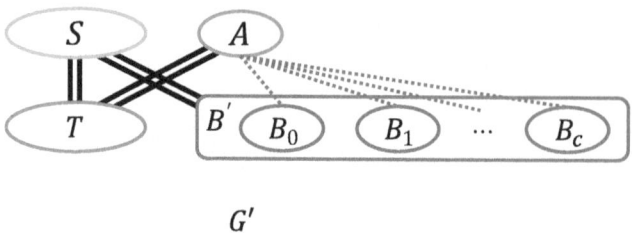

G'

Fig. 4. Illustration of our construction of G'. The double lines represent all possible edges between the two vertex sets. The dotted line between A and B_i represents all edges in G^*.

Theorem 5 (∗). *Let G be an even-hole-free graph and let I and J be two independent sets of G. Then, I and J are reconfigurable under TS if and only if I and J are reconfigurable under k-TS.*

ISR under TS is known to be PSPACE-complete even for split graphs [2] and solvable in polynomial time for interval graphs [4], where these graphs are even-hole-free. Therefore, from Theorem 5, we obtained the following corollary.

Corollary 6. *For any integer k with $1 \le k \le |I|$, ISR under k-TS is PSPACE-complete even for split graphs and solvable in polynomial time for interval graphs, where I is an initial independent set of a given instance.*

5   Conclusion

In this paper, to extensively investigate how the extension of reconfiguration rules affects the complexity of reconfiguration problems, we researched a comprehensive reconfiguration problem, namely the INDUCED SUBGRAPH ISOMORPHISM RECONFIGURATION problem. We presented two meta-theorems that include both negative and positive results for this problem. Our insight would be useful for future research on reconfiguration problems under extended rules.

Acknowledgements. We are grateful to Naoki Domon and Takahiro Suzuki for valuable discussions with them. We thank anonymous referees for their valuable comments and suggestions, which greatly helped improve this paper's presentation.

References

1. Agrawal, A., Allumalla, R.K., Dhanekula, V.T.: Refuting FPT algorithms for some parameterized problems under Gap-ETH. In: IPEC. LIPIcs, vol. 214, pp. 2:1–2:12. Schloss Dagstuhl - Leibniz-Zentrum für Informatik (2021)
2. Belmonte, R., Kim, E.J., Lampis, M., Mitsou, V., Otachi, Y., Sikora, F.: Token sliding on split graphs. Theory Comput. Syst. **65**(4), 662–686 (2021)

3. de Berg, M., Jansen, B.M.P., Mukherjee, D.: Independent-set reconfiguration thresholds of hereditary graph classes. Disc. Appl. Math. **250**, 165–182 (2018)
4. Bonamy, M., Bousquet, N.: Token sliding on chordal graphs. In: Bodlaender, H.L., Woeginger, G.J. (eds.) WG 2017. LNCS, vol. 10520, pp. 127–139. Springer, Cham (2017). https://doi.org/10.1007/978-3-319-68705-6_10
5. Bonsma, P.S.: The complexity of rerouting shortest paths. Theor. Comput. Sci. **510**, 1–12 (2013)
6. Bonsma, P.S.: Independent set reconfiguration in cographs and their generalizations. J. Graph Theory **83**(2), 164–195 (2016)
7. Bonsma, P., Kamiński, M., Wrochna, M.: Reconfiguring independent sets in claw-free graphs. In: Ravi, R., Gørtz, I.L. (eds.) SWAT 2014. LNCS, vol. 8503, pp. 86–97. Springer, Cham (2014). https://doi.org/10.1007/978-3-319-08404-6_8
8. Brianski, M., Felsner, S., Hodor, J., Micek, P.: Reconfiguring independent sets on interval graphs. In: MFCS. LIPIcs, vol. 202, pp. 23:1–23:14. Schloss Dagstuhl - Leibniz-Zentrum für Informatik (2021)
9. Chalermsook, P., et al.: From Gap-exponential time hypothesis to fixed parameter tractable inapproximability: clique, dominating set, and more. SIAM J. Comput. **49**(4), 772–810 (2020)
10. Christen, R., et al.: PARIS: planning algorithms for reconfiguring independent sets. In: ECAI. Frontiers in Artificial Intelligence and Applications, vol. 372, pp. 453–460. IOS Press (2023)
11. Chudnovsky, M., Robertson, N., Seymour, P., Thomas, R.: The strong perfect graph theorem. Ann. Math. **164**, 51–229 (2002)
12. Demaine, E.D., et al.: Linear-time algorithm for sliding tokens on trees. Theor. Comput. Sci. **600**, 132–142 (2015)
13. Dinur, I.: Mildly exponential reduction from gap 3SAT to polynomial-gap labelcover. Electron. Colloquium Comput. Complex. **TR16-128** (2016)
14. Domon, N., Suzuki, A., Tamura, Y., Zhou, X.: The shortest path reconfiguration problem based on relaxation of reconfiguration rules. In: WALCOM. Lecture Notes in Computer Science, vol. 14549, pp. 227–241. Springer, Heidelberg (2024). https://doi.org/10.1007/978-981-97-0566-5_17
15. Gajjar, K., Jha, A.V., Kumar, M., Lahiri, A.: Reconfiguring shortest paths in graphs. Algorithmica **86**(10), 3309–3338 (2024)
16. Garey, M.R., Johnson, D.S.: Computers and Intractability: A Guide to the Theory of NP-Completeness. Freeman W. H, New York City (1979)
17. Grötschel, M., Lovász, L., Schrijver, A.: Polynomial algorithms for perfect graphs. In: Topics on Perfect Graphs, North-Holland Mathematics Studies, vol. 88, pp. 325–356. North-Holland (1984)
18. Hanaka, T., et al.: Reconfiguring spanning and induced subgraphs. Theor. Comput. Sci. **806**, 553–566 (2020)
19. Hearn, R.A., Demaine, E.D.: PSPACE-completeness of sliding-block puzzles and other problems through the nondeterministic constraint logic model of computation. Theor. Comput. Sci. **343**(1–2), 72–96 (2005)
20. van den Heuvel, J.: The complexity of change. In: Surveys in Combinatorics, London Mathematical Society Lecture Note Series, vol. 409, pp. 127–160. Cambridge University Press (2013)
21. Hoang, D.A., Uehara, R.: Sliding tokens on a cactus. In: ISAAC. LIPIcs, vol. 64, pp. 37:1–37:26. Schloss Dagstuhl - Leibniz-Zentrum für Informatik (2016)
22. Ito, T., et al.: On the complexity of reconfiguration problems. Theor. Comput. Sci. **412**(12–14), 1054–1065 (2011)

23. Ito, T., et al.: ZDD-based algorithmic framework for solving shortest reconfiguration problems. In: CPAIOR. Lecture Notes in Computer Science, vol. 13884, pp. 167–183. Springer, Heidelberg (2023). https://doi.org/10.1007/978-3-031-33271-5_12

24. Ito, T., Ono, H., Otachi, Y.: Reconfiguration of cliques in a graph. Disc. Appl. Math. **333**, 43–58 (2023)

25. Kaminski, M., Medvedev, P., Milanic, M.: Shortest paths between shortest paths. Theor. Comput. Sci. **412**(39), 5205–5210 (2011)

26. Kaminski, M., Medvedev, P., Milanic, M.: Complexity of independent set reconfigurability problems. Theor. Comput. Sci. **439**, 9–15 (2012)

27. Lokshtanov, D., Mouawad, A.E.: The complexity of independent set reconfiguration on bipartite graphs. ACM Trans. Algor. **15**(1), 7:1–7:19 (2019)

28. Mahajan, M., Raman, V.: Parameterizing above guaranteed values: MaxSat and MaxCut. J. Algor. **31**(2), 335–354 (1999)

29. Manurangsi, P., Raghavendra, P.: A birthday repetition theorem and complexity of approximating dense CSPs. In: ICALP. LIPIcs, vol. 80, pp. 78:1–78:15. Schloss Dagstuhl - Leibniz-Zentrum für Informatik (2017)

30. Nishimura, N.: Introduction to reconfiguration. Algorithms **11**(4), 52 (2018)

31. Toda, T., Ito, T., Kawahara, J., Soh, T., Suzuki, A., Teruyama, J.: Solving reconfiguration problems of first-order expressible properties of graph vertices with boolean satisfiability. In: ICTAI, pp. 294–302. IEEE (2023)

32. Wrochna, M.: Reconfiguration in bounded bandwidth and tree-depth. J. Comput. Syst. Sci. **93**, 1–10 (2018)

33. Yamada, Y., Banbara, M., Inoue, K., Schaub, T., Uehara, R.: Combinatorial reconfiguration with answer set programming: algorithms, encodings, and empirical analysis. In: WALCOM. Lecture Notes in Computer Science, vol. 14549, pp. 242–256. Springer, Heidelberg (2024). https://doi.org/10.1007/978-981-97-0566-5_18

A Unified Model of Congestion Games with Priorities

Two-Sided Markets with Ties, Finite and Non-affine Delay Functions, and Pure Nash Equilibria

Kenjiro Takazawa[✉][iD]

Hosei University, Tokyo 184-8584, Japan
takazawa@hosei.ac.jp

Abstract. The study of equilibrium concepts in congestion games and two-sided markets with ties has been a primary topic in game theory, economics, and computer science. Ackermann, Goldberg, Mirrokni, Röglin, Vöcking (2008) gave a common generalization of these two models, in which a player more prioritized by a resource produces an infinite delay on less prioritized players. While presenting several theorems on pure Nash equilibria in this model, Ackermann et al. posed an open problem of how to design a model in which more prioritized players produce a large but finite delay on less prioritized players. In this paper, we present a positive solution to this open problem by combining the model of Ackermann et al. with a generalized model of congestion games due to Bilò and Vinci (2023). In the model of Bilò and Vinci, the more prioritized players produce a finite delay on the less prioritized players, while the delay functions are of a specific kind of affine functions, and all resources have the same priorities. By unifying these two models, we achieve a model in which the delay functions may be finite and non-affine, and the priorities of the resources may be distinct. We prove some positive results on the existence and computability of pure Nash equilibria in our model, which extend those for the previous models and support the validity of our model.

Keywords: Algorithmic game theory · Congestion game · Two-sided market · Pure Nash equilibrium · Potential game · Matroid

1 Introduction

The study of equilibrium concepts in noncooperative games is a primary topic in the fields of game theory, economics, and computer science. In particular, the models of *congestion games* and *two-sided markets with ties* have played important roles in the literature.

Congestion games, introduced by Rosenthal [34] in 1973, represent situations in which congestion results in more delay. Each *player* chooses a strategy, which is a set of *resources*. If a resource is shared by many players, then much delay is imposed on those players. The objective of a player is to minimize the total delay of the resources in her strategy.

ⓒ The Author(s), under exclusive license to Springer Nature Singapore Pte Ltd. 2025
S. Nakano and M. Xiao (Eds.): WALCOM 2025, LNCS 15411, pp. 361–376, 2025.
https://doi.org/10.1007/978-981-96-2845-2_23

Table 1. Results of Ackermann et al. [1]. "NPS" stands for "Non-Player-Specific," while "PS" stands for "Player-Specific." "Polynomial BR Dynamics" means that there exists a sequence of a polynomial number of best responses reaching a pure Nash equilibrium. Theorems on two-sided markets are omitted in this extended abstract.

	Consistent Priorities	**Inconsistent Priorities**
NPS	**Polynomial BR Dynamics** - Singleton Game (Theorem 6) - Matroid Game (Theorem 9)	**Potential Function** - Singleton Game (Theorem 7) - Matroid Game (Theorem 10)
PS	—	**Potential Function** - Two-Sided Singleton Market - Two-Sided Matroid Market **Polynomial Algorithm** - Singleton Game (Theorem 8) - Matroid Game (Theorem 11)

Rosenthal [34] proved that every congestion game is an *exact potential game*, and hence possesses a pure Nash equilibrium: see Sect. 2.1 for definition. Moreover, Monderer and Shapley [32] proved the converse: every exact potential game can be represented as a congestion game. On the basis of these results, congestion games are recognized as a fundamental model in the study of pure Nash equilibria in noncooperative games (see, e.g., [7,35]).

A *two-sided market* consists of *agents* and *markets*, which have preferences over the other side. Typical special cases are the stable matching and Hospitals/Residents problems. Since the pioneering work of Gale and Shapley [14], analyses on equilibria have been a primary subject, and a large number of generalized models have been proposed. In particular, a typical generalization of allowing *ties* in the preferences [24] critically changes the difficulty of the analyses (see [16,29]), and has attracted intensive interests [15,17,21,22,25–27,30,41].

In 2008, Ackermann, Goldberg, Mirrokni, Röglin, and Vöcking [1] introduced a model which commonly generalizes congestion games and two-sided markets with ties, referred to as *congestion games with priorities*. Each resource e has priorities (preferences) with ties over the players. Among the players choosing e in their strategies, only the players most prioritized by e receive a finite delay from e, and the less prioritized players receive an infinite delay. In other words, only the most prioritized players are accepted. For several classes of their model, Ackermann et al. [1] presented some positive results on the existence and computability of pure Nash equilibria. These results are summarized in Table 1 and will be formally described in Sect. 2.2. In the class of *consistent priorities*, all resources have the same priorities over the players, while it is not the case with that of *inconsistent priorities*. In *player-specific congestion games*, each resource e has a specific delay function $d_{i,e}$ for each player i. In a *singleton congestion game*, every strategy of every player consists of a single resource. In a *matroid congestion game*, the strategies of each player are the bases of a matroid, which we refer to as *inconsistent priorities*.

Table 2. Summary of Our Results. "NPS" stands for "Non-Player-Specific," while "PS" stands for "Player-Specific." "Polynomial BR Dynamics" means that there exists a sequence of a polynomial number of better responses reaching a pure Nash equilibrium. "PNE" stands for "Pure Nash Equilibrium." Theorems on two-sided markets are omitted in this extended abstract.

	Consistent Priorities	Inconsistent Priorities
NPS	**Polynomial BR Dynamics** - Singleton Game (Theorem 13) - Matroid Game (Theorem 16) **Existence of a PNE** - General Game (Theorem 19)	**Potential Function** - Singleton Game (Theorem 14) - Matroid Game (Theorem 17)
PS	Polynomial BR Dynamics - Singleton Game (Theorem 13) - Matroid Game (Theorem 16)	**Potential Function** - Two-Sided Singleton Market - Two-Sided Matroid Market **Existence of a PNE** - Singleton Game (Theorem 15) - Matroid Game (Theorem 18)

Meanwhile, Ackermann et al. [1] posed an open question of how to design a model in which the less prioritized players receive a large but finite delay. We appreciate the importance of this question because such a model provides a closer connection to the many-to-many models of stable matchings [4,5,9,12,36,37] (see also [29]). In the many-to-many models of stable matchings, a market can accept agents with different preferences. Thus, the question in [1] is crucial to attain a reasonable generalization of congestion games with priorities corresponding to the many-to-many stable matching problem.

A main contribution of this paper is to design a model which gives a positive and full answer to the open problem of Ackermann et al. [1]. Before presenting our model, we first point out that a generalized model of congestion games by Bilò and Vinci [6] partially answers the question. In their model, the players more prioritized by a resource indeed produce a finite delay on the less prioritized players. Meanwhile, this model only covers a special case of the model of [1] in which the delay functions are of a specific kind of affine functions and the priorities of the resources are consistent.

We then present our model, which unifies those of Ackermann et al. [1] and Bilò and Vinci [6]. Its characteristic feature is that the more prioritized players produce a finite delay on the less prioritized players, the delay function may be non-affine, and the resources may have inconsistent priorities. We prove some positive results on the existence and computability of pure Nash equilibria in our model, which extend those for the previous models [1,6] and support the validity of our model. Our technical results are listed in Table 2.

The rest of the paper is organized as follows. We review previous results in Sect. 2, and describe our model in Sect. 3. In Sect. 4, we present some positive results on pure Nash equilibria in our model. Finally, in Sect. 5, we deal with more general classes which are not of singleton games. Due to the page limitation, the description of two-sided markets with ties is deferred to the full version of this paper.

2   Preliminaries

Let $\mathbb{Z}$ denote the set of the integers, and $\mathbb{R}$ that of the real numbers. Subscripts $+$ and $++$ represent that the set consists of nonnegative numbers and positive numbers, respectively. Further, let $\bar{\mathbb{R}}_+$ denote $\mathbb{R}_+ \cup \{+\infty\}$.

2.1   Congestion Games

A congestion game is described by a tuple $(N, E, (\mathcal{S}_i)_{i \in N}, (d_e)_{e \in E})$. Here, $N = \{1, \ldots, n\}$ denotes the set of the players and E that of the resources. Each player $i \in N$ has her *strategy space* $\mathcal{S}_i \subseteq 2^E$, and chooses a *strategy* $S_i \in \mathcal{S}_i$. The collection $(S_1, \ldots, S_n)$ of the chosen strategies is called a *strategy profile*. For a resource $e \in E$ and a strategy profile $S = (S_1, \ldots, S_n)$, let $N_e(S) \subseteq N$ denote the set of players whose strategy includes e, and let $n_e(S) \in \mathbb{Z}_+$ denote the size of $N_e(S)$, i.e., $N_e(S) = \{i \in N : e \in S_i\}$ and $n_e(S) = |N_e(S)|$.

Each resource $e \in E$ has its *delay function* $d_e \colon \mathbb{Z}_{++} \to \mathbb{R}_+$. In a strategy profile S, the function value $d_e(n_e(S))$ represents the delay of a resource $e \in E$. The objective of each player $i \in N$ is to minimize her cost $\gamma_i(S) = \sum_{e \in S_i} d_e(n_e(S))$.

Let $S = (S_1, \ldots, S_n)$ be a strategy profile and $i \in N$. Let S_{-i} denote a collection of the strategies in S other than S_i, namely $S_{-i} = (S_1, \ldots, S_{i-1}, S_{i+1}, \ldots, S_n)$. A *better response* of i in S is to change her strategy from S_i to S_i' such that $\gamma_i(S_{-i}, S_i') < \gamma_i(S)$. In particular, a better response from S_i to S_i' is a *best response* if S_i' minimizes $\gamma_i(S_{-i}, S_i')$. A *pure Nash equilibrium* is a strategy profile in which no player has a better response.

An *exact potential game* admits an *exact potential function* Φ, which is defined on the set of the strategy profiles and satisfies $\Phi(S_{-i}, S_i') - \Phi(S) = \gamma_i(S_{-i}, S_i') - \gamma_i(S)$ for each strategy profile S, each player $i \in N$, and each strategy $S_i' \in \mathcal{S}$. More generally, a *potential function* is one which maps the set of the strategy profile to an ordered set so that $\Phi(S_{-i}, S_i') \prec \Phi(S)$ if $\gamma_i(S_{-i}, S_i') < \gamma_i(S)$. A game admitting a potential function is referred to as a *potential game*, and a potential game possesses a pure Nash equilibrium because a strategy profile minimizing the potential function must be a pure Nash equilibrium.

The following theorem is a primary result on congestion games, stating that each congestion game is an exact potential game and vice versa.

Theorem 1 ([32,34]).   *A congestion game is an exact potential game, and hence possesses a pure Nash equilibrium. Moreover, every exact potential game is represented as a congestion game.*

Study on congestion games from the viewpoint of *algorithmic game theory* [7,33,35] has appeared since around 2000. For singleton congestion games, Ieong, McGrew, Nudelman, Shoham, and Sun [23] proved the following theorem.

Theorem 2 ([23]). *In a singleton congestion game, starting from an arbitrary strategy profile, a pure Nash equilibrium is attained after a polynomial number of better responses.*

This theorem is followed by a large number of extensions. Recall that a *player-specific congestion game* is one in which each resource $e \in E$ has a delay function $d_{i,e} \colon \mathbb{Z}_{++} \to \mathbb{R}_+$ specific to each player $i \in N$.

Theorem 3 ([31]). *In a player-specific singleton congestion game, there exists a sequence of polynomial number of best responses starting from an arbitrary strategy profile and reaching a pure Nash equilibrium.*

Note that Theorem 3 differs from Theorem 2 in that not any sequence of best responses reaches to a pure Nash equilibrium.

A significant work which has employed the discrete structure of *matroids* into congestion games is due to Ackermann, Röglin, and Vöcking [2,3]. For a finite set E and its subset family $\mathcal{S} \subseteq 2^E$, the pair $(E, \mathcal{S})$ is a *matroid* if $\mathcal{S} \neq \emptyset$ and

$$\text{if } S, S' \in \mathcal{S} \text{ and } e \in S \backslash S', (S \backslash \{e\}) \cup \{e'\} \in \mathcal{S} \text{ for some } e' \in S' \backslash S. \quad (1)$$

A set in $\mathcal{S}$ is referred to as a *base*. It follows from (1) that all bases in $\mathcal{S}$ have the same cardinality, which is referred to as the *rank* of the matroid $(E, \mathcal{S})$.

A congestion game $(N, E, (\mathcal{S}_i)_{i \in N}, (d_e)_{e \in E})$ is referred to as a *matroid congestion game* if $(E, \mathcal{S}_i)$ is a matroid for every player $i \in N$. It is straightforward to see that a singleton congestion game is a special case of a matroid congestion game. Ackermann, Röglin, and Vöcking [2,3] proved the following extensions of Theorems 2 and 3.

Theorem 4 ([2]). *In a matroid congestion game, starting from an arbitrary strategy profile, a pure Nash equilibrium is attained after a polynomial number of best responses.*

Theorem 5 ([3]). *In a player-specific matroid congestion game, there exists a sequence of polynomial number of better responses starting from an arbitrary strategy profile and reaching a pure Nash equilibrium.*

Since these results, matroid congestion games have been recognized as a well-behaved class of congestion games, and study on more generalized and related models followed. In the models of *congestion games with mixed objectives* [11] and *congestion games with complementarities* (or *congestion games with non-additive aggregation*) [10,39,40], the cost on a player is not necessarily the sum of the delays in her strategy. A *budget game* [8] is a variant of a congestion game, and their common generalization is proposed in [28]. A *resource buying game* [19,38] is another kind of a noncooperative game in which the players share the resources. In all of the above models, the fact that $(E, \mathcal{S}_i)$ is a matroid for each player i plays a key role to guaranteeing the existence of a pure Nash equilibrium. A further generalized model in which the strategy space is represented by a *polymatroid* is studied in [18,20]. A different kind of relation between matroids and congestion games is investigated in [13].

2.2 Congestion Games with Priorities

Hereafter, we assume that each delay function d_e $(e \in E)$ is monotonically nondecreasing, i.e., $d_e(x) \leq d_e(x')$ if $x < x'$.

Ackermann et al. [1] offered a model which commonly generalizes congestion games and a certain class of two-sided markets with ties, referred to as a *congestion game with priorities* and described by a tuple $(N, E, (\mathcal{S}_i)_{i \in N}, (p_e)_{e \in E}, (d_e)_{e \in E})$.

In this model, each resource $e \in E$ has a *priority function* $p_e \colon N \to \mathbb{Z}_{++}$. If $p_e(i) < p_e(j)$ for players $i, j \in N$, then the resource e prefers i to j. Recall that a special case in which all resources have the same priority function is called a *congestion game with consistent priorities*, and the general model is referred to as a *congestion game with inconsistent priorities*.

In a strategy profile $S = (S_1, \ldots, S_n)$, the delay of e imposed on each player in $N_e(S)$ is determined in the following way. Define $p_e^*(S) \in \mathbb{Z}_{++} \cup \{+\infty\}$ by

$$p_e^*(S) = \begin{cases} \min\{p_e(i) \colon i \in N_e(S)\} & \text{if } N_e(S) \neq \emptyset, \\ +\infty & \text{if } N_e(S) = \emptyset. \end{cases}$$

For a positive integer q, define $N_e^q(S) \subseteq N_e(S)$ and $n_e^q(S) \in \mathbb{Z}_+$ by

$$N_e^q(S) = \{i \in N_e(S) \colon p_e(i) = q\}, \qquad n_e^q(S) = |N_e^q(S)|. \tag{2}$$

Now the delay imposed on a player $i \in N_e(S)$ by the resource e is defined as

$$\begin{cases} d_e\left(n_e^{p_e^*(S)}(S)\right) & \text{if } p_e(i) = p_e^*(S), \\ +\infty & \text{if } p_e(i) > p_e^*(S). \end{cases}$$

It is straightforward to see that a congestion game reduces to a congestion game with priorities. As mentioned above, congestion games with priorities also include *correlated two-sided markets with ties*: see the full version for details.

For singleton congestion games with consistent priorities, Ackermann et al. [1] proved the following theorem.

Theorem 6 ([1]). *In a singleton congestion game with consistent priorities, there exists a sequence of a polynomial number of best responses starting from an arbitrary strategy profile and reaching a pure Nash equilibrium.*

To the best of our knowledge, a theorem on player-specific singleton congestion games with consistent priorities is missing in the literature, and will be presented in a more generalized form in Sect. 4.1.

Ackermann et al. [1] further proved the following theorems for inconsistent priorities. Let n denote the number of the players and m that of the resources.

Theorem 7 ([1]). *A singleton congestion game with inconsistent priorities is a potential game, and hence possesses a pure Nash equilibrium.*

Theorem 8 ([1]). *A player-specific singleton congestion game with inconsistent priorities possesses a pure Nash equilibrium, which can be computed in polynomial time with $O(n^3m)$ strategy changes.*

Finally, Ackermann et al. [1] provided the following extensions of Theorems 6, 7, and 8 from singleton games to matroid games. For a matroid game, define its *rank* r as the maximum rank of the matroids forming the strategy spaces of all players. A better response of a player $i \in N$ in a strategy profile S from a strategy S_i to another strategy S_i' is referred to as a *lazy better response* if there exists a sequence $(S_i^0, S_i^1, ..., S_i^k)$ of strategies of i such that $S_i^0 = S_i$, $S_i^k = S_i'$, $|S_i^{k'+1} \backslash S_i^{k'}| = 1$ and the cost on i in a strategy profile $(S_{-i}, S_i^{k'+1})$ is strictly smaller than that in $(S_{-i}, S_i^{k'})$ for each $k' = 0, 1, \ldots, k-1$. A *potential game with respect to lazy better responses* is a game admitting a potential function which strictly decreases by a lazy better response.

Theorem 9 ([1]). *In a matroid congestion game with consistent priorities, there exists a sequence of a polynomial number of best responses starting from an arbitrary strategy profile and reaching a pure Nash equilibrium.*

Theorem 10 ([1]). *A matroid congestion game with inconsistent priorities is a potential game with respect to lazy better responses, and hence possesses a pure Nash equilibrium.*

Theorem 11 ([1]). *A player-specific matroid congestion game with inconsistent priorities possesses a pure Nash equilibrium, which can be computed in polynomial time with $O(n^3mr)$ strategy changes.*

2.3 Priority-Based Affine Congestion Games

We refer to the model of Bilò and Vinci [6] as a *priority-based affine congestion game with consistent priorities*, which is described as $(N, E, (S_i)_{i \in N}, p, (\alpha_e, \beta_e)_{e \in E})$. Note that all resources have the same priority function $p \colon N \to \mathbb{Z}_{++}$.

What is specific is that each resource $e \in E$ is associated with two nonnegative real numbers $\alpha_e, \beta_e \in \mathbb{R}_+$, which determine the delay function of e in the following manner. For a positive integer $q \in \mathbb{Z}_+$, define $n_e^q(S) \in \mathbb{Z}_+$ as in (2), in which p_e is replaced by p. Similarly, define $n_e^{<q}(S) \in \mathbb{Z}_+$ by

$$n_e^{<q}(S) = |\{i \in N_e(S) \colon p(i) < q\}|. \tag{3}$$

Now the delay imposed on a player $i \in N_e(S)$ by e is defined as

$$\alpha_e \cdot \left(n_e^{<p(i)}(S) + \frac{n_e^{p(i)}(S) + 1}{2} \right) + \beta_e. \tag{4}$$

Note that $(n_e^{p(i)}(S){+}1)/2$ in (4) is the expected number of the players in $N_e^{p(i)}(S)$ more or equally prioritized than i when the ties of the players in $N_e^{p(i)}(S)$ are broken uniformly at random.

Theorem 12 ([6]). *A priority-based affine congestion game with consistent priorities possesses a pure Nash equilibrium.*

3   Our Model

We first point out that the model of Bilò and Vinci [6] partially answers to the open question of Ackermann et al. [1]. Indeed, the delay (4) of a player $i \in N_e(S)$ is finitely affected by the more prioritized players in $N_e(S)$. Meanwhile, compared to the model of Ackermann et al. [1], the delay (4) is specific in that it is a particular affine function of $n_e^{<p(i)}(S)$ and $n_e^{p(i)}(S)$, and the priorities of the resources are consistent. Below we resolve these points by providing a common generalization of the two models, which provides a full answer to the open question in [1].

Our model is represented as $(N, E, (\mathcal{S}_i)_{i \in N}, (p_e)_{e \in E}, (d_e)_{e \in E})$, which is often abbreviated as $(N, E, (\mathcal{S}_i), (p_e), (d_e))$. Let $S = (S_1, \ldots, S_n)$ be a strategy profile, $i \in N$, and $e \in S_i$. Reflecting the delay function (4), our delay function $d_e \colon \mathbb{Z}_+ \times \mathbb{Z}_{++} \to \bar{\mathbb{R}}_+$ is a bivariate function with variables $n_e^{<p_e(i)}(S)$ and $n_e^{p_e(i)}(S)$. Namely, the delay imposed on i by e is

$$d_e \left(n_e^{<p_e(i)}(S), n_e^{p_e(i)}(S) \right). \tag{5}$$

We assume that each delay function d_e ($e \in E$) has the following properties:

$$d_e(x, y) \le d_e(x', y) \qquad \text{(if } x < x'\text{)}, \tag{6}$$
$$d_e(x, y) \le d_e(x, y') \qquad \text{(if } y < y'\text{)}, \tag{7}$$
$$d_e(x, y) \le d_e(x + y - 1, 1) \quad \text{(for each } x \in \mathbb{Z}_+ \text{ and } y \in \mathbb{Z}_{++}\text{)}. \tag{8}$$

Properties (6) and (7) mean that the delay function d_e is nondecreasing with respect to $n_e^{<p_e(i)}(S)$ and $n_e^{p_e(i)}(S)$, respectively. Property (8) means that the cost on i increases if the $n_e^{p_e(i)}(S) - 1$ players in $N_e(S)$ with the same priority as i are replaced by the same number of more prioritized players. This property captures the characteristic of the models of [1,6] that prioritized players produce more delays than those with the same priority.

We refer to our model as a *priority-based congestion game with inconsistent priorities*, or *priority-based congestion game* for short. If the resources have the same priority function, then the game is referred to as a *priority-based congestion game with consistent priorities*. The model of a *priority-based player-specific congestion game* can be defined by replacing the delay functions d_e ($e \in E$) with player-specific delay functions $d_{i,e}$ ($i \in N, e \in E$).

A priority-based affine congestion game $(N, E, (\mathcal{S}_i)_{i \in N}, p, (\alpha_e, \beta_e)_{e \in E})$ with consistent priority [6] is represented as a priority-based congestion game with consistent priorities p and delay function $d_e \colon \mathbb{Z}_+ \times \mathbb{Z}_{++} \to \bar{\mathbb{R}}_+$ ($e \in E$) defined as in (4), namely

$$d_e \left(n_e^{<p_e(i)}(S), n_e^{p_e(i)}(S) \right) = \alpha_e \left(n_e^{<p(i)}(S) + \frac{n_e^{p(i)}(S) + 1}{2} \right) + \beta_e. \qquad (9)$$

It is clear that the delay function d_e in (9) satisfies the properties (6)–(8).

The model of [1] is also a special case of a priority-based congestion game. Given a congestion game $(N, E, (\mathcal{S}_i)_{i \in N}, (p_e)_{e \in E}, (d_e)_{e \in E})$ with priorities, define a delay function $d'_e \colon \mathbb{Z}_+ \times \mathbb{Z}_{++} \to \bar{\mathbb{R}}_+$ of a priority-based congestion game by

$$d'_e(x, y) = \begin{cases} +\infty & (x \geq 1), \\ d_e(y) & (x = 0). \end{cases} \qquad (10)$$

Again, the delay function d'_e in (10) satisfies the properties (6)–(8) if d_e is a nondecreasing function. The properties (6) and (7) are directly derived. The property (8) follows from the fact that $d'_e(x+y-1, 1) \neq +\infty$ only if $(x, y) = (0, 1)$, and in that case both $d'_e(x, y)$ and $d'_e(x + y - 1, 1)$ is equal to $d'_e(0, 1) = d_e(1)$.

In the subsequent sections, we present our technical contributions. Some of the proofs are omitted due to the page limitation.

4 Priority-Based Singleton Congestion Games

4.1 Consistent Priorities

The following theorem is not only an extension of Theorem 6, concerning pure Nash equilibria in non-player-specific singleton congestion games with consistent priorities, but also implies the existence of pure Nash equilibria in player-specific congestion games with consistent priorities (Corollary 1), which is missing in the literature.

Theorem 13. *In a priority-based player-specific singleton congestion game $G = (N, E, (\mathcal{S}_i), p, (d_{i,e}))$ with consistent priorities, there exists a sequence of polynomial number of better responses starting from an arbitrary strategy profile and reaching a pure Nash equilibrium.*

Corollary 1. *In a player-specific singleton congestion game with consistent priorities, there exists a sequences of polynomial number of better responses starting from an arbitrary strategy profile and reaching a pure Nash equilibrium.*

4.2 Inconsistent Priorities

We first prove the following extension of Theorem 7.

Theorem 14. *A priority-based singleton congestion game $(N, E, (\mathcal{S}_i), (p_e), (d_e))$ with inconsistent priorities is a potential game, and hence possesses a pure Nash equilibrium.*

Proof. For each strategy profile $S = (e_1, \ldots, e_n)$, define its potential $\Phi(S) \in (\bar{\mathbb{R}}_+ \times \mathbb{Z}_{++})^n$ as follows. Let $e \in E$ be a resource, and let $Q_e(S) = \{q_1, \ldots, q_{k^*}\}$ be a set of integers such that $Q_e(S) = \{q \colon n_e^q(S) > 0\}$ and $q_1 < \cdots < q_{k^*}$. The resource $e \in E$ contributes the following $n_e(S)$ vectors in $\bar{\mathbb{R}}_+ \times \mathbb{Z}_{++}$ to $\Phi(S)$:

$$
\begin{aligned}
&(d_e(0,1),\, q_1), \ldots, (d_e(0, n_e^{q_1}(S)),\, q_1), \\
&(d_e(n_e^{q_1}(S), 1),\, q_2), \ldots, (d_e(n_e^{q_1}(S), n_e^{q_2}(S)),\, q_2), \ldots, \\
&\left(d_e\left(n_e^{<q_k}(S), 1\right),\, q_k\right), \ldots, \left(d_e\left(n_e^{<q_k}(S), n_e^{q_k}(S)\right),\, q_k\right), \ldots, \\
&\left(d_e\left(n_e^{<q_{k^*}}(S), 1\right),\, q_{k^*}\right), \ldots, \left(d_e\left(n_e^{<q_{k^*}}(S), n_e^{q_{k^*}}(S)\right),\, q_{k^*}\right).
\end{aligned}
\tag{11}
$$

For two vectors $(x,y), (x',y') \in \bar{\mathbb{R}}_+ \times \mathbb{Z}_{++}$, we define a lexicographic order $(x,y) \preceq_{\mathrm{lex}} (x',y')$ if

$$
x < x', \quad \text{or} \quad x = x' \text{ and } y \le y'.
$$

The strict relation $(x,y) \prec_{\mathrm{lex}} (x',y')$ means that $(x,y) \preceq_{\mathrm{lex}} (x',y')$ and $(x,y) \ne (x',y')$ hold. The potential $\Phi(S)$ is obtained by ordering the n vectors contributed by all resources in the lexicographically nondecreasing order. We remark that the order in (11) is lexicographically nondecreasing, which can be derived from (6)–(8) as follows. It follows from (7) that

$$
d_e(n_e^{<q_k}(S), y) \le d_e(n_e^{<q_k}(S), y+1) \quad (k = 1, \ldots, k^*,\, y = 1, \ldots, q_{k-1}),
\tag{12}
$$

and from (6) and (8) that

$$
d_e(n_e^{<q_k}(S), n_e^{q_k}(S)) \le d_e(n_e^{<q_{k+1}}(S) - 1, 1) \le d_e(n_e^{<q_{k+1}}(S), 1)
$$
$$
(k = 1, \ldots, k^* - 1).
\tag{13}
$$

We then define a lexicographic order over the potentials. For strategy profiles S and S', where

$$
\Phi(S) = ((x_1, y_1), \ldots, (x_n, y_n)), \quad \Phi(S') = ((x_1', y_1'), \ldots, (x_n', y_n')),
$$

define $\Phi(S') \preceq_{\mathrm{lex}} \Phi(S)$ if there exists an integer ℓ with $1 \le \ell \le n$ such that $(x_{\ell'}', y_{\ell'}') = (x_{\ell'}, y_{\ell'})$ for each $\ell' < \ell$ and $(x_\ell', y_\ell') \prec_{\mathrm{lex}} (x_\ell, y_\ell)$. The strict relation $\Phi(S') \prec_{\mathrm{lex}} \Phi(S)$ means that $\Phi(S') \preceq_{\mathrm{lex}} \Phi(S)$ and $\Phi(S') \ne \Phi(S)$ hold.

Suppose that $i \in N$ has a better response in a strategy profile S, changing her strategy from e to e'. Let $S' = (S_{-i}, e')$. Below we show $\Phi(S') \prec_{\mathrm{lex}} \Phi(S)$, completing the proof.

Let $p_e(i) = q$ and $p_{e'}(i) = q'$. Since the delay imposed on i becomes smaller due to the better response, it holds that

$$
d_{e'}(n_{e'}^{<q'}(S), n_{e'}^{q'}(S) + 1) < d_e(n_e^{<q}(S), n_e^q(S)).
\tag{14}
$$

Note that e' contributes a vector

$$
(d_{e'}(n_{e'}^{<q'}(S), n_{e'}^{q'}(S) + 1),\, q')
\tag{15}
$$

to $\Phi(S')$ but not to $\Phi(S)$. To prove $\Phi(S') \prec_{\text{lex}} \Phi(S)$, it suffices to show that a vector belonging to $\Phi(S)$ but not to $\Phi(S')$ is lexcographically larger than the vector (15).

First, consider the vectors in $\Phi(S)$ contributed by e. Let $Q_e(S) = \{q_1, \ldots, q_{k^*}\}$, where $q_1 < \cdots < q_{k^*}$. The better response of i changes the vectors in $\Phi(S)$ whose second component is larger than q, because the first argument of the delay function d_e decreases by one. If $q = q_{k^*}$, then those vectors do not exist and thus we are done. Suppose that $q = q_k$ for some $k < k^*$. Among those vectors, the lexicographically smallest one is $(d_e(n_e^{<q_{k+1}}(S), 1), q_{k+1})$. Recall (13), saying that $d_e(n_e^{<q_k}(S), n_e^{q_k}(S)) \leq d_e(n_e^{<q_{k+1}}(S), 1)$, and thus $d_{e'}(n_{e'}^{<q'}(S), n_{e'}^{q'}(S) + 1) < d_e(n_e^{<q_{k+1}}(S), 1)$ follows from (14). Hence, we conclude that $(d_{e'}(n_{e'}^{<q'}(S), n_{e'}^{q'}(S) + 1), q') \prec_{\text{lex}} (d_e(n_e^{<q_{k+1}}(S), 1), q_{k+1})$.

Next, consider the vectors in $\Phi(S)$ contributed by e'. Without loss of generality, suppose that there exists a positive integer q'' such that $q'' \in Q_{e'}(S)$ and $q'' > q'$. Let q'' be the smallest integer satisfying these conditions. The lexicographically smallest vector in $\Phi(S)$ contributed by e' and changed by the better response of i is $(d_{e'}(n_{e'}^{<q''}(S), 1), q'')$. It follows from the property (8) that $d_{e'}(n_{e'}^{<q'}(S), n_{e'}^{q'}(S) + 1) \leq d_{e'}(n_{e'}^{<q''}(S), 1)$, and thus $(d_{e'}(n_{e'}^{<q'}(S), n_{e'}^{q'}(S) + 1), q') \prec_{\text{lex}} (d_{e'}(n_{e'}^{<q''}(S), 1), q'')$, completing the proof. $\square$

The next theorem corresponds to Theorem 8, while it does not include a polynomial bound. We use a term *state* to refer to a collection of the strategies of some of the players in N, and let $N(S)$ denote the set of the players contributing to a state S. For a state S and a resource $e \in E$, let $N_e(S)$ denote the set of players in $N(S)$ choosing e as the strategy, and let $|N_e(S)| = n_e(S)$.

Theorem 15. *A priority-based player-specific singleton congestion game with inconsistent priorities possesses a pure Nash equilibrium, which can be computed with a finite number of strategy changes.*

Proof. We prove this theorem by presenting an algorithm for computing a pure Nash equilibrium of a priority-based player-specific singleton congestion game $(N, E, (S_i), (p_e), (d_{i,e}))$. The algorithm constructs a sequence $S_0, S_1, \ldots, S_k$ of states in which $N(S_0) = \emptyset$, $N(S_k) = N$, and

$$\text{each player in } N(S_{k'}) \text{ has no incentive to change her strategy} \tag{16}$$

for each $k' = 0, 1, \ldots, k$, implying that S_k is a pure Nash equilibrium.

It is clear that (16) is satisfied for $k' = 0$. Below we show how to construct $S_{k'+1}$ from $S_{k'}$ under an assumption that $S_{k'}$ satisfies (16) and $N(S_{k'}) \subsetneq N$.

Take a player $i \in N \setminus N(S_{k'})$, and let i choose a resource $e \in E$ imposing the minimum cost on i if i is added to $N_e(S_{k'})$. We construct the new state $S_{k'+1}$ by changing the strategy of each player $j \in N_e(S_{k'})$ in the following way. The other players do not change their strategies.

When i is added to $N_e(S_{k'})$, we have the following cases A and B.

Case A. No player in $N_e(S_{k'})$ comes to have a better response.

Case B. Some players in $N_e(S_{k'})$ comes to have a better response.

In Case A, we do not change the strategies of the players in $N_e(S_{k'})$. In Case B, if a player $j \in N_e(S_{k'})$ comes to have a better response, it must hold that $p_e(j) \geq p_e(i)$. We further separate Case B into the following two cases.

Case B1. A player $j \in N_e(S_{k'})$ with a better response satisfies $p_e(j) = p_e(i)$.
Case B2. Each player $j \in N_e(S_{k'})$ with a better response satisfies $p_e(j) > p_e(i)$.

In each case, the strategies are changed as follows.

Case B1. Only one player $j \in N_e(S_{k'})$ having a better response and satisfying $p_e(j) = p_e(i)$ changes her strategy by discarding it. Namely, $j \notin N(S_{k'+1})$. The other players do not change their strategies.
Case B2. Each player $j \in N_e(S_{k'})$ with a better response discards her strategy.

We have now constructed the new state $S_{k'+1}$. It is straightforward to see that the state $S_{k'+1}$ satisfies (16). We complete the proof by showing that this algorithm terminates within a finite number of strategy changes.

For a resource $e \in E$, let $q_e^* = \max\{p_e(i) : i \in N\}$. For each state $S_{k'}$ appearing in the algorithm, define its potential $\Phi(S_{k'}) \in \left(\times_{e \in E} \mathbb{Z}_+^{q_e^*} \right) \times \mathbb{Z}_{++}$ in the following manner. For each resource $e \in E$, define a vector $\phi_e \in \mathbb{Z}_+^{q_e^*}$ by

$$\phi_e(q) = n_e^q(S_{k'}) \quad (q = 1, 2, \ldots, q_e^*),$$

which is a contribution of e to the first component of $\Phi(S_{k'})$. The first component of $\Phi(S_{k'})$ is constructed by ordering the vectors ϕ_e ($e \in E$) in the lexicographically nondecreasing order.

For a resource $e \in E$ and a player $i \in N_e(S_{k'})$, define $\mathrm{tol}(i, S_{k'}) \in \mathbb{Z}_{++}$ as the maximum number $y \in \mathbb{Z}_{++}$ such that e is an optimal strategy for i if i shares e with y players having the same priority as e, i.e.,

$$d_{i,e}(n_e^{<p_e(i)}(S_{k'}), y) \leq d_{i,e'}(n_{e'}^{<p_{e'}(i)}(S_{k'}), n_{e'}^{p_{e'}(i)}(S_{k'}) + 1)$$

for each e' with $e' \neq e$ and $\{e'\} \in S_i$. Note that i herself is counted in y, and hence $\mathrm{tol}(i, S_{k'}) \geq 1$ for each $i \in N(S_{k'})$. Now the second component of the potential $\Phi(S_{k'})$ is defined as $\sum_{i \in N(S_{k'})} \mathrm{tol}(i, S_{k'})$.

We prove that the potential $\Phi(S_{k'})$ increases lexicographically monotonically during the algorithm. Let a state $S_{k'+1}$ be constructed from $S_{k'}$ and the involvement of a player $i \in N \backslash N(S_{k'})$ choosing a resource $e \in E$. It is straightforward to see that $\phi_{e'}$ is unchanged for each $e' \in E \backslash \{e\}$. Consider the vector ϕ_e.

Case A. The unique change of ϕ_e is that $\phi_e(p_e(i))$ increases by one, implying that the first component of $\Phi(S_{k'+1})$ is lexicographically larger than that of $\Phi(S_{k'})$.
Case B1. Let $j^* \in N_e(S_{k'})$ denote the unique player who discard her strategy. Recall that $p_e(j^*) = p_e(i)$. It follows that ϕ_e is unchanged, and hence the first

component of $\Phi(S_{k'+1})$ is the same as that of $\Phi(S_{k'})$. The second component of $\Phi(S_{k'+1})$ is strictly larger than that of $\Phi(S_{k'})$, because

$$\mathrm{tol}(j, S_{k'+1}) = \mathrm{tol}(j, S_{k'}) \quad \text{for each } j \in N(S_{k'}) \backslash \{i, j^*\},$$

$$\mathrm{tol}(i, S_{k'+1}) \geq n_e^{p_e(i)}(S_{k'}) + 1, \text{ and } \mathrm{tol}(j^*, S_{k'}) = n_e^{p_e(i)}(S_{k'}).$$

Case B2. It holds that $n_e^q(S_{k'+1}) = n_e^q(S_{k'})$ for each $q < p_e(i)$ and $n_e^{p_e(i)}(S_{k'+1}) = n_e^{p_e(i)}(S_{k'}) + 1$. Thus, the first component of Φ lexicographically increases. $\square$

5 Extension Beyond Singleton Games

We first discuss matroid games. Theorems 13, 14, and 15 are extended as follows.

Theorem 16. *In a priority-based player-specific matroid congestion game with consistent priorities, there exists a sequences of polynomial number of better responses starting from an arbitrary strategy profile and reaching a pure Nash equilibrium.*

Theorem 17. *A priority-based matroid congestion game with inconsistent priorities is a potential game, and hence possesses a pure Nash equilibrium.*

Theorem 18. *A priority-based player-specific matroid congestion game with inconsistent priority possesses a pure Nash equilibrium, which can be computed with a finite number of strategy changes.*

We next prove the existence of pure Nash equilibria in games with consistent priorities and arbitrary strategy spaces.

Theorem 19. *A priority-based congestion game with consistent priorities possesses a pure Nash equilibrium.*

6 Conclusion

We have presented a common generalization of the models of congestion games by Ackermann et al. [1] and Bilò and Vinci [6]. This generalization gives a positive and full answer to the open question posed by Ackermann et al. [1]. We then proved some theorems on the existence of pure Nash equilibria, extending those in [1,6]. Once the existence of pure Nash equilibria is established, a possible direction of future work is to design an efficient algorithm for finding a pure Nash equilibrium in our model. Analyses on the price of anarchy and the price of stability in our model are also of interest, as are intensively done for the model of Bilò and Vinci [6]. Finally, investigation on the connection of our model and the many-to-many model of stable matchings is to be expected.

Acknowledgement. The author is partially supported by JSPS KAKENHI Grant Numbers JP20K11699, JP24K02901, JP24K14828, Japan.

References

1. Ackermann, H., Goldberg, P.W., Mirrokni, V.S., Röglin, H., Vöcking, B.: A unified approach to congestion games and two-sided markets. Internet Math. **5**(4), 439–457 (2008). https://doi.org/10.1080/15427951.2008.10129171

2. Ackermann, H., Röglin, H., Vöcking, B.: On the impact of combinatorial structure on congestion games. J. ACM **55**(6), 25:1–25:22 (2008). https://doi.org/10.1145/1455248.1455249

3. Ackermann, H., Röglin, H., Vöcking, B.: Pure Nash equilibria in player-specific and weighted congestion games. Theor. Comput. Sci. **410**(17), 1552–1563 (2009). https://doi.org/10.1016/j.tcs.2008.12.035

4. Baïou, M., Balinski, M.: Many-to-many matching: stable polyandrous polygamy (or polygamous polyandry). Discrete Appl. Math. **101**(1–3), 1–12 (2000). https://doi.org/10.1016/S0166-218X(99)00203-6

5. Bansal, V., Agrawal, A., Malhotra, V.S.: Polynomial time algorithm for an optimal stable assignment with multiple partners. Theor. Comput. Sci. **379**(3), 317–328 (2007). https://doi.org/10.1016/J.TCS.2007.02.050

6. Bilò, V., Vinci, C.: Congestion games with priority-based scheduling. Theor. Comput. Sci. **974**, 114094 (2023). https://doi.org/10.1016/j.tcs.2023.114094

7. Bilò, V., Vinci, C.: Coping with Selfishness in Congestion Games—Analysis and Design via LP Duality. Monographs in Theoretical Computer Science. An EATCS Series, Springer (2023). https://doi.org/10.1007/978-3-031-30261-9

8. Drees, M., Feldotto, M., Riechers, S., Skopalik, A.: Pure Nash equilibria in restricted budget games. J. Comb. Optim. **37**(2), 620–638 (2019). https://doi.org/10.1007/s10878-018-0269-7

9. Eirinakis, P., Magos, D., Mourtos, I., Miliotis, P.: Finding all stable pairs and solutions to the many-to-many stable matching problem. INFORMS J. Comput. **24**(2), 245–259 (2012). https://doi.org/10.1287/IJOC.1110.0449

10. Feldotto, M., Leder, L., Skopalik, A.: Congestion games with complementarities. In: Fotakis, D., Pagourtzis, A., Paschos, V.T. (eds.) CIAC 2017. LNCS, vol. 10236, pp. 222–233. Springer, Cham (2017). https://doi.org/10.1007/978-3-319-57586-5_19

11. Feldotto, M., Leder, L., Skopalik, A.: Congestion games with mixed objectives. J. Comb. Optim. **36**(4), 1145–1167 (2018). https://doi.org/10.1007/s10878-017-0189-y

12. Fleiner, T.: On the stable b-matching polytope. Math. Soc. Sci. **46**(2), 149–158 (2003). https://doi.org/10.1016/S0165-4896(03)00074-X

13. Fujishige, S., Goemans, M.X., Harks, T., Peis, B., Zenklusen, R.: Congestion games viewed from M-convexity. Oper. Res. Lett. **43**(3), 329–333 (2015). https://doi.org/10.1016/j.orl.2015.04.002

14. Gale, D., Shapley, L.: College admissions and the stability of marriage. Am. Math. Mon. **69**, 9–15 (1962)

15. Goko, H., Makino, K., Miyazaki, S., Yokoi, Y.: Maximally satisfying lower quotas in the hospitals/residents problem with ties. In: Berenbrink, P., Monmege, B. (eds.) 39th International Symposium on Theoretical Aspects of Computer Science, STACS 2022. LIPIcs, vol. 219, pp. 31:1–31:20 (2022). https://doi.org/10.4230/LIPICS.STACS.2022.31

16. Gusfield, D., Irving, R.W.: The Stable Marriage Problem—Structure and Algorithms. Foundations of Computing Series, MIT Press, Cambridge (1989)

17. Hamada, K., Miyazaki, S., Yanagisawa, H.: Strategy-proof approximation algorithms for the stable marriage problem with ties and incomplete lists. In: Lu,

P., Zhang, G. (eds.) 30th International Symposium on Algorithms and Computation, ISAAC 2019. LIPIcs, vol. 149, pp. 9:1–9:14 (2019). https://doi.org/10.4230/LIPICS.ISAAC.2019.9

18. Harks, T., Klimm, M., Peis, B.: Sensitivity analysis for convex separable optimization over integral polymatroids. SIAM J. Optim. **28**(3), 2222–2245 (2018). https://doi.org/10.1137/16M1107450

19. Harks, T., Peis, B.: Resource buying games. In: Schulz, A.S., Skutella, M., Stiller, S., Wagner, D. (eds.) Gems of Combinatorial Optimization and Graph Algorithms, pp. 103–111. Springer, Cham (2015). https://doi.org/10.1007/978-3-319-24971-1_10

20. Harks, T., Timmermans, V.: Uniqueness of equilibria in atomic splittable polymatroid congestion games. J. Comb. Optim. **36**(3), 812–830 (2018). https://doi.org/10.1007/s10878-017-0166-5

21. Huang, C.-C., Iwama, K., Miyazaki, S., Yanagisawa, H.: A tight approximation bound for the stable marriage problem with restricted ties. In: Garg, N., Jansen, K., Rao, A., Rolim, J.D.P. (eds.) Approximation, Randomization, and Combinatorial Optimization. Algorithms and Techniques, APPROX/RANDOM 2015. LIPIcs, vol. 40, pp. 361–380 (2015). https://doi.org/10.4230/LIPICS.APPROX-RANDOM.2015.361

22. Huang, C.-C., Kavitha, T.: Improved approximation algorithms for two variants of the stable marriage problem with ties. Math. Program. **154**(1–2), 353–380 (2015). https://doi.org/10.1007/S10107-015-0923-0

23. Ieong, S., McGrew, R., Nudelman, E., Shoham, Y., Sun, Q.: Fast and compact: a simple class of congestion games. In: Veloso, M.M., Kambhampati, S. (eds.) 20th Annual AAAI Conference on Artificial Intelligence, AAAI 2005, pp. 489–494 (2005). http://www.aaai.org/Library/AAAI/2005/aaai05-077.php

24. Irving, R.W.: Stable marriage and indifference. Discrete Appl. Math. **48**(3), 261–272 (1994). https://doi.org/10.1016/0166-218X(92)00179-P

25. Kamiyama, N.: Stable matchings with ties, master preference lists, and matroid constraints. In: Hoefer, M. (ed.) SAGT 2015. LNCS, vol. 9347, pp. 3–14. Springer, Heidelberg (2015). https://doi.org/10.1007/978-3-662-48433-3_1

26. Kamiyama, N.: Many-to-many stable matchings with ties, master preference lists, and matroid constraints. In: Elkind, E., Veloso, M., Agmon, N., Taylor, M.E. (eds.) 18th International Conference on Autonomous Agents and MultiAgent Systems, AAMAS 2019, pp. 583–591. IFAAMAS (2019). http://dl.acm.org/citation.cfm?id=3331743

27. Kavitha, T.: Stable matchings with one-sided ties and approximate popularity. In: Dawar, A., Guruswami, V. (eds.) 42nd IARCS Annual Conference on Foundations of Software Technology and Theoretical Computer Science, FSTTCS 2022. LIPIcs, vol. 250, pp. 22:1–22:17 (2022). https://doi.org/10.4230/LIPICS.FSTTCS.2022.22

28. Kiyosue, F., Takazawa, K.: A common generalization of budget games and congestion games. J. Comb. Optim. **48**, 24 (2024). https://doi.org/10.1007/S10878-024-01218-7

29. Manlove, D.F.: Algorithmics of Matching Under Preferences. Series on Theoretical Computer Science, vol. 2. World Scientific (2013). https://doi.org/10.1142/8591

30. Marx, D., Schlotter, I.: Parameterized complexity and local search approaches for the stable marriage problem with ties. Algorithmica **58**(1), 170–187 (2010). https://doi.org/10.1007/S00453-009-9326-Z

31. Milchtaich, I.: Congestion games with player-specific payoff functions. Games Econ. Behav. **13**, 111–124 (1996)

32. Monderer, D., Shapley, L.S.: Potential games. Games Econ. Behav. **14**, 124–143 (1996). https://doi.org/10.1006/game.1996.0044

33. Nisan, N., Roughgarden, T., Tardos, É., Vazirani, V.V. (eds.): Algorithmic Game Theory. Cambridge University Press, Cambridge (2007)

34. Rosenthal, R.W.: A class of games possessing pure-strategy Nash equilibria. Int. J. Game Theory **2**, 65–67 (1973). https://doi.org/10.1007/BF01737559

35. Roughgarden, T.: Twenty Lectures on Algorithmic Game Theory. Cambridge University Press, Cambridge (2016)

36. Sotomayor, M.: The lattice structure of the set of stable outcomes of the multiple partners assignment game. Int. J. Game Theory **28**(4), 567–583 (1999). https://doi.org/10.1007/S001820050126

37. Sotomayor, M.: Three remarks on the many-to-many stable matching problem. Math. Soc. Sci. **38**(1), 55–70 (1999)

38. Takazawa, K.: Generalizations of weighted matroid congestion games: pure Nash equilibrium, sensitivity analysis, and discrete convex function. J. Comb. Optim. **38**(4), 1043–1065 (2019). https://doi.org/10.1007/s10878-019-00435-9

39. Takazawa, K.: Pure Nash equilibria in weighted congestion games with complementarities and beyond. In: Dastani, M., Sichman, J.S., Alechina, N., Dignum, V. (eds.) Proceedings of the 23rd International Conference on Autonomous Agents and Multiagent Systems, AAMAS 2024, pp. 2495–2497. ACM (2024). https://doi.org/10.5555/3635637.3663205

40. Takazawa, K.: Pure Nash equilibria in weighted matroid congestion games with non-additive aggregation and beyond. Discrete Appl. Math. **361**, 226–235 (2025). https://doi.org/10.1016/j.dam.2024.10.017

41. Yokoi, Y.: An approximation algorithm for maximum stable matching with ties and constraints. In: Ahn, H., Sadakane, K. (eds.) 32nd International Symposium on Algorithms and Computation, ISAAC 2021. LIPIcs, vol. 212, pp. 71:1–71:16 (2021). https://doi.org/10.4230/LIPICS.ISAAC.2021.71

Dag-Like Unit Refutations in UTVPI Constraint Systems

Piotr Wojciechowski and K. Subramani[(⊠)]

LDCSEE, West Virginia University, Morgantown, WV, USA
{pwojciec,k.subramani}@mail.wvu.edu

Abstract. In this paper, we investigate the problem of finding the shortest dag-like unit refutation of a system of Unit Two Variable Per Inequality (UTVPI) constraints. Unit refutations in difference constraints (both tree-like and dag-like) have been studied in the literature. Unit refutations are important from the perspective of identifying restrictions on variables which cause inconsistencies. They are also useful from the perspective of visualizing the cause of inconsistency in a constraint system. Unit resolution refutations form the backbone of logic programming engines. Here, we investigate unit refutations in polyhedral constraint systems. In particular, our focus is on optimal length dag-like unit refutations. Previous research has established that this problem is **NP-hard**. In the current work, we design a pseudo-polynomial time algorithm and an approximation algorithm for this problem. It is unusual for a problem to be both **APX-complete** and solvable in pseudo-polynomial time.

1 Introduction

This paper is concerned with the problem of finding the shortest dag-like unit refutation of a UTVPI constraint system (UCS). A UTVPI constraint is a linear relationship of the form: $\pm x_i \pm x_j \leq b_{ij}$. A conjunction of such constraints is called a UCS and can be expressed in matrix form as: $\mathbf{A} \cdot \mathbf{x} \leq \mathbf{b}$. In this paper, we use n to denote the number of variables in a UCS and m to denote the number of constraints. UCSs find applications in a number of domains such as program verification [2], array bounds checking [6], abstract interpretation [8] and scheduling [9]. Several decision procedures for linear and integer feasibility problems have been developed over the years [4,6]. Some of these procedures are certifying, whereas others are not. A decision procedure is said to be certifying if it provides certificates which attest to the correctness of its output [7]. Certificates enhance the reliability of the decision procedure and make it more trustworthy. A certificate which attests to the correctness of "yes"-instances is called a positive certificate. Likewise, a certificate which attests to the correctness of "no"-instances is called a negative certificate. This paper is concerned with refutations, which are a specialized class of negative certificates. Furthermore, we focus on unit refutations, which are a subclass of general refutations. A unit refutation must use the absolute constraints present in the input UCS. Thus, a

S. Nakano and M. Xiao (Eds.): WALCOM 2025, LNCS 15411, pp. 377–392, 2025.
https://doi.org/10.1007/978-981-96-2845-2_24

unit refutation of a UCS depends on the domains of the variables in the UCS. On the other hand, unrestricted refutations can utilize only the relative constraints in the input UCS. This means that a non-unit refutation may be domain agnostic. For this reason, studying unit refutations helps in determining the structure of domain specific refutations. Previous work has examined unit refutations for difference constraints [10]. Additionally, a specific form of unit refutation, known as unit read-once refutation (UROR), has been studied for DCSs [12]. Refutations of this form have an additional restriction, namely that each constraint is used at most once.

When analyzing refutations, the typical focus is on existence. A secondary focus is on finding optimal length refutations in specialized forms. In previous work on UCSs, we have focused on efficiently finding solutions and linear refutations [11]. In this paper we focus specifically on dag-like unit refutations, as opposed to the unrestricted refutations examined previously. In particular, we investigate the problem of finding the "shortest" dag-like unit refutations (see Sect. 2).

The principal contributions of this paper are as follows:

1. An $O(m \cdot n)$ algorithm for determining if a UCS has a unit refutation (Sect. 4).
2. An $O(m \cdot n^2 \cdot ||\mathbf{b}||_\infty)$ time algorithm for finding a shortest dag-like unit refutation of a UCS (Sect. 5).
3. A 2-approximation algorithm for a shortest dag-like unit refutation of a UCS (Sect. 6).

2 Statement of Problems

In this section, we formally describe the problems under consideration.

Definition 1. *A difference constraint is a constraint of the form* $a_i \cdot x_i + a_j \cdot x_j \leq b_k$ *where* $a_i \neq a_j \in \{1, 0, -1\}$ *and* $b_k \in \mathbb{Z}$.

We focus on a type of constraint known as a Unit Two Variable Per Inequality (UTVPI) constraint.

Definition 2. *A Unit Two Variable Per Inequality (UTVPI) constraint is a constraint of the form* $a_i \cdot x_i + a_j \cdot x_j \leq b_k$ *where* $a_i, a_j \in \{1, 0, -1\}$, a_i *and* a_j *are not both 0, and* $b_k \in \mathbb{Z}$.

Note that every difference constraint is also a UTVPI constraint. In a UTVPI constraint or difference constraint, we refer to b_k as the defining constant. Additionally, the terms x_i and $-x_i$ are referred to as literals.

If a UTVPI constraint or difference constraint has only one non-zero coefficient, then it is called an absolute constraint.

Definition 3. *A conjunction of difference constraints is known as a Difference Constraint System (DCS).*

Definition 4. *A conjunction of UTVPI constraints is known as a UTVPI Constraint System (UCS).*

Example 1. System (1) is a UCS with 2 constraints over 3 variables.

$$x_1 - x_2 \leq 5 \quad x_2 + x_3 \leq -2 \tag{1}$$

A UCS can be represented as a matrix $\mathbf{A} \cdot \mathbf{x} \leq \mathbf{b}$. In this representation, $\mathbf{b}$ is referred to as the defining constant vector. In this paper, we use n to denote the number of variables in a UCS and m to denote the number of constraints.

When finding linear refutations of UCSs, there is a single inference rule used. This rule is Rule (2).

$$\text{ADD} : \frac{\sum_{i=1}^{n} a_i \cdot x_i \leq b_1 \qquad \sum_{i=1}^{n} a'_i \cdot x_i \leq b_2}{\sum_{i=1}^{n} (a_i + a'_i) \cdot x_i \leq b_1 + b_2} \tag{2}$$

We refer to Rule (2) as the **ADD rule**. It is easy to see that Rule (2) is sound since any assignment that satisfies the hypotheses also satisfies the consequent. From Farkas' lemma [3], it follows that the rule is also complete. We refer to any constraint derived using an inference rule as a derived constraint.

Definition 5. *A sequence of applications of the ADD rule that results in a contradiction is known as a linear refutation.*

In this paper, we study a restricted version of the ADD rule, known as the unit-ADD (UADD) rule. In the UADD rule, at least one of the constraints must be an absolute constraint. Rule (3) is the UADD rule.

$$\text{UADD} : \frac{a_{i,1} \cdot x_i \leq b_1 \qquad \sum_{j=1}^{n} a_{j,2} \cdot x_j \leq b_2}{(a_{i,1} + a_{i,2}) \cdot x_i + \sum_{j \neq i} a_{j,2} \cdot x_j \leq b_1 + b_2} \tag{3}$$

A linear refutation using only the UADD rule is called a unit refutation. The problem of finding such a refutation is called the unit refutability problem (URP).

It is important to note that, unlike the ADD rule, the UADD rule is incomplete. This means that there are UCSs with no linear solutions that do not have a refutation using only the UADD rule.

In this paper, we study dag-like unit refutations.

Definition 6. *A* **dag-like** *refutation is a refutation in which each constraint, both in the original system and derived, can be used multiple times.*

In particular, we focus on dag-like unit refutations.

Definition 7. *A* **dag-like unit** *refutation is a unit refutation in which each constraint, both in the original system and derived, can be used multiple times.*

Example 2. Let $\mathbf{U}$ be the UCS in System (4).

$$- x_1 \leq -1 \quad x_1 + x_2 \leq 0 \quad - x_2 + x_3 \leq 1 \quad - x_2 - x_3 \leq 0 \tag{4}$$

A dag-like unit refutation of $\mathbf{U}$ is shown in Fig. 1.

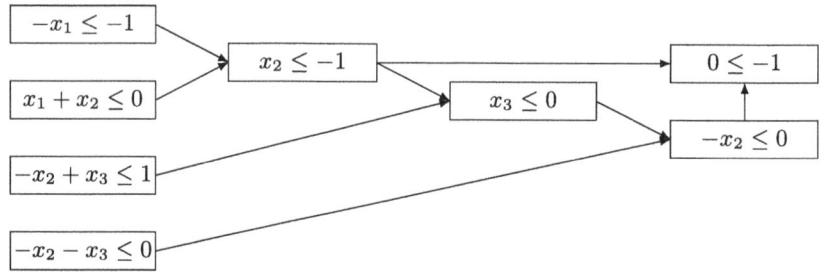

Fig. 1. Unit Refutation

Note that the derived constraint $x_2 \leq -1$ is reused.

This paper also focuses on short refutations. Thus, we introduce the notion of refutation length.

Definition 8. *The* **length** *of a refutation is the number of inferences in that refutation.*

We focus on the problem of finding a shortest dag-like unit refutation.

Definition 9. *The Optimum Length Dag-like Unit Refutation (ODLUR) Problem: Given a UCS $\mathbf{U}$ find a dag-like unit refutation of $\mathbf{U}$ with the shortest length.*

3   Constraint Network Representation

In this paper, we utilize a constraint network to represent UTVPI constraints. Note that this constraint network was first described in [11].

Let $\mathbf{U} : \mathbf{A} \cdot \mathbf{x} \leq \mathbf{b}$ denote a UTVPI constraint system and let $\mathbf{X}$ denote the set of all solutions to $\mathbf{U}$. Corresponding to this constraint system, we construct the constraint network $\mathbf{G} = \langle \mathbf{V}, \mathbf{E}, \mathbf{b} \rangle$ as follows. For each variable x_i create a node in $\mathbf{V}$. For ease of reference, both the variable and its corresponding node are denoted as x_i. Constraints are represented as edges using the following rules:

1. A constraint of the form $x_i - x_j \leq b_{ij}$ is represented as an undirected "gray" edge, $(x_j \overset{b_{ij}}{\blacksquare} x_i)$, or $(x_i \overset{b_{ij}}{\blacksquare} x_j)$, with cost b_{ij}.
2. A constraint of the form $-x_i - x_j \leq b_{ij}$ is represented by an undirected "black" edge, $(x_i \overset{b_{ij}}{\blacksquare} x_j)$, with cost b_{ij}.
3. A constraint of the form $x_i + x_j \leq b_{ij}$ is represented by an undirected "white" edge, $(x_i \overset{b_{ij}}{\square} x_j)$, with cost b_{ij}.

Finally, we add a node x_0 to the network. This node permits the addition of absolute constraints. Each absolute constraint $x_i \leq b_i$ is replaced by a pair of constraints $x_i + x_0 \leq b_i$ and $x_i - x_0 \leq b_i$. The corresponding edges $x_0 \overset{b_i}{\square} x_i$ and $x_0 \overset{b_i}{\blacksquare} x_i$ are added to the constraint network.

We now recall the definitions of paths and cycles from [11].

Definition 10. *A k-path is a sequence of $(k+1)$ nodes, $x_1, x_2, \ldots x_{k+1}$, and k edges $e_1, e_2, \ldots e_k$, such that e_i is the edge corresponding to one of the constraints between x_i and x_{i+1} in* **U**.

Definition 11. *A k-path is considered* valid *if it has the following property: For every i from 2 to k, the coefficients of x_i in the constraints corresponding to the edges e_i and e_{i-1} have opposite signs.*

Definition 12. *The* cost *of a path is the sum of the costs of the edges along that path.*

Definition 13. *A cycle is a valid k-path for which $x_1 = x_{k+1}$.*

Example 3. Suppose we have the system of constraints and corresponding constraint network provided in Fig. 2:

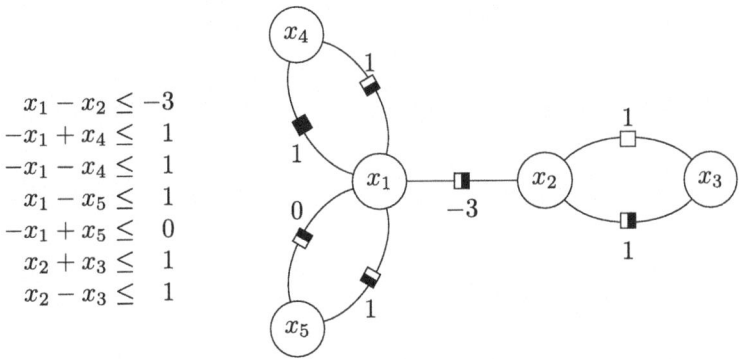

Fig. 2. Example Constraint System and Network (without node x_0)

From Fig. 2, we can see that the 8-path

$$(x_1 \overset{-3}{\blacksquare} x_2 \overset{1}{\square} x_3 \overset{1}{\blacksquare} x_2 \overset{-3}{\blacksquare} x_1 \overset{0}{\blacksquare} x_5 \overset{1}{\blacksquare} x_1 \overset{1}{\blacksquare} x_4 \overset{1}{\blacksquare} x_1)$$

forms a cycle even though the vertices x_1 and x_2 and the edge $(x_2 \overset{-3}{\blacksquare} x_1)$ are used multiple times.

We also utilize the edge reductions from [11].

Definition 14. *An edge reduction is an operation that determines a single edge equivalent to a two-edge path and represents the addition of the two UTVPI constraints which correspond to the edges in question. If this addition results in a UTVPI constraint, the reduction is said to be* valid.

We use definition 14 to define path types.

Definition 15. *We say that a path has type t, if it can be reduced to a single edge of type t, where $t \in \{\square, \blacksquare, \square, \square\}$ by a series of valid edge reductions.*

Using this, we can define a negative cost gray cycle.

Definition 16. *A negative cost gray cycle is a path which can be reduced to an edge $x_i \overset{b_i}{\square} x_i$ (or $x_i \overset{b_i}{\square} x_i$) for some node x_i such that $b_i < 0$.*

Example 4. Consider the following system:

$$
\begin{array}{lll}
l_1 : x_1 - x_2 \leq -3 & l_2 : -x_1 + x_4 \leq 1 & l_3 : -x_1 - x_4 \leq 1 \\
l_4 : x_2 + x_3 \leq 1 & l_5 : x_2 - x_3 \leq 1 &
\end{array}
\tag{5}
$$

First, observe that l_1 is the only constraint with a negative defining constant. Hence, it must be included in *any* refutation. In order to eliminate x_1, we must include both l_2 and l_3 in the refutation. Otherwise, x_4 is not eliminated. Similarly, to eliminate $-x_2$, we must include both l_4 and l_5. Otherwise, x_3 is not eliminated. In other words, *read-once* refutations are not complete for UTVPI constraint systems.

This dag-like refutation corresponds to cycle $(x_1 \overset{1}{\square} x_4 \overset{1}{\blacksquare} x_1 \overset{-3}{\square} x_2 \overset{1}{\square} x_3 \overset{1}{\square} x_2 \overset{-3}{\square} x_1)$ in Fig. 3.

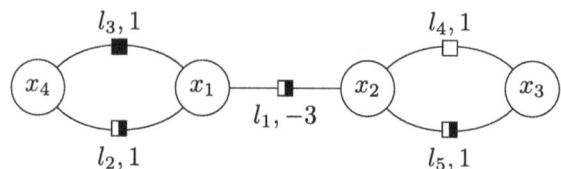

Fig. 3. Example Constraint Network (without node x_0)

This cycle reduces to the edge $(x_1 \overset{-2}{\square} x_1)$, thus it is a negative cost gray cycle.

Our results utilize the following theorem from [13].

Theorem 1. *A system of UTVPI constraints **U** has a dag-like refutation of length l, if and only if, the corresponding constraint network **G** has a negative cost gray cycle of length l.*

4 Unit Refutations in UTVPI Constraint Systems

In this section, we examine the problem of finding unit refutations of UCSs. We utilize the network construction described in Sect. 3.

We first show that the length of a dag-like unit refutation can be exponential in the size of the UCS. Thus, we cannot construct the refutation inference by inference.

Theorem 2. *There exists a UCS with n variables for which the ODLUR has length $(2 \cdot n - 2) \cdot (2 \cdot f(n) + 1)$ for an arbitrary function $f(n)$.*

Proof. Let **U** be the UCS in System (6):

$$\begin{aligned}
x_1 &\leq f(n) & x_1 - x_2 &\leq 0 & -x_1 - x_2 &\leq 0 \\
x_2 - x_3 &\leq 0 & \cdots & & x_{n-1} - x_n &\leq -1 \\
x_{n-1} + x_n &\leq 0
\end{aligned} \tag{6}$$

Let R be a dag-like unit refutation of **U**. Since $x_1 \leq f(n)$ is the only absolute constraint in **U**, R must use this constraint. Note that the constraint $x_{n-1} - x_n \leq -1$ is the only constraint in **U** with a negative right-hand side. Thus, R must use at least $(f(n) + 1)$ copies of this constraint. Otherwise, the right-hand side of the final derived constraint will not be negative. To be able to use the constraint $x_{n-1} - x_n \leq -1$ as part of a unit refutation, we need to derive a constraint of the form $-x_{n-1} \leq b$. To derive this constraint from $x_1 \leq f(n)$, R needs to use the constraints $-x_1 - x_2 \leq 0$ through $x_{n-2} - x_{n-1} \leq 0$. This uses $(n-2)$ applications of the UADD rule, and derives the constraint $-x_{n-1} \leq f(n)$. Applying the UADD rule to this constraint and $x_{n-1} - x_n \leq -1$ results in the constraint $-x_n \leq f(n) - 1$. Then, deriving the constraint $x_1 \leq f(n) - 1$ requires using the constraints $x_{n-1} + x_n \leq 0$ through $x_1 - x_2 \leq 0$. This is an additional $(n-1)$ applications of the UADD rule. In total, deriving the constraint $x_1 \leq f(n) - 1$ requires $(2 \cdot n - 2)$ applications of the UADD rule.

Consequently, deriving $x_2 \leq -f(n) - 1$ requires a total of $((2 \cdot n - 2) \cdot (2 \cdot f(n) + 1) - 1)$ applications of the UADD rule. To derive the final contradiction, we need to use the derived constraint $-x_2 \leq f(n)$ through an additional use of the UADD rule. Thus, any unit refutation of **U** has length at least $(2 \cdot n - 2) \cdot (2 \cdot f(n) + 1)$. □

Note that a similar result was obtained for DCSs in [10]. However, the result in this paper derives a different bound on the length of unit refutations due to the different constraint system.

We now show that a UCS **U** has a unit refutation, if and only if, the corresponding constraint network **G** has a negative weight gray cycle that uses the vertex x_0. From Sect. 3, the vertex x_0 is the vertex in the constraint network used to handle absolute constraints.

Lemma 1. *A UCS **U** has a unit refutation, if and only if, the corresponding constraint network **G** has a negative weight gray cycle using x_0.*

Note that Lemma 1 deals with the existence of unit refutations while Theorem 1 is about the length of unrestricted dag-like refutations.

Proof. First, suppose that $\mathbf{U}$ has a unit refutation R. Recall that each application of the UADD rule has the following form:

$$\frac{a_i \cdot x_i \leq b_i \qquad\qquad -a_i \cdot x_i + a_j \cdot x_j \leq b_{ij}}{a_j \cdot x_j \leq b_i + b_{ij}}$$

Note that no two-variable constraints can be derived in this way. If a non-absolute constraint is derived, then it has to be of the form $0 \leq b$. If $b < 0$, then we have derived a contradiction and thus produced a refutation. If $b \geq 0$, then the remaining portion of R must be a unit refutation. Thus, without loss of generality, we can assume that only the final application of the UADD rule in R derives a non-absolute constraint.

Since constraint addition is associative, we can assume without loss of generality that the first inference in R is

$$\frac{a_i \cdot x_i \leq b_i \qquad\qquad a_k \cdot x_k - a_i \cdot x_i \leq b_{ki}}{a_k \cdot x_k \leq b_i + b_{ki}}$$

and that the last inference is

$$\frac{a_j \cdot x_j \leq b - b_j \qquad\qquad -a_j \cdot x_j \leq b_j}{0 \leq b}.$$

From R, we can construct a negative weight gray cycle C in $\mathbf{G}$ as follows:

1. Start the constraint $a_i \cdot x_i \leq b_i$. If $a_i = 1$, then by construction of $\mathbf{G}$, the edge $x_0 \overset{b_i}{\square} x_i$ is in $\mathbf{G}$. Add this edge to C.

   If $a_i = -1$, then by construction of $\mathbf{G}$, the edge $x_0 \overset{b_i}{\blacksquare} x_i$ is in $\mathbf{G}$. Add this edge to C.

2. Let $a_k \cdot x_k - a_i \cdot x_i \leq b_{ki}$ be the other constraint used by the first application of the UADD rule. Add the corresponding edge of $\mathbf{G}$ to C.

3. Let $a_{i'} \cdot x_{i'} + a_{j'} \cdot x_{j'} \leq b_{i'j'}$ be the a constraint used by the rth application of the UADD rule. Add the corresponding edge of $\mathbf{G}$ to C.

4. Finally, consider the constraint $-a_j \cdot x_j \leq b_j$. If $a_j = 1$, then by construction of $\mathbf{G}$, the edge $x_j \overset{b_j}{\blacksquare} x_0$ is in $\mathbf{G}$. Add this edge to C.

   If $a_i = -1$, then by construction of $\mathbf{G}$, the edge $x_j \overset{b_j}{\square\!\blacksquare} x_0$ is in $\mathbf{G}$. Add this edge to C.

From the structure of R, the constraint introduced by the rth inference must cancel the variable introduced by the $(r-1)^{th}$ inference. Thus, these constraints share a variable $x_{i'}$ and the coefficients of $x_{i'}$ in these constraints have opposite signs. These constraints correspond to the $(r+1)^{th}$ and r^{th} edges in C respectively. Thus, C is a valid path from x_0 to itself.

The first edge of C is either $x_0 \overset{b_i}{\square} x_i$ or $x_0 \overset{b_i}{\square\!\blacksquare} x_i$, and the last edge of C is either $x_j \overset{b_j}{\blacksquare} x_0$ or $x_j \overset{b_j}{\square\!\blacksquare} x_0$. Thus, C can be reduced to a single gray edge from x_0 to itself. Thus, C is a gray cycle through x_0.

By construction, the total weight of all the edges in C is $b < 0$. Thus, C is a negative weight gray cycle through x_0.

Now suppose that $\mathbf{G}$ has a negative weight gray cycle C through the vertex x_0. We can reverse the process used to construct C to construct a unit refutation R. A detailed description of the process will be included in the full version of this paper. □

Now we need a way to detect if $\mathbf{G}$ contains a negative weight gray cycle through the vertex x_0. Recall that the length of this walk can be exponential in the size of the input.

This can be done in polynomial time by looking at the vertices of $\mathbf{G}$ reachable from x_0 by white paths and the vertices of $\mathbf{G}$ reachable by black paths. Let $S^{\square}$ be the set of vertices in $\mathbf{G}$ reachable from x_0 by a white path. Additionally, let $S^{\blacksquare}$ be the set of vertices reachable from $\mathbf{G}$ by a black path.

Lemma 2. *If $\mathbf{U}$ has a unit refutation, then the graph $\mathbf{G}'$ induced by $S^{\square} \cap S^{\blacksquare}$ contains a negative weight gray cycle.*

Proof. Assume that $\mathbf{U}$ has a unit refutation. From Theorem 1, $\mathbf{G}$ contains a negative weight gray cycle C through the vertex x_0. Let x_i be a vertex on this cycle.

By the construction of C in Theorem 1, the constraints $x_i \leq b_i$ and $-x_i \leq b_i'$ must be derivable from $\mathbf{U}$ by unit refutation. If we repeat the process used to construct C using the derivation of $x_i \leq b_i$, we obtain a white path between x_0 and x_i. Thus, $x_i \in S^{\square}$. If we repeat the process used to construct C using the derivation of $-x_i \leq b_i'$, we obtain a black path between x_0 and x_i. Thus, $x_i \in S^{\blacksquare}$. Consequently, $x_i \in S^{\square} \cap S^{\blacksquare}$.

Since x_i was an arbitrary vertex on C, every vertex on C must be in $S^{\square} \cap S^{\blacksquare}$. Thus, $\mathbf{G}'$ contains a negative weight gray cycle. □

Lemma 3. *If the graph $\mathbf{G}'$ induced by $S^{\square} \cap S^{\blacksquare}$ contains a negative weight gray cycle, then $\mathbf{U}$ has a unit refutation.*

Proof. Assume that $\mathbf{G}'$ contains a negative weight gray cycle C' of weight $b_{C'} < 0$. Let x_i be a vertex on this cycle. Since x_i is in $S^{\square} \cap S^{\blacksquare}$, there must be a white path p_1 between x_0 and x_i. Let b_1 be the weight of this path. Similarly there must be a black path p_2 between x_0 and x_i. Let b_2 be the weight of this path.

Algorithm 4.1 Algorithm for constructing the subgraph of vertices reachable from x_0.

Function GET-SUBGRAPH($\mathbf{U}$)

1: Let $\mathbf{G}$ be the constraint network corresponding to $\mathbf{U}$.

2: Let $S^{\square} \leftarrow \{x_0\}$ be the set of vertices reachable from x_0 by a white path.

3: Let $S^{\blacksquare} \leftarrow \{x_0\}$ be the set of vertices reachable from x_0 by a black path.

4: Let $E^{\blacksquare}$ be the set of gray edges leaving $S^{\square}$.

5: Let $E^{\square}$ be the set of white edges leaving $S^{\blacksquare}$.

6: Let $E^{\blacksquare}$ be the set of gray edges entering $S^{\blacksquare}$.

7: Let $E^{\blacksquare}$ be the set of black edges leaving $S^{\square}$.

8: **while** ($E^{\blacksquare} \cup E^{\square} \cup E^{\blacksquare} \cup E^{\blacksquare} \neq \emptyset$) **do**

9:     **for** (each edge $(x_i \overset{b_k}{\square} x_j) \in E^{\square}$) **do** ▷ Assume without loss of generality that $x_i \in S^{\blacksquare}$.

10:         Add x_j to $S^{\square}$.

11:     **for** (each edge $(x_i \overset{b_k}{\blacksquare} x_j) \in E^{\blacksquare}$) **do** ▷ Assume without loss of generality that $x_i \in S^{\square}$.

12:         Add x_j to $S^{\square}$.

13:     **for** (each edge $(x_i \overset{b_k}{\blacksquare} x_j) \in E^{\blacksquare}$) **do** ▷ Assume without loss of generality that $x_i \in S^{\square}$.

14:         Add x_j to $S^{\blacksquare}$.

15:     **for** (each edge $(x_i \overset{b_k}{\blacksquare} x_j) \in E^{\blacksquare}$) **do** ▷ Assume without loss of generality that $x_i \in S^{\blacksquare}$.

16:         Add x_j to $S^{\blacksquare}$.

17:     Let $E^{\blacksquare}$ be the set of gray edges leaving $S^{\square}$.

18:     Let $E^{\square}$ be the set of white edges leaving $S^{\blacksquare}$.

19:     Let $E^{\blacksquare}$ be the set of gray edges entering $S^{\blacksquare}$.

20:     Let $E^{\blacksquare}$ be the set of black edges leaving $S^{\square}$.

21: **return** $S^{\square} \cap S^{\blacksquare}$.

Thus, we construct a negative weight gray cycle C as follows:

1. Add the white path p_1 to C.
2. Add $\left\lfloor \frac{b_1+b_2}{-b_C} + 1 \right\rfloor$ copies of the negative weight gray cycle C' to C.

3. Add the black path p_2 to C.

By construction, C is a gray cycle through the vertex x_0. We now show that C is a negative weight gray cycle.

Note that C has a weight of:

$$b_1 + b_2 + \left\lfloor \frac{b_1 + b_2}{-b_C} + 1 \right\rfloor \cdot b_C \le b_1 + b_2 + \left(\frac{b_1 + b_2}{-b_C} + 1 \right) \cdot b_C = b_C < 0.$$

Thus C is a negative weight gray cycle through x_0 as desired.   □

Figure 4 shows the structure of a unit refutation found in this way.

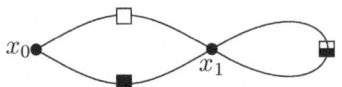

Fig. 4. Structure of a unit refutation

From Lemmas 2 and 3, we have the following result.

Theorem 3. U *has a unit refutation, if and only if, the graph* **G'** *induced by* $S^{\square} \cap S^{\blacksquare}$ *contains a negative weight gray cycle.*

Proof. If **U** has a unit refutation, then by Lemma 2, **G'** contains a negative weight gray cycle. If **G'** contains a negative weight gray cycle, then by Lemma 3, **U** has a unit refutation.   □

We can construct $S^{\square} \cap S^{\blacksquare}$ using Algorithm 4.1.

Theorem 4. *Algorithm 4.1 correctly constructs both* $S^{\square}$ *and* $S^{\blacksquare}$.

Proof. Let **U** be a UCS and let **G** be the corresponding constraint network.

Let x_i be a vertex in $S^{\square}$ and let x_j be a vertex in $S^{\blacksquare}$. We will show that Algorithm 4.1 will correctly add x_i to $S^{\square}$ and x_j to $S^{\blacksquare}$. This will be done by induction on the lengths of the paths between x_0 and both x_i and x_i.

Base case: Suppose x_i is reachable from x_0 by a white path of length 1. Thus, by construction there exists an edge $x_0 \square x_i$ in **G**. Initially, when $S^{\blacksquare} = x_0$, this is a white edge leaving $S^{\blacksquare}$. Thus, x_i is added to $S^{\square}$ as desired.

Similarly, if x_j is reachable from x_0 by a black path of length 1, then x_j is added to $S^{\blacksquare}$ as desired.

Inductive Step: Assume that all vertices reachable from x_0 by a white path with k edges are added to $S^{\square}$ by Algorithm 4.1. Additionally, assume all vertices reachable from x_0 by a black path with k edges are added to $S^{\blacksquare}$ by Algorithm 4.1.

Suppose x_i is reachable from x_0 by a white path P with $(k+1)$ edges. By construction, the last edge of P is either $x_{i'} \,\square\, x_i$ or $x_{i'} \,\blacksquare\, x_i$ in $\mathbf{G}$.

If the last edge of P is $x_{i'} \,\square\, x_i$, then the remaining edges of P form a gray path from x_0 to $x_{i'}$. Consider the first edge of this path. By construction, this edge must be either $x_0 \,\square\, x_l$ or $x_0 \,\blacksquare\, x_l$ for some vertex x_l. If this edge is $x_0 \,\square\, x_l$, then by construction of $\mathbf{G}$, $\mathbf{G}$ must also contain the edge $x_0 \,\blacksquare\, x_l$. Similarly, if this edge is $x_0 \,\blacksquare\, x_l$, then by construction of $\mathbf{G}$, $\mathbf{G}$ must also contain the edge $x_0 \,\blacksquare\, x_l$. Let P' be the path made by replacing the edge $x_0 \,\square\, x_l$ (or $x_0 \,\blacksquare\, x_l$) with the edge $x_0 \,\blacksquare\, x_l$ or $(x_0 \,\blacksquare\, x_l)$. Note that P' is a black path between x_0 and $x_{i'}$. Since P' has k edges $x_{i'}$ was added to $S^{\blacksquare}$ by Algorithm 4.1. After adding $x_{i'}$ to $S^{\blacksquare}$, the edge $x_{i'} \,\square\, x_i$ is a white edge leaving $S^{\blacksquare}$. Thus, x_i is added to $S^{\square}$ by Algorithm 4.1.

The case where the last edge of P is $x_{i'} \,\blacksquare\, x_i$ is handled similarly, as is the case where x_j is reachable from x_0 by a black path P with $(k+1)$ edges.

Thus, by the first principle of induction, $S^{\square}$ contains all vertices reachable from x_0 by white paths, and $S^{\blacksquare}$ contains all vertices reachable from x_0 by black paths. □

From Lemmas 2 and 3 and Theorem 4, we have the following result.

Theorem 5. *The problem of checking if a UCS $\mathbf{U}$ with m constraints over n variables has a unit refutation can be solved in $O(m \cdot n)$ time.*

Proof. From Lemmas 2 and 3 we know that $\mathbf{U}$ has a unit refutation, if and only if, the graph $\mathbf{G}'$ induced by $S^{\square} \cap S^{\blacksquare}$ contains a negative weight gray cycle.

Observe that each edge in $\mathbf{G}$ is processed by Algorithm 4.1 at most twice. Thus, Algorithm 4.1 finds $S^{\square} \cap S^{\blacksquare}$ in $O(m+n)$ time. Once $S^{\square} \cap S^{\blacksquare}$ is found checking if the graph induced by $S^{\square} \cap S^{\blacksquare}$ contains a negative weight gray cycle can be done in $O(m \cdot n)$ time [11]. □

5   Shortest Dag-Like Unit Refutations

In this section, we provide a pseudopolynomial time algorithm for the ODLUR problem for UCSs.

We utilize the structure described in Lemma 3 to prove upper and lower bounds on the lengths of dag-like unit refutations of UCSs. In the theorems in this section, b_{max} represents the largest defining constant in a UCS.

Theorem 6. *For each even n, there exists a UCS $\mathbf{U}$ with n variables such that the ODLUR of $\mathbf{U}$ has length $(2 \cdot n^2 \cdot b_{max} + 2 \cdot n)$.*

Proof. Let $\mathbf{U}$ be the UCS in System (7).

$$
\begin{array}{lll}
x_{\frac{n}{2}-1} \le b_{max} & x_1 - x_2 \le b_{max} & -x_1 - x_2 \le b_{max} \\
x_2 - x_3 \le b_{max} & \cdots & x_{\frac{n}{2}-2} - x_{\frac{n}{2}-1} \le b_{max} \\
x_{\frac{n}{2}-1} + x_{\frac{n}{2}} \le b_{max} & x_{\frac{n}{2}-1} - x_{\frac{n}{2}} \le b_{max} & x_{\frac{n}{2}} - x_{\frac{n}{2}+1} \le 0 \\
-x_{\frac{n}{2}} - x_{\frac{n}{2}+1} \le 0 & x_{\frac{n}{2}+1} - x_{\frac{n}{2}+2} \le 0 & \cdots \\
x_{n-2} - x_{n-1} \le 0 & x_{n-1} - x_n \le 0 & x_{n-1} + x_n \le -1
\end{array}
\tag{7}
$$

Let $\mathbf{G}$ be the constraint network corresponding to $\mathbf{U}$. By construction, the only negative cost gray cycle C' in $\mathbf{G}$ is

$$(x_{\frac{n}{2}} \; \blacksquare \; \overset{0}{} \; x_{\frac{n}{2}+1} \ldots x_{n-1} \; \blacksquare \; \overset{0}{} \; x_n \; \square \; \overset{-1}{} \; x_{n-1} \; \blacksquare \; \overset{0}{} \; x_{n-1} \ldots x_{\frac{n}{2}+2} \; \blacksquare \; \overset{0}{} \; x_{\frac{n}{2}+1} \; \blacksquare \; \overset{0}{} \; x_{\frac{n}{2}}).$$

This cycle has n edges and total cost -1.

Additionally by construction, a shortest way to reach a vertex on C using both a white path and a black path are the following two paths p_1 and p_2:

$$(x_0 \; \square \; \overset{b_{max}}{} \; x_{\frac{n}{2}-1} \; \blacksquare \; \overset{b_{max}}{} \; x_{\frac{n}{2}-2} \ldots x_2 \; \blacksquare \; \overset{b_{max}}{} \; x_1 \; \blacksquare \; x_2 \; \blacksquare \; \overset{b_{max}}{} \; x_3 \ldots x_{\frac{n}{2}-2} \; \blacksquare \; \overset{b_{max}}{} \; x_{\frac{n}{2}-1} \; \square \; \overset{b_{max}}{} \; x_{\frac{n}{2}}).$$

$$(x_0 \; \blacksquare \; \overset{b_{max}}{} \; x_{\frac{n}{2}-1} \; \blacksquare \; \overset{b_{max}}{} \; x_{\frac{n}{2}-2} \ldots x_2 \; \blacksquare \; \overset{b_{max}}{} \; x_1 \; \blacksquare \; x_2 \; \blacksquare \; \overset{b_{max}}{} \; x_3 \ldots x_{\frac{n}{2}-2} \; \blacksquare \; \overset{b_{max}}{} \; x_{\frac{n}{2}-1} \; \blacksquare \; \overset{b_{max}}{} \; x_{\frac{n}{2}}).$$

Each of these paths uses n edges and has a total cost of $n \cdot b_{max}$.

A shortest dag-like refutation C of $\mathbf{U}$ corresponds to the path p_1 followed by $\left\lfloor \frac{b_1+b_2}{-b_{C'}} + 1 \right\rfloor$ copies of C' followed by the path p_2. Thus, wee need $(2 \cdot n \cdot b_{max} + 1)$ copies of C'. Note that p_2 consists of the derived constraint $-x_{\frac{n}{2}-1} \le (n-1) \cdot b_{max}$ and the constraint $x_{\frac{n}{2}-1} - x_{\frac{n}{2}} \le b_{max}$. Thus, we can think of p_2 as using only two constraints in a dag-like refutation.

In total C uses $(2 \cdot n^2 \cdot b_{max} + 2 \cdot n + 1)$ edges. Thus, the ODLUR of $\mathbf{U}$ has length $(2 \cdot n^2 \cdot b_{max} + 2 \cdot n)$. $\qquad\square$

Theorem 7. *Let $\mathbf{U}$ be a UCS with n variables. The ODLUR of $\mathbf{U}$ has length at most $(8 \cdot n^2 \cdot b_{max} + 6 \cdot n)$.*

Proof. Let $\mathbf{U}$ be a UCS with a unit refutation. Additionally, let $\mathbf{G}$ be the constraint network corresponding to $\mathbf{U}$. From Lemma 2, the graph induced by $S^{\square} \cap S^{\blacksquare}$ contains a negative weight gray cycle.

From Lemma 3, $\mathbf{U}$ has a unit refutation that corresponds to a white path p_1, a possibly repeated negative cost gray cycle C' and a black path p_2. Recall that each of these uses no vertex more than twice. Thus, p_1, p_2 and C' each have at most $2 \cdot n$ edges.

Since b_{max} is the largest edge weight in $\mathbf{G}$. The weight b_1 of p_1 is at most $2 \cdot n \cdot b_{max}$. Similarly, the weight b_2 of p_2 is at most $2 \cdot n \cdot b_{max}$. Since, C' is a negative cost gray cycle the weight $b_{C'}$ of C' is at most -1.

To, make a negative weight gray cycle C through x_0, we need $\left\lfloor \frac{b_1+b_2}{-b_{C'}} + 1 \right\rfloor$ copies of C'. Note that $\left\lfloor \frac{b_1+b_2}{-b_{C'}} + 1 \right\rfloor \le 4 \cdot n \cdot b_{max} + 1$. Thus, the number of edges in C is at most $2 \cdot n + 2 \cdot n + 2 \cdot n \cdot (4 \cdot n \cdot b_{max} + 1) = 8 \cdot n^2 \cdot b_{max} + 6 \cdot n$

This means that $\mathbf{U}$ has a dag-like unit refutation of length at most $(8 \cdot n^2 \cdot b_{max} + 6 \cdot n)$. Thus, the ODLUR of $\mathbf{U}$ has length at most $(8 \cdot n^2 \cdot b_{max} + 6 \cdot n)$. $\square$

We now present an $O(m \cdot n^2 \cdot ||\mathbf{b}||_\infty)$ time pseudo-polynomial time algorithm for the ODLUR problem for UCSs.

From Theorem 7, an ODLUR of a UCS has length at most $(8 \cdot n^2 \cdot ||\mathbf{b}||_\infty + 6 \cdot n)$, it can be found in $O(m \cdot n^2 \cdot ||\mathbf{b}||_\infty)$ time using the algorithm in [11].

Theorem 8. *An ODLUR of a UCS can be found in $O(m \cdot n^2 \cdot ||\mathbf{b}||_\infty)$ time.*

Proof. Let $\mathbf{U}$ be a UCS and let $\mathbf{G}$ be the corresponding constraint network. From Theorem 7, an ODLUR of $\mathbf{U}$ has length l^* which is at most $(8 \cdot n^2 \cdot ||\mathbf{b}||_\infty + 6 \cdot n)$. Thus, there is a negative weight gray cycle in $\mathbf{G}$ through x_0 with at most $(8 \cdot n^2 \cdot ||\mathbf{b}||_\infty + 6 \cdot n)$ edges. Thus, we can find a shortest negative weight gray cycle through x_0 as follows:

1. After k iterations of the algorithm in [11], we will find the least weight gray path in $\mathbf{G}$ with at most k edges from x_0 to each vertex.
2. Since $\mathbf{U}$ has a unit refutation of length l^*, $\mathbf{G}$ has a negative weight gray cycle from x_0 to itself of length l^*.
3. After $(8 \cdot n^2 \cdot ||\mathbf{b}||_\infty + 6 \cdot n) \geq l^*$ iterations of the algorithm in [11], we will find a negative weight gray cycle from x_0 to itself.
4. This takes $O(m \cdot n^2 \cdot ||\mathbf{b}||_\infty)$ time.   □

6   Approximability

We now show that the problem of finding the length of an ODLUR for a UCS can be approximated to within a factor of 2 in polynomial time. This algorithm focuses on a restricted form of dag-like unit refutation.

Definition 17. *A* **simple dag-like unit refutation** *is a dag-like unit refutation that can be decomposed into either:*

1. *A white path between x_0 and some vertex x_i,*
2. *a black path between x_i and some vertex x_j,*
3. *a gray path from x_i to x_j, and*
4. *a negative weight gray cycle through the vertex x_j, possibly repeated.*

or:

1. *A black path between x_0 and some vertex x_i,*
2. *a white path between x_i and some vertex x_j,*
3. *a gray path from x_j to x_i, and*
4. *a negative weight gray cycle through the vertex x_j, possibly repeated.*

First, we show that every UCS with a dag-like unit refutation has a simple dag-like unit refutation.

Lemma 4. *Let $\mathbf{U}$ be an infeasible UCS and let R be a dag-like unit refutation of $\mathbf{U}$. $\mathbf{U}$ has a simple dag-like unit refutation.*

Proof. From Lemma 1, we know that R corresponds to a negative weight gray walk through x_0. This walk can be divided into:

1. A white walk w_1 from x_0 to some vertex x_j.
2. A closed gray walk w_2 through the vertex x_j, possibly repeated.
3. A black walk w_3 from x_j to x_0.

We now show that for some vertex x'_j, none of these paths use a vertex more than twice.

Assume for the sake of contradiction, that the walk w_1 uses a vertex more than twice. Thus, it contains a gray cycle C' [11]. If this is not a negative weight gray cycle, then we can remove this cycle from w_1 and shorten the corresponding refutation. Thus, we can assume that C' has a weight $-b_{C'} < 0$.

Consider the average weight of this cycle $\frac{-b_{C'}}{|C'|}$, and compare it to the average weight of the gray cycle w_2 ($\frac{-b_{w_2}}{|w_2|}$). If the average weight of C' is lower than that of w_2, then we can replace repetitions of w_2 with repetitions of C' until w_2 no longer appears in the walk corresponding to R. Similarly, if the average weight of w_2 is lower than that of C', then we can replace repetitions of C' with repetitions of w_2. In each case, remove one of these two cycles from the refutation.

After this process, w_1 has no gray cycles, and so uses each vertex at most twice. We can similarly show that w_2 and w_3 must also use each vertex at most twice.

Note that the white and black paths between x_0 and x_j may be the same up until a vertex x_i. This gives us a simple dag-like unit refutation.   □

From Lemma 4, we know that every UCS $\mathbf{U}$ with a dag-like unit refutation has a simple dag-like refutation. We now show that the shortest simple dag-like unit refutation is at most twice the length of the ODLUR of $\mathbf{U}$ and that it can be found in polynomial time.

Theorem 9. *The problem of approximating the length of an ODLUR R of a UCS $\mathbf{U}$ with m constraints over n variables to within a factor of 2 can be solved in $O(n^6)$ time.*

The proof of Theorem 9 has been omitted for space reasons and will be found in the full version of this paper.

Theorem 9 establishes that this problem is in **APX**. From [10], the ODLUR problem for DCSs is **APX-complete**. Consequently, the ODLUR problem for UCSs is **APX-hard** as well, since UTVPI constraints subsume difference constraints. It follows that the ODLUR problem for DCSs is **APX-complete**.

From [1], the Min 2SAT problem cannot be approximated to within $(\frac{15}{14} - \epsilon)$ for any $\epsilon > 0$. Note that the reduction in [10] preserves this approximation ratio. Consequently, the length of an ODLUR of a DCS, and thus a UCS, cannot be approximated to within $(\frac{15}{14} - \epsilon)$ for any $\epsilon > 0$.

Note that the ODLUR problem for UCSs is both **APX-complete** and can be solved with a pseudo-polynomial time algorithm. Such problems are unusual. However, there are other problems with similar properties. For example, the 2-dimensional Knapsack problem has a pseudo-polynomial time algorithm but no FPTAS [5]. From the **APX-completeness** result and the pseudo-polynomial time algorithm, we have the following dichotomy:

1. If $||\mathbf{b}||_\infty$ is polynomial in n and m, then the ODLUR problem for UCSs can be solved in polynomial time.
2. If $||\mathbf{b}||_\infty \geq n \cdot 2^m - 2$, then the ODLUR problem for UCSs is **APX-complete**.

References

1. Avidor, A., Zwick, U.: Approximating MIN 2-sat and MIN 3-sat. Theory Comput. Syst. **38**(3), 329–345 (2005)
2. Cousot, P., Cousot, R.: Abstract interpretation: a unified lattice model for static analysis of programs by construction or approximation of fixpoints. In: POPL, pp. 238–252 (1977)
3. Farkas, G.: Über die Theorie der Einfachen Ungleichungen. Journal für die Reine und Angewandte Mathematik **124**(124), 1–27 (1902)
4. Jaffar, J., Maher, M.J., Stuckey, P.J., Yap, H.C.: Beyond finite domains. In: Proceedings of the Second International Workshop on Principles and Practice of Constraint Programming (1994)
5. Korte, B., Schrader, R.: On the existence of fast approximation schemes. In: Mangasarian, O.L., Meyer, R.R., Robinson, S.M. (eds.) Nonlinear Programming (1981)
6. Lahiri, S.K., Musuvathi, M.: An efficient decision procedure for UTVPI constraints. In: Proceedings of the 5th International Workshop on the Frontiers of Combining Systems, Vienna, Austria, 19–21 September 2005, pp. 168–183. Springer, New York (2005)
7. McConnell, R.M., Mehlhorn, K., Näher, S., Schweitzer, P.: Certifying algorithms. Comput. Sci. Rev. **5**(2), 119–161 (2011)
8. Miné, A.: The octagon abstract domain. Higher-Order Symb. Comput. **19**(1), 31–100 (2006)
9. Subramani, K.: On the complexity of selected satisfiability and equivalence queries over Boolean formulas and inclusion queries over hulls. J. Appl. Math. Decis. Sci. (JAMDS) **1–18**, 2009 (2009)
10. Subramani, K., Wojciechowski, P.: Unit refutations of difference constraint systems. In: Gal, K., Nowé, A., Nalepa, G.J., Fairstein, R., Radulescu, R. (eds.) ECAI 2023 - 26th European Conference on Artificial Intelligence, Kraków, Poland, 30 September–4 October 2023 - Including 12th Conference on Prestigious Applications of Intelligent Systems (PAIS 2023). Frontiers in Artificial Intelligence and Applications, vol. 372, pp. 2226–2233. IOS Press (2023)
11. Subramani, K., Wojciechowski, P.J.: A combinatorial certifying algorithm for linear feasibility in UTVPI constraints. Algorithmica **78**(1), 166–208 (2017)
12. Subramani, K., Wojciechowski, P.J.: Analyzing unit read-once refutations in difference constraint systems. In: Faber, W., Friedrich, G., Gebser, M., Morak, M. (eds.) Logics in Artificial Intelligence - 17th European Conference, JELIA 2021, Virtual Event, 17–20 May 2021, Proceedings. LNCS, vol. 12678, pp. 147–161. Springer (2021)
13. Wojciechowski, P.J., Subramani, K., Williamson, M.D.: Polynomial time algorithms for optimal length tree-like refutations of linear infeasibility in UTVPI constraints. Discrete Appl. Math. **305**, 272–294 (2021)

Online Contention Resolution Schemes for Size-Stochastic Knapsacks

Toru Yoshinaga[✉][iD] and Yasushi Kawase[✉][iD]

The University of Tokyo, Hongo 7-3-1, Bunkyo, Tokyo, Japan
yoshinaga-toru106@g.ecc.u-tokyo.ac.jp, kawase@mist.i.u-tokyo.ac.jp

Abstract. Online contention resolution schemes (OCRSs) are effective rounding techniques for online combinatorial optimization problems with stochastic inputs. These schemes randomly and sequentially round a fractional solution to a relaxed problem that can be formulated in advance. In this study, we propose OCRSs for online stochastic knapsack problems and, more generally, online stochastic generalized assignment problems. In our setup, each item arriving sequentially is inserted into one of multiple knapsacks or discarded. Its size, which follows a known distribution, is revealed only after insertion. The goal of the problem is to maximize the *acceptance probability*, which is the smallest probability among the items being placed in the knapsack. Since the item sizes are unknown beforehand, a violation of capacity constraints may occur. Thus, we consider two distinct settings: the hard constraint setting, where items that cause such violations are rejected, and the soft constraint setting, where these items are accepted. Under the hard constraint setting, we present an algorithm with an acceptance probability of 1/3 and show that no algorithm can achieve an acceptance probability greater than 3/7. Under the soft constraint setting, we propose an algorithm with an acceptance probability of 1/2 and demonstrate that this is best possible.

Keywords: Online contention resolution schemes · Stochastic knapsack problem · Online algorithms

1 Introduction

The *Online Stochastic Generalized Assignment Problem (OSGAP)* is an online selection problem that represents several significant online problems, such as online knapsack problems [19,26], web advertising problems [13,25], and online scheduling problems [9,21]. In the OSGAP, multiple knapsacks with unit capacity are given first, and then a sequence of items with varying sizes and values arrives in an online manner. Upon the arrival of an item, it must either be packed into a knapsack or discarded. Although distributional information on the size of each item is available from the beginning, its actual size is revealed only after it is placed in a knapsack. The objective of the problem is to maximize the total value of the packed items. This problem applies to a scheduling scenario where

© The Author(s), under exclusive license to Springer Nature Singapore Pte Ltd. 2025
S. Nakano and M. Xiao (Eds.): WALCOM 2025, LNCS 15411, pp. 393–408, 2025.
https://doi.org/10.1007/978-981-96-2845-2_25

parallel machines process a sequence of jobs. In this context, each knapsack corresponds to a machine with a maximum load, and each item corresponds to a job with processing time and profit. Although the processing time for each job can be predicted, the actual processing time is determined only after the job is completed.

A promising approximation technique for online selection problems is to utilize an *Online Contention Resolution Scheme (OCRS)*. This scheme initially constructs a fractional solution to a relaxed problem using prior information and then stochastically and sequentially rounds it into a solution to the original online problem. This concept was introduced by Feldman et al. [14,15] and has been applied to various problems, such as prophet inequalities [20,22], posted price mechanisms [6,15], and online matchings [12,17,23].

Before Feldman et al. [14,15] established the concept of OCRSs, Alaei et al. [3] proposed an OCRS for an OSGAP, where the size of any arriving item does not exceed $1/k$ of the knapsack capacity, with k being a prefixed positive integer. For the rounding process, they designed an optimization problem called the *Generalized Magician Problem*. This problem is a stochastic online knapsack problem in which a unit-capacity knapsack is given initially, followed by a sequence of items with varying sizes arriving online. Upon an item's arrival, it must be packed into the knapsack or discarded. Here, items can be placed in the knapsack only when at least $1/k$ of the knapsack capacity remains. The goal of the problem is to maximize the *acceptance probability*, which is the smallest probability among the items of being placed in the knapsack. If the acceptance probability for the generalized magician problem is γ, their scheme guarantees a γ-fraction of the optimal value in expectation for the OSGAP. Alaei et al. [3] provided the γ-conservative algorithm, which achieves the acceptance probability of $(1 - 1/\sqrt{k})$. However, when $k = 1$, no algorithm can guarantee a positive acceptance probability.

An OSGAP algorithm based on the generalized magician problem attempts to put an item in a knapsack only when there is no risk of exceeding capacity. However, in applications such as online scheduling, it is reasonable to try to put an item in a knapsack if there is a chance it will fit (i.e., try to process a job even if there is a possibility that the job will not be completed by the deadline). Motivated by this, we introduce a slightly different problem, which we refer to as the *size-stochastic knapsack OCRS (KOCRS)* problem. Unlike the generalized magician problem, the KOCRS problem assumes that each item can take any size, allowing attempts to place items in knapsacks regardless of remaining capacity.

1.1   Our Results

We analyze the KOCRS problem and give algorithms with constant acceptance probabilities. Since a capacity violation may occur in our setup, we consider two settings: *hard* and *soft*. In the hard constraint setting, any item that causes a violation is rejected, while in the soft constraint setting, the item is accepted despite the violation. Such setups are commonly observed in stochastic knapsack

Table 1. Summary of the acceptance probability of the best algorithm for the KOCRS problem.

Setting	Item size	Acceptance Probability	
		Lower bound	Upper bound
Hard constraint	$[0, 1]$	1/3 (Theorem 3)	3/7 (Theorem 2)
Hard constraint	$\{\epsilon, 1\}$	3/7 (Theorem 5)	3/7 (Theorem 2)
Soft constraint	$[0, 1]$	1/2 (Theorem 7)	1/2 (Theorem 6)
Soft constraint	$\{\epsilon, 1\}$	1/2 (Theorem 7)	1/2 (Theorem 6)

problems [5, 8, 9, 26, 27] and can be interpreted in online scheduling as follows. In the hard constraint setting, jobs that are not completed within a certain period are rejected and bring no profit. On the other hand, in the soft constraint setting, jobs that start processing within the allotted time are accepted, regardless of whether they are completed by the time or not.

For the hard constraint setting, we provide an algorithm with an acceptance probability of $1/3$ (Theorem 3). We also devise an algorithm with an acceptance probability of $3/7$ when the item sizes are limited to a binary distribution on 1 and a sufficiently small positive real ϵ. Further, we prove that the acceptance probability is at most $3/7$ for any algorithm, even when the item sizes are limited to $\{\epsilon, 1\}$ (Theorem 2). For the soft constraint setting, we present an algorithm with the acceptance probability of $1/2$ (Theorem 7) and show that this is best possible even when item sizes are restricted to $\{\epsilon, 1\}$ (Theorem 6). Our results are summarized in Table 1.

In Sect. 3, we discuss how algorithms for the KOCRS problem can be incorporated into the rounding framework for OSGAP, similar to those for the generalized magician problem by Alaei et al. [3]. For the soft constraint setting, our OCRS implies an online algorithm for the OSGAP that gives a $1/2$-fraction of the optimal value of a relaxation problem in expectation. For the hard constraint setting, our OCRS guarantees at least $1/3$-fraction of the optimal value if the values of the items are determined in advance.

1.2 Related Work

The online knapsack problem is one of the hardest online problems, and it often requires additional assumptions to solve. In the first study of online knapsacks by Marchetti-Spaccamela and Vercellis [24], they showed that the competitive ratio of any algorithm becomes arbitrarily small for the general case. Consequently, it is common to impose restrictions on the online knapsack problem based on real-world situations. For example, in the context of scheduling and web advertising, it is often assumed that the value and size of each item follow known probability distributions [2, 29].

In (offline) stochastic knapsack problems, hard and soft constraint settings are popular approaches to handle capacity violation [4, 9, 18, 27]. In the problems, we are given a knapsack with a unit capacity and a known number of items. Each item has a deterministic value and a stochastic size that follows a given distribution. The size of an item is determined after its insertion. The objective of the decision-maker is to sequentially insert items into the knapsack, aiming to maximize the total value of the fitted items.

Our problem setting is closely related to the magician problem [2] and the generalized magician problem [3]. The magician problem is a special case of the generalized magician problem, in which the size of each item is either $1/k$ or 0. Alaei [2] presented the γ-conservative algorithm, which attains the acceptance probability of $1 - 1/\sqrt{k+3}$ for the magician's problem. Note that, in the magician's problem and the generalized magician problem, each item can be accepted only when there is no possibility of capacity violation, which differs from our setting.

Contention resolution schemes (CRSs) were introduced by Chekuri et al. [7] as an approximation scheme for (offline) submodular function maximization. After their work, Feldman et al. [14,15] extended the concept of CRSs for adversarial order online optimization problems as OCRSs. In recent years, OCRSs have been applied to a variety of online optimization problems [6,12,15,16,20,22], and they have been further studied in random order online optimization problems [1,10,17,23].

As with other online allocation problems, OCRSs have been proposed for online knapsack problems in various settings [1,11,15]. A typical technique for constructing OCRSs for online knapsack problems is to place each arriving item into the knapsack with a certain probability. We also adopt this approach, and from this perspective, our research is relatively close to the work by Alaei et al. [3] explained in the introduction and the work by Jiang et al. [20]. Jiang et al. [20] studied the knapsack OCRS problem, assuming that the item size follows a probability distribution. However, unlike our setting, the size of each item is revealed before its insertion. Jiang et al. [20] proposed an algorithm that places each realized item with probability $1/(3 + e^{-2}) \approx 0.319$. Our algorithms just guarantee the acceptance probability with respect to only items rather than pairs of items and their size realizations. Consequently, our setup is more challenging as the size realization is determined only after the insertion of each item, but it is easier in terms of the acceptance probability guarantee not being dependent on the realized size of items.

2   Preliminaries

In this section, we formally define the KOCRS problem. In this problem, we are initially given a knapsack with a unit capacity, and then n items arrive sequentially. Upon the arrival of the ith item, its size distribution F_i is revealed, and we must immediately decide whether to insert the item into the knapsack or irrevocably discard it. The size of each item is determined only after it has been inserted. If the item fits the knapsack, it is permanently placed inside and cannot be removed. As long as there is available space in the knapsack, we can insert items even if they might cause a capacity violation.

We consider two settings for handling the item that causes a capacity violation, which we refer to as a *critical item*. In the hard constraint setting, the critical item is discarded, and the capacity consumed by the critical item is not restored. In the soft constraint setting, the critical item is accepted without any penalty for exceeding capacity. In either setting, all items following the critical item are discarded.

The goal is to design an algorithm that successfully inserts every item with probability at least γ, where $\gamma \in [0, 1]$ is as large as possible. Here, we assume that $F_1, \ldots, F_n$ are independent probability distributions on $[0, 1]$.[1] In addition, it is promised that the total expected size of the items is less than or equal to the capacity of the knapsack, that is, $\sum_{i=1}^{n} \mathbb{E}[S_i] \leq 1$.[2]

The assumption that the total expected size of the items is at most 1 does not provide a guarantee that all items can always be successfully inserted into the knapsack. We demonstrate this fact with the following example, which also clarifies how the process differs between hard and soft constraint settings.

Example 1. Suppose that there are three items: the first item has a size of $1/2$ with probability 1, and the second and third items have sizes of 1 with probabilities of $1/4$ and 0 with probabilities of $3/4$, i.e.,

$$S_1 = 1/2 \;\; \text{w.p. } 1, \qquad S_2 = \begin{cases} 1 & \text{w.p. } 1/4, \\ 0 & \text{w.p. } 3/4, \end{cases} \qquad S_3 = \begin{cases} 1 & \text{w.p. } 1, \\ 0 & \text{w.p. } 3/4. \end{cases}$$

Note that the total expected size of the items is $1/2 + 1/4 + 1/4 = 1$.

Let us consider the strategy of greedily inserting incoming items into the knapsack. Since the knapsack is initially empty, the first item is successfully inserted into it, and its size is revealed to be $1/2$ with a probability of 1.

When attempting to insert the second item, it is revealed to be of size 1 and exceeds the capacity of the knapsack with probability $1/4$. Therefore, in the hard constraint setting, the second item is discarded with probability $1/4$. In the soft constraint setting, the second item is always accepted, but no further items can be accepted with probability $1/4$.

Finally, the procedure successfully inserts the third item only with probability $3/4 \cdot 3/4 = 9/16$ and $3/4$ in the hard and soft constraint settings, respectively. Thus, placing all the items with probability 1 is not always possible.

---

[1] The assumption that the size of each item is at most 1 is not necessary in the soft constraint setting, but it is crucial in the hard constraint setting. Indeed, consider an input sequence consisting of a single item with the following size distribution with a positive real ϵ: the size is $1 + \epsilon$ with probability $1 - \epsilon$ and 0 with probability ϵ. The expected size of the item is $1 - \epsilon^2$ (< 1). However, for the input sequence only with this item, the acceptance probability of any algorithm under the hard constraint setting is at most ϵ because the item fits only when its size is 0. Thus, by setting ϵ to be arbitrarily small, the acceptance probability becomes arbitrarily small.

[2] This assumption follows from Alaei et al. [3]. If $\sum_{i=1}^{n} \mathbb{E}[S_i] \leq \alpha$ for a real α, our algorithms remain feasible, but the guaranteed acceptance probability is scaled by a factor of $1/\alpha$.

We propose an algorithm that guarantees a constant acceptance probability for any input sequence. The algorithm determines the probability of inserting items by tracking the probability distribution of the knapsack, which can be computed by using the size distributions of the items that have arrived.

It should be noted that an OCRS by Jiang et al. [20] addressed a setting where the size of each item is revealed before deciding whether to insert the item into the knapsack. They introduced an algorithm that inserts each item with a probability depending on its realized size. However, this algorithm cannot be directly applied to our setting because the size is revealed only after insertion.

3   Rounding Framework for the OSGAP

We describe how the KOCRS problem can be employed to round a fractional solution to a relaxed problem of the OSGAP. Specifically, we demonstrate that, by utilizing an algorithm for the KOCRS problem of acceptance probability γ, we can construct an algorithm for the OSGAP that achieves at least γ-fraction of the value of the fractional solution. For each integer k, let $[k] := \{1, 2, \ldots, k\}$.

The OSGAP is defined as follows. Initially, we are given m knapsacks, each with a capacity of 1. Then, n items are presented sequentially. Each item has a pair of a value and a size distribution called a *type*. Item i has r_i possible types, and its actual type is initially known only as a distribution. Upon the arrival of item $i \in [n]$, its type is determined to be $t \in [r_i]$ with probability p_{it}. Then, the item must be inserted into one of the knapsacks or discarded immediately and irrevocably. If the capacity constraint is violated in this process, the acceptance/rejection of the critical item is determined according to the setting: the critical item is accepted in the soft constraint setting and rejected in the hard constraint setting. In both settings, items cannot be inserted into a knapsack after the critical item appears for it. If item i of type t is inserted into knapsack j, its value is v_{itj}, and its size S_{itj} is determined according to distribution F_{itj} on $[0, 1]$, after the insertion. We assume that all information on r_i, p_{it}, v_{itj}, and F_{itj} (for each $i \in [n]$, $t \in [r_i]$, and $j \in [m]$) is known from the beginning.

Recall that Alaei et al. [3] studied the OSGAP problem under an additional assumption called the large-capacity assumption: no item takes a size larger than $1/k$ of the knapsack capacity, where k is a prefixed positive integer. In contrast, our setting allows items to take any size. In addition, Alaei et al. [3] assumed that items are only placed in knapsacks with a remaining capacity of at least $1/k$, whereas we make no such assumption.

We construct an online algorithm for the problem by using a linear programming (LP) relaxation problem derived from prior information. The algorithm sequentially and stochastically rounds the optimal solution of the LP through the algorithms for the size-stochastic knapsack. We analyze the expected instance of the problem where everything happens as per expectation. Specifically, for each item i of type t and knapsack j, we denote the expected size of the item as $\tilde{s}_{itj} = \mathbb{E}[S_{itj}]$, which is computable before the arrival of items. The linear programming is formalized as follows:

$$\begin{aligned}
\text{max. } & \sum_{i\in[n]}\sum_{t\in[r_i]}\sum_{j\in[m]} v_{itj}x_{itj} \\
\text{s.t. } & \sum_{i\in[n]}\sum_{t\in[r_i]} \tilde{s}_{itj}x_{itj} \le 1 && \forall j \in [m], \\
& \sum_{j\in[m]} x_{itj} \le p_{it} && \forall i \in [n],\, t \in [r_i], \\
& x_{itj} \in [0,1] && \forall i \in [n],\, t \in [r_i],\, j \in [m].
\end{aligned} \tag{1}$$

We interpret the optimal solution x_{itj}^* of the LP as the probability of inserting an item. Note that the expected total size of the item set sent to each knapsack is at most 1. Ideally, items should be inserted according to the relaxed problem's assignment, but this may cause a capacity violation, similar to what we observed in Example 1. To avoid this, we use the KOCRS problem to realize the assignment with probability $\gamma \cdot x_{itj}^*$, and obtain γ-fraction in expectation. To be precise, we execute the following procedure: item i of type t is sent to knapsack j with probability x_{itj}^*/p_{it}, and an algorithm for the KOCRS problem determines whether actually to place the item into the knapsack or discard it. We formally describe our procedure in Algorithm 1.

Algorithm 1: Rounding Framework for the OSGAP with a KOCRS algorithm

---

Input: $\gamma \in [0,1]$, $(F_{itj})_{i\in[n],t\in[r_i],j\in[m]}$, an algorithm for the KOCRS problem
Output: Item allocation in m knapsacks and their size realization
1 Solve LP (1) and get the optimum x^*;
2 For each knapsack $j \in [m]$, construct an instance I_j of the KOCRS problem;
3 **for** $i \leftarrow 1$ **to** n **do**
4   Let t be the type of item i and select a knapsack (or nothing) $j^* \in [m] \cup \{\varnothing\}$ at random such that each knapsack $j \in [m]$ is chosen with probability x_{itj}^*/p_{it};
5   **for** $j \leftarrow 1$ **to** m **do**
6   Let F_{ij} be the CDF with
$$F_{ij}(s) = \left(1 - \sum_{t'\in[r_i]} x_{it'j}^*\right) + \sum_{t'\in[r_i]} x_{it'j}^* F_{it'j}(s);$$
7   Feed F_{ij} to I_j as the CDF of the ith item size;
8   **if** *Item i is accepted in the algorithm for the KOCRS problem of j^** **then**
9   Assign item i to knapsack j^*;
10   **else** Discard item i ;
11   Let the realization of the ith item be 0 for I_j with $j \ne j^*$ and S_{itj^*} for I_{j^*};

The combination of Algorithm 1 with an algorithm for the KOCRS problem of acceptance probability γ has the following guarantee for the OSGAP.

Theorem 1. *For the soft constraint setting, Algorithm 1 obtains at least γ-fraction of the optimal value of the LP (1) if the acceptance probability of the utilized KOCRS problem is at least γ. For the hard constraint setting, the same statement holds if the value of each item is independent of its type (i.e., $v_{itj} = v_{ij}$ for all i, j, t).*

Proof. The algorithm sent item i of type t to knapsack j with probability x_{itj}^*/p_{it}. In the soft constraint setting, the item is successfully placed in the knapsack with a probability of at least γ. Thus, the contribution of item i to the objective value is $\sum_{t \in [r_i]} \sum_{j \in [m]} v_{itj} p_{it}(x_{itj}^*/p_{it}) \cdot \gamma = \gamma \sum_{t \in [r_i]} \sum_{j \in [m]} v_{itj} x_{itj}^*$.

In the hard constraint setting, item i of type t sent to knapsack j may be accepted with a probability less than γ because the probability of capacity violation depends on the type. Nevertheless, since the probability that item i sent to knapsack j fits is at least γ in total, the contribution of item i to the objective value is $\sum_{j \in [m]} v_{ij} \sum_{t \in [r_i]} p_{it}(x_{itj}^*/p_{it}) \cdot \gamma = \gamma \sum_{t \in [r_i]} \sum_{j \in [m]} v_{ij} x_{itj}^*$ if the value of the item is independent of its type (i.e., $v_{itj} = v_{ij}$ for all i, j, t).

By summing up all the contributions, we obtain $\gamma \sum_{i \in [n]} \sum_{j \in [m]} \sum_{t \in [r_i]} v_{itj} x_{itj}^*$, which is equal to the optimal value of the LP (1) multiplied by γ. $\square$

4   Analysis for the Hard Constraint Setting

This section analyzes the KOCRS problem with the hard constraint setting. We first provide an instance of the hard constraint KOCRS problem, in which, for any algorithm, there exists an item with an acceptance probability at most $3/7$ (≈ 0.429). We then provide an algorithm that accepts each item with probability at least $1/3$ (≈ 0.333).

4.1   Impossibility

We demonstrate that no algorithm can accept every arriving item with a probability greater than $3/7$. The fact implies that our problem belongs to an essentially different class from the magician problem with $k = 1$, as we can ensure the acceptance probability of $1/2$ for the magician problem with $k = 1$.

Theorem 2. *For any positive real δ, there exists an instance of the hard constraint KOCRS problem where no algorithm has the acceptance probability of at least $3/7 + \delta$. This impossibility holds even when item sizes are restricted to 1 or a sufficiently small positive constant.*

Proof. Let ϵ be a sufficiently small positive real. We will show that no algorithm has the acceptance probability of greater than $3/7$ for the following instance with four items as ϵ approaches 0:

- the size of the first item S_1 is ϵ with probability 1,
- the sizes of the second and the third items S_2 and S_3 are 1 with probability $1/2 - \sqrt{\epsilon}$ and ϵ with probability $1/2 + \sqrt{\epsilon}$, and
- the size of the last item S_4 is ϵ with probability 1.

Note that the total expected size of the items is

$$\sum_{i=1}^{4} \mathbb{E}[S_i] = \epsilon + 2 \cdot (1 \cdot (1/2 - \sqrt{\epsilon}) + \epsilon \cdot (1/2 + \sqrt{\epsilon})) + \epsilon$$

$$= 1 - 2 \cdot \epsilon^{1/2} + 3 \cdot \epsilon + 2 \cdot \epsilon^{3/2} < 1 - 2 \cdot \epsilon^{1/2} + \frac{3}{2} \cdot \epsilon^{1/2} + \frac{2}{4} \cdot \epsilon^{1/2} = 1,$$

where the inequality holds if $\epsilon < 1/4$.

We will investigate the existence of an algorithm that accepts each item with a probability of at least γ. To this end, it suffices to consider the existence of an algorithm that accepts any item with exactly a probability of γ.

Since this algorithm accepts the first item with probability γ, the size of the item inside the knapsack after processing the first item becomes ϵ with probability γ, and 0 with probability $1 - \gamma$ (see Fig. 1a).

Suppose the algorithm tries to insert the second item with probability $\alpha \in [0, 1]$ under the condition that the first item is placed in the knapsack. If it tries to insert and the second item is revealed to be of size 1, then the second item is rejected since it violates the knapsack capacity. Hence, the probability that the first item is placed in the knapsack and the second item is also placed in the knapsack is $\gamma \cdot \alpha(1/2 + \sqrt{\epsilon})$. To accept the second item with probability γ overall, the probability that the first item is not placed but the second item is placed in the knapsack is $\gamma - \gamma \cdot \alpha(1/2 + \sqrt{\epsilon})$. The total size of the items in the knapsack, just after processing the second item, can be summarized as follows: the probability of exceeding 1 (violating the capacity) is $\gamma \cdot \alpha/2 + O(\sqrt{\epsilon})$, the probability of being exactly 1 is $\gamma(1/2 - \alpha/4) + O(\sqrt{\epsilon})$, the probability of being ϵ or 2ϵ is $\gamma(3/2 - 3\alpha/4) + O(\sqrt{\epsilon})$, and the probability of being 0 is $1 - \gamma(2 - \alpha/2) + O(\epsilon)$ (see Figs. 1b and 1c). Here, we use the notation $O(\sqrt{\epsilon})$ to represent a value in the range $[-10\sqrt{\epsilon}, 10\sqrt{\epsilon}]$ for each $\epsilon \geq 0$. For the third item, assume that the algorithm tries to insert it with probability $\beta \in [0, 1]$, under the condition that the knapsack utilization is ϵ or 2ϵ. Similar to the insertion of the second item, the algorithm also tries to insert the third item into the knapsack with a utilization rate of 0, as appropriate, to ensure that the third item can be inserted with an overall probability γ. The total size of the items in the knapsack, just after processing the third item, is in $\{\epsilon, 2\epsilon, 3\epsilon\}$ with probability $\gamma(2 - 3\alpha/4 - 9\beta/8 + 9\alpha\beta/16) + O(\sqrt{\epsilon})$ and 0 with probability $1 - \gamma(3 - \alpha/2 - 3\beta/4 + 3\alpha\beta/8) + O(\sqrt{\epsilon})$ (see Fig. 1d).

Here, the probability $1 - \gamma(3 - \alpha/2 - 3\beta/4 + 3\alpha\beta/8) + O(\sqrt{\epsilon})$ must be non-negative. By considering $\epsilon \to 0$, we have

$$1 - \gamma \left(3 - \frac{\alpha}{2} - \frac{3\beta}{4} + \frac{3\alpha\beta}{8} \right) \geq 0. \tag{2}$$

For the last item, it can be accepted with probability at most

$$1 - \gamma \left(3 - \frac{\alpha}{2} - \frac{3\beta}{4} + \frac{3\alpha\beta}{8} \right) + O(\sqrt{\epsilon}) + \gamma \left(2 - \frac{3\alpha}{4} - \frac{9\beta}{8} + \frac{9\alpha\beta}{16} \right) + O(\sqrt{\epsilon})$$

$$= 1 - \gamma \left(1 + \frac{\alpha}{4} - \frac{3\beta}{8} + \frac{3\alpha\beta}{16} \right) + O(\sqrt{\epsilon}). \tag{3}$$

To accept the last item with probability γ, the value (3)o must be at least γ. Thus, by setting $\epsilon \to 0$, we have

$$1 - \gamma \left(1 + \tfrac{\alpha}{4} - \tfrac{3\beta}{8} + \tfrac{3\alpha\beta}{16}\right) \geq \gamma. \qquad (4)$$

By summing up inequality (4) twice and inequality (2), we have $3 - 5\gamma \geq 2\gamma$. Hence, the upper bound of γ is $3/7$, and this implies that no algorithm can accept every item with a probability of greater than $3/7$.   □

4.2   Algorithm

Next, we prove that each item can be placed in the knapsack with a probability of at least $1/3$ by the following γ-*aggressive algorithm*. Let W_k and N_k be the random variables representing the total size and the number of items in the knapsack just before processing the kth item, respectively. Then, the algorithm tries to insert the kth item into the knapsack with the following probabilities:

- if $W_k > 1$, the probability is 0,
- if $W_k \leq 1$ and $N_k \geq 1$, the probability is 1, and
- if $N_k = 0$, the probability is $(\gamma - \Pr[W_k + S_k \leq 1 \text{ and } N_k \geq 1])/\Pr[N_k = 0]$.

For a given instance, if the γ-aggressive algorithm is feasible (i.e., the probability $(\gamma - \Pr[W_k + S_k \leq 1 \text{ and } N_k \geq 1])/\Pr[N_k = 0]$ is in the range $[0,1]$ for every k), the algorithm successfully inserts every item with a probability of

$$\Pr[N_k = 0] \cdot \frac{\gamma - \Pr[W_k + S_k \leq 1 \text{ and } N_k \geq 1]}{\Pr[N_k = 0]}$$
$$+ \Pr[W_k \leq 1 \text{ and } N_k \geq 1] \cdot \Pr[W_k + S_k \leq 1 \mid W_k \leq 1 \text{ and } N_k \geq 1],$$

which is equal to γ. Consequently, we have $\Pr[W_k \leq 1 \text{ and } N_k \geq 1] = \gamma$ for $k > 1$. By showing that the γ-aggressive algorithm is feasible for $\gamma = 1/3$, we demonstrate that the γ-aggressive algorithm can guarantee each item is accepted with a probability of $1/3$.

Theorem 3. *For any $\gamma \in (0, 1/3]$, the γ-aggressive algorithm is feasible for any instance of the hard constraint KOCRS problem.*

Proof. To analyze our algorithm, we consider the following procedure:

1. Draw $X_1, \ldots, X_n$ and $X'_1, \ldots, X'_n$ independently from the item size distributions $F_1, \ldots, F_n$.
2. Set T to 1. If $\sum_{i=1}^{n} X_i \leq 1$, return T. Otherwise, let k be the smallest index such that $\sum_{i=1}^{k} X_i > 1$.
3. Increment T by 1. If $X'_k + \sum_{i=k+1}^{n} X_i \leq 1$, return T. Otherwise, let ℓ be the smallest index such that $X'_k + \sum_{i=k+1}^{\ell} X_i > 1$.
4. Set k to ℓ and repeat Step 3.

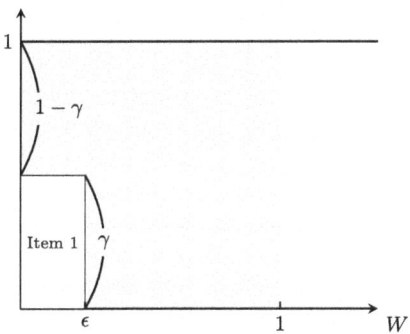

(a) The utilization of the knapsack after processing the first item. The horizontal axis represents the utilization of the knapsack, and the vertical axis represents the probability of each event. This figure represents that the knapsack utilization is ϵ with probability γ and 0 with probability $1 - \gamma$.

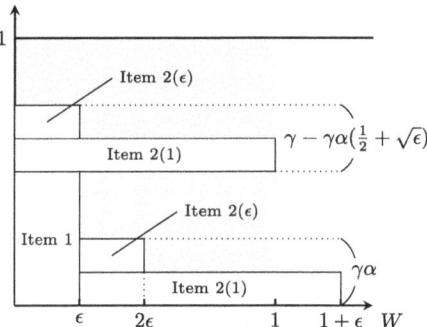

(b) The utilization of the knapsack after processing the second item. For the second item, we denote it as Item 2(1) if its size realization is 1 and Item 2(ϵ) if the realization is ϵ. The item colored in gray is not accepted due to exceeding the capacity.

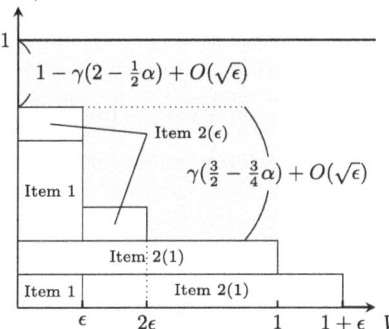

(c) The utilization of the knapsack after processing the second item, sorted by the knapsack utilization. The utilization of ϵ and 2ϵ can be considered equivalent in the sense that the items that can be put in the future are the same.

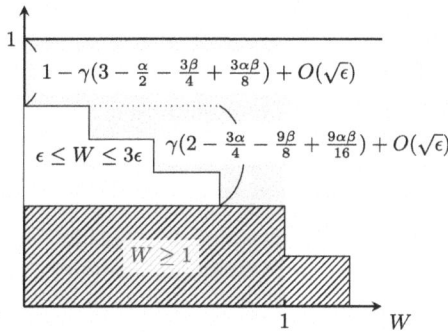

(d) The sorted utilization of the knapsack after processing the third item. The last item cannot be added if the utilization is at least 1 (hatched area).

Fig. 1. The behavior of an algorithm for the instance that accepts every item with probability exactly γ where W is a random variable representing the utilization rate of the knapsack.

We shall explain the correspondence between this procedure and our algorithm. The scenarios where the first item is inserted, occurring with probability γ, correspond to "Set T to 1" in Step 2.

For each $k > 1$, the algorithm always tries to insert the kth item if there is an item in the knapsack and the capacity is not violated (i.e., $W_k \leq 1$ and $N_k \geq 1$). If there is a possibility of capacity violation with such insertion (i.e., $\Pr[W_k + S_k \leq 1 \mid W_k \leq 1 \text{ and } N_k \geq 1] < 1$), the algorithm may insert the kth item under the condition that no item is currently in the knapsack. This is done with a probability that compensates for the probability of capacity violation. In the procedure, Step 2 with $\sum_{i=1}^{k} X_i > 1$ corresponds to such a capacity violation at the kth item. Step 3 corresponds to the insertion into the knapsack with no item. The value T represents the number of knapsacks required to process the items when $(X_1, \ldots, X_n, X_1', \ldots, X_n')$ occurs. Thus, $\gamma \cdot \mathbb{E}[T]$ corresponds to the probability that the algorithm inserts at least one item. It is sufficient to keep this value at most 1 to guarantee the feasibility.

We give an upper bound of $\mathbb{E}[T]$. The variable T is set to 1 at the beginning of Step 2. During the procedure, the value T is incremented by 1 only when $\sum_{i=1}^{k} X_i > 1$ in Step 2 or $X_k' + \sum_{i=k+1}^{\ell} X_i > 1$ in Step 3 occurs. Hence, for a fixed $(X_1, \ldots, X_n, X_1', \ldots, X_n')$, the number of times such an increment occurs is at most $\lfloor X_1 + \cdots + X_n + X_1' + \cdots + X_n' \rfloor$. Thus, we obtain

$$
\begin{aligned}
\mathbb{E}[T] &\leq 1 + \mathbb{E}\big[\lfloor X_1 + \cdots + X_n + X_1' + \cdots + X_n' \rfloor\big] \\
&\leq 1 + \mathbb{E}[X_1 + \cdots + X_n] + \mathbb{E}[X_1' + \cdots + X_n'] \leq 3,
\end{aligned}
$$

where the last inequality holds by the assumption that the total expected size of the items is at most 1. Therefore, if we fix $\gamma \leq 1/3$, the algorithm is feasible for any instance since $\mathbb{E}[T] \cdot \gamma \leq 1$. □

Furthermore, we observe that our analysis in Theorem 3 is best possible.

Theorem 4. *For any $\gamma > 1/3$, there exists an instance such that the γ-aggressive algorithm is not feasible.*

The proof is provided by considering a specific instance consisting of three items (see the full version of this paper [28]).

5 $\{\epsilon, 1\}$-Sized KOCRS in the Hard Constraint Setting

In this section, we examine a special case of the hard constraint KOCRS problem where the size distributions are binary, either 1 or ϵ ($\leq 1/n$). We call this problem the $\{\epsilon, 1\}$-*sized KOCRS problem*. For this problem, we present an algorithm that successfully puts every item with probability 3/7. Since the instance in the proof of Theorem 2 belongs to this special case, the probability 3/7 is best possible.

Our algorithm, which we refer to as the γ-*reserved* algorithm, keeps the probability that the usage of the knapsack is in $(0, 1)$ at γ. For simplicity in the analysis, we assume that the algorithm initially puts a virtual item of size ϵ (deterministically) into the knapsack with probability γ. We treat this virtual item as the 0th item. Let W_i be the random variable denoting the usage of the knapsack just before processing the ith item, including the virtual item. Also, let p_i be the probability that the size of the ith item is 1. Then, the γ-reserved algorithm tries to insert the ith item into the knapsack with the following probabilities depending on the realization W_i:

- if $W_i \geq 1$, then the probability is 0 (in this case, we cannot put any more items),

- if $0 < W_i < 1$, then the probability is $\frac{1-p_i}{1-p_i+p_i^2}$, and
- if $W_i = 0$, then the probability is $\frac{\gamma \cdot p_i}{1-p_i+p_i^2} \cdot \frac{1}{\Pr[W_i=0]}$.

We call the algorithm feasible if $\Pr[W_i = 0] \geq \frac{\gamma \cdot p_i}{1-p_i+p_i^2}$.

We show by induction that the algorithm keeps $\Pr[0 < W_{i+1} < 1] = \gamma$ if it is feasible. The base case $\Pr[0 < W_1 < 1] = \gamma$ clearly holds due to the treatment of the virtual item. For the inductive step, we get the probability $\Pr[0 < W_{i+1} < 1]$ as

$$\gamma \cdot \frac{1-p_i}{1-p_i+p_i^2} \cdot (1-p_i) + \gamma \cdot \left(1 - \frac{1-p_i}{1-p_i+p_i^2}\right) + \frac{\gamma \cdot p_i}{1-p_i+p_i^2} \cdot (1-p_i) = \gamma.$$

Here, on the left-hand side of the above equation, the three terms represent the probabilities of (i) $0 < W_i < 1$, the ith item is attempted for insertion, and its realized size is ϵ, (ii) $0 < W_i < 1$, the ith item is discarded, and (iii) $W_i = 0$, the ith item is attempted for insertion, and its realized size is ϵ, respectively. Moreover, the algorithm successfully puts the ith item with probability γ if it is feasible. In fact, the probability that the ith item is accepted is

$$\gamma \cdot \frac{1-p_i}{1-p_i+p_i^2} \cdot (1-p_i) + \frac{\gamma \cdot p_i}{1-p_i+p_i^2} = \gamma,$$

where the first term is the probability that $0 < W_i < 1$, the ith item is attempted for insertion, and its realized size is ϵ, and the second term is the probability that $W_i = 0$ and the ith item is attempted for insertion, respectively.

We prove that the γ-reserved algorithm can guarantee each item to be accepted with a probability of $3/7$. The proof can be found in the full version of this paper [28].

Theorem 5. *For any $\gamma \in (0, 3/7]$, the γ-reserved algorithm is feasible for any instance of the $\{\epsilon, 1\}$-sized KOCRS problem in the hard constraint.*

6 Analysis for the Soft Constraint Setting

In this section, we analyze the soft constraint KOCRS problem.

6.1 Impossibility

We demonstrate that no algorithm can guarantee acceptance probability greater than $1/2$. The proof can be found in the full version of this paper [28].

Theorem 6. *For any positive real δ, there exists an instance of the soft constraint KOCRS problem, where no algorithm has the acceptance probability of at least $1/2 + \delta$. This impossibility holds even when item sizes are restricted to 1 or a small positive constant ϵ.*

6.2   Algorithm

We next prove that an aggressive algorithm can put every item into the knapsack with a probability of at least $1/2$, which is best possible by Theorem 6. For the soft constraint setting, the γ-aggressive algorithm is an algorithm that tries to insert a new item into the knapsack if the items in the knapsack exist and the total size is less than 1, ensuring that the probability of each item being placed is exactly γ. Let W_i be the random variable representing the total size of items in the knapsack just before processing the ith item. Then, the γ-aggressive algorithm for the soft constraint setting inserts the ith item into the knapsack with the following probabilities, according to the value of W_i:

- if $W_i \geq 1$, then the probability is 0 (in this case, we cannot put any more items in the knapsack),
- if $0 < W_i < 1$, then the probability is 1, and
- if $W_i = 0$, then the probability is $(\gamma - \Pr[0 < W_i < 1])/\Pr[W_i = 0]$.

For a given instance, if the γ-aggressive algorithm is feasible, the algorithm successfully inserts every item with probability $\Pr[0 < W_i < 1] + \Pr[W_i = 0] \cdot (\gamma - \Pr[0 < W_i < 1])/\Pr[W_i = 0] = \gamma$. We show that the $1/2$-aggressive algorithm is feasible for any instance of the soft constraint KOCRS problem.

Theorem 7. *For any $\gamma \in (0, 1/2]$, the γ-aggressive algorithm successfully inserts every item with probability γ for any instance of the soft constraint KOCRS problem.*

Proof. We prove this by mathematical induction. The base case holds because the first item is inserted with probability γ by $\Pr[W_1 = 0] = 1$. Assume that each of the items up to the $(i-1)$st one has been successfully placed in the knapsack with probability γ. The expected total size of items in the knapsack just before processing the ith item is $\mathbb{E}[W_i] = \sum_{j=1}^{i-1} \gamma \cdot \mathbb{E}[S_j]$. By using the assumptions that $\sum_{j=1}^{n} \mathbb{E}[S_j] \leq 1$ and $\gamma \leq 1/2$, we have $\mathbb{E}[W_i] \leq 1/2$. Consequently, the probability that W_i is less than one is at least

$$\Pr[W_i < 1] = 1 - \Pr[W_i \geq 1] \geq 1 - \mathbb{E}[W_i] \geq 1/2,$$

where the first inequality holds by Markov's inequality.   □

Acknowledgment. This work was supported by JST SPRING Grant Number JPMJSP2108, JST ERATO Grant Number JPMJER2301, JST PRESTO Grant Number JPMJPR2122, JSPS KAKENHI Grant Number JP20K19739, and Value Exchange Engineering, a joint research project between Mercari, Inc. and the RIISE.

References

1. Adamczyk, M., Włodarczyk, M.: Random order contention resolution schemes. In: Proceedings of the 59th Annual Symposium on Foundations of Computer Science, pp. 790–801 (2018)

2. Alaei, S.: Bayesian combinatorial auctions: expanding single buyer mechanisms to many buyers. SIAM J. Comput. **43**(2), 930–972 (2014)
3. Alaei, S., Hajiaghayi, M.T., Liaghat, V.: The online stochastic generalized assignment problem. In: Proceedings of the 16th International Workshop on Approximation Algorithms for Combinatorial Optimization, pp. 11–25 (2013)
4. Bhalgat, A., Goel, A., Khanna, S.: Improved approximation results for stochastic knapsack problems. In: Proceedings of the 22nd Annual ACM-SIAM Symposium on Discrete Algorithms, pp. 1647–1665 (2011)
5. Carraway, R.L., Schmidt, R.L., Weatherford, L.R.: An algorithm for maximizing target achievement in the stochastic knapsack problem with normal returns. Nav. Res. Logist. **40**(2), 161–173 (1993)
6. Chawla, S., Miller, J.B.: Mechanism design for subadditive agents via an ex ante relaxation. In: Proceedings of the 17th ACM Conference on Economics and Computation, pp. 579–596 (2016)
7. Chekuri, C., Vondrák, J., Zenklusen, R.: Submodular function maximization via the multilinear relaxation and contention resolution schemes. In: Proceedings of the 43rd Annual ACM Symposium on Theory of Computing, pp. 783–792 (2011)
8. De, A.: Boolean function analysis meets stochastic optimization: an approximation scheme for stochastic knapsack. In: Proceedings of the 29th Annual ACM-SIAM Symposium on Discrete Algorithms, pp. 1286–1305 (2018)
9. Dean, B.C., Goemans, M.X., Vondrák, J.: Approximating the stochastic knapsack problem: the benefit of adaptivity. Math. Oper. Res. **33**(4), 945–964 (2008)
10. Dughmi, S.: Matroid secretary is equivalent to contention resolution. In: Proceedings of the 13th Innovations in Theoretical Computer Science Conference, vol. 215, pp. 58:1–58:23 (2022)
11. Dütting, P., Feldman, M., Kesselheim, T., Lucier, B.: Prophet inequalities made easy: stochastic optimization by pricing nonstochastic inputs. SIAM J. Comput. **49**(3), 540–582 (2020)
12. Ezra, T., Feldman, M., Gravin, N., Tang, Z.G.: Online stochastic max-weight matching: prophet inequality for vertex and edge arrival models. In: Proceedings of the 21st ACM Conference on Economics and Computation, pp. 769–787 (2020)
13. Feldman, J., Korula, N., Mirrokni, V., Muthukrishnan, S., Pál, M.: Online ad assignment with free disposal. In: Proceedings of the 5th Workshop of Internet and Network Economics, pp. 374–385 (2009)
14. Feldman, M., Svensson, O., Zenklusen, R.: Online contention resolution schemes. In: Proceedings of the 27th Annual ACM-SIAM Symposium on Discrete Algorithms, pp. 1014–1033 (2016)
15. Feldman, M., Svensson, O., Zenklusen, R.: Online contention resolution schemes with applications to Bayesian selection problems. SIAM J. Comput. **50**(2), 255–300 (2021)
16. Feng, Y., Niazadeh, R., Saberi, A.: Near-optimal Bayesian online assortment of reusable resources. In: Proceedings of the 23rd ACM Conference on Economics and Computation, pp. 964–965 (2022)
17. Fu, H., Tang, Z.G., Wu, H., Wu, J., Zhang, Q.: Random order vertex arrival contention resolution schemes for matching, with applications. In: Proceedings of the 48th International Colloquium on Automata, Languages, and Programming, pp. 68:1–68:20 (2021)
18. Gupta, A., Krishnaswamy, R., Molinaro, M., Ravi, R.: Approximation algorithms for correlated knapsacks and non-martingale bandits. In: Proceedings of the 52nd Annual Symposium on Foundations of Computer Science, pp. 827–836 (2011)

19. Han, X., Kawase, Y., Makino, K.: Randomized algorithms for online knapsack problems. Theoret. Comput. Sci. **562**, 395–405 (2015)
20. Jiang, J., Ma, W., Zhang, J.: Tight guarantees for multi-unit prophet inequalities and online stochastic knapsack. In: Proceedings of the 33rd Annual ACM-SIAM Symposium on Discrete Algorithms, pp. 1221–1246 (2022)
21. Kleinberg, J., Rabani, Y., Tardos, É.: Allocating bandwidth for bursty connections. In: Proceedings of the 29th Annual ACM Symposium on Theory of Computing, pp. 664–673 (1997)
22. Lee, E., Singla, S.: Optimal online contention resolution schemes via ex-ante prophet inequalities. In: Proceedings of the 26th Annual European Symposium on Algorithms, pp. 57:1–57:14 (2018)
23. MacRury, C., Ma, W., Grammel, N.: On (random-order) online contention resolution schemes for the matching polytope of (bipartite) graphs. Oper. Res. (2024)
24. Marchetti-Spaccamela, A., Vercellis, C.: Stochastic on-line knapsack problems. Math. Program. **68**(1–3), 73–104 (1995)
25. Mehta, A., Saberi, A., Vazirani, U., Vazirani, V.: Adwords and generalized online matching. J. ACM **54**(5), 22–31 (2007)
26. Papastavrou, J.D., Rajagopalan, S., Kleywegt, A.J.: The dynamic and stochastic knapsack problem with deadlines. Manag. Sci. **42**(12), 1706–1718 (1996)
27. Yang, S., Khuller, S., Choudhary, S., Mitra, S., Mahadik, K.: Correlated stochastic knapsack with a submodular objective. In: Proceedings of the 30th Annual European Symposium on Algorithms, ESA 2022, pp. 91:1–91:14 (2022)
28. Yoshinaga, T., Kawase, Y.: Size-stochastic knapsack online contention resolution schemes. arXiv preprint arXiv:2305.08622 (2023)
29. Zhou, Y., Naroditskiy, V.: Algorithm for stochastic multiple-choice knapsack problem and application to keywords bidding. In: Proceedings of the 17th International Conference on World Wide Web, pp. 1175–1176 (2008)

The Connected k-Vertex One-Center Problem on Graphs

Jingru Zhang$^{(\boxtimes)}$

Cleveland State University, Cleveland, OH 44115, USA

Abstract. We consider a generalized version of the (weighted) one-center problem on graphs. Given an undirected graph G of n vertices and m edges and a positive integer $k \leq n$, the problem aims to find a point on G so that the maximum (weighted) distance from it to k connected vertices on its shortest path tree(s) is minimized. No previous work has been proposed for this problem except for the case $k = n$, that is, the classical graph one-center problem. In this paper, an $O(mn \log n \log mn + m^2 \log n \log mn)$-time algorithm is proposed for the weighted case, and an $O(mn \log n)$-time algorithm is presented for the unweighted case, provided that the distance matrix is given. When G is a tree graph, we give an $O(n \log^2 n \log k)$-time algorithm for the weighted case and improve it to $O(n \log^2 n)$ for the unweighted case.

Keywords: Algorithms · Data Structures · Facility Locations · k-Vertex One-Center · Graphs · Connectivity

1 Introduction

The one-center problem is a classical problem in facility locations which aims to compute the best location of a single facility on a graph network to serve customers such that the maximum (weighted) distance between the facility and all customers is minimized [3,13,15–19,24]. Due to the resource limits, it is quite natural to consider the partial version where the facility serves only k neighboring customers with the minimized maximum transportation cost, that is, this connected k-vertex one-center problem.

Let $G = (V, E)$ be an undirected graph of n vertices and m edges where each vertex $v \in V$ has a weight $w_v > 0$ and each edge $e \in E$ is of length $l(e) > 0$. For any two vertices $u, v \in V$, let $e(u, v)$ be the edge between them. By considering $e(u, v)$ as a line segment of length $l(e(u, v))$, we can talk about "points" on it. Formally, a point $p = (u, v, t(p))$ on edge $e(u, v)$ is characterized by being located at a distance of $t(p) \leq l(e(u, v))$ from the vertex u. We say that p is interior of $e(u, v)$ if $0 < t(p) < l(e(u, v))$. For any two points x, y of G, the distance $d(x, y)$ between them is defined as the length of their shortest path(s) $\pi(x, y)$ in G.

This research was supported in part by U.S. National Science Foundation under Grant CCF-2339371.

A point x of G may have multiple shortest path trees, and denote their set by $G(x)$. Let T represent a tree graph and T^k denote a tree of size k. A subgraph of G is called a *k-subtree* if and only if it is a tree of size k. Let $G^k(x)$ be the set of all distinct k-subtrees of trees in $G(x)$. Denote by $V(G')$ the subset of all vertices in a subgraph G' of G.

Define $\phi(x, G)$ as $\min_{T^k \in G^k(x)} \max_{v \in V(T^k)} w_v d(v, x)$. The problem aims to compute a point x^* on G, called the *partial center*, so as to minimize $\phi(x, G)$. Note that x^* might be interior of an edge in G.

If G is a tree graph, every point of G has a unique shortest path tree, i.e., G itself. Clearly, in this situation, the problem is equivalent to the problem of finding a k-subtree of minimum (weighted) radius where x^* is the (weighted) center of an optimal k-subtree.

When $k = n$, x^* is exactly the (weighted) center of G with respect to V. Provided that the distance matrix is given, the center can be found in $O(mn \log n)$ time and the unweighted case can be addressed in $O(mn + n^2 \log n)$ time [18]. Additionally, for G being a tree, the (weighted) one-center problem can be solved in $O(n)$ time [20].

As far as we are aware, however, this connected k-vertex one-center problem has not received any attention even when G is a tree. In this paper, we solve this problem in $O(mn \log n \log mn + m^2 \log n \log mn)$ time and address the unweighted case more efficiently in $O(mn \log n)$ time with the given distance matrix. For G being a tree, an $O(n \log^2 n \log k)$-time algorithm is proposed for the weighted case and an $O(n \log^2 n)$-time approach is presented for the unweighted case.

1.1 Related Work

As introduced above, when $k = n$, Kariv and Hamiki [18] proposed an $O(mn \log n)$-time algorithm for the weighted case and an $O(mn + n^2 \log n)$-time algorithm for the unweighted case, provided that the distance matrix is given. Megiddo [20] solved the (weighted) one-center problem on trees in linear time by the prune-and-search techniques.

Although this partial version of the one-center problem has not been studied before, some other partial variants of the general p-center problem have been explored in literature. Megiddo et al. [23] gave an $O(n^2 p)$-time algorithm to solve the maximum coverage problem which aims to place p facilities on tree networks to cover maximum customers within their covering range. Berman et al. [5] considered another variant that places p facilities to minimize the maximum distance between them and customers within their covering range. See other partial versions of the p-center problem and variances [6,7,10].

Another most related problem is the graph maximum t-club problem where the goal is to find the maximum-cardinality subgraph of diameter no more than value t. Bourjolly et al. [8] revealed the NP-Hardness of this problem. Asahiro et al. [2] proposed an approximation algorithm of $O(n^{\frac{1}{2}})$ ratio, which was proved to be optimal for any $t > 2$. Additionally, a constant ratio was achieved for this problem on unit disk graphs [1].

1.2   Our Approach

Denote by λ^* the minimized objective value $\phi(x^*, G)$. We show that λ^* belongs to a finite set that includes the following values w.r.t. every edge e: the one-center objective values of every two distinct vertices with the *local* constraint where their centers must be on e, and the (weighted) distance of every vertex to its *semicircular* point on e which is the point on e such that the vertex has two shortest paths to it without any common intermediate vertex. For the unweighted version or the version where G is a tree, however, we observe that λ^* is only relevant to those constrained one-center objective values w.r.t. each edge.

For the weighted version, by forming this finite set as a set of y-coordinates of intersections between $O(mn)$ lines, we can adapt the line arrangement search technique [9] to find λ^* with the assistance of our feasibility test that determines for any given value λ, whether $\lambda \geq \lambda^*$. Obviously, $\lambda \geq \lambda^*$ if there exists a point on G such that it covers a k-subtree in its shortest path tree(s) under λ (that is, the weighted distance from it to each vertex of a k-subtree in its shortest path tree(s) is no more than λ). Otherwise, $\lambda < \lambda^*$, so λ is not feasible.

Our feasibility test is motivated by a critical observation: λ is feasible if and only if there exists a point in G such that the largest *self-inclusive* subtree covered by it in its shortest path tree(s) is of size at least k. Hence, our algorithm examines every edge e of G to decide the existence of such a point by algorithmically constructing a function that computes for every point on e the size of the largest self-inclusive subtree covered by the point, which is one of our main contributions. By determining the breakpoints of these functions, the feasibility of λ can be known in $O(mn \log n + m^2 \log n)$ time.

For the unweighted version, our key observation, that λ^* is decided by the smallest one among the k-th shortest path lengths of all vertices, implies that finding the *local* partial center on every edge is equivalent to a geometry problem that computes the lowest vertex on the *k-th level* of $O(n)$ x-monotone polygonal chains of complexity $O(1)$. We develop an $O(n \log n)$-time algorithm for this geometry problem, which is our another main contribution, and the unweighted version is thus addressed in $O(mn \log n)$ time.

When G is a tree, we develop several data structures so that for any given point of G, the counting query on the largest self-inclusive subtree covered by the point can be answered in $O(\log n \log k)$ time. In addition, instead of examining every edge, we observe that only $O(n)$ points on the tree need to be examined in order to decide the feasibility of λ. These lead an $O(n \log n \log k)$-time feasibility test for the tree version. Then, by implicitly forming the set of the one-center objective values for all pairs of vertices, λ^* can be computed in $O(n \log^2 n \log k)$ time. For the unweighted case, $O(n \log n)$ sorted sets of intervertex distances are implicitly formed by employing the tree decomposition technique [21], so λ^* can be found in $O(n \log^2 n)$ time.

2   Preliminary

When all vertex weights are same, without loss of generality, we assume their weights are one. Let T^* represent an optimal k-subtree in shortest path tree(s) of x^* so that $\phi(x^*, G) = \max_{v \in V(T^*)} w_v d(x^*, v)$. Although the optimal solution may not be unique, the following observation helps figure out one of them.

Observation 1. *There exists an optimal solution where x^* is a point of T^*. For the unweighted case, T^* is induced by the k closest vertices of x^* on G.*

Proof. Suppose no such optimal solutions exist. Let x^* and T^* be any optimal solution. Consider the closest vertex v' in $V(T^*)$ to x^*. In fact, T^* is a k-subtree in a shortest path tree of v'. Let T' be the shortest path tree of x^* containing T^*. It is because for each vertex $v \in V(T^*)$, the (shortest) path between v and x^* on T' is the concatenation of the subpath between v and v' and the subpath between v' and x^*; both subpaths are their shortest paths in G. Due to $d(v', x^*) > 0$, we have $\max_{v \in V(T^*)} w_v d(x^*, v) > \max_{v \in V(T^*)} w_v d(v', v)$. Further, due to $\phi(v', G) \leq \max_{v \in V(T^*)} w_v d(v', v)$, a contradiction where $\phi(x^*, G) > \phi(v', G)$ occurs. Hence, the first statement is proved.

Moreover, for any point x in G, x and its k closest vertices in V must induce a connected k-subtree in x's shortest path tree(s). This implies the second statement. □

Let $e(r, s)$ be an arbitrary edge of G. Let x be any point on $e(r, s)$, which is at distance $t(x)$ to r along $e(r, s)$. For each vertex $v \in V$, we use $D(v, x)$ to represent

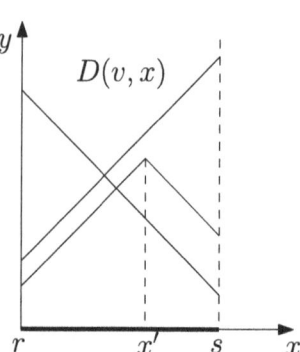

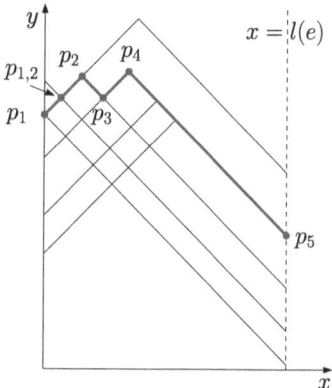

Fig. 1. Illustrating the three cases of the (weighted) distance function $D(v, x)$ for $x \in e(r, s)$: As x moves from r to s on e, at rate w_v, $D(v, x)$ increases, or decreases, or first increases until v's semicircular point x' and then decreases.

Fig. 2. Illustrating the 6-level closure (the heavy chain) of seven functions $D(v, x)$, which has five vertices p_1, p_2, p_3, p_4, p_5. It cannot turns at the intersection $p_{1,2}$.

the (weighted) distance from v to x. Figure 1 shows the three cases of $D(v,x)$. Denote by $I_{e(r,s)}(y,y')$ the path (segment) along $e(r,s)$ between two points y, y' on $e(r,s)$. If there is a point $x' \in e(r,s)$ so that $d(v,r) + t(x') = d(v,s) + l(e(r,s)) - t(x')$, then v has two shortest paths to x' respectively containing $I_{e(r,s)}(r,x')$ and $I_{e(r,s)}(x',s)$. We refer to x' as the semicircular point of v on $e(r,s)$ and also say that v is a *neutral* vertex of x'. Notice that every vertex has at most one semicircular point on each edge of G.

Denote by $\bar{V}(x)$ the set of all neutral vertices of a point x of G. W.r.t. $x \in e(r,s)$, V can be partitioned into three subsets $\bar{V}(x)$, $V_r(x)$, and $V_s(x)$: $V_r(x)$ is composed of all vertices in $V - \bar{V}(x)$ satisfying the condition $d(v,r) + t(x) < d(v,s) + l(e(r,s)) - t(x)$, and $V_s(x)$ contains all remaining in $V - \bar{V}(x)$ with $d(v,r) + t(x) > d(v,s) + l(e(r,s)) - t(x)$. Indeed, $V_r(x)$ (resp., $V_s(x)$) contains all vertices in V whose shortest paths to x each contains $I_{e(r,s)}(r,x)$ (resp., $I_{e(r,s)}(s,x)$) on $e(r,s)$.

In the situation where x is interior of $e(r,s)$, we consider x as a dummy vertex on its shortest path tree(s) (rather than G). Let the vertex containing x be the root of its shortest path tree(s). We have the following useful properties.

Observation 2. *In any shortest path tree of x, every vertex of $\bar{V}(x)$ is either a leaf or an internal node whose descendants are all in $\bar{V}(x)$. Further, for any two points y, y' on $e(r,s)$ with $t(y) < t(y')$, $\bar{V}(y') \in V_r(y)$ and $V_r(y') \in V_r(y)$, and symmetrically, $\bar{V}(y) \in V_s(y')$ and $V_s(y) \in V_s(y')$.*

For the case $k = n$, as proved in [18], λ^* is in the set of the (weighted) one-center objective value of every two vertices by constraining their center lying on each edge of G, that is, the set of the y-coordinates of the intersections between every two functions $D(v,x)$ w.r.t. each edge where their slopes are of opposite signs. Denote by Λ this set. Additionally, let Λ' be the set of values $D(v,x)$ of each $v \in V$ at all $O(mn)$ semicircular points in G. Clearly, $|\Lambda| = O(mn^2)$ and so is $|\Lambda'|$. For the general case $k \leq n$, the following observation holds.

Observation 3. $\lambda^* \in \Lambda \cup \Lambda'$.

For the unweighted case or the (weighted) tree version, we have the following observation.

Observation 4. *For the unweighted case or the (weighted) tree version, $\lambda^* \in \Lambda$, and x^* is the (weighted) center of T^* w.r.t. $V(T^*)$.*

As in [11], the (weighted) diameter $W(G)$ of G is defined as $\max_{v,u \in V} \frac{w_v w_u d(u,v)}{w_v + w_u}$; when all weights are same, $W(G)$ is exactly one half of the diameter of G. Regarding to our problem, the observation below reveals the equivalency between our problem and the problem of finding the k-subtree of minimum (weighted) diameter in a graph.

Observation 5. *For the unweighted case, T^* is of minimum diameter among all k-subtrees of G if and only if $W(T^*) = \lambda^*$. When G is a tree, T^* is a k-subtree of minimum diameter.*

Furthermore, the following corollary can be utilized to determine whether any given k-subtree T^k of G is of minimum diameter among all its k-subtrees.

Corollary 1. *For the unweighted case or the weighted tree version, T^k is of minimum diameter if and only if $W(T^k) = \lambda^*$.*

In the following, when we talk about a point x on an edge, we use x to denote $t(x)$ for convenience. If x is interior of the edge, we consider x as a dummy vertex in its shortest path tree(s) (not G) except when we say its shortest path tree(s) in G. In addition, for any point $p \in \mathbb{R}^2$, we use $x(p)$ and $y(p)$ to denote its x- and y-coordinates, respectively; if the context is clear, for a point x on the x-axis, we directly use x to denote its x-coordinate.

3   The Problem on a Vertex-Weighted Graph

In this section, we shall present our algorithm for the weighted version on undirected graphs. To introduce our algorithm, we first give the result for the feasibility test that determines for any given value λ, whether $\lambda \geq \lambda^*$, and defer its proof in Sect. 3.1.

Lemma 1. *Given any value λ, we can decide whether $\lambda \geq \lambda^*$ in $O(mn \log n + m^2 \log n)$ time.*

Our idea for computing λ^* is as follows: First, we compute a set of $O(mn)$ lines in the x, y-coordinate plane generated by extending line segments on graphs of functions $y = D(v, x)$ of all vertices w.r.t. every edge of G and including vertical lines through incident vertices of every edge on x-axis. Clearly, the set $\Lambda \cup \Lambda'$ belongs to the set of y-coordinates of all intersections between these obtained lines, and λ^* is the y-coordinate of the lowest intersection with a feasible y-coordinate. Next, we compute λ^* by finding that lowest one among all intersections by utilizing the line arrangement search technique [9] with the assistance of our feasibility test in Lemma 1. The line arrangement search technique is reviewed as follows.

Suppose L is a set of N lines in the plane. Denote by $\mathcal{A}(L)$ the arrangement of lines in L. In $\mathcal{A}(L)$, every intersection of lines defines a vertex of $\mathcal{A}(L)$ and vice versa. Let $v_1(L)$ be the lowest vertex of $\mathcal{A}(L)$ whose y-coordinate $y(v_1(L))$ is a feasible value, and let $v_2(L)$ be the highest vertex of $\mathcal{A}(L)$ whose y-coordinate $y(v_2(L))$ is smaller than $y(v_1(L))$. By the definitions, $y(v_2(L)) < \lambda^* \leq y(v_1(L))$ and no vertices in $\mathcal{A}(L)$ have y-coordinates in range $(y(v_2(L)), y(v_1(L)))$. Lemma 2 was given in [9] to find the two vertices.

Lemma 2. *[9] Both vertices $v_1(L)$ and $v_2(L)$ can be computed in $O((N + \tau) \log N)$ time, where τ is the running time of the feasibility test.*

Regarding to our problem, if these $O(mn)$ lines are known, with the assistance of Lemma 1, we can adapt Lemma 2 to compute λ^* in $O(mn \log n \log mn + m^2 \log n \log mn)$ time.

It remains now to compute these $O(mn)$ lines. We use a list L to store all these lines, which is empty initially. For every edge of G, we perform the following $O(n)$-time routine: Suppose we are about to process edge $e(r, s)$. Consider $e(r, s)$ being on the x-axis with vertex r at the origin and vertex s at the point of x-coordinate $l(e(r, s))$. First, we join into L the two vertical lines respectively through vertices r, s on x-axis. Next, for each $v \in V$, we determine in $O(1)$ time function $y = D(v, x)$ w.r.t. $x \in e(r, s)$ with the provided distance matrix, and then insert into L the lines containing all $O(1)$ line segments on its graph.

As a result, a set of $O(mn)$ lines stored in L is obtained in $O(mn)$ time. Clearly, the set $\Lambda \cup \Lambda'$ belongs to the set of y-coordinates of intersections between lines in L. Now we can employ Lemma 2 to find the lowest vertex with a feasible y-coordinate in the line arrangement $\mathcal{A}(L)$ by applying Lemma 1 to decide the feasibility of every tested y-coordinate. The y-coordinate of that lowest vertex is exactly λ^*. Accordingly, x^* and T^* can be found by applying Lemma 1 to λ^*. Thus, we have the following theorem.

Theorem 1. *The weighted connected k-vertex one-center problem can be solved in $O(mn \log n \log mn + m^2 \log n \log mn)$ time.*

3.1 The Feasibility Test

For any given value λ, we say that a subgraph G' of G can be covered by a point x in G (under λ) if and only if the maximum (weighted) distance of x to $V(G')$ is no more than λ. Indeed, the feasibility test asks for the existence of a point in G that covers a k-subtree of its shortest path tree(s) in G (without joining a dummy vertex for x into its shortest path trees). For any point x of G, there is a subtree of maximum cardinality in G that contains x, is covered by x, and belongs to a shortest path tree of x in G; we refer to this subtree as the *largest covered self-inclusive subtree* of x, and denote it by $T_\lambda(x)$. The following key observation leads our decision algorithm.

Observation 6. *There exists a point x' in G with $|V(T_\lambda(x'))| \geq k$ if and only if λ is feasible.*

Proof. It suffices to show that if $\lambda \geq \lambda^*$ then such a point must exist. Since λ is feasible, there must be a point on G so that it covers a k-subtree of its shortest path trees in G. Let x' be such a point. Clearly, the statement is true if a k-subtree of its shortest path trees in G covered by x' contains x'.

Otherwise, the largest self-inclusive subtree of its every shortest path tree in G covered by x' is of size less than k. But the largest one covered by x' excluding x' contains at least k vertices of V. Let T^k be such a largest subtree in a shortest path tree of x'. Suppose T^k is rooted at vertex u in that shortest path tree. (Recall that each shortest path tree of x' is rooted at the dummy or real vertex containing x'.) Because for each vertex $v \in V(T^k)$, the path of v to u on T^k is exactly its shortest path to u on G. T^k is thus a subtree of a shortest path tree of u in G; additionally, it includes u and is covered by u under λ. Due to $|V(T^k)| \geq k$, the observation holds. □

The underlying idea of our feasibility test is: For every edge e of G, we determine $|V(T_\lambda(x))|$, i.e., the size $|T_\lambda(x)|$ of $T_\lambda(x)$, for points on e where $|V(T_\lambda(x))|$ changes. In the process, if a point is found so that its $V(T_\lambda(x))$ is of size at least k, then λ is feasible and so we immediately return. Otherwise, no such points exist and thus $\lambda < \lambda^*$.

Let S be a subset of V. For any point $x \in G$, we refer to a vertex as a *descendant* of subset S w.r.t. x if its every shortest path to x contains vertices in S. We say a vertex is a *heavy* vertex of x if it cannot be covered by x. Clearly, $T_\lambda(x)$ includes neither any heavy vertex of x nor any descendant of the set $H(x)$ of all its heavy vertices. Otherwise, a vertex is called a *light* vertex of x (if it is neither a vertex in $H(x)$ nor a descendant of $H(x)$).

The following observation sets the base for determining $|V(T_\lambda(x))|$.

Observation 7. $T_\lambda(x)$ *is induced by all light vertices of* x.

Proof. It is sufficient to show that x must have a shortest path tree where the path from x to its every light vertex contains no (heavy) vertex in $H(x)$.

On the one hand, every vertex adjacent to x in G is in $H(x)$. It follows that if x is interior of an edge in G then x has no light vertices, and otherwise, the one containing x is its only light vertex. Hence, the statement is true in this situation.

On the other hand, x is adjacent to at least one of its light vertices. Let G' be the subgraph generated by removing $H(x)$ and all descendants of $H(x)$ from G. Note that if x is interior of an edge and only one of its two adjacent vertices is light, then x is not in G', which contains its only light adjacent vertex though; otherwise, x is in G'. For the former case, to maintain the reachability, we join a dummy vertex for x into G' by connecting x and its only light adjacent vertex with an edge of the same length as their segment length along the edge of G containing x.

By the definition, every light vertex of x has at least one shortest path to x on G where every vertex is light w.r.t. x. This implies the following properties: (1) G' is connected; (2) G' contains such shortest path(s) in G from x to its every light vertex. Hence, there must be a shortest path tree w.r.t. x where the path from x to its every light vertex contains only its light vertices.

Therefore, the observation holds. □

Furthermore, a problem needs to be addressed for determining $|V(T_\lambda(x))|$: Given any subset S of V, the goal is to find all descendants of S w.r.t. x on G and the subgraph generated by removing S and its descendants from G. The following lemma gives the result.

Lemma 3. *With* $O((m + n) \log n)$*-time preprocessing work, given any subset S of V, all descendants of S w.r.t. x and the subgraph generated by removing S and its descendants from G can be obtained in* $O(n' + m' \log n)$ *time where n' is the total number of vertices in S and its descendants, and m' is their total degrees.*

We now present our algorithm for determining values $|V(T_\lambda(x))|$ at necessary points on an arbitrary edge $e(r, s)$. We simply use e to denote edge $e(r, s)$. For any point $x \in e$, we define $f(e, x) = |V(T_\lambda(x))|$. Recall that x is at distance $t(x)$ to r along e. $f(e, x)$ is indeed a function w.r.t. $t(x)$. The following Lemma can be employed to construct $f(e, x)$.

Lemma 4. $f(e, x)$ *is a piece-wise constant function of complexity $O(n)$. The ordered set of all its breakpoints can be computed in $O((m + n) \log n)$ time.*

Recall that the feasibility test is to decide if a point exists in G so that its largest covered self-inclusive $T_\lambda(x)$ is of size no less than k, i.e., it has at least k light vertices. We can decide the existence of such a point by applying Lemma 4 to every edge of G. During the procedure, once such a point is found (so that $f(e, x)$ at this point is at least k), λ is known to be feasible and so we immediately return. Otherwise, no such points exist and hence λ is infeasible. As a result, the feasibility of any given value λ can be known in $O(m^2 \log n + mn \log n)$ time.

4   The Problem on a Vertex-Unweighted Graph

In this section, we introduce the algorithm for the unweighted case where every vertex is of weight one.

For any edge e of G, the point on e minimizing $\phi(x, G)$ among its all points is called the *local* partial center on e; denote it by x_e^*. x_e^* may not be unique on e but let x_e^* represent any of them. To find x^* on G, our strategy is to compute for every edge of G its local partial center x_e^* and the objective value at x_e^* such that x^* is the one with the smallest objective value among them.

Consider the problem of computing the local partial center on an arbitrary edge $e(r, s)$ of G, which is denoted by e for simplicity. As analyzed in Sect. 2, in the unweighted case, at any point x on G, $\phi(x, G)$ equals to the distance of x to its k-th closest vertex, so the k-subtree T' of its shortest path trees in G with $\phi(x, G) = \max_{v \in V(T')} d(x, v)$ is induced by its k closest vertices. It thus follows that finding the local partial center x_e^* on e requires to compute the distance of every point on e to its k-th closest vertex. As we shall show below, it is equivalent to solve the problem of computing the *k-th level* of a set of x-monotone polygonal chains.

Consider function $y = d(v, x)$ of each $v \in V$ for $x \in e$ in the x, y-coordinate plane. Function $y = d(v, x)$ defines a x-monotone polygon chain C_v whose left-most and rightmost endpoints are respectively on the vertical line $x = 0$ and $x = l(e)$. Specifically, either C_v is a line segment of slope $+1$ or -1, or it consists of two line segments where the left one is of slope $+1$ and the right one is of slope -1. We refer to the segment of slope $+1$ on C_v as its *x-segment*, and its segment of slope -1 as its *y-segment*. (Because rotating the x, y-plane by 45 degree along the positive x-axis causes that the segment of slope $+1$ becomes horizontal and the segment of slope -1 becomes vertical.)

Let C be the set of the n polygonal chains of all vertices w.r.t. e. C can be obtained in $O(n)$ time by the given distance matrix. If values $d(v, r)$ of all vertices are distinct and so are values $d(v, s)$, then any two chains in C intersect at most once. If so, x_e^* is of the same x-coordinate as the lowest point on the *k-th level* of an arrangement of C that is the closure of the set of all points that lie on chains of C such that the open downward-directed vertical ray emanating from this closure intersects exactly k chains of C; this k-th level of C can be computed in $O(n \log^2 n + nk)$ time [12].

Generally, chains of C may overlap with each other. Specifically, the x-segments (resp., the y-segment) of two vertices with same values $d(v, r)$ (resp., $d(v, s)$) have the same left (resp., right) endpoint, so they overlap partially or fully from their common left (resp., right) endpoint. This implies that the k-th level of C may not exist since it requires exactly k chains not above its closure. Hence, we give a more generalized definition: The (general) k-th level of C is the closure of a set of all points that lie on chains of C such that the open downward-directed vertical ray emanating from this closure intersects fewest but at least k chains in C. See Fig. 2 for an example.

Clearly, the (general) k-th level of C always exists and x_e^* is decided by its lowest point. Let C^k denote the k-th level of C. We can utilize the following lemma to construct C^k.

Lemma 5. C^k *is a x-monotone polygon chain of complexity $O(n)$, and it can be constructed in $O(n \log n)$ time.*

Based on the above analysis, x^* and λ^* can be computed as follows: For each edge e of G, we first determine in $O(n)$ time the set C of functions $d(v, x)$ of each $v \in V$, and then apply Lemma 5 to find the lowest point on C^k in $O(n \log n)$ time. Among all obtained points, we keep the one with the smallest y-coordinate since λ^* is its y-coordinate and x^* is its projection to the corresponding edge. The total running time is $O(mn \log n)$.

Theorem 2. *The unweighted connected k-vertex one-center problem can be solved in $O(mn \log n)$ time.*

5   Solving the Problem on a Tree

In this section, we propose two faster algorithms for the problem on tree graphs respectively in the weighted and unweighted case, where the algorithm for the unweighted case is presented in the proof of Theorem 4. Note that the intervertex distance matrix of the tree graph is not given.

Let T represent the given tree graph, and $R(T)$ be its root. So, $|T| = O(n)$. In the preprocessing, we compute and maintain the distance of every vertex to $R(T)$, and then apply the lowest common ancestor data structure [4] to T so that given any two points y, y' on T, $d(y, y')$ can be known in $O(1)$ time. The preprocessing time is $O(n)$.

Below, we shall first give a faster feasibility test for tree graphs, and then present an algorithm based on tree decomposition techniques for computing λ^*.

5.1 A Faster Feasibility Test

Recall that on a general graph, to decide the feasibility of any given λ, Lemma 1 is applied to every edge to find a point whose largest self-inclusive subtree covered by it (under λ) is of size at least k. On tree graphs, however, the following analysis shows that only $O(n)$ points on T need to be considered.

For each vertex v, $D(v, x)$ increases as x moves away from v along any path. There is a point x' on path $\pi(v, R(T))$ so that $D(v, x') = \lambda$ if $D(v, R(T)) \leq \lambda$ and otherwise, $D(v, x)$ achieves its maximum at x', i.e., $x' = R(T)$. We refer to x' as the *critical* point of v (w.r.t. $R(T)$). Denote by Q_λ the set of all n critical points. We have the following observation for Q_λ.

Observation 8. *There must be a point in Q_λ so that the largest self-inclusive subtree covered by it is of size at least k if and only if λ is feasible.*

Our algorithm decides the feasibility of λ as follows: Traverse T to compute set Q_λ. Next, for each $x \in Q_\lambda$, we determine the size of the largest x-inclusive subtree $T_\lambda(x)$ covered by x. If a point in Q_λ has $|T_\lambda(x)| \geq k$, then λ is feasible. Otherwise, it is infeasible.

To compute Q_λ, we traverse T from $R(T)$ in the post-order, and during the traversal, we maintain the path from $R(T)$ to the current vertex by employing a stack. For each vertex encountered, we perform a binary search on its path to $R(T)$ to compute its critical point. Since the distance of any two points on T can be obtained in $O(1)$ time, the critical point of every vertex can be figured out in $O(\log n)$ time. Hence, computing Q_λ takes $O(n \log n)$ time (Fig. 3).

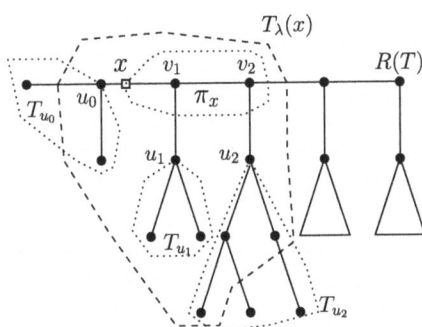

Fig. 3. Illustrating the counting and reporting query on $T_\lambda(x)$ for T being a balanced binary search tree.

It remains to solve the query problems of counting and reporting $V(T_\lambda(x))$ for any given point $x \in T$. If T is a balanced binary tree, which can be verified in $O(n)$ time, Lemma 6 can be employed to construct a data structure $\mathcal{A}_1$ in $O(n \log n)$ time that answers the two queries in $O(\log^2 n)$ and $O(\log^2 n + |V(T_\lambda(x))|)$ time, respectively. In general, T is a general tree graph.

Then, Lemma 7 can be applied to build a data structure $\mathcal{A}_2$ in $O(n \log n)$ time that counts and reports $V(T_\lambda(x))$ for any $x \in T$ respectively in $O(\log^2 n)$ and $O(\log^2 n + |V(T_\lambda(x))|)$ time. Note that compared to $\mathcal{A}_2$, the construction of $\mathcal{A}_1$ and the query on it are much simpler.

Lemma 6. *For T being a balanced binary tree, we can build a data structure $\mathcal{A}_1$ in $O(n \log n)$ time that answers the counting query of $V(T_\lambda(x))$ in $O(\log^2 n)$ time, and reports $V(T_\lambda(x))$ in $O(\log^2 n + K)$ time, where $K = |V(T_\lambda(x))|$.*

Lemma 7. *For T being a general tree, we can build a data structure $\mathcal{A}_2$ in $O(n \log n)$ time that answers the counting query on $V(T_\lambda(x))$ in $O(\log^2 n)$ time, and reports $V(T_\lambda(x))$ in $O(\log^2 n + K)$ time.*

Regarding to our problem, the goal is to decide for any given point $x \in T$ whether $|V(T_\lambda(x))| \geq k$. By maintaining partial but enough information on $\mathcal{A}_1$ and $\mathcal{A}_2$, the counting query on $V(T_\lambda(x))$ can be answered in $O(\log n \log k)$ time. This improvement is demonstrated in Corollary 2. Note that Corollary 2 does not support reporting $V(T_\lambda(x))$.

Corollary 2. *For any given point x on T, with $O(n \log n)$-time preprocessing work, we can decide in $O(\log n \log k)$ whether $T_\lambda(x)$ is of size at least k.*

By Corollary 2, we have the following lemma for the feasibility test.

Lemma 8. *The feasibility test on trees can be solved in $O(n \log n \log k)$ time.*

5.2   Computing λ^*

Observation 4 in Sect. 2 shows that λ^* is in the set Λ that consists of values generated by solving $D(v, x) = D(u, x)$ for every pair of vertices u, v w.r.t. $x \in \pi(u, v)$. Unlike the graph version, we employ the centroid decomposition [21] to implicitly enumerate every pair of vertices such that only $O(n \log n)$ linear functions $y = D(v, x)$ are needed so that Λ belongs to the set of the y-coordinates of all their intersections.

The *centroid* of a tree is a vertex at which the tree can be decomposed into three or fewer subtrees with only this common vertex such that each of them is of size at most half of the tree. The centroid of a tree and these subtrees can be found in linear time [15, 22]. As shown in [21], by recursively decomposing T at its centroid, a decomposition tree of height $O(\log n)$ can be constructed in $O(n \log n)$ time: T is stored in the root; for each internal node, the subtrees of T that are stored in all its (at most three) children nodes are generated by decomposing the subtree in this node at the centroid; every leaf maintains a vertex of T uniquely.

Similarly, we find the centroid c of T and decompose it into three subtrees T_1, T_2, and T_3. Consider c at the origin of the x, y-coordinate plane. For each vertex $v \in V$, let v_l and v_r be the two points on the x-axis at distance $d(c, v)$ to c (the origin) respectively on its left and right, and we construct the distance function $y = w_v \cdot (x - x(v_l))$ for v_l and $y = w_v \cdot (x(v_r) - x)$ for v_r. The set of the y-coordinates of all intersections between the obtained $2n$ lines contains all values in Λ caused by every pair of vertices from different subtrees of T_1, T_2, and T_3.

We recursively decompose T_1, T_2, and T_3 respectively at their centroids, and construct distance functions for vertices w.r.t. each centroid. Thus, in $O(n \log n)$ time, $O(n \log n)$ linear functions are derived so that Λ belongs to the set of the y-coordinates of their intersections.

At this point, we can adapt Lemma 2 with the assistance of Lemma 8 to find λ^* among the y-coordinates of all intersections between the $O(n \log n)$ lines, which runs totally in $O(n \log^2 n \log k)$ time. The following result is thus obtained.

Theorem 3. *The weighted connected k-vertex one-center problem on trees can be solved in $O(n \log^2 n \log k)$ time.*

When the vertices of T are all weights one, the following theorem computes λ^* and x^* in $O(n \log^2 n)$ time.

Theorem 4. *The unweighted connected k-vertex one-center problem on trees can be solved in $O(n \log^2 n)$ time.*

Proof. Observation 4 and 5 imply that the unweighted tree version aims to compute a k-subtree of smallest diameter so that its diameter, which is its longest path length, is exactly $2\lambda^*$, and x^* is the center of this longest path. Hence, computing λ^* is equivalent to solving the problem of finding the vertex so that the distance from it to its k-th closest vertex is smallest.

Our algorithm is simple and consists of three steps: First, we implicitly form the intervertex distance subsets for each vertex by employing the centroid decomposition [21]. Second, we find the length of that k-th shortest path for every vertex. Last, we compute the smallest value of them, which is exactly $2\lambda^*$, and find the center x^*.

The intervertex distance subsets for each vertex of V are implicitly formed in a similar way as in [21], which is for computing the k-th longest path on a tree. More specifically, T is decomposed at its centroid c into three or fewer subtrees, e.g., T_1, T_2, and T_3. Then, three sorted subsets L_1, L_2 and L_3 are explicitly formed so that for each $1 \le i \le 3$, L_i is the ordered set of the distances from c to every other vertex of T_i. Next, for each vertex v of $T_1/\{c\}$, a sorted subset is implicitly formed for v's distance to every vertex in $T_2/\{c\}$ by adding $d(v, c)$ to L_2, and another sorted subset is implicitly formed for v's distances to every vertex of $T_3/\{c\}$ by adding $d(v, c)$ to L_3. Additionally, for each vertex of $T_2/\{c\}$, two sorted subsets are implicitly generated for its distances to vertices respectively in $T_1/\{c\}$ and $T_3/\{c\}$. Similarly, two sorted subsets are implicitly formed for each vertex in $T_3/\{c\}$. Clearly, at most $3n$ sorted subsets are generated but the total storage space is $O(n)$. Due to the sorting work on L_1, L_2, and L_3, the time complexity is $O(n \log n)$.

We proceed to recursively decompose each of the three subtrees T_1, T_2, and T_3 at their own centroids, and form sorted subsets as the above for each subtree.

To the end, $O(n \log n)$ sorted subsets are generated, which implicitly enumerate the intervertex distances for every vertex of V but take $O(n \log n)$ space in total. Note that the value of every entry in any subset can be known in constant time.

Further, we compute for every vertex the length of its k-th shortest path. More specifically, for each vertex v, let n_v represent the number of all intervertex distance subsets of v. Since every subset of v is sorted and each entry can be accessed in constant time, its k-th shortest path length can be found in $O(n_v \log n)$ time by the algorithm [14], which is for finding the k-th smallest value of multiple sorted arrays. Among all obtained values, we set λ^* as the smallest one. The total running time is $O(\sum_{v \in V} n_v \log n)$, which is $O(n \log^2 n)$.

By the corresponding entry of λ^* in the subsets, the two vertices and their path that decide λ^* can be obtained in $O(n)$ time. Consequently, x^*, which is exactly the center of this path, can be found in $O(n)$ time. It follows that the optimal k-subtree T^*, induced by the k closest vertices of x^*, can be obtained in $O(n)$ time by reporting k vertices within distance λ^* to x^* during the pre-order traversal on T from x^*.

As a summary, we can find the partial center x^* of T and T^* in $O(n \log^2 n)$ time. The theorem thus follows.   □

References

1. Abu-Affash, A., Carmi, P., Maheshwari, A., Morin, P., Smid, M., Smorodinsky, S.: Approximating maximum diameter-bounded subgraph in unit disk graphs. Discrete Comput. Geom. **66**, 1401–1414 (2021)
2. Asahiro, Y., Doi, Y., Miyano, E., Samizo, K., Shimizu, H.: Optimal approximation algorithms for maximum distance-bounded subgraph problems. Algorithmica **80**, 1834–1856 (2018)
3. Ben-Moshe, B., Bhattacharya, B., Shi, Q., Tamar, A.: Efficient algorithms for center problems in cactus networks. Theor. Comput. Sci. **378**, 237–252 (2007)
4. Bender, M., Farach-Colton, M.: The level ancestor problem simplified. Theor. Comput. Sci. **321**, 5–12 (2004)
5. Berman, O.: The p maximal cover - p partial center problem on networks. Eur. J. Oper. Res. **72**, 432–442 (1994)
6. Berman, O., Drezner, Z., Wesolowsky, G.: The maximal covering problem with some negative weights. Geogr. Anal. **41**, 30–42 (2009)
7. Berman, O., Krass, D.: The generalized maximal covering location problem. Comput. Oper. Res. **29**(6), 563–581 (2002)
8. Bourjolly, J., Laporte, G., Pesant, G.: An exact algorithm for the maximum k-club problem in an undirected graph. Eur. J. Oper. Res. **138**(1), 21–28 (2002)
9. Chen, D., Wang, H.: A note on searching line arrangements and applications. Inf. Process. Lett. **113**, 518–521 (2013)
10. Daskin, M., Owen, S.: Two new location covering problems: the partial p-center problem and the partial set covering problem. Geogr. Anal. **31**(3), 217–235 (1999)
11. Dearing, P., Francis, R.: A minimax location problem on a network. Transp. Sci. **8**(4) (1974)
12. Everett, H., Robert, J.M., van Kreveld, M.: An optimal algorithm for the ($\leq k$)-levels and with applications to separation and transversal problems. Int. J. Comput. Geom. Appl. **6**(3), 247–261 (1996)

13. Foul, A.: A 1-center problem on the plane with uniformly distributed demand points. Oper. Res. Lett. **34**(3), 264–268 (2006)
14. Frederickson, G., Johnson, D.: The complexity of selection and ranking in $X + Y$ and matrices with sorted columns. J. Comput. Syst. Sci. **24**(2), 197–208 (1982)
15. Goldman, A.: Minimax location of a facility in a network. Transp. Sci. **6**, 407–418 (1972)
16. Handler, G.: Minimax location of a facility in an undirected tree graph. Transp. Sci. **7**, 287–293 (1973)
17. Hu, R., Kanani, D., Zhang, J.: Computing the center of uncertain points on cactus graphs. In: Proceedings of the 34th International Workshop on Combinatorial Algorithms, pp. 233–245 (2023)
18. Kariv, O., Hakimi, S.: An algorithmic approach to network location problems. I: the p-centers. SIAM J. Appl. Math. **37**(3), 513–538 (1979)
19. Lan, Y., Wang, Y., Suzuki, H.: A linear-time algorithm for solving the center problem on weighted cactus graphs. Inf. Process. Lett. **71**(5), 205–212 (1999)
20. Megiddo, N.: Linear-time algorithms for linear programming in R^3 and related problems. SIAM J. Comput. **12**(4), 759–776 (1983)
21. Megiddo, N., Tamir, A., Zemel, E., Chandrasekaran, R.: An $O(n \log^2 n)$ algorithm for the k-th longest path in a tree with applications to location problems. SIAM J. Comput. **10**, 328–337 (1981)
22. Megiddo, N., Zemel, E.: An $O(n \log n)$ randomizing algorithm for the weighted Euclidean 1-center problem. J. Algorithms **7**, 358–368 (1986)
23. Megiddo, N., Zemel, E., Hakimi, S.: The maximum coverage location problem. SIAM J. Algebraic Discrete Methods **4**(2), 253–261 (1983)
24. Wang, H., Zhang, J.: Computing the center of uncertain points on tree networks. Algorithmica **78**(1), 232–254 (2017)

Author Index